Power to Explore

NASA SP–4313

Power to Explore

A History of Marshall Space Flight Center 1960–1990

Andrew J. Dunar
and
Stephen P. Waring

The NASA History Series

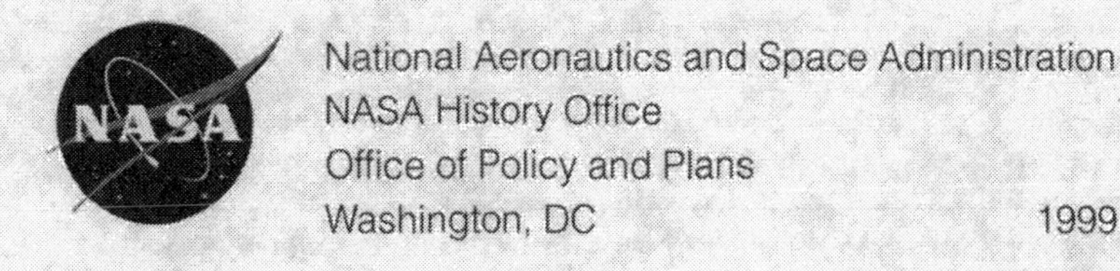

National Aeronautics and Space Administration
NASA History Office
Office of Policy and Plans
Washington, DC 1999

Library of Congress Cataloging in Publications Data

Dunar, Andrew J.
Power to explore : NASA's Marshall Space Flight Center, 1960–1990 / by Andrew J. Dunar and Stephen P. Waring.
p. cm.—(NASA historical series)
Includes bibliographical references and index.
1. George C. Marshall Space Flight Center—History. I. Title. II. Series III. Waring, Stephen P.

TL862.G4 D86 1999
629.4'0973 21–dc21 99-043539

For sale by the U.S. Government Printing Office
Superintendent of Documents, Mail Stop: SSOP, Washington, DC 20402-9328

To Johanna N. Shields

Mentor, Colleague, and Friend

Contents

Acknowledgments

This history was a complex project that evolved over a long period and involved diverse sources and multiple perspectives. In the course of the research and writing, we made many new friends and became indebted to many people. With their help, our understanding became more sophisticated and extensive. Without the advice and advocacy of some of these people, the project never would have reached completion.

This book is based upon work supported by the National Aeronautics and Space Administration (NASA) under Contract NAS8–36955. Any opinions, findings, and conclusions expressed herein are those of the authors and do not necessarily reflect the views of NASA. Several people at the Marshall Space Flight Center contributed to the project's success; they include J. R. Thompson, Thomas J. Lee, Wayne Littles, Susan Smith, JD Horne, Bill Ornburn, Bob Phillips, and Bob Sheppard. We owe special thanks to Mike Wright and Annette Tingle for their work in assembling sources, overseeing the review process, and managing our contract from beginning to end. Tom Gates, Sarah McKinley, Jessie Whalen, Alan Grady, Laura Ballentine, and Angela Fanning also helped with sources, interviews, and chronologies. Pamela Vaughn copy edited the manuscript, Halley Little prepared the layout, and Robert Jaques assisted in assembly of the photographs.

In addition, our book would not have been possible without the help of several people from the NASA History Division at NASA Headquarters in Washington, DC; they include Sylvia Fries Kraemer, Lee Saegasser, and J. D. Hunley. Roger Launius, the chief NASA historian throughout most of the project, provided an invaluable intellectual perspective and shepherded the project through many obstacles.

Marshall assembled a review panel whose members offered comments on our draft manuscript. The historians on the panel, James Hansen and Roger Bilstein, helped us see the relationships between Marshall's history and NASA as a whole. The technical reviewers, mainly Marshall veterans and employees, were a rich source of suggestions, corrections, and insights; particularly helpful were James Kingsbury, George McDonough, and Charles R. Chappell. We wish to express gratitude to Ernst Stuhlinger for his lengthy comments and detailed suggestions. In addition to the technical reviewers, we benefited from numerous oral history interviews with Marshall and NASA retirees and employees; William Lucas, James Kingsbury, Thomas J. Lee, and Charles R. Chappell, and Ernst Stuhlinger granted us more than one interview that revealed the richness of the Center's past.

We benefited from conversations and comments from several scholars at the National Air and Space Museum; we appreciate the help of Martin Collins, David Devorkin, Robert Smith, and especially Michael Neufeld. Pamela Mack at Clemson University gave us useful comments on preliminary versions of our work. Our research at the Johnson Space Center would not have been successful without the assistance of Janet Kovacevich and Joey Pellarin. In Huntsville, Nick Shields offered us insider insights and plenty of cold drinks. Cathie Dunar gave support and encouragement throughout.

Throughout our project, we received tremendous support from the University of Alabama in Huntsville (UAH). Among the administrators and staff members who helped at points in the project were John White, Robert Rieder, Michael Spearing, Dan Rochowiak, Glenna Colclough, Jerry Mebane, Sue Kirkpatrick, and Roy Meek. Offering important behind-the-scenes support were Sue Weir, Jack Ellis, and Frank Franz. Throughout the project and often at critical junctures, Ken Harwell and Charles Lundquist gave us sound advice and crucial backing. Several graduate research assistants collected and filed materials; they included Tony Helton, Dan Gleason, Sarah Walker, and especially Jo Gartrell. Beverley Robinson provided assistance whenever we wished, and our colleagues in the History Department at UAH gave us backing and sympathy. During the five years that she worked with us, Sarah Kidd began as a graduate assistant, became our staff assistant, and brought industry, dedication, and levity. Ann Lee began as our staff assistant, abandoned us midstream, but came to our rescue at the end. Finally, and most importantly, we wish to thank our colleague and friend Johanna Shields; at the beginning of the project, she served as principal investigator and even after leaving this position, she continued to offer us wise counsel and encouragement. We are pleased to dedicate the book to her.

Introduction

Since its inception in 1960, the Marshall Space Flight Center in Huntsville, Alabama has been at the center of the American space program. The Center built the rockets that powered Americans to the Moon, developed the propulsion system for Space Shuttle, and managed the development of *Skylab*, the Hubble Space Telescope, and Spacelab. It is one of NASA's most diversified field Centers, with expertise in propulsion, spacecraft engineering, and human systems and multitudinous space sciences.

Yet the Center's role in American space exploration has often been obscure. Americans following the major space flights of the Mercury, Gemini, and Apollo Programs in the 1960s, *Skylab* in the 1970s, and the Shuttle in the 1980s focused most of their attention on the launch site in Florida or mission control in Houston. Popular histories of the space program accentuate astronauts. When accounts of the early space program do examine Marshall's role, they tend to highlight the dominating presence of Wernher von Braun, the Center's first director, rather than the institution itself. The Center's achievements have often been behind-the-scenes, and if they have not always captured public attention, they have frequently been at the center of NASA's triumphs.

The present work explores Marshall's evolution at the center of NASA, from its origins as an Army missile development organization through its participation in major American space programs. We have employed a generally chronological approach, exploring in topical chapters Marshall's contributions to NASA's major programs. In each chapter, we have traced the Center's contributions to the program and the ways in which the Center's participation shaped the institution itself.

Our own inclinations and the scope and requirements of the NASA contract under which we wrote this book have led us to examine Marshall's history differently from previous treatments. Most previous studies of Marshall's contributions to the space program have been products of what British aerospace scholar Rip Bulkeley called the "Huntsville school" of American space historians,[1] a group that included von Braun himself and several of his associates, most prominently Frederick I. Ordway. Works of this school have chronicled the technical achievements of early space projects in Huntsville, focusing on the role of von Braun and his German team. The Huntsville school took a narrow approach and minimized the social and political context of technological history. The most significant work on Marshall's contributions that is not a product of the Huntsville school is Roger Bilstein's *Stages to Saturn* 1980, a detailed technological history of the Saturn family of launch vehicles.

Technological achievements are the heart of the Marshall story. The Center's accomplishments in engineering and technology have not only contributed to most of NASA's major efforts in human space flight, but have included an array of automated spacecraft that have made breakthroughs in space science, and provided platforms for researchers from other Centers, universities, and private industry.

Nonetheless, the story of the Center cannot be understood apart from its social and political context. Often the Center and its technical efforts developed as much because of political pressures—both from within NASA and from the outside—as because of the technological imperatives of space exploration. The NASA contract under which we worked in fact mandated that we explore Marshall's contributions toward, and responses to, changes in its social, political, and technological environment. While research was underway, several Marshall veterans reviewing our manuscript questioned the social and political approach even to the point that the Center canceled the contract under which we were working. Ultimately, however, NASA and the Center confirmed an approach to MSFC's history that extended beyond technology and reinstated the original contract and its research design.

A broad approach to the Center's history is necessary because Marshall has always been complex, even enigmatic. In six years of research we have talked to people at Marshall and elsewhere in NASA, and have heard interpretations of the Center that are often strikingly contradictory. Some outsiders criticize Marshall as having a closed culture, impervious to penetration from the outside; most Marshall veterans see their Center as open, seeking interaction with other groups at every opportunity. Outsiders sometimes describe Marshall's management as authoritarian; insiders typically see top officials as responsive to ideas from lower-ranking experts. Some see Marshall's history as a prosaic tale of bureaucratic growth and inertia, common to NASA; others see a story of unique organizational culture. Howard McCurdy's recent book *Inside NASA* examines NASA's evolution and shows how early dynamism fell victim to increasingly complex limitations and tightening budgets. Not surprisingly many of his interviewees were Marshall veterans. Yet Marshall's team of German rocket experts and American engineers was unique in the annals of space pioneering, and the Center's first 30 years led to space science and engineering achievements of unparalleled breadth.

Marshall has been at the forefront of the frontier of space, but it has also been a center of controversy. In its first three decades, NASA had three major crises: the Apollo fire in 1967, the *Challenger* disaster in 1986, and a crisis of confidence in the late 1980s in which initial shortcomings of the Hubble Space Telescope and questions about Space Station planning and funding focused national attention on NASA's uncertain future. Marshall was at the margins of the Apollo fire investigation, but at the center of the crises of the 1980s.

One of our major goals then has been to show the complexity of Marshall's history and culture. Moreover, the story of the Center sheds light on the contemporary history of the government-industrial complex, the management of technological endeavors, and the evolving networks of engineers and researchers in "big science." In addition, anyone who hopes to understand NASA's future must come to terms with Marshall's past, for the Center has been a microcosm of the Agency. The major themes of NASA's development over its first 30 years extend through Marshall's history.

The Federal Government assumed responsibility to fund technological research and development tasks in the years after World War II, and by the late 1950s it became apparent that a new federal agency, the National Aeronautics and Space Administration, would be one of the major recipients of federal money. President Kennedy made that commitment a national quest when he directed the new agency to land a man on the Moon by the end of the decade. With that mission NASA emerged as one of the most visible federal agencies. Marshall was one of the three major NASA installations involved in Apollo, and the Center was the largest recipient of NASA funds and had the largest workforce in the early 1960s. Marshall's expertise in rocketry made fulfillment of Kennedy's challenge possible.

The aftermath of Apollo ushered in a new era for Marshall and for NASA. Marshall was the first NASA installation to experience the impact of tightening budgets, cutbacks, and readjusted schedules as Apollo wound down. As one of NASA's two largest field Centers and the one with the most entrenched tradition of in-house production, Marshall was at the center of NASA's shift from the arsenal organization, capable of internal development of hardware to contractor production. Marshall and its surrounding community learned that federal money does not come unencumbered, and the government used the Center to pressure Alabama to reform its pattern of racial segregation. When the government determined that NASA's mission would broaden to include international participation in its programs, Marshall was again in the forefront, managing development of Spacelab with the European Space Agency and incorporating multinational participation in Space Station and other programs. Post-Apollo cutbacks forced the Center to compete with other NASA Centers for business. NASA fostered competition, convinced it promoted creativity, and certain that the benefits of resourcefulness outweighed the costs of Center rivalry. Marshall proved an able competitor, and in the late 1960s began extensive diversification that restructured the Center. Marshall now began to supplement its work on NASA's major human space flight programs with work in space science, which involved both piloted and robotic space technology. The Center worked on technology supporting all types of missions, and in the process developed a scientific and technological diversity unmatched at other Centers.

Marshall in 1990 was a very different institution than it had been in the 1960s. The changes reflected the vision, will, and talent of the people who have worked there through its first three decades, and the external environment in which they worked. No longer merely a propulsion Center, it developed a vast capacity to develop new generations of space vehicles and to lead research investigations in emerging fields of space science. For 30 years the Marshall Space Flight Center indeed remained at the center of NASA's quest to explore space.

---

1 Rip Bulkeley, *The Sputniks Crisis and Early United States Space Policy: A Critique of the Historiography of Space* (Bloomington: Indiana University Press, 1991), pp. 204–205.

Chapter I

Origins of Marshall Space Flight Center

On his way to dedication ceremonies for the George C. Marshall Space Flight Center on the morning of 8 September 1960, President Dwight D. Eisenhower paused before a test stand holding an enormous Saturn I booster. He turned to Wernher von Braun, the director of the new Center, and said that he had never seen anything like it. "They come into my office and say it has eight engines. I didn't know if they put one on top of the other or what," he told von Braun.

Wernher von Braun describes the Saturn I to President Eisenhower.

The President was not the only American who was impressed but somewhat mystified by what had been going on in Huntsville, Alabama. Indeed, when Eisenhower addressed the 20,000 people who assembled at the ceremonies later that morning, he acknowledged a decade of achievements in rocketry at the Army's Redstone Arsenal in Huntsville. There the Army had developed the Redstone and Jupiter missiles, and with the assistance of the Jet Propulsion Laboratory (JPL) created America's first Earth satellite, Explorer I.[1] The dedication signified a change of command, as the Development Operations Division of the Army Ballistic Missile Agency (ABMA) transferred from military authority to the civilian direction of the National Aeronautics and Space Administration (NASA).

The dedication of the Marshall Center marked the absorption of a talented group of 100 German rocket experts and 4,570 American engineers and technicians into the ranks of the civilian agency. The Germans had been working together for more than two decades, and their experience and leadership gave the new Center cohesion. They had trained the Americans to continue their legacy in rocketry. In the years that followed, Marshall would be at the heart of the American space program, one of NASA's two largest field Centers, proud of its many achievements in technology and exploration.[2]

Chemical munitions work at Redstone Arsenal during World War II.

The political struggles that culminated in transfer of ABMA's Development Operations Division to NASA left a legacy that affected Marshall's role in the space program. ABMA was at the center of key debates over national space policy in the late 1950s. An Army agency, it forced consideration of Army-Air Force rivalry in military missile development. A military organization, it prompted the Eisenhower administration to seek a balance between civilian and military space programs. A research and development enterprise of such versatility that it was virtually a space program unto itself, it opened debate over whether experimental work on rocketry should be contracted to private business or conducted by government specialists. A leader in propulsion and high technology, it stimulated contention over the division of labor between NASA Centers. None of these questions would have final answers by the time of the establishment of Marshall. They would reverberate through Marshall's early years, and carry implications that would affect the Center for decades.

If Marshall's future would be tied to the fortunes of the American space program, its origins rested on an improbable coalition: a small southern town, an obscure federal agency, expatriate rocket experts, and young American engineers. Tracing those origins leads inevitably to World War II, when the circumstances developed that would bring the coalition together. Huntsville, Alabama, was an agricultural community in the 1940s, an unlikely site for space research, but the wartime activation of an ordnance plant at the outskirts established the future site of the Army's Redstone Arsenal and the Marshall Center. During the war the National Advisory Committee for Aeronautics (NACA), a small federal agency and the forerunner of NASA, broadened its research base beyond the interest in aerodynamics on which its reputation rested.[3] Another part of the coalition was comprised of German rocket scientists and engineers who, during the war, worked under the direction of von Braun on a remote island in the Baltic Sea developing missiles for Adolph Hitler's army.

Huntsville Before the Space Age

Huntsville, a small town a dozen miles north of the southern-most bend in the Tennessee River, welcomed the arrival of defense plants in 1941 as a solution to economic woes. A compact site of four square miles, Huntsville had seen prosperous days, as blocks of ante-bellum houses east of the Courthouse Square attested. By the late 1920s, Huntsville had become the textile center of Alabama, and Madison County the state's leading cotton producer. But even before the Great Depression its single-crop economy fluctuated with the vagaries of cotton prices. Huntsville's leading citizens yearned for economic growth. Two new twelve-story "skyscrapers" revealed ambitions to be more than a small cotton town. One businessman emblazoned his building with the slogan "Great is the Power of Cash," and the Chamber of Commerce declared Huntsville the "Watercress Capital of the World," the "Biggest Town on Earth for its Size."

To the west and south spread a broad plain of cotton fields dotted with mill villages. Mill wages remained low even when cotton prices rose. Many African Americans left Madison County to seek jobs in northern cities; the black population was lower in 1940 than it had been at the turn of the century, even though the total population increased by fifty percent. The Depression made conditions worse. Mill strikes in 1934 hastened decline, ending Huntsville's domination of the state's textile industry.

The infusion of federal money into the economy during World War II lifted Huntsville out of the Depression, and permanently altered the community. During the Fourth of July weekend in 1941, the Chemical Weapons Service announced plans to establish a chemical weapons plant in Huntsville, and 500 people applied for jobs by the following Monday. The Huntsville Arsenal manufactured toxic agents and incendiary material, and packed them in shells, grenades, and bombs supplied by the Ordnance Department. Three months later, Redstone Ordnance Plant began operations on adjacent land southwest of the city. Redstone manufactured and assembled ammunition for Ordnance. Construction costs for the two arsenals totaled $81.5 million; peak employment exceeded 11,000 civilians. For the duration of the war, Madison County prospered.

The end of the war brought fears of renewed depression, and within months the Huntsville economy seemed on the verge of collapse. Jobs disappeared, and despite efforts to encourage diversification, another "bust" period in Madison County's cyclical economic fortunes seemed imminent. The Department of the Army declared the 35,000 acres of Huntsville Arsenal surplus property, and offered it for sale.[4]

The war nonetheless proved more than a temporary economic surge for Huntsville. The presence of the federal government in Madison County established a foundation for continued prosperity. North Alabama, beneficiary first of the Tennessee Valley Authority, then of Huntsville's defense plants, would see an increasing infusion of federal funds. The twin arsenals, whose futures were uncertain in 1945, would become the launching pads of future growth when the Army chose the site for its missile development team.

NACA: Forerunner of NASA

The war also influenced NACA, which would become the second component of the Marshall coalition, and enhanced its reputation as a research institution. Founded in 1915, NACA supported the aircraft industry with basic research and investigations suited to specific aeronautical problems. With the coming of war in Europe, NACA expanded to new facilities at Moffett Field in California in 1939 and in Cleveland in 1940.

Wartime demands limited NACA to a support role for military requirements. After the war, NACA shed its conservative image, adding new facilities at Wallops Island, Virginia, and at Edwards Air Force Base in California and branching into new fields of research. Hugh L. Dryden, who became director in 1947, initiated research into rocket propulsion, upper atmosphere exploration, hypersonic flight, and other fields previously ignored by NACA. Still minimally funded, but no longer bound by an emphasis on aeronautics, NACA had already begun the transition by the late 1940s that would lead to the formation of a national space agency a decade later.[5]

Peenemünde and Marshall's German Roots

The third component of the Marshall coalition was a talented team of German specialists who developed the V–2 rocket used against Britain and Allied positions on the European continent in the last years of the war. During World War II, German rocketry advanced beyond that of any other nation. The story of the American acquisition of German rocket expertise, intertwined in the origins of the Cold War, has been controversial ever since.[6]

German rocketry originated with the pioneering efforts of the Rumanian Hermann Oberth and the experimentation of amateur rocket societies in the 1920s and 1930s. Among the members of one such society in Berlin in 1929 was von Braun, a recent high school graduate from the town of Wirsitz in Posen, territory along the Oder River that became part of Poland after World War II.[7]

Rocketry changed from a hobby to a profession in the late 1920s when the German army became interested in using it as a means to take advantage of a loophole in the Versailles Treaty. The treaty forbade Germany to build long-range guns, but included no prohibition against rocketry.[8] The Army wanted to develop a liquid-fueled rocket that could be produced inexpensively and surpass existing guns in range.

Von Braun became a civilian army employee in 1932. Beginning with only one mechanic to assist him, von Braun began to build a team of researchers, drawing from amateur rocket societies, universities, and industry. They began work at Kummersdorf near Berlin and by 1936 began moving to Peenemünde. The army provided von Braun with whatever equipment he needed. The Center concentrated all phases of research and development at one location, a concept that von Braun's military supervisor Captain Walter Dornberger described as "everything under one roof." Von Braun first resisted the notion, arguing that he had no experience in production, but later embraced it.[9] Researchers were available if problems arose during production. Test launch sites were only two miles from manufacturing facilities. Dornberger compared the organization at Peenemünde to "a large private research institute combined with a production plant."[10]

The need for secrecy limited cooperation with industry. Rocket technology was too arcane in the early years for industry to desire participation, and conventional arms contracts offered more money. Ernst Stuhlinger recalled that the arsenal concept took hold in Peenemünde simply because "nobody could build rockets at that time in Germany. Nobody knew how to build rocket motors. We had to develop it, and von Braun had gotten the team together. We did it in our Peenemünde laboratories and became the experts before anybody else was an expert."[11]

Formal cooperation with industry and academia increased as the Peenemünde operation matured, but by then the in-house approach was established. Von Braun sought cooperation with universities, especially for research and recruitment. "The main professors, the lead investigators, became our laboratory directors," Georg von Tiesenhausen recounted. Von Braun preferred direct private contacts to the more rigid structures of the German bureaucracy. "We worked closely with universities all over the country. We gave them the list of problems and they had to solve them," von Tiesenhausen explained.[12]

Von Braun established a flexible management system that could respond to external constraints. He envisioned major projects on a vertical axis, technical support laboratories superimposed on a horizontal axis. Every project manager had direct access to all laboratory facilities. Technical departments were not dependent on the fortunes of any given project, yet had the flexibility to adapt to changing demands.[13]

The research team assembled at Peenemünde included men of exceptional talent. Many of them had advanced degrees and practical experience in industry before joining von Braun. Few had worked in rocketry, but expertise in fields like physics, chemistry, mechanical engineering, and electrical engineering suited them to work on various aspects of rocket development.

Not that everything went smoothly at Peenemünde. Early rocketry was an inexact science, with progress registered through trial and error. Von Braun recalled that "Our main objective for a long time was to make it more dangerous to be in the target area than to be with the launch crew."[14] Hundreds of test firings from 1938 to 1942 brought improvements in stability, propulsion, gas stream rudders used for steering, the wireless guidance communication system, and instruments to plot flight paths.[15]

British intelligence discerned that rocket research was underway at Peenemünde as early as May 1943. On the night of 17 August, British bombers staged a large raid that killed 815 people, destroyed test stands, and disrupted transportation. The raid did little to disrupt V–2 production plans, but nonetheless precipitated changes in plans—most significantly the decision that no production would take place at Peenemünde.[16]

Labor for V–2 production became a pressing problem in 1943. In April Arthur Rudolph, chief engineer of the Peenemünde factory, learned of the availability of concentration camp prisoners, enthusiastically endorsed their use, and helped win approval for their transfer. The first prisoners began working in June. Hitler's concern for V–2 development after July 1943 peaked the interest of Heinrich Himmler, the commander of the SS, who conspired to take control of the rocket program and research activities at Peenemünde as a means to expand his power base. When Dornberger and von Braun resisted his advances, the SS arrested von Braun, charging that he had tried to sabotage the V–2 program. Himmler cited as evidence remarks that von Braun had made at a party suggesting developing the V–2 for space travel after the war. Dornberger's intercession won von Braun's release, but Himmler had made his point. Von Braun's defenders cite his arrest as proof of his differences with the Nazi Party and his distance from the use of slave labor. Von Braun's relationship to the Nazi Party is complex; although he was not an ardent Nazi, he did hold rank as an SS officer. His relationship to slave labor is likewise complicated, for his distance from direct responsibility for the use of slave labor must be balanced by the fact that he was aware of its use and the conditions under which prisoners labored.[17]

Atrocities perpetrated at V–2 production facilities at Nordhausen and the nearby concentration camp at Dora—where some 20,000 died as a result of execution, starvation, and disease—stimulated controversy that plagued the rocket pioneers who left Germany after the war. The most important V–2 production sites were the central plants, called Mittelwerk, in the southern Harz Mountains near Nordhausen, where an abandoned gypsum mine provided an underground cavern large enough to house extensive facilities in secrecy. Slave labor from Dora carved out an underground factory in the abandoned mine, which extended a mile into the hillside. Foreign workers under the supervision of skilled German technicians assumed an increasing burden; at Mittelwerk, ninety percent of the 10,000 laborers were non-Germans.[18]

The oft-delayed V–2 production program staggered into low gear in the fall of 1943. Production built steadily through the early months of 1944, peaking in late 1944 and early 1945 at rates of between 650 and 850 V–2s per month.[19] But the V–2 was a military disappointment. As many as two-thirds of the rockets exploded in mid-air before reaching targets. The campaign against England perhaps did more to rally the British people than to inflict damage. So disappointing was the campaign that Nazi officials regretted the decision to concentrate on the V–2 at the expense of the anti-aircraft rockets.[20]

Project Paperclip: American Acquisition of German Rocket Experts

By the beginning of 1945, the advance of the Russian army into Pomerania threatened Peenemünde, and an Allied victory appeared inevitable. With an Allied victory imminent, von Braun and his associates agreed that their future would be brightest with the Americans, who had suffered the least from the war and might be able to afford to support rocket research. Evacuation of Peenemünde began late in January. Workers destroyed records that could not be evacuated and detonated remaining facilities to keep them out of Russian hands. Von Braun moved his organization to the Harz Mountains near Mittelwerk, where he worked on improving V–2 accuracy and eliminating mid-air explosions.[21]

Work ceased only when the advance of Allied troops forced another move. By early April, 400 key members of the von Braun group scattered in villages near Oberammergau. Anticipating the advance of Allied troops, von Braun directed his men to hide research documents from Peenemünde. They hid 14 tons of numbered crates in an abandoned mine, then sealed the opening to the mine with a dynamite explosion.[22]

Research at a standstill, the Germans waited for the arrival of the Allies. On 2 May, two days after Hitler's suicide in his Berlin bunker, American forces moved into the vicinity of Oberammergau. Von Braun and his group surrendered to the Americans.[23]

The destiny of von Braun's rocket experts, now severed from the fate of Hitler's Reich, passed into the crosscurrents of a new international struggle between the United States and the Soviet Union. The meeting of President Franklin D. Roosevelt, British Prime Minister Winston Churchill, and Soviet leader Joseph Stalin at Yalta in February exposed tension between the wartime Allies.

Consideration of what to do with captured scientists and engineers succumbed to emerging Cold War attitudes, as Washington measured hostility toward an old adversary against fear of a new one.

Colonel Gervais William Trichel, the chief of the Rocket Branch of U.S. Army Ordnance, was one of the few Americans who had pondered the disposition of German rocket experts prior to their surrender. He sent Major Robert Staver to London to work with British intelligence developing a list of German rocket technicians, ranking them in order of significance. Wernher von Braun's name headed the list. Trichel negotiated a contract with General Electric late in 1944 for Project Hermes, an agreement for the development of long-range guided missiles. He anticipated using V–2 rockets in his research, and in March 1945 he directed Colonel Holger Toftoy, chief of Ordnance Technical Intelligence, to locate 100 operational V–2s and ship them to an Army range in White Sands, New Mexico.[24]

As soon as Toftoy learned about the Allied discovery of the V–2 plant at Mittelwerk, he sent Staver to Nordhausen to investigate. After verifying the astounding discovery of rows of partially assembled V–2s in the underground facilities, Staver met with members of von Braun's staff and learned of the hidden cache of Peenemünde documents. The peace agreement stipulated that the Soviet Union would occupy Nordhausen, and Britain would control Dornten before the end of May, so Toftoy and Staver had to improvise quickly. Toftoy sent Major James P. Hamill to Nordhausen, where in nine days he supervised shipment of 341 rail cars containing 100 V–2s to Antwerp in preparation for shipment to the United States. Staver convinced the Germans to help him find the hidden documents. He shipped 14 tons of the Peenemünde cache out of Dornten even as the British were erecting roadblocks prior to assuming control.[25]

The question of what to do with German technicians in American custody was laden with political, military, and moral overtones. Some feared that allowing them to continue their research might allow for a rebirth of German militarism. Secretary of the Treasury Henry Morgenthau sought a punitive policy toward Germany, with no room for coddling weapons developers.[26] The most compelling moral argument hinged on the involvement of the Germans with either the Nazi Party or slave labor at Mittelwerk.

Many German academics, scientists, and technicians had been members of the Nazi Party, often because party membership brought benefits such as research grants and promotions. The Party often bestowed honorary rank as a reward. Heinrich Himmler personally awarded an honorary SS rank to von Braun in May 1940, which von Braun accepted only after he and his colleagues agreed that to turn it down might risk Himmler's wrath. Party membership alone seemed an inadequate criteria, and advocates of using German scientists suggested distinguishing "ardent" Nazis from those who joined the Party out of expediency.[27]

Similar ambiguities clouded the issue of responsibility for the slave labor at Nordhausen. Manufacture facilities were far from Peenemünde, under the supervision of Himmler's SS. Himmler and SS-General Kammler dictated production schedules and allocated V–2s for deployment and for testing. Neither Dornberger nor von Braun had direct authority over Mittelwerk, but both men visited the plant several times and observed conditions. Dornberger—and von Braun—could influence V–2 production only indirectly, by lobbying for greater resources.[28]

In the years after the war, when von Braun and other Peenemünde veterans had risen to responsible positions in the American space program, accusations regarding their role in the Mittelwerk slave labor production rose occasionally. Responding to charges leveled by former inmates of the Dora-Ellrich concentration camps in the mid-1960s, von Braun gave his most detailed response. He admitted that he had indeed visited Mittelwerk on several occasions, summoned there in response to attempts by Mittelwerk management to hasten the V–2 into production. He insisted that his visits lasted only hours, or at most one or two days, and that he never saw a prisoner beaten, hanged, or otherwise killed. He conceded that in 1944 he learned that many prisoners had been killed, and that others had died from mistreatment, malnutrition, and other causes, that the environment at the production facility was "repulsive."[29]

In later years some members of the von Braun group countered criticism by explaining that the Germans at Peenemünde were more interested in the scientific potential of rocketry than weapons, and that they often spent evenings discussing space travel. Some stories, repeated many times, became part of the legend of the von Braun group after its successful work on the Saturn rocket. Several stories revolved around the first successful V–2 test of 3 October 1942, when Dornberger proclaimed the birth of the space age.[30] Von Braun's discussion of the potential of the V–2 as a step toward space travel had given Himmler the pretense for his arrest in 1944. Eberhard Rees, von Braun's closest lieutenant, put the issue in perspective years later, saying, "Let us be very honest. In Peenemünde we did not work in the field of space flight whatsoever. We worked directly on rockets and guided missiles, and only privately we talked in the evening about space flight. . . . A lot of people have talked about how strongly we worked in space flight and that just simply is not so."[31]

After V-E Day, concern with the background of the Germans gave way to the Cold War preoccupation with the Soviet Union. American strategists argued that the Germans might help bring the war in the Pacific to an end, and pressured the Truman administration to support a program of exploitation of German scientific expertise. Russian and British interest in German scientists raised concern that the United States might miss a historic opportunity. Truman had no reservations about using German expertise as long as the program could be kept secret. On 6 July, the Joint Chiefs of Staff responded by initiating Project Overcast—later renamed Project Paperclip—a top secret program authorizing recruitment of up to 350 experts in specialties of interest to American military.[32]

Interrogation of von Braun's inner circle, now ensconced in Witzenhausen in the American zone, gave way to negotiations over terms for consultation services. Colonel Toftoy requested authority to bring 300 rocket experts to the United States, and received permission to transfer 100. Von Braun had insisted that the smallest group that could be transferred was 520, but he helped pare the list to 127, ensuring that they represented a cross-section of his organization.

Negotiations did not always proceed smoothly. Questions rose over whether transfers would be permanent, if they could be renewed, whether wives could accompany their husbands, what salary they would be paid—none of which had clear-cut answers, given the ad hoc nature of the program. Persistent French, British, and Russian interest in exploitation gave the Germans some leverage. In the end, the von Braun group remained together and stayed with the Americans as the least undesirable alternative. "We despised the French, we were mortally afraid of the Soviets, we did not believe the British could afford us, so that left the Americans," one member of the group explained.[33]

Time in the Desert

In September 1945, seven Germans including von Braun traveled to the United States.[34] All except von Braun went to Aberdeen Proving Ground in Maryland, where they helped organize and translate the cache of Peenemünde documents. Von Braun traveled cross-country by train with Major Hamill to Fort Bliss in El Paso, Texas, where Colonel Toftoy planned to reassemble "the world's only experienced supersonic ballistic missile team."[35] Nearby White Sands Proving Ground, 25 miles north of Las Cruces, New Mexico, offered a vast desert expanse for testing.

By the spring of 1946, most of the Germans selected by Toftoy had arrived at Fort Bliss. The Germans knew little of the desert terrain of the American southwest other than what they had read in the westerns of Karl May, a popular German novelist who set some of his stories in El Paso. An isolated enclave at Fort Bliss, the Germans were never more than a marginal part of the El Paso community. They were still wards of the Army in 1946, subject to many restrictions, living behind a fence in converted barracks, required to have an American escort if they left the base. Those involved in testing at White Sands had fewer restrictions because of its remote location, but their isolation was greater. At first, none of the Germans had much contact with Americans other than those they met in their official duties.[36]

General Toftoy's principal purpose in bringing the Germans to Fort Bliss was Project Hermes, the test firing of the Mittelwerk V–2s, a project intended to give Americans experience in rocket research, testing, and development. The V–2 parts were in disarray, having been packed by soldiers, shipped to New Orleans, reloaded on freight cars, repacked once again on trucks, and finally left in the open on the desert at White Sands. Working with General Electric as the prime contractor, the Germans reassembled rockets, tested engines, and fired the first American V–2 on 16 April 1946.[37]

For the remainder of the decade, the Germans served as consultants to the Army, Navy, and private contractors. Forty-five of the sixty-eight V–2s fired performed successfully, yielding aerodynamic data, information on the composition of the upper atmosphere, and launching American rocketry research. Major achievements included launching a V–2 from the deck of the USS Midway, and firing a Bumper-Wac (a modified V–2 first stage with a Wac Corporal second stage)

The original Peenemünde team shortly after their arrival at Fort Bliss.

from the White Sands Proving Grounds to a record altitude of 250 miles, the first object to be sent outside the Earth's atmosphere.[38]

The years at Fort Bliss were a literal time in the desert for von Braun's rocket experts. Unlike the Peenemünde years before or the Saturn years later, no clear goal unified them. They were consultants to American military and industrial researchers, advisers to the dreams of other men. But the period was crucial, for at Fort Bliss the members of the von Braun group began to view themselves as members of a team. Dornberger and von Braun had fostered cooperative enterprise, of course; but no corresponding sense of collective identity emerged from the military-industrial-university complex supporting Peenemünde.[39]

The peculiar circumstances of life at Fort Bliss reinforced the sense of a team. New to a foreign country in which many had at best a cursory understanding of the language, separated from their families, sharing professional interests, viewed with suspicion by the people of El Paso, the Germans drew together. They hiked in the nearby Organ Mountains, played chess and read, and played ball games on a makeshift field between the barracks.[40] Pranks reflected a boarding-school atmosphere, as when Major Hamill reprimanded von Braun: "The wall of Mr. Weisemann's [sic] room has been broken through. This matter was not reported to this office. The pieces of the wall have evidently been distributed to various occupants of Barracks Number 1."[41] The elite nature of the group that led to charges of arrogance created another common front; one American described them as "a president and 124 vice presidents."[42]

The president, of course, was von Braun. Not only did the other Germans accept him unequivocally as their leader, but von Braun insisted on his prerogatives. Relations with Hamill were often prickly. Von Braun resented it when

Hamill questioned his subordinates, issued orders, or transferred personnel without working through him, and threatened to resign several times. Hamill ignored the threats, but acceded to von Braun's control of the team.[43]

Relations between von Braun and Colonel Toftoy remained on a higher plane. Toftoy exerted a calming influence on the group, and worked to meet their needs. Within a year, he had won the right for the Germans to begin bringing their families. In the spring of 1948, Toftoy and Hamill devised a scheme to overcome a legal technicality that troubled the group. Since they had entered the United States without passports or visas, their immigration status was in doubt. They crossed into Mexican territory and returned the same day with papers listing Ciudad Juarez as their port of debarkation, El Paso their port of arrival.[44]

The Transfer to Huntsville

In 1949, General Toftoy began to search for a new location at which to conduct Army rocket research, thus initiating the chain of events that would lead to the establishment of the Marshall Center. The commander of Fort Bliss rejected Toftoy's plans for expansion, and insufficient funds forced cancellation of research projects.[45] Toftoy believed rocket research had become too decentralized. In August, he visited Redstone Arsenal and neighboring Huntsville Arsenal, then listed for sale by the Army Chemical Corps. Toftoy liked the site. Senator John Sparkman, a Huntsville resident and chair of the city's Industrial Expansion Committee, lent support after the city lost a bid for an Air Force aeronautical research laboratory to Tullahoma, Tennessee. After a personal appeal to General Matthew B. Ridgway, Toftoy won approval in October 1949 to incorporate Huntsville Arsenal into Redstone Arsenal and transfer the von Braun group to Alabama.[46]

Toftoy's shift to Redstone Arsenal began the economic, cultural, and political transformation of Madison County, Alabama. The first small contingent of Germans arrived in March 1950, and others soon followed. The move to Huntsville involved not only the German rocket experts, but 800 others, including General Electric and Civil Service employees, and 500 military personnel. By June 1951, more than 5,000 people worked at the Arsenal.[47] Huntsville's population would triple by the end of the decade, and much of the growth was due to the infusion of federal money for the Arsenal.

When the Germans began the move to Huntsville in April 1950, they did so with some trepidation. Unlike the isolation at Fort Bliss, they would live in the community, and some worried that resentment from the war, which had risen occasionally in Texas, might be a problem. "We had fears," Hertha Heller remembered, recalling especially warnings that Alabama ranked near the bottom in state expenditures for education.[48]

The contrast to the restrictions and bleak terrain of Fort Bliss, however, left the Germans enthusiastic about their new home. "Our freedom began for us," Stuhlinger recalled. "We could live where we wanted to, we could buy or rent houses, buy property. We could send the children to any school we wanted to. We could go to church." Hertha Heller recalled that "we liked Huntsville because it was green and reminded us of Germany."[49]

Huntsville, although a small cotton town, was better prepared to accept its highly educated new residents than might have been expected. "Huntsville was not just a 'hick' town," recalled Ruth von Saurma, who arrived with her husband shortly after the Fort Bliss contingent. "As you can see from the Twickenham District and the ante-bellum homes, there were a good number of educated and prominent families who lived in Huntsville." At first a natural reticence characterized relations between the Germans and native Huntsvillians, and each side perceived clannishness on the part of the other. The Germans lived in clusters, some on Monte Sano, others in downtown Huntsville. Some Huntsvillians were not sure they wanted the Army back, and were not sure what to make of the Germans. But as von Saurma remembered, "Most of the people in Huntsville knew that this was not a group that had just come from nowhere, but that the majority of them were people with a very good professional background." Over time, individuals established friendships, and interaction brought the groups closer. After the Heller's house burned, people contributed clothing, furniture, and money to help the family recover. "The generosity was unbelievable," Heller recalled. "Americans are extremely generous and start immediately. They are 'action-pushed' in America. 'Let's do something!'"[50]

The Germans participated in Huntsville's civic life; one observer claimed "they plunged into community affairs with a proprietary interest."[51] When they arrived, the single bookstore in Huntsville only sold textbooks for public schools; soon a new bookstore opened in response to the new demand. The Germans supported a campaign to build a new public library. They helped found a symphony orchestra, and several performed with the group. Von Braun and a few others helped form a local astronomical society. Walt Wiesman, the only non-technical person in the group, became president of the Junior Chamber of Commerce in his second year in Huntsville. On 15 April 1955, von Braun and 40 members of his team and their families assembled in Huntsville High School to take the oath as American citizens.[52]

The American Engineers

The Germans provided leadership for an Army rocket development team that included military, civil service, and contractor personnel. Many of those who came to work for the Ordnance Guided Missile Center and its successor organizations at Redstone Arsenal later became second-generation leaders at Marshall. The Army drafted people with professional experience during the Korean War, and they provided a rich pool of talent for Redstone Arsenal.

Charles Lundquist, an assistant professor of engineering research at Penn State University, recalled being drafted into a basic training unit that included lawyers, CPAs, and other professors before he received his orders to Huntsville. "There were lots of people brought in to augment the von Braun team by that process," he explained. They were "sort of a second echelon under the German folks." Robert Lindstrom, who managed Marshall's Space Shuttle Projects Office in the 1970s, came to Redstone via the draft.[53] So did James Kingsbury, who stayed for 36 years and eventually headed the Science and Engineering Directorate. A college graduate with an electrical engineering degree, Kingsbury remembered being pulled out of the ranks and sent to Huntsville in 1951 when his unit shipped out to Korea. "My first job was to take a warehouse that stored chemical weapons during World War II and convert it into a laboratory," he recalled.[54] Henry Pohl, who spent most of his career at Houston, came first to Huntsville as a draftee with a new engineering degree. His first job was at the test layout, where a supervisor told him he would have to watch a Redstone missile launch. "This huge massive building that we were in—you could feel a quiver from the power of that thing," he recalled. "I was hooked. I would have given my $75 a month to work there!"[55]

Not all who came to Redstone with the Army were draftees. Joe Lombardo, a graduate of MIT, enlisted in order to complete his military obligation, and later asked for a transfer to Redstone Arsenal after "reading an article about this team of German scientists that was working on rockets in a place called Huntsville, Alabama."[56] Stan Reinartz, called to active duty after participating in ROTC at the University of Cincinnati, received orders to Redstone Arsenal and soon found himself working in the Project Control Office.[57] Lee James, a West Point graduate and a World War II veteran who later served as a program manager on Saturn stages, had a unique perspective. "Guided missiles were something I had been introduced to," James recalled. "I had occasion to be in London when the V–2s were landing." When he was in Germany, "the V–1s would go over so low you could read the chalk marking written on them by the soldiers."[58]

Other young engineers came to work at the Arsenal as employees of contractors. Richard A. Marmann, who later managed payload development for Spacelab, first worked for Chrysler Corporation doing weight engineering on many of the early missiles before moving over to work for the government.[59] Jack Waite worked for a contractor as a research design engineer at Redstone Arsenal after graduating from the University of Alabama with a degree in mechanical engineering.[60] John Robertson came to Redstone Arsenal after being laid off from his work on bomber contracts for the Air Force.[61] A few people transferred to Redstone Arsenal from other government agencies. Leland Belew began working with the Tennessee Valley Authority in 1951, but found that the work was not challenging. "Most of the work there was replication of work that had already been done," he explained. He soon took a job with the von Braun group, and later helped manage work on Saturn and *Skylab*.[62]

Some new employees came to Huntsville directly from college or graduate school. William R. Lucas, who would have the longest tenure of any Center director in Marshall's first three decades, was a graduate student at Vanderbilt University when he learned about the missile work at Redstone Arsenal from a professor who was working as a consultant in Huntsville.[63] William Snoddy, who came to Huntsville in 1958 with a degree in physics from the University of Alabama, was another of the dozens of graduates from southern universities who took jobs in Huntsville.[64]

Graduates of southern universities predominated among new employees in Huntsville, but people came from around the nation. Art Sanderson, who made recruiting trips as part of his responsibilities in the personnel office, recalled that "They wanted top-notch engineers and we had a charter to go all over this country to get them."[65] Snoddy, a die-hard Crimson Tide fan, said that the diverse origins of his fellow workers became most noticeable during football season. "It was really strange to be in Alabama and yet work around people that didn't care," he laughed. "They had these weird teams they were cheering for. Some of them were even Yankee teams [from] places I'd never heard of like North Dakota."

Redstone Test Stand—the "poor man's test stand."

The young American engineers were a brash, irreverent, talented group, who after serving in apprenticeship to the Germans during the 1960s, would emerge as Marshall's leaders in the Center's second and third decades. Snoddy remembered that in his first summer, he, Robert Naumann, and three others rented a lodge on the back side of Monte Sano. "We'd sit out on the back, Bob and myself and others, and drink beer and throw the cans off the back of the mountain," Snoddy recalled. Von Braun had organized a brainstorming group called the Redstone Technical Society. "We formed a counterpart we called the Rednose Technical Society," Snoddy remembered. "We had some really senior level folks that came, [including] the manager of Thiokol in Huntsville at the time, and the head of Research. We'd get quite a group up there, and we had some darned good discussions. One night in the heat of the discussion, there was this tremendous display of the Northern Lights. It was really wondrous; there's never been any thing like it in this part of the country in recent times. . . . So that was a great summer—and the ranger found the beer cans and made us go pick them all up."[66]

Army Missile Development in Huntsville

The German-American team set to work developing missiles for the Army. Within months after arrival in Huntsville, General Toftoy's Ordnance Guided Missile Center won approval to develop the Redstone, a new surface-to-surface missile intended to augment the Army's Corporal and Hermes. Army requirements to use existing components where possible led some of the Germans to consider the Redstone simply another redesign of the V–2. But the development plan contained considerable flexibility. Not only did the Redstone become a reliable vehicle, but its development provided answers to pressing problems in rocketry and served as a foundation for the Jupiter.[67]

The Redstone gave the Germans a project of their own, and Toftoy's confidence in von Braun gave the group latitude they had not known at Fort Bliss. In 1952, the Army established the Ordnance Missile Laboratories at Redstone Arsenal, with von Braun as chief of its Guided Missile Development Division. He began to employ the principles that would be the hallmark of rocket development in Huntsville for the next two decades. "When the Redstone came upon us, we were prepared," Stuhlinger remembered. "We could go right to work."[68]

The "arsenal system" was the heart of von Braun's approach. The system was not uniquely German. It was well understood in the United States, employed first at the arsenal and armory at Harper's Ferry, West Virginia, in the 19th century, and endorsed by the Army ever since. The circumstances under which the von Braun team had matured intensified its commitment to the system, however, and by the time an interservice debate developed in the 1950s over the relative merits of in-house versus contractor development, the group had come to epitomize the arsenal approach. Its principles had been applied at Peenemünde. American engineers concentrated on design and contracted others to execute; German training emphasized hands-on experience, enabling the German engineers to execute a project from design and development to construction. Karl Heimburg, director of von Braun's test laboratory, noted that in Germany "you are not admitted to any technical college or university if you do not have some practical time."[69] Thus training reinforced the German commitment to in-house work, and von Braun's approach meshed well with the Army's own reliance on the arsenal system. Ultimately, the arsenal system would be caught in the whipsaw of a debate over military procurement, with the Air Force and aerospace industrial firms pushing to increase reliance on contractors.

The Army's continued reliance on the arsenal system in its Huntsville rocket program was also a response to budgetary constraints imposed by the beginning of the Korean War. The Army terminated its Hermes program and reduced funding to Redstone. Work could often be accomplished internally at a much lower cost than could be done by a contractor. After he received a bid of $75 thousand to build a static test stand to test rocket motors, Heimburg had his own people build a "poor man's" test stand for only $1 thousand in materials.[70]

Reliability testing became an adjunct to the arsenal system, a response both to conservative engineering practices among the German group and the Army's insistence on better than 90 percent reliability on Redstone. Dr. Kurt Debus proposed a system for monitoring reliability in February 1952. Soon adopted in all laboratories, it became the basis for later management systems. "The proposal derived from analyzing guided missile systems and concluded that any part could be classified as 'parallel' or 'series' in operation," Debus explained. "Failure of a 'parallel' part would probably not result in failure of the system since its function could be taken over by another part. Failure of a 'series' item, on the other hand, would ultimately result in total failure."[71]

In addition to work on hardware, top officials in the missile team also advanced a vision of future space exploration. In a series of articles in Collier's magazine in 1952, von Braun propounded his ideas about prospects for space travel, suggesting that a Moon landing could take place within the next quarter century.[72] The articles established him as one of the foremost American spokesmen on space. His ability to communicate complex ideas in simple terms and his appeal as a speaker made him an attractive public figure.

Von Braun formulated proposals for the initial steps that might make his speculations a reality. In 1953 he proposed using existing hardware to orbit an Earth satellite.[73] The next year the Army suggested an interservice satellite project, which became the basis for a joint Army-Navy proposal known as Project Orbiter. The Air Force and Naval Research Laboratories also proposed independent satellite programs. The Defense Department formed a panel to evaluate these proposals, and in August 1955 ruled in favor of the Naval Research Laboratory's Project Vanguard, apparently ending Redstone Arsenal's space aspirations.[74] Some suspected that sentiment in the Defense Department that the first American satellite should not be launched by a German team influenced the decision.[75]

The Army Ballistic Missile Agency

Organizational changes and new assignments nonetheless demonstrated that Huntsville would remain at the center of military rocketry. The Army reorganized its missile development program, establishing the ABMA at Redstone Arsenal. The new organization incorporated the Guided Missile Center and the Redstone missile project. Redstone's Ordnance Missile Laboratory also received authorization to begin development of an Intermediate Range Ballistic Missile (IRBM), a single-stage liquid-fuel vehicle expected to have a range of 1,500 miles. Designated the Jupiter, the new missile was to exploit Redstone technology.

General John B. Medaris, who assumed command of ABMA in February 1956, was a no-nonsense commander. "He had an iron fist," Helmut Hoelzer recalled, but he was "an excellent, outstanding man." Medaris's direct, demanding approach suited the high expectations the Army had for ABMA. Medaris was "very blunt" according to Erich Neubert, but "it was a time to be blunt." Using the high priority granted him by the Army, Medaris expanded operations. He brought in top military and civilian personnel, tripling the number of employees to 5,000.[76]

The optimistic, "can-do" mood that visitors noticed at ABMA in 1956 was tempered by restrictions preventing Jupiter from competing with Project Vanguard as the American satellite program. Medaris submitted proposals to the Defense Department requesting authority to develop Jupiter as an alternate means of launching a satellite, only to be rebuffed. "We at Huntsville knew that our rocket technology was fully capable of satellite application and could quickly be implemented," von Braun later reflected. When ABMA launched its first Jupiter–C on 20 September 1956, the Defense Department sent observers to ensure that the Army did not activate a dummy fourth stage and orbit a booster before Vanguard.[77]

Jupiter research proceeded in competition not with Vanguard, but with the Air Force's Thor. The greater altitude to be achieved by the new generation of missiles nonetheless allowed ABMA to study problems related to space flight. One of the most puzzling questions was how to deal with the heat generated by re-entry of missiles into the Earth's atmosphere. The Air Force favored a heat-sink concept in which nosecone materials would absorb heat; ABMA preferred ablation, in which materials shielding the nosecone would melt and peel away, carrying off excessive heat. Ablation had the advantage of dissipating more heat and allowing more accuracy, and came to be the preferred technique. Jupiter–C launches in 1956 and 1957 tested the feasibility of ablation, and allowed ABMA to demonstrate the capabilities of the new vehicle by exceeding an altitude of 600 miles.[78] Reentry studies also gave ABMA a means to skirt Defense Department range restrictions. William R. Lucas remembered that in spite of these restrictions, "we went ahead and developed a launch vehicle anyway and justified it on the basis of testing nose cones."[79]

Explorer Project Leaders: Dr. Rees, Major General Medaris, Dr. von Braun, Dr. Stuhlinger, and (in back) Mr. Mrazek, and Dr. Haeussermann.

From Sputnik I to Explorer I

Until the autumn of 1957, the United States had no coherent space program except as an adjunct to military missile research. The launch of Sputnik I by the Russians on 4 October prompted a reevaluation of the national role in space research. Neil McElroy, the incoming secretary of defense, was visiting Redstone Arsenal when he received news of Sputnik. At dinner that evening von Braun and Medaris sat on either side of McElroy; von Braun insisted that ABMA could launch a satellite into orbit within 60 days, Medaris cautioned that 90 might be more realistic. Three days later, Secretary of the Army Wilbur M. Brucker urged the secretary of defense to allow ABMA to use a Jupiter–C to launch a satellite, promising a launch within four months of approval. Only after the Soviet Union launched a 1,120 pound Sputnik II with the dog Laika aboard on 3 November did the Department of Defense agree. At the request of the Army, Defense set a launch date of 29 January. After Vanguard exploded on its launch pad on 6 December, ABMA became the focus of American hopes to recoup some of the prestige lost to the Soviet Union.[80]

Frantic activity at Huntsville and the Atlantic Missile Range at Cape Canaveral, Florida, characterized the 84 days between authorization and launch of ABMA's satellite. President Eisenhower, trying to avoid being pushed into a race with the Russians, refused to approve a mission without a scientific satellite that could contribute to the International Geophysical Year (IGY).[81] Dr. William H. Pickering of the JPL at the California Institute of Technology developed Explorer I, a 34-inch-long satellite, 6 inches in diameter, weighing just over 18 pounds. Dr. James A. Van Allen of the University of Iowa contributed instruments to measure cosmic radiation. ABMA fashioned a launch vehicle, designated Juno 1, by attaching a cluster of solid propellant rockets atop a Jupiter–C. Explorer I was ready for launch on schedule, but weather forced postponement for two days. On 31 January 1958, Explorer I lifted into an orbit with an apogee of 1,594 miles.[82]

The Establishment of NASA and the Fate of ABMA

In the harried atmosphere of panic following Sputnik, the Defense Department, Congress, and the Eisenhower administration all generated proposals from which a national space policy would emerge. In the balance were crucial decisions: Would the space program be civilian or military? How would the military services divide responsibility for missile development? Should space research be dominated by manned programs or unmanned satellites?

Since the American space program before Sputnik had been exclusively military, the Defense Department became the principal target of post-Sputnik criticism. Some was facile, such as the allegation that the Russians had gotten the better Germans after the war. More substantive critiques charged duplication in the Army's Jupiter and the Air Force's Thor, bureaucratic delays, and interservice rivalry. Even before Sputnik, Defense apportioned the military program by limiting the Army to land-based IRBMs with ranges up to 200 miles (the range of Redstone), and giving the Air Force longer range Intercontinental Ballistic Missiles (ICBMs). A week after the launch of Explorer I, Secretary of Defense McElroy sought greater coordination of military space programs by establishing an Advanced Research Project Agency (ARPA), and appointing General Electric vice president Roy W. Johnson as its director. The Agency had authority to initiate space projects approved by the President for one year, and Johnson soon received proposals to put a man in space from ABMA (Project Adam) and the Air Force (Man-in-Space-Soonest).[83]

Congress, awakened to public pressure, entered the debate. Senator Lyndon B. Johnson chaired hearings that treated Sputnik as "a technological Pearl Harbor," and Congressmen began filing proposals for a national space policy.[84]

The Eisenhower administration refused to be stampeded into a space race. Eisenhower transferred the Office of Defense Mobilization Science Advisory Committee to the White House staff, and named James R. Killian, Jr. as its chairman and as special assistant to the President for Science and Technology. Killian agreed with the President that space research should not be approached as a measure of national prestige, but rather as one of many avenues for scientific inquiry, each of which should be evaluated solely on the basis of its potential contribution to scientific progress. Eisenhower directed them to prepare two reports, a policy statement on space research and a recommendation for national space policy. Late in February, the Presidential Science Advisory Committee (PSAC) submitted a proposal to use the NACA as a foundation for a new agency to direct national research on astronautics. In a message to Congress on 2 April, Eisenhower proposed establishment of a National Aeronautics and Space Agency that would absorb the NACA. American space exploration, the President insisted, should be conducted "under the direction of a civilian agency except for those projects primarily associated with military requirements."[85]

While Congress debated the President's proposal, von Braun kept alive ABMA hopes for a role in space by supporting projects managed by ARPA. Another Jupiter–C (Juno 1) failed to put Explorer II in orbit when the fourth stage failed to ignite on 5 March, but the same configuration succeeded in orbiting Explorer III later that month. By the end of the Juno 1 series in October, ABMA had launched three satellites and failed in three other attempts.[86]

In August, ARPA approved an ABMA proposal to develop a multi-stage rocket with a clustered-engine first stage. Although originally called Juno 5, the new project envisioned a rocket much larger than those used in the Juno/Explorer program, powerful enough to generate 1.5 million pounds thrust—enough to lift payloads weighing tons into orbit. Later called the Saturn I, it soon became ABMA's most important project.[87]

ABMA also proposed using a Redstone as a booster for a manned suborbital flight. Project Adam advocated sealing a man in a cylindrical capsule for a flight of 150 miles in altitude and 150 miles range. Ridiculed as the equivalent of firing a person from a circus cannon, the proposal died aborning, the victim of Air Force opposition and uncertainty over plans for a civilian space agency. Despite such criticism, the early suborbital Mercury flights were much like Project Adam.[88]

The civilian space agency became a reality when President Eisenhower signed the National Aeronautics and Space Act on 29 July 1958. Dr. T. Keith Glennan became the first Administrator of NASA. NASA went into operation on 1 October, absorbing NACA's 8,000 personnel and five laboratories.[89] The Space Act also assigned the Navy's Vanguard project and several Air Force projects to NASA, as well as three of ABMA's satellite projects and two of its lunar probes.[90]

Although the Space Act gave some ABMA projects to NASA, it did not specify whether the von Braun team should remain with the Army or transfer to NASA. By the middle of October, Glennan requested transfer of more that half of the Ordnance Missile Command (von Braun's group) to NASA. Medaris was enraged at the prospect of losing the heart of ABMA and by the lack of support from Assistant Secretary of Defense Donald A. Quarles, who seemed to accept the prospect of transfer with undue equanimity. Von Braun opposed transfer, fearing that it might lead to dispersal of his team. He owed Medaris loyalty and feared that NASA might not be as supportive of in-house development.[91] He and some of his lieutenants told of lucrative offers from private industry and threatened to resign from government service if the team was divided.[92]

Eisenhower held a meeting of the National Aeronautics and Space Council on 29 October, and made it clear that he expected NASA and the Department of Defense to resolve the dispute. Five weeks later, Defense and NASA announced an agreement that transferred JPL to NASA. Von Braun's team would remain intact under Army control, but would be "continually responsive to NASA requirements." Neither side was satisfied. NASA considered the compromise a victory for the Army, since von Braun's Ordnance Missile Command was the more important facility. The Army resented loss of JPL. Although NASA Director Glennan insisted "this agreement is a final agreement," some in the Army suspected that NASA considered the arrangement only a deferred decision, not a resolution.[93]

NASA was disappointed with the failure to acquire the von Braun team, but its appraisal of ABMA was ambivalent. NASA administrators respected the achievements of the Germans at Redstone Arsenal, but harbored misgivings about their way of doing business. Glennan's staff warned him that the Aircraft Industries Association considered the arsenal system to be "hopelessly outmoded," and suggested that if NASA were to absorb ABMA, "it should be made plain beyond any possibility of mistake that what is being taken over are the ABMA personnel and facilities, not the ABMA way of doing business."[94] After reading an article by Walter Dornberger on the lessons of Peenemünde, Deputy Administrator Hugh L. Dryden concluded "I have been generally familiar with the V–2 operation, and I have talked with many of the scientists and engineers involved. The general principles of the required management are well known; it seems difficult to get them adopted in a democracy."[95]

But ABMA was too important to ignore. NASA had to depend on the Army for boosters, and Saturn was a key to civilian space exploration. Glennan respected his agreement not to try to absorb ABMA, but his subordinates had other ideas. "We should move in on ABMA in the strongest possible way," his assistant Wesley L. Hjornevik argued, urging Glennan to seek "a beachhead on the big cluster." Hjornevik, however, worried that ABMA might not "play ball right down the line," and suggested "making clear to ABMA that we don't propose to delegate control or responsibility."[96]

The Army and NASA nonetheless began to work under their ambiguous relationship. Medaris and Glennan maintained proper but cool relations. Glennan rejected Medaris's suggestion to add ABMA representatives to NASA research advisory committees, and dispatched a NASA representative to Huntsville.[97] NASA contracted with ABMA to provide eight Redstones for early Project Mercury suborbital flights; reconfigured Mercury-Redstones would be the workhorses of the early manned space program. ABMA continued work on the clustered Saturn booster, which figured prominently in NASA's long-range plans.

Development of the first stage H–1 engine, which would be clustered to power the first stage, proceeded as ABMA considered proposed configurations for other stages.[98]

Project Saturn elicited controversy from the start, and was the catalyst that led to the transfer of ABMA to NASA. ABMA's position became increasingly untenable, its mission at odds with its capabilities. Project Saturn's large boosters offered power far beyond anything needed by the Army under Department of Defense directives for military missile programs. So while the Air Force and NASA needed large boosters, their capabilities in this field were less than those of the Army, which was forbidden to use them. The Air Force used this logic in proposing the transfer of the von Braun team to its cognizance.[99]

Herbert F. York, the Defense Department's director of Research and Engineering, posed a more serious challenge. York believed that big boosters should be developed under NASA, and that Saturn was becoming both a distraction and a financial drain on DOD's resources. "Von Braun, Medaris, and ABMA were and had been seriously interfering with the ability of the Army to accomplish its primary mission," York recalled. "Whenever the Army was given another dollar, Secretary Brucker put it into space rather than into supporting the Army's capability for ground warfare."[100] In April, York issued an order to cancel Saturn, arguing that there was "no military justification" for the large booster.[101]

York's decision cast doubts on the future not only of Saturn, but of ABMA itself. In bitter memoirs, Medaris described what he considered a well-orchestrated plan by "project snatchers" to sever von Braun's group from the Army. He described the dilemma: "By this time it was crystal clear to both von Braun and myself that we were faced with a Solomon's choice—either we could hold firm in an attempt to keep the von Braun group in the Army, being sure that in doing so we were guaranteeing that their space capabilities would die on the vine, or we could support the effort to take the von Braun organization out of the Army and hope that a fond and wealthy foster parent could be found."[102]

The only potential foster parents were the Air Force and NASA. The Air Force, which would have fallen under York's strictures in any case, was an anathema to Medaris and von Braun. Von Braun feared that Air Force reliance on contractors, and aircraft industry hostility to major in-house activities operated by the government, would have led to decay of his team under the Air Force. NASA had drawbacks, too. Eisenhower and his science advisors favored a civilian space program, but one in which space would have to compete with other scientific research programs for federal dollars, so funding could be limited.[103] In contrast, pressures of the Cold War, which by now included allegations of a missile gap between the United States and the Soviet Union, seemed to promise a continued military program. Nonetheless, to Medaris and von Braun, NASA seemed the lesser evil.

Discussions between Defense and NASA continued through the summer and into the autumn of 1959. York, who later claimed that he was "largely responsible" for the transfer of the von Braun group, approached Glennan and proposed another attempt. Glennan agreed, although York admitted "there was more push on my part than there was pull on his part." York conferred with McElroy and the President, and won their concurrence.[104] By 6 October, negotiators hammered out an agreement to transfer von Braun's Development Operations Division of ABMA to NASA, and to assign NASA "responsibility for the development of space booster vehicle systems of any generations beyond those based upon IRBM and ICBM missiles as first stages."[105]

Medaris and von Braun attacked the agreement. Medaris announced that he would retire, and von Braun threatened to do the same.[106] Brucker privately assured von Braun that his team could stay together and continue to work on Saturn under NASA, and later claimed that von Braun "expressed to me at the time not only a willingness, but finally a desire" for the transfer.[107]

From ABMA to the George C. Marshall Space Flight Center

President Eisenhower met with Glennan, McElroy, Dryden, York, and his top science advisers on 21 October and approved the transfer.[108] Glennan suggested that the new NASA facility be named for General George C. Marshall because of his "image of a military man greatly dedicated to the cause of peace." Marshall's Nobel Peace Prize, initiation of the Marshall Plan, and service as secretary of state obviated concerns about the propriety of naming a civilian space center after a military man. Eisenhower agreed, saying "I can think of no one whom I would more wish to honor."[109]

The President forwarded a formal transfer plan to Congress on 14 January 1960. Under the terms of the 1958 Space Act, the transfer would become effective in 60 days unless Congress adopted a resolution opposing it. Joint Army-NASA support made opposition unlikely, but rumors persisted that von Braun had been "clubbed" or "blackmailed," that communications between Defense and NASA had broken down. The Senate Committee on Aeronautical and Space Sciences held hearings in February to determine if there were difficulties that might impede transfer. General Medaris, by then retired, offered the most volatile testimony, explaining that "With the army's total inability to secure from the Department of Defense sufficient money or responsibility to do the space job properly, we found ourselves in the position of either agreeing with the transfer of the team or watching it be destroyed by starvation and frustration." But even Medaris conceded that "this transfer is the least bad solution that can be found, and I therefore support it."[110]

Nothing rose in hearings in either the House or Senate that threatened to derail the plan. The House even passed a resolution urging immediate implementation. The Senate failed to follow suit, however, and the plan became effective on 14 March after the expiration of the 60-day statutory waiting period. President Eisenhower issued an executive order on 15 March making the action official.

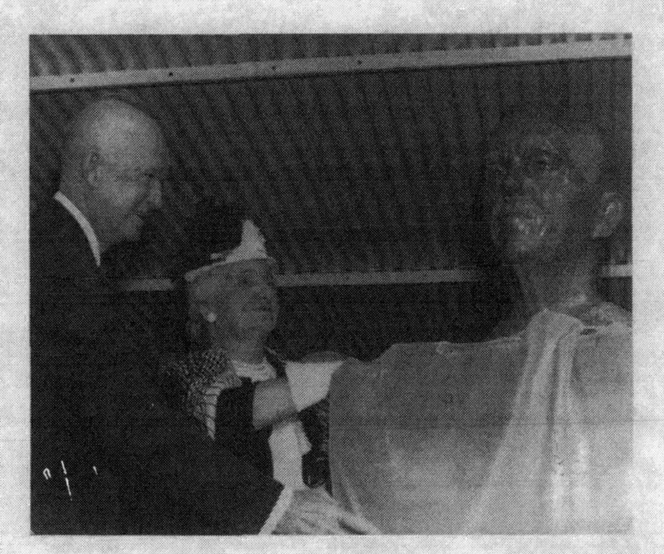

President Eisenhower and Mrs. George C. Marshall unveiling the bust of General Marshall at MSFC dedication.

The transfer would be effective on 1 July to coincide with the start of a new fiscal year, allowing time to work out final details of the arrangement. NASA received all unobligated Saturn funds immediately, although it did not assume full responsibility for Saturn until July.[111]

Von Braun remained at the head of his organization and became the director of the George C. Marshall Space Flight Center. The transfer shifted 4,670 people to NASA. NASA took control of 1,200 acres at Redstone Arsenal under a 99-year, non-revocable, renewable use permit, and received facilities of the Development Operations Division of ABMA valued at $100 million, of which $14 million was at Cape Canaveral. ABMA's Missile Firing Laboratory at the Cape became the Launch Operations Directorate under NASA, with Debus of the von Braun team retained as its director. The operational laboratories under ABMA's Development Operations Division became the new divisions of the new space center.[112]

George C. Marshall Space Flight Center became a reality in a quiet ceremony on 1 July. Major General August Schomburg, commander of the Army Ordnance Missile Command, said he felt like the father of the bride, commenting that the Army had provided a sizable dowry. "And I don't mean to imply that this is a shotgun wedding," he joked.[113] On 8 September, President Eisenhower dedicated the Center in a ceremony attended by Marshall's widow, and highlighted by the unveiling of a granite bust of the general which now stands in the lobby of the Marshall Center headquarters.

Marshall was now a full-fledged unit of NASA. For most employees, the change made little difference. Kingsbury remembered that on 1 July, "about 4,000 of us were told, 'You now work for somebody else. Your check will have a green stripe down the middle.' That was the only difference."[114]

But the year of controversy preceding transfer of the Development Operations Division had ramifications. Von Braun's decision to stay with the Army kept his team together, but also kept it out of NASA during the Agency's formative first year, limiting its role in the early development of the American civilian space program. During that year a small group of engineers from Langley, designated the Space Task Group (STG), assumed a role at the center of NASA planning for manned space flight. Comprised of only 35 members at NASA's founding, STG's numbers swelled to 350 by July 1959.[115] Suspicion of ABMA's approach—arsenal system, reliability testing, engineering conservatism—took hold among NASA administrators. One account of the Apollo program claimed that von Braun's people "had missed their chance to run the whole mission when they had stayed with the Army for the first year after NASA was founded."[116]

Other uncertainties clouded Marshall's future. The new Center had responsibility for "research and development of large launch vehicle systems" under NASA; Saturn would remain its major project. But would NASA allow Marshall to broaden its mission beyond propulsion? NASA recognized its new acquisition as "a team of outstanding experts who are capable not only of 'in-house' research and development of large launch vehicles, but also of providing, as needed, the responsible technical monitoring and direction of the various industrial contractors who assist in the engineering and production of such launch vehicles."[117] Would Marshall maintain this in-house capability under NASA? In 1960, even the extent of the national commitment to space was not clear, nor had the military relinquished interest in space. Eisenhower's visit to Huntsville to dedicate Marshall took place just two months before the 1960 presidential election. The questions surrounding the new Center's future would be decided under a new administration.

---

1 *Huntsville Times*, 8 September 1960.

2 The Manned Spacecraft Center in Houston (later renamed the Johnson Space Center) was approximately the same size as Marshall in terms of personnel in the 1960s and 1970s.

3 James R. Hansen, *Engineer in Charge: A History of the Langley Aeronautical Laboratory*, 1917–1958 (Washington: NASA SP–4305), pp. 187–88; Alex Roland, "Model Research: The National Advisory Committee for Aeronautics, 1915–1958" (Washington: NASA SP–4103, 1985); Loyd S. Swenson, Jr., James M. Grimwood, and Charles C. Alexander, *This New Ocean: A History of Project Mercury* (Washington: NASA SP–4201, 1966), pp. 6, 8.

4 Helen Brents Joiner, "The Redstone Arsenal Complex in the Pre-Missile Era: A History of Huntsville Arsenal, Gulf Chemical Warfare Depot, and Redstone Arsenal, 1941–1949" (Redstone Arsenal, Alabama: Historical Division, Army Missile Command, 22 June 1966), passim.; Elise Hopkins Stephens, *Historic Huntsville: A City of New Beginnings* (Woodland Hills, California: Windsor Publications, 1984), pp. 82–113.

5 Hansen, pp. 187–217; Arnold S. Levine, *Managing NASA in the Apollo Era* (NASA SP–4102, 1982), pp. 9–11; Swenson, Grimwood and Alexander, pp. 9–10.

6 The two most recent comprehensive treatments of the German roots of the group of rocket experts that accompanied Wernher von Braun to Huntsville in the 1950s, and especially of their experiences at Peenemünde, are Michael J. Neufeld, *The Rocket and the Reich: Peenemünde and the Coming of the Ballistic Missile Era* (New York: The Free Press, 1995); and Ernst Stuhlinger and Frederick I. Ordway III, *Wernher von Braun, Crusader for Space: A Biographical Memoir* (Malabar, Florida: Krieger Publishing Company, 1994).

7 Ernst Stuhlinger and Frederick I. Ordway III, *Wernher von Braun, Crusader for Space: A Biographical Memoir* (Malabar, Florida: Krieger Publishing Company, 1994), p. 15; Walter A. McDougall, *... the Heavens and the Earth: A Political History of the Space Age*

(New York: Basic Books, Inc., 1985), pp. 20–40. Stuhlinger and Ordway's book is the most comprehensive biography of von Braun.

8 Historian Michael J. Neufeld has argued, however, that the lack of a clause forbidding rocketry in the Versailles Treaty has been overrated as factor in the army's interest in rockets since the Army pursued development of other prohibited weapons, such as poison gas. Neufeld, "The Guided Missile and the Third Reich: Peenemünde and the Forging of a Technological Revolution," in Monika Remmeberg and Mark Walker, eds., Science, *Technology and National Socialism* (Cambridge: Cambridge University Press, 1994), p. 56.

9 Cited in Michael J. Neufeld, "Peenemünde-Ost: The State, the Military, and Technological Change in the Third Reich," paper presented at the International Congress of the History of Science, Hamburg, West Germany, 2 August 1989, p. 3. Neufeld cites von Braun team member Gerhard Reisig in attributing the phrase to Dornberger.

10 Walter R. Dornberger, "The Lessons of Peenemünde," *Astronautics* (March 1958), p. 18.

11 Neufeld, "Peenemünde-Ost: The State, the Military, and Technological Change in the Third Reich," pp. 4–6, 10–11; Ernst Stuhlinger, Oral History Interview by Stephen P. Waring and Andrew J. Dunar (hereafter OHI by SPW and AJD), 24 April 1989, Huntsville, Alabama.

12 Georg von Tiesenhausen, OHI by AJD and SPW, 29 November 1988, Huntsville, Alabama; Neufeld, pp. 10–11.

13 Stuhlinger, OHI, 24 April 1989.

14 Major General John B. Medaris with Arthur Gordon, *Countdown for Decision* (New York: G. P. Putnam's Sons, 1960), pp. 37–38.

15 General Walter Dornberger, "Rockets," interview by F. Zwicky, 1 October 1945, Von Braun Interrogations in Germany folder, NASA History Division Documents Collection; Von Braun, "An Historical Essay."

16 Dieter K. Huzel, *Peenemünde to Canaveral* (Englewood Cliffs, New Jersey: Prentice-Hall, 1962; reprinted by Greenwood Press, 1981), pp. 57–63; Michael J. Neufeld, *The Rocket and the Reich: Peenemünde and the Coming of the Ballistic Missile Era*, (New York: The Free Press, 1995), pp. 198–200; Ordway and Sharpe, pp. 111–23.

17 James McGovern, *Crossbow and Overcast* (New York: William Morrow & Co., 1964), pp. 46–54; Albert Speer, *Inside the Third Reich* (New York: The Macmillan Company, 1970), pp. 371–72.

18 "History of German Guided Missiles;" Christopher Simpson, *Blowback: America's Recruitment of Nazis and Its Effects on the Cold War* (New York: Weidenfeld & Nicolson, 1988), pp. 28–31.

19 Ordway and Sharpe, pp. 405–408. Appendix A in this volume discusses sources for estimates of V–2 production.

20 "History of German Guided Missiles;" Speer, pp. 365–66.

21 "History of German Guided Missiles."

22 McGovern, pp. 110–17; Dieter K. Huzel, *Peenemünde to Canaveral* (Westport, Connecticut: Greenwood Press, 1962), pp. 151–65; Ordway and Sharpe, pp. 261–67.

23 McGovern, pp. 141–45.

24 McGovern, pp. 96, 100–03; Ordway and Sharpe, pp. 277–78. Toftoy's position was chief of Ordnance Technical Intelligence in Europe; Staver was assigned to the Ordnance Technical Division in London. Trichel would soon retire, but his contribution to Army

rocketry and to the assembly of the von Braun team was crucial. Staver and Toftoy (and Toftoy subordinate James Hamill) played key roles in Project Paperclip and in the transfer of V–2s to the United States, and Robert Porter of Project Hermes was one of the first Americans to interrogate the Germans. Toftoy credited Trichel with "planning and working out the original Army missile program," and for doing the ground work from which the program developed. Cited in Shirley Thomas, *Men of Space*, Vol. 3 (Philadelphia: Chilton Company, 1961), p. 220.

25 McGovern, pp. 154–75. The Army shipped the V–2 parts, which filled the hulls of sixteen Liberty ships, to the newly commissioned White Sands Proving Ground in New Mexico, storing them on the open desert. Major General H.N. Toftoy and Colonel J.P. Hamill, "Historical Summary on the Von Braun Missile Team," 29 September 1959, p. 2, University of Alabama in Huntsville (UAH) Saturn Collection.

26 Clarence A. Lasby, *Project Paperclip: German Scientists and the Cold War* (New York: Atheneum, 1971), pp. 51–61.

27 Christopher Simpson, *Blowback: America's Recruitment of Nazis and Its Effects on the Cold War* (New York: Weidenfeld & Nicolson, 1988), pp. 32–36. Regarding von Braun's SS award, Ernst Stuhlinger explained that Himmler tried to lure von Braun from the army to the SS, presenting the honorary rank in 1940 and a promotion in 1943. Von Braun's refusal contributed to Himmler's decision to imprison him in a Gestapo prison in 1944. Stuhlinger, undated comments to authors on chapter draft; Stuhlinger and Ordway, pp. 32–33.

28 Simpson, pp. 28–30. The case of Arthur Rudolph was another matter, for Rudolph served as production manager at Mittelwerk. Rudolph, who earned honors as the chief of the Saturn booster program in the 1960s, became the center of controversy in the 1980s when a Justice Department investigation reopened his case. First investigated in 1947, Rudolph was at that time cleared for entry into the United States. For information on Dornberger's and von Braun's influence on production schedules, see Stuhlinger to Michael Wright, comments on draft chapter, 1990, MSFC history office.

29 Stuhlinger and Ordway, pp. 46–53.

30 David Irving, *The Mare's Nest* (Boston: Little, Brown and Company, 1964, 1965), p. 21; McGovern, pp. 19–20; Ordway and Sharpe, pp. 41–42.

31 Eberhard Rees, OHI by Donald Tarter and Konrad Dannenberg, Huntsville, Alabama, 1985.

32 Simpson, pp. 32–34; McGovern, p. 78.

33 Lasby, pp. 88–92, 102, 114–16, 124–25; McGovern, pp. 73–74.

34 The six to accompany von Braun were Wilhelm Jungert, Erich Neubert, Theodore Poppel, Eberhard Rees, Wilhelm Schulze, and Walter Schwidetzki. William Joseph Stubno, Jr., "The Impact of the Von Braun Board of Directors on the American Space Program," (MA thesis, University of Alabama in Huntsville, 1980), p. 23.

35 Ordway and Sharpe, pp. 310–14; quotation is from Toftoy and Hamill, "Historical Summary on the Von Braun Missile Team."

36 Ordway and Sharpe, pp. 346–49; Stuhlinger, OHI, 24 April 1989; Erich Neubert, OHI by AJD, 27 July 1989, Huntsville, Alabama.

37 Colonel Holger N. Toftoy, press release, 10 May 1946, Press Release by Col. Toftoy file, Dr Wernher von Braun, White Sands/Fort Bliss 1945 (Sept.)—1950 (April) Drawer, Von

Braun Collection, ASRC; Stuhlinger, OHI, 24 April 1989; E. H. Krause, "General Introduction" to "Upper Atmosphere Research Report No. I," Naval Research Laboratory, (1 October 1946), pp. 1–2.

38 "What We Have Learned from V–2 Firings," *Aviation Week* (26 November 1951), 23 ff; David S. Akens, "Historical Origins of the George C. Marshall Space Flight Center," MSFC Historical Monograph No. 1 (Huntsville: MSFC, 1960), pp. 30–35. The launching that won the most publicity was an errant V–2 that strayed over the Mexican border toward Ciudad Juarez. A tragedy was averted when the rocket landed harmlessly on a hillside, and the Germans could later joke that they had launched the first American missile against a foreign nation.

39 Michael J. Neufeld, who is conducting a study of Peenemünde, has found no term corresponding to "team" in the Peenemünde period, and suggests that the closest equivalent might be the term "Arbeitsgemeinschaft Vorhaben Peenemünde" (roughly translated as Peenemünde Project Working Group or Working Community), a label used to signify some of the university-technical links that relied on personal ties as well. Neufeld to SPW, 29 August 1989.

40 Stuhlinger and Ordway, pp. 74–75.

41 Hamill to von Braun, memorandum, 7 February 1946, Personnel: Regulations, Restrictions, 1946 Fort Bliss File, Dr. Wernher von Braun White Sands-Fort Bliss, 1945 (Sept.)-1950 (April) Drawer, Von Braun Papers, U.S. Space and Rocket Center, Huntsville, Alabama.

42 Ordway and Sharpe, p. 351.

43 Von Braun to Hamill, 12 January 1948, Correspondence—Special Letters File, White Sands-Fort Bliss, 1945 (Sept.)-1950 (April) Drawer, Von Braun Papers, U.S. Space and Rocket Center; Ordway and Sharpe, p. 351.

44 Thomas, p. 223; Ordway and Sharpe, pp. 358–59.

45 Among the canceled projects was a 1947 proposal developed by von Braun, Hamill, and Eberhard Rees to develop a 200-ton thrust rocket motor. Major General H.N. Toftoy and Colonel J.P. Hamill, "Historical Summary on the Von Braun Missile Team," p. 10, UAH Saturn Collection.

46 Erik Bergaust, *Wernher von Braun* (Washington, D.C.: National Space Institute, 1975), pp. 151–53; Peter Cobun, "Toftoy's Foresight Led to Huntsville's Rebirth," *Huntsville Times*, 4 June 1989; Toftoy and Hamill, p. 10; Stuhlinger and Ordway, p. 97.

47 Helen Brents Joiner and Elizabeth C. Jolliff, "The Redstone Arsenal Complex in its Second Decade, 1950–1960" (Army Missile Command, Redstone Arsenal, 28 May 1969), p. 29.

48 Hertha Heller, OHI by AJD, 16 February 1989, Huntsville, Alabama; Stuhlinger, OHI, 24 April 1989; Ordway and Sharpe, pp. 357–59.

49 Stuhlinger, OHI, 24 April 1989; Heller, OHI.

50 Ruth von Saurma, OHI by AJD, 21 July 1989, Huntsville, Alabama; Hertha Heller, OHI; "From Cotton to Space," *Huntsville Times*, 8 October 1989.

51 Paul O'Neill, "The Splendid Anachronism of Huntsville," *Fortune* 65 (June 1962), p. 155.

52 Ordway and Sharpe, pp. 367–69, 376–77; Walter Wiesman, OHI by AJD and SPW, 13 April 1989, Huntsville, Alabama; Hertha Heller, OHI. The first members of the von Braun group to become citizens took their oath in Birmingham in November 1954. Weisman, undated note to authors on chapter draft.

53 Charles R. Lundquist, OHI by SPW, 21 August 1990, Huntsville, Alabama, p. 1.
54 James Kingsbury, OHI by SPW, 22 August 1990, Huntsville, Alabama, pp. 1–2.
55 Henry Pohl, OHI by AJD and SPW, 13 July 1990, Houston, Texas, pp. 1–2.
56 Joe Lombardo, OHI by Jessie Whalen, 12 May 1989, Huntsville, Alabama, p. 4.
57 Stan Reinartz, OHI by SPW, 10 January 1991, Huntsville, Alabama, p. 1.
58 Lee B. James, OHI by SPW, 1 December 1989, Huntsville, Alabama, p. 1.
59 Richard A. Marmann, OHI by AJD, 19 July 1994, Huntsville, Alabama.
60 Jack Waite, OHI by Sarah A. Kidd, 5 August 1992, Huntsville, Alabama, p. 1.
61 Sylvia Doughty Fries, *NASA Engineers and the Age of Apollo* (Washington, D.C.: NASA SP–4104), pp. 102–03.
62 Leland Belew, OHI by SPW, 1 September 1990, Huntsville, Alabama, pp. 1–2.
63 William R. Lucas, OHI by AJD and SPW, 19 June 1989, Huntsville, Alabama, pp. 1–2.
64 William C. Snoddy, OHI by AJD and SPW, 22 July 1992, p. 1.
65 Arthur E. Sanderson, OHI by AJD, 20 April 1990, Huntsville, Alabama, pp. 1–2.
66 Snoddy, OHI, pp. 3–5.
67 Toftoy and Hamill, pp. 10–13; Akens, pp. 36–38; Ordway and Sharpe, pp. 370–74.
68 Stuhlinger, OHI, 24 April 1989.
69 Karl Heimburg, OHI by AJD and SPW, 2 April 1989, Huntsville, Alabama.
70 Akens, p. 37; Heimburg, OHI; Ordway and Sharpe, pp. 372. James Kingsbury, former director of science and engineering at MSFC, note that the $75,000 figure is not truly a fair basis for comparison, since it included salaries while the $1,000 figure did not. Kingsbury, undated note to authors on chapter draft.
71 Kurt H. Debus, "From A4 to Explorer I," paper presented at 24th International Astronautical Congress, Baku, USSR, 8 October 1973, Debus/1973/Redstone/Pershing/Jupiter folder, NASA History Division Documents Collection, Washington, D.C.
72 Wernher von Braun, "Man on the Moon: The Journey," *Collier's* (18 October 1952), pp. 52–59. The *Collier's* series continued with occasional articles by von Braun and other authorities on space travel in 1953 and 1954. Another von Braun article (27 June 1953) advocated the establishment of a Space Station.
73 Von Braun's proposal, titled "A Minimum Satellite Vehicle Based Upon Components Available From Missile Development of the Army Ordnance Corps," advocated using the Redstone for a launch vehicle.
74 Akens, pp. 38–39.
75 Ordway and Sharpe, p. 376.
76 Helmut Hoelzer, OHI by AJD, 25 July 1989, Huntsville, Alabama; Neubert, OHI; Debus, p.33; Medaris, p. 72.
77 Patricia Yingling White, "The United States Enters Space, 1945–1958: A Study of National Priorities and the Decision-Making Process in the Artificial Satellite Program," (MA Thesis, Ohio State University, 1969), pp. 90–93; Akens, pp. 42–43.
78 Debus, p. 36; Medaris, pp. 142–44; William Lucas, OHI by AJD and SPW, 19 June 1989, Huntsville, Alabama. H. Julian (Harvey) Allen of Ames Research Center, who developed the blunt-body theory used for Project Mercury capsule re-entry, also conducted pioneering experiments on ablation in the 1950s independent of the work at ABMA. Muenger, Elizabeth, *Searching the Horizon: A History of Ames Research Center*, 1940–1976 (Washington: NASA SP–4304, 1985), pp. 66–67, 131–32.

79 William Lucas, OHI by SPW, 4 April 1994, Huntsville, Alabama, p. 7.

80 White, p. 95; Medaris, pp. 151–90; Ordway and Sharpe, pp. 382–83; Akens, pp. 44–47.

81 Robert A. Divine's *The Sputnik Challenge: Eisenhower's Response to the Soviet Satellite* (New York: Oxford University Press, 1993) is a comprehensive examination of Eisenhower's attempts to respond calmly in the midst of national panic after the launch of Sputnik. The IGY was a 1957 program of international cooperation in Earth science research.

82 Ordway and Sharpe, pp. 382–86; Debus, pp. 42–54; Medaris, pp. 207–26; Stuhlinger and Ordway, pp. 134–40.

83 Swenson, Grimwood, and Alexander, pp. 25–28; White, pp. 11–112; John M. Logsdon, *The Decision to Go to the Moon: Project Apollo and the National Interest* (Cambridge, Mass.: MIT Press, 1970), pp. 28–29; Alison Griffith, *The National Aeronautics and Space Act: A Study of the Development of Public Policy* (Washington, DC: Public Affairs Press, 1962), pp. 10–11.

84 White, pp. 114–20; McDougall, pp. 151–52.

85 White, pp. 110–11; Swenson, Grimwood, and Alexander, pp. 82–83; Griffith, p. 14; Akens, p. 67.

86 J. Boehm, H. J. Fichtner, and Otto A. Hoberg, "Explorer Satellites Launched by Juno 1 and Juno 2 Vehicles," in *Peenemünde to Outer Space*, edited by Ernst Stuhlinger, et. al. (Huntsville: Marshall Space Flight Center, 1962), pp. 163–65.

87 ABMA initially proposed Saturn in December 1957 in a report titled "Proposal for a National Integrated Missile and Space Vehicle Development Program." Akens, pp. 58–60; Wernher von Braun, "The Redstone, Jupiter, and Juno," *Technology and Culture*, IV(Fall 1963).

88 ABMA, "Development Proposal for Project Adam," (17 April 1958), pp. 1–3; Norman L. Baker, "Air Force Won't Support Project Adam," *Missiles and Rockets* (June 1958), pp. 20–21; Swenson, Grimwood, and Alexander, pp. 99–101, 177.

89 NASA assumed control of the following NACA facilities: Wallops Island Station and Langley Research Center in Virginia; Lewis Research Center in Cleveland; Ames Research Center at Moffett Field and the Flight Research Center at Edwards Air Force Base in California.

90 Richard L. Smoke, "Civil-Military Relations in the American Space Program, 1957–60" (BA Honors Thesis, Harvard University, 1965), pp. 70–71.

91 Smoke, pp. 73–75; Medaris, pp. 242–45; Hoelzer, OHI.

92 *Washington Post*, 16 October 1958; Jim G. Lucas, "Army Expects to Lose Von Braun," *New York World-Telegram & Sun*, 31 October 1958; "The Periscope," *Newsweek*, 10 November 1958.

93 Smoke, pp. 75–76; Medaris, pp. 244–47, 264; T. Keith Glennan and Donald A. Quarles, Press Conference, 3 December 1958, Folder 4.8.2, NASA History Division Documents Collection, Washington, D.C.

94 Memorandum, Walter T. Bonney to Glennan, 30 September 1958, NASA-Army (ABMA) folder, NASA History Division Documents Collection, Washington, D.C.

95 Dryden to Lieutenant General Arthur G. Trudeau, 25 February 1959, Krafft Ehricke—The Peenemünde Rocket Center folder, Vertical file, JSC History Office. Dornberger had

come to the United States as part of Project Paperclip, but not with the von Braun group. He worked for Bell Aircraft Corporation in Buffalo, New York. The article to which Dryden responded was "The Lessons of Peenemünde" *Astronautics* (March 1958), pp. 18–20, 58–60.

96 Memorandum, Wesley L. Hjornevik to Glennan, 20 January 1959, Folder 4.8.2 ABMA, NASA History Division Documents Collection, Washington, D.C.

97 Medaris to Glennan, 18 March 1959, and Glennan to Medaris, 26 March 1959, Folder 4.8.2 ABMA; Memorandum, Glennan, 29 April 1959, NASA History Division Documents Collection, Washington, D.C.

98 Von Braun, "The Redstone, Jupiter and Juno;" Roger E. Bilstein, *Stages to Saturn: A Technological History of the Apollo/Saturn Launch Vehicles* (Washington: NASA SP–4206, 1980), pp. 35–37. ABMA also continued to work on ARPA projects, including several satellite missions and deep space probes launched by the Juno 2 series of rockets. Successful missions in this series in 1959 included the launch of Pioneer 4, a deep space probe which went into orbit around the Sun on 3 March; and the launch of Explorer VII, which contained x-ray and cosmic ray experiments on 13 October. The most acclaimed mission of the year was the launch (using a Jupiter booster) and recovery of two monkeys, Able and Baker, on 28 May 1959. Finally, ABMA supervised development of the Pershing, a solid fuel missile, by the Martin Company of Orlando, Florida. Akens, pp. 52–58; U.S. Army Ordnance Missile Command press release, 26 October 1959.

99 Smoke, pp. 78–79.

100 York to Eugene Emme, 10 June 1974, UAH Saturn Collection.

101 Eugene M. Emme, "Historical Perspectives on Apollo," *Journal of Spacecraft and Rockets* 5(April 1968), p. 372. York later acknowledged legitimate Defense interest in Saturn, but not until after approval of the ABMA transfer. At a press conference on 23 October 1959, he granted that "people in Defense regard it as probable that a need will develop, may develop over the next few years. So we do have a positive interest in Saturn because we feel that despite the fact we have no clear-cut need this is a rapidly changing world, space is a new environment, and under no conditions could we foreclose on the possibility that we would need one." Herbert F. York, NASA press conference, 23 October 1959, Washington, D.C., Folder 4.8.2 ABMA, NASA History Division Documents Collection, Washington, D.C.

102 Medaris, p. 266.

103 John M. Logsdon, *Decision to Go to the Moon: Apollo and the National Interest*, (Cambridge, Mass.: MIT Press, 1970), pp. 17–18, 33.

104 York to Eugene Emme, 2 May 1973, UAH Saturn Collection.

105 Draft Memorandum, NASA Administrator and Secretary of Defense to the President, 6 October 1959, NASA-Army (ABMA) folder, NASA History Division Documents Collection, Washington, D.C.

106 *Washington Post*, 21 October 1959.

107 U.S., Congress, House, Committee on Science and Astronautics, hearing, "Transfer of the Development Operations Division of the Army Ballistic Missile Agency to the National Aeronautics and Space Administration: Hearings pursuant to H. J. Res. 567," 86th Cong., 2d sess., 3 February 1960, p. 12.

108 Eisenhower statement, 21 October 1959, ABMA-Transfer to NASA folder, NASA History Division Documents Collection, Washington, D.C.

109 Glennan to Major General Wilton B. Persons, 17 December 1959, Transfer from ABMA file, UAH Saturn Collection.

110 "Transfer Plan Making Certain Transfers from the Department of Defense to the National Aeronautics and Space Administration," 14 January 1959, NASA-Army (ABMA) folder, NASA History Division Documents Collection, Washington, D.C.; Smoke, pp. 88–89; U.S., Congress, Senate, NASA Authorization Subcommittee of the Committee on Aeronautical and Space Sciences, "Transfer of Von Braun Team to NASA: Hearings pursuant to H. J. Res. 567," 86th Cong., 2d sess., 18 February 1960, pp. 2–3, 8–9, 39.

111 Akens, pp. 76–77.

112 ABMA facilities transferred to NASA included: Computation Laboratory (58,465 square feet); Aeroballistics Laboratory (38,860 square feet), including two wind tunnels; Fabrication and Assembly Engineering Laboratory (348,411 square feet), including two missile assembly shops and the Structural Fabrication Building; Guidance and Control Laboratory (306,475 square feet); Test Laboratory (186,614 square feet), including the blockhouse and Static Test Tower; Research Projects Laboratory (7,000 square feet); Missile Firing Laboratory (135,000 square feet), of which the major facilities were located at Cape Canaveral, Florida, and which became the Launch Operations Directorate on 14 June. "Army-NASA Transfer Plan," 11 December 1959, Appendix C in Akens; Akens, pp. 77–80; "Facilities Given NASA Worth $100,000,000," *Redstone Rocket* (6 July 1960), pp. 9–10. Akens details the Army-NASA agreements reached during the negotiations between 14 March and 1 July.

113 *Huntsville Times*, 1 July 1960.

114 James Kingsbury, OHI by SPW, 22 August 1990, p. 2.

115 Swenson, Grimwood, and Alexander, p. 149.

116 The account continued, "They could have come in, after NASA's civilian supremacy was conceded, if they had been willing to desert the Army unceremoniously, but von Braun had not been willing to do that and he had reconciled himself to the consequences." Charles Murray and Catherine Bly Cox, *Apollo: The Race to the Moon* (New York: Simon and Schuster, (1989), p. 136.

117 Albert F. Siepert, "Relating to the Transfer of the ABMA Development Operations Division to NASA," 3 February 1960, NASA Release No. 60–121.

Chapter II

The Center in the Saturn Years: Culture, Choice, and Change

When Huntsville's rocketeers transferred from the Army, they brought a unique organizational culture to NASA. Marshall's laboratories had a technical ethos which sought control over all phases of a space project, from design, development, manufacture, and testing, all the way to launch. The labs could, and did, manufacture anything from subscale engineering prototypes to Redstone missiles. The Center's contract managers already had experience in directing missile development. Heading the team was von Braun, one of the most charismatic leaders of any American organization.

In its first decade in NASA, the Marshall Center helped make American space plans, and those plans in turn reshaped the Center. The Center influenced decisions to undertake a manned lunar landing, select the Saturn launch vehicles, and choose a mode for going to the Moon, and in the process formed patterns of interaction with NASA Headquarters and other field centers. The plans and the subsequent work on the Saturn boosters changed Marshall in various ways, leading it to add personnel and facilities, enhance its capabilities in project management and systems engineering, and help NASA create a launch center. Indeed, it would be no exaggeration to say that the Apollo Program shaped Marshall's first decade.

Dirty Hands

In 1960 NASA's newest field center was fundamentally a rocketry research organization with a professional engineering code that sought hands-on control over all phases of booster development and operation. The foundation of Marshall's organization and culture in 1960 was the "Army arsenal system" in which Civil Servants performed all types of technical work. Rather than being primarily supervisors of contractors, Center personnel were hands-on designers, testers, manufacturers, and operators.

The arsenal approach was a legacy from the German and American military but was similar to the laboratory culture of NASA's other field centers. Government research organizations, whether military or civilian, evolved because business initially had limited interest and expertise in rocketry or aerospace research. Moreover, in the 1950s, rocketry was still relatively unexplored technology, and pioneers in the field faced many unanticipated problems that made contracting problematic. As Dr. Ernst Stuhlinger, the chief of the Center's Research Projects laboratory, recalled, "it is very difficult to tell them [industry] just exactly what to build, because we don't know ourselves before we have begun with some experiments."[1] Dr. William Lucas, a materials specialist in the Structures and Mechanics Lab and later Marshall's Center director, remembered that "in the early days, we could go from the idea to the proving ground," because there were "not [industry] people who wanted to do this or were able to do it."

The ABMA experience with the Redstone missile illustrated the problem. When ABMA asked industry to make bids for the project, no business responded, and the Department of Defense had to convince Chrysler Corporation to take on the job. Even so ABMA was the innovator; its labs designed and built the first 17 Redstones, trained Chrysler personnel, and only then turned the work over to the company. Lucas explained "it wasn't a matter of going to the contractor and saying 'do this for us,'" and then assigning the firm a task it had done before. Marshall had to find contractors and say "here's what we want you to do" and then show them how to do it.[2]

The arsenal system showed in various ways. Despite Marshall's location among wooded hills and lush valleys, the physical appearance of the Center was industrial and was in stark contrast to some other NASA field centers that looked like college campuses. The center's layout displayed a functional character, with areas for management, engineering, manufacturing, and testing. The architecture also looked industrial, with utilitarian office buildings, cavernous factory structures, and huge test facilities, all linked by a web of electrical wires and above-ground pipes.

Marshall's original organization was also industrial and much like a large aerospace company. Each of the Center's eight laboratories had a functional specialty and its own technical facilities; together they could design, test, and build rockets or almost any other kind of aerospace hardware. The Aeroballistics Laboratory, later called Aero-Astrodynamics, used wind tunnels and vacuum

Drafting specialists from the Propulsion and Vehicle Engineering Lab work in the Huntsville Industrial Center building.

chambers to study air flows on vehicles and developed programs to control them. The Guidance and Control Laboratory, later named Astrionics, developed systems and components for communications, guidance and control, and electrical power. Its facilities and equipment ranged from standard bench equipment like oscilloscopes to specialized test equipment, telemetry instruments, and simulators. The Research Projects Laboratory, later called Space Sciences, used smaller "plug-in" equipment for scientific research in physics, astrophysics, and thermodynamics; the lab also provided scientific support for engineering projects, helping develop several spacecraft in the Explorer series of satellites. The Computation Laboratory's computers helped administer the Center and supported research activities in the other labs.

The SA–2 booster is in final assembly stages at the NASA Marshall Space Flight Center, Huntsville, AL.

The Structures and Mechanics Lab, later called Propulsion and Vehicle Engineering, had broad capabilities in rocketry, with specialties in structural and mechanical design, materials analysis, and systems engineering. It could conduct heat transfer research, chemical and radiation analyses, cryogenic tests, and fluid and hydraulic studies. With its capability to make prototypes and test components, the Structures and Mechanics lab in itself had capabilities comparable to a rocketry corporation. The Manufacturing Engineering Laboratory could manufacture large prototypes and had high bay structures with cranes, large access doors, and machine shops. The Test Laboratory operated the huge test stands that handled the smoke-and-fire rocket tests. The Quality Laboratory also tested vehicle systems and subsystems, and had facilities ranging from high bay buildings to small bench equipment for electronic calibration tests on flight components. The Launch Operations lab had facilities in Huntsville and at Merritt Island, Florida. All in all, Marshall's laboratories had nearly comprehensive capabilities in propulsion and aerospace engineering; the Center was almost a space agency in miniature.[3]

Center officials believed in the arsenal system. Convinced that it should be more than a transitionary step in the maturation of aerospace industry, they argued that the system improved quality, accelerated progress, and contained costs. Von Braun argued that in-house design and manufacturing capability attracted engineers and specialists who wanted to build things rather than shuffle paper. It also trained young engineers fresh out of college, who had more theoretical than practical knowledge, and gave them industrial experience.[4]

Marshall engineers also believed that the arsenal system improved quality and reduced red tape. They appreciated working with in-house machinists and craftsmen. Typical of their views were the comments of Peter Broussard, an engineer in the Sensor Branch of the Guidance and Control Lab whose team developed the navigation system for the lunar roving vehicle. In an arsenal system, Broussard said, "you can work hand in glove with the man that is doing it. He could call you and say, 'I don't understand this; come over and talk to me.'" Later contracting methods, he believed, were "far more expensive and far less efficient" and even after a slow process "you may not get what you contracted for."[5]

In addition, the arsenal system and the technical depth of the labs helped the Center direct its contractors. Marshall officials often contrasted the arsenal system to the Air Force system which gave business contractors much wider scope.

Saturn I booster checkout in 1961.

Lee B. James, Saturn I and Saturn V project manager, said that "the difference in managing a program at Marshall has always been the laboratories, which give our Center unusual depth." Marshall's engineers had detailed knowledge which allowed for meticulous design requirements in their contracts. In some cases, like the Redstone and the first stages for the Saturn I and V, Center personnel invented manufacturing methods and built full-scale prototypes to accelerate progress. Moreover, knowledge of engineering and manufacturing detail allowed Marshall to evaluate contractors. Building prototypes was especially effective because Marshall learned about costs, creating a "yardstick" to measure contractor prices. Karl Heimburg, chief of the Test Lab, recalled that "what industry didn't like was, since we made it ourselves here, we knew what it would cost. They would come out with a flat sum that was three times as high as it should cost. We said 'if you do it this way, we will manufacture it ourselves.' So you see they didn't like it at all that we dictated it."[6]

The intimacy with hardware produced by arsenal practice and laboratory culture affected nearly everything at the Center. Marshall developed customs of conservative engineering, meticulous quality control, testing-to-failure, dirty-hands management, matrix organization, automatic responsibility, and open communications.

Conservative engineering was a natural lesson from rocketry experience. Rockets put extreme stresses on technology, and propulsion pioneers often faced fiery failures. Lucas recalled watching his first Redstone launch. "It got about thirty feet off the ground and fell back and exploded." During any launch or test, he noted, "there were thousands of things that could go wrong," and "we knew at any time that one lousy little twenty-five cent part somewhere could cost you the whole ball game."[7]

Center engineers developed a habit of conservatism in engineering, preferring things simple and sturdy, tried and true. James Odom, chief engineer for the Saturn S–II stage, recalled that Marshall designed its hardware to be "stout," often to the point of being "over-stout." Conservative design led to technology with high margins of safety and reliability.[8] Conservatism showed in an "incremental" approach to innovation; rather than designing from scratch, Marshall preferred to build on proven concepts. For instance, the Saturn rocket engines and stages, while innovative in size, materials, and manufacturing processes, drew on the engineering knowledge and research programs of military rocketry. Even more telling, the Center used successful technology in new ways, most famously helping conceive the conversion of a Saturn S–IVB rocket stage into the *Skylab* space station. Flight tests of rockets were also conservative; under the Center's original stage-by-stage approach, first stages flew first without upper stages, and only after successful flights were live upper stages added.

Marshall used rigorous quality control and test practices. Again rocketry experience had taught Center personnel that quality had to be built into hardware from the beginning. As von Braun observed, it was "better to build a rocket in the factory than on the launch pad." The Center, especially its Quality and Reliability Assurance Lab, taught contractors how to ensure quality products and monitored their manufacturing and test procedures. Part of this was what Dieter Grau, the lab's chief, called a "rigid inspection program" in which all Center personnel, rather than only designated quality inspectors, were responsible for quality.[9]

When Center people applied this approach to contractors, they called the practice "penetration." Marshall believed in giving contractors specific design requirements and then observing their operations closely to ensure that the requirements were met. The Center's resident manager offices were key tools of penetration. Located at major contractors' plants, each had a staff of administrators and engineers who monitored work and acted as liaison between the contractor and Marshall's labs. Center specialists carefully watched the manufacturing process, discussed problems with contractor personnel, and as a result often knew more about the corporation and its products than the corporation's own management. During the resource-rich Saturn years, Marshall assigned as much as one-tenth of its workforce to resident offices. One Center manager admitted that penetration was often "traumatic" for the company at first, especially for those accustomed to working under Air Force supervision. Compared to Marshall, one contractor pointed out, the Air Force was "not in your pants all the time."[10]

One Marshall project official noted that during the Saturn program the Center would "penetrate down to excruciating detail on a continuous basis. Engineer to engineer. Designer to designer." Headquarters sometimes questioned such practices and wanted Marshall to trust its contractors more. During a visit by NASA Administrator James Webb, Center engineers showed him a rag they had found in a rocket engine and explained that such problems revealed why they mistrusted contractors.[11]

Center personnel contrasted their method of monitoring contracts with the methods used by the Air Force. When Marshall replaced the Air Force as monitor of the Centaur rocket contract, the difference became clear. The Air Force had assigned 8 officials to the project, while Marshall assigned 140. One Center engineer noted that aerospace contractors wanted Marshall to manage like the Air Force: "they [the government] give you [the contractor] the money; you go away; you deliver a product; they buy it." Marshall, he noted, did not work like this because the Center did not want to get "taken to the cleaners."[12]

Marshall people also contrasted their quality practices with those of private industry. For most of its hardware, aerospace industry and the military relied on mass production. In mass production, cheapness compensated for defects, and when a customer complained about product quality, he would receive a replacement. But NASA's launch vehicles were not mass produced, and a failure in the propulsion system could be catastrophic rather than merely inconvenient. As Grau explained, "you cannot put a man on a [launch vehicle] and say 'if it fails, and if you get killed, take the next one.'" Consequently Marshall had to change the mentality of its contractors from "mass production with acceptable errors" to "craftsmanship—do it right the first time—with no error."[13]

Marshall also questioned the statistical risk assessment methods used by aerospace contractors and the military. With mass production, engineers could use random tests and statistical measures to isolate defects and predict reliability. But since NASA built only a few vehicles and required that each work flawlessly, random tests and statistical measures of reliability seemed questionable to Marshall engineers. In 1961, Eberhard Rees, Marshall's deputy technical director, observed that NASA rules required reliability statistics, but that he did not trust the numbers; his attitude was "if they [Headquarters managers] are happy with the figures let them have it." According to Marshall lore, Headquarters asked von Braun for a reliability figure on a Saturn stage and he replied by saying it was 0.99999 reliable. The figure, the Center director said, came from calling his lab directors and asking them if the stage would cause trouble. Von Braun called five directors and they replied in turn "Nein," "Nein," "Nein," "Nein," "Nein."[14]

Marshall's confidence in its hardware resulted from rigorous testing. All the labs performed tests, and two labs, the Test Lab and the Quality Assurance and Reliability Laboratory, independently checked the work of the other labs and contractors. The two labs, remembered Walter Haeussermann, chief of the Astrionics Lab, sought to prevent the "camouflaging of short-comings." Heimburg believed that experience in rocketry had convinced Center leaders that safety and economy depended on thorough tests on the ground; with severe tests, engineers could detect and correct problems and thus minimize costly, or even deadly, launch mishaps and failures. In a response to questions from a NASA propulsion committee in 1961, Heimburg explained the Marshall policy that "each sensitive event, component, subassembly, and stage should be subjected to design evaluation testing." The tests should be realistic, using full-scale flight equipment rather than subscale models, and should occur at "exaggerated environmental conditions." The practice allowed Marshall engineers to discover failures and flaws. The goal, Lucas recalled, was to "test until we wear it out" in order to understand weaknesses. Marshall insisted that its contractors bring their hardware to Huntsville for tests, even after that hardware had already been tested at contractor facilities.

Thorough tests were of course expensive. Tests accounted for one-half of the Saturn project's total cost as measured in man-hours and material resources. Heimburg justified these costs in 1961, arguing that "a shortage of funds means a minimum of ground testing, below the optimum, which means increased mission failures. The money temporarily saved, and more, will be spent later in repetition of testing." Lucas noted that NASA reduced testing in the 1970s and based its decision on "so-called economics." Reducing tests to save money, he believed, was "one of the costliest mistakes" that NASA ever made; "maybe we overdid it [testing] on the Saturn program, but we clearly underdid it on everything since then."[15]

Organizational and managerial patterns also evolved from Marshall's arsenal practices and research culture. The key organizational custom was "automatic responsibility." Konrad Dannenberg, a Center veteran, explained that the labs, regardless of whether they had formal authority, were automatically responsible for problems in their specialty. They could not, he said, "sit in the corner" and "wait until something went wrong and say 'I told you so.'" James believed that the practice helped expedite problem solving because the lab experts "feel responsible [and] they bring these things to the program manager's attention without being asked."[16]

Automatic responsibility helped produce a matrix organization based on interdisciplinary groups. The practice, which von Braun called "teamwork," evolved from the complex tasks of aerospace and rocketry engineering. Because problems overlapped engineering specialties, no single discipline could design, develop, and evaluate an entire launch vehicle or even major subsystems. Success depended on the cooperation of specialists from many labs.[17] Moreover, as Dr. Mathias P. Siebel, deputy director of Marshall's Manufacturing Engineering Laboratory, observed, the Center was making "small quantities of high cost articles" that had to work "the first time." This meant, Siebel added, that each vehicle was a research project based on continuous innovation in response to unpredictable technical problems and program changes. Solving the problems systematically required teams with experts in design, manufacturing, quality control, testing, and operations.[18]

Accordingly, Marshall had many task-specific, interdisciplinary teams. At the beginning of each project, lab chiefs and project managers formed temporary teams with members drawn from several labs. The project managers had responsibility for budgets and schedule, and the lab chiefs had authority over technical problems. Each team and its contractor counterpart worked on a specific problem until it was resolved. For example, specialists from several labs and contractors cooperated closely on the guidance and control systems for the Saturn V. The Astrionics Lab designed the guidance and control processors and built prototypes, IBM manufactured the flight models, the Quality Lab tested the processors, the Aeroballistics Lab developed guidance equations for the processors, and the Computation Lab simulated flights in its computers and generated data for the guidance equations.

Leland Belew remembered that teamwork meant that "you would see every major decision treated by the total organization. It was a fishbowl type operation. It was 20/20 visibility from the outside in." The systematic approach to engineering that Marshall used in the 1960s, Belew believed, anticipated 1980s innovations associated with management guru W. Edwards Deming or systems like Total Quality Management.[19]

Another central feature of the laboratory culture was that Center managers were intimately involved in technical matters. In-house research and development, von Braun said, helped top officials "keep their knowledge up to date and judgment sharp by keeping their hands dirty at the work bench." He believed that managers with "dirty hands" were both planners and doers, and consequently were more effective leaders.[20]

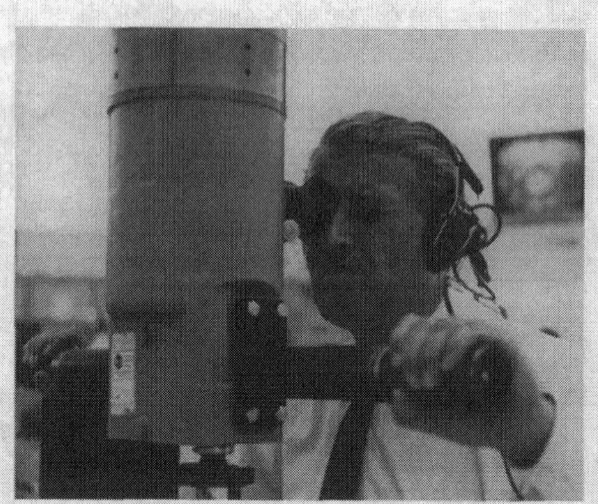

Von Braun watches a Saturn launch.

Von Braun was the model of the dirty-hands manager and his persona and management style have generated much comment. One commentator described von Braun as the "managerial lord" of Marshall's "feudal order." He ruled over German "vassals," each of whom had rights in their fiefs and responsibilities to their lord. The Marshall leader, the novelist and pundit Norman Mailer wrote, was "the deus ex machina of the big boosters" who corporate managers worshipped as the "high priest" of innovative organizations.[21]

Marshall colleagues recalled von Braun's charisma. Dannenberg noted that von Braun inspired each employee to feel like he was "the second most important man" in the world working for the most important man. Ruth von Saurma, who as a member of the public affairs staff often helped out with international correspondence, recalled that "there was hardly anyone who did not like him and look up to him, although he never looked down on anyone. He always seemed to be on the same level as the person he would be talking to. What was fantastic was that individuals grew tremendously under his leadership and performed so much more for him as a group than they ever would have been able to do individually."[22] "Wernher von Braun was not a dictator—he didn't have to be," Georg von Tiesenhausen insisted. "His personality was such, his authority was such, that everyone did what he wanted anyway." Von Braun had confidence in his ability to pick the right person for a job, and delegated responsibility.[23] His dynamism challenged people. "Von Braun was always overflowing with ideas," according to Dannenberg.[24]

Von Braun, Stuhlinger remembered, "never said any disparaging word or derogatory word about anyone." This habit encouraged the openness and cooperation necessary for problem-solving. Center veterans recollected how von Braun had responded to a young engineer who admitted an error. The man had violated a launch rule by making a last-minute adjustment to a control device on a Redstone, and thereby had caused the vehicle to fly out of control. Afterwards the engineer admitted his mistake, and von Braun, happy to learn the source of the failure and wanting to reward honesty, brought the man a bottle of champagne.[25]

Wernher von Braun suited up for conducting tests in Marshall's Neutral Buoyancy Simulator.

Marshall's first leader was also the Agency's master publicist and lobbyist. In addition to appearances before congressional committees, von Braun averaged nearly 150 articles and speeches a year, and kept two full-time writers busy in Marshall's Public Affairs Office. Between 1963 and 1973 he contributed monthly articles to the magazine *Popular Science*. His topics were diverse and included anticommunism, Christianity, and Creationism, but the vast majority promoted space exploration and research. Recognizing that space projects needed public support, his motto was "Early to bed, early to rise, work like hell and advertise!" Such boostership made von Braun, in the words of Amos Crisp, one of his writers, "Mister Space" in the 1960s.[26] Norman Mailer observed that von Braun was the only NASA manager known to the public and was "the real engineer, the spiritual leader, the inventor, the force, the philosopher, the genius! of America's Space Program."[27]

In his space speeches and articles, Marshall's director made NASA projects, plans, and technology understandable to the public. More importantly he sold the excitement, significance, and benefits of space exploration. Von Braun pointed out technological spinoffs and scientific discoveries, but mainly argued that the greatest benefit of the space program was in generating new challenges. Spurred by space exploration, scientists, engineers, and technicians innovated faster and teachers educated students better. In the long-term, he thought, meeting the challenges of space boosted economic growth.[28]

Rees, von Braun's deputy since Peenemünde, complemented his chief's leadership style. Von Braun was the visionary, Rees the practical manager; von Braun inspired people to conceive new ideas, Rees drove them to complete old tasks. His direct supervision became more important as von Braun's public appearances absorbed more of the Center director's time. Rees "paid attention to minor details. He was the technical man, but von Braun always floated with his feet above the ground," von Tiesenhausen explained. "Dr. Rees would say to Wernher, 'Now simmer down.'"[29]

Von Braun expounded a philosophy of management, and some of its elements became parts of Marshall's culture. Teamwork in a research and development organization, he argued, depended on a proper balance between centralized management and decentralized specialists. Without centralization, the team could not set common goals and harmonize differences. But managers in an ivory tower could not command cooperation or solve technical problems "in a high-handed fashion." Without decentralization, specialized technicians could not develop knowledge and work together. For von Braun, managing teamwork required "communication" between managers and specialists; and communication depended on "a kind of four-way stretch: up and down the organizational chart, and laterally in both directions."[30]

Two of von Braun's methods of communication, "board meetings" and "weekly notes," became Marshall traditions. The Marshall director had weekly meetings of his "board" of top Center officials, laboratory directors, project managers, and invited specialists. The meetings had formal presentations, but their primary feature was the free, often heated, discussion of problems and policies. Von Braun presided over the discussion without dominating the exchange.

In board meetings and in other Center-level meetings he showed his skill as a systems engineer and manager. Subordinates marveled at von Braun's vision of space exploration, understanding of arcane technical and scientific issues, and ability to recall details and fit them into patterns. They wondered at his ability to summarize complex and confused presentations in a few sentences, translate technical jargon, and integrate conflicting opinion. One colleague recalled how experts "would be talking almost like in unknown tongues" and "finally von Braun would take over and explain what was being said in terms that everybody could understand." Another remembered that "von Braun's gift was, after listening to each one, to join all the information into one package that each one agreed to." The consensus and clear policy that emerged at the top helped give Marshall a very disciplined organization.[31]

While meetings were common in research organizations, von Braun's "weekly notes" were unique to Marshall. Under his direction, the Center's laboratory chiefs and project managers submitted a single page weekly summary of their activities and problems. Von Braun scribbled comments and recommendations in the margins and circulated copies to all top officials. Marshall people eagerly read the notes and used them as a forum for discussing technical problems, arguing policy issues, complaining about inadequate resources and cooperation, and discussing solutions. The benefits multiplied because many superiors generated information for their "Monday Notes" by having subordinates submit "Friday Notes." In the process of learning about the problems and ideas of other officials, Marshall's managers could develop a holistic view of the Center and determine how to synthesize their part with the whole. Later Center directors continued von Braun's weekly notes, imitating his use of communication networks as tools for managing teamwork.[32]

The Marshall team's arsenal practices and laboratory culture were sources of strength during the 1960s and early 1970s. Although much of the original culture persisted, the Center's participation in the Apollo Program would impose political and managerial pressures that led in new directions.

Planning and Propulsion

When members of the Development Operations Division of ABMA became NASA employees in 1960, America's civilian space policy was still in flux. Over the next few years, American leaders and NASA officials made important decisions, eventually choosing the Apollo lunar landing mission and giving Marshall its task of producing the Saturn launch vehicles. These discussions and decisions mixed scientific and technical issues with strategic and political ones. Lucas recalled that "some of the most significant decisions made in the Saturn program had little to do with engineering. They were mostly political. To be successful in a major project like that, you have to have a national commitment to it, you have to have a defined goal, you have to have a timetable, and you have to have resources."[33]

In the late 1950s American space plans developed in the political context of the Cold War and competition with the Soviet Union. Many Americans feared the military threat of apparent Soviet supremacy in rocketry after the success of the Sputnik satellite in October 1957. The Eisenhower administration had photos from U–2 spy planes to show that no "missile gap" existed, but refused to release this information and compromise its source. Consequently fears persisted, and politicians, public officials, journalists, and scientists debated alternative ways to promote American progress in space.

While still in the Army, the rocket group in Huntsville participated in the national discussions about future space missions and launch vehicles. In early 1958 von Braun stood in the spotlight of Explorer I's success and appeared before Congress to lobby for more space exploration and for a trip to the Moon. In June 1959, General Medaris had ABMA release a "Project Horizon" plan which proposed to establish a permanent, 12-person lunar outpost by 1966.[34]

ABMA also contributed to planning of new launch vehicles. In 1957 the team proposed construction of a clustered-engine booster with 1.5 million pounds of thrust. By August the following year the ARPA of the Department of Defense had agreed to provide research and development funding for the new vehicle, called the Juno V and later the Saturn I, and in December ABMA began working on the vehicle as a subcontractor to NASA. Concurrently ABMA worked with military and NASA planners in choosing advanced vehicle designs and upper-stage configurations appropriate to missions in Earth orbit or lunar voyages. ABMA's engineers examined concepts using space planes, solid-fuel rockets, or various liquid fuels. By 1960, NASA's propulsion planning committee, chaired by Abe Silverstein, formerly of Lewis Research Center, had selected liquid hydrogen, a relatively new but powerful fuel, for the upper stage. By late in the year, NASA and Marshall had begun preliminary design of an even more powerful Saturn. Later called the Saturn V, its first stage would use a cluster of F–1 engines, originally developed by Rocketdyne for the Air Force, each with 1.5 million pounds of thrust.[35]

In the spring of 1961, the new administration of John F. Kennedy chose a lunar landing as the primary task of space exploration. Although the choice rested on technical data from NASA committees and special space policy groups, it depended more on political considerations. The Kennedy administration wanted to ease the anxieties of the American public and bolster national prestige by achieving a dramatic first in space exploration. Staging such a drama would demonstrate the superiority of the American system of enterprise, management, technology, and science. The Kennedy people defined space as a "new frontier" and believed that exploring it would promote progress. Accordingly in his State of the Union message on 25 May 1961, President Kennedy asked for a national commitment to "landing a man on the Moon, and returning him safely to the Earth" before the decade was out. Congress endorsed his request, and NASA created the Apollo Program to put "man-on-the-Moon."[36]

With a clear mission and timetable, NASA and science planners within the Kennedy administration now began studying methods for getting to the Moon. This "mode" decision was difficult because the method had to be economical in time and money, technically feasible, and acceptable within NASA.

The Agency made this decision based on consultations between NASA Headquarters and its field organizations. The groups responsible for human space flight—Marshall and the STG—were especially influential. The Agency had formed the STG, composed of aeronautical engineers from the Langley Research Center and led by Robert Gilruth, to manage the manned satellite program called Project Mercury. By late in 1961 NASA had redesignated the group as the Manned Spacecraft Center (MSC), given it responsibility for manned spacecraft, astronauts, and mission operations, and selected Houston, Texas, as its permanent site. Over the decades the history of the MSC and Marshall would be intertwined; although partners who worked well together, they were sometimes competitors who struggled for resources and control over projects.[37]

From 1960 to 1962, NASA conducted studies of various lunar mission modes, evaluating each plan according to weight margins, guidance accuracy, communications, reliability, development complexity, schedules, costs, flexibility, growth potential, and military usefulness. Marshall personnel investigated two modes, "direct ascent" and "Earth orbital rendezvous." Direct ascent would limit the number of vehicles and launches. A Nova booster, a sort of super-Saturn, would launch one heavy spacecraft, which would travel to the Moon, land on the surface, lift off, and return to Earth. Earth orbital rendezvous, referred to as EOR, could be traced to von Braun's 1952 articles in *Collier's* and had two versions, each depending on Saturn V boosters rather than a Nova. One "connecting" version of EOR would divide the heavy spacecraft in two parts, launch each separately, and integrate them in Earth orbit. The other "fueling" mode would launch the heavy spacecraft with one Saturn booster and its fuel in another, then transfer the fuel in Earth orbit.[38]

The direct ascent mode fell out of favor by the spring 1962. Although officials at Headquarters, the MSC, and Marshall believed that a powerful Nova booster would be useful for a space station, a lunar base, or interplanetary exploration, planners concluded that Nova was too big a leap beyond existing technology and doubted that it could be ready by the end of the decade. Preliminary designs called for the Nova to be twice as powerful as the Saturn V and to have 10 F–1 engines for its first stage. It would be so big—50 feet in diameter in contrast to the Saturn V's 35 feet—that it would not fit test stands and assembly buildings. Moreover, Marshall expected that Nova would be even more technically difficult to develop than Saturn, and they doubted that they could develop two super-boosters at one time, especially if each siphoned money away from the other.[39]

Marshall's dire forecasts about the Nova led to criticism of the Center's commitment to liquid fuels. The criticism focused on Marshall's plans for a liquid-fueled version and failure to study a potentially less expensive and more powerful solid-fuel rocket. Maxime Faget of the MSC later contended that Marshall engineers were "liquid-fuel people" who did not "trust" solid fuels and "tried to think of everything wrong with solids they could." At the time, Marshall did not seriously consider solid rockets because Center propulsion engineers doubted their safety for human flight. Dannenberg pointed out that solid-rocket engines kept burning once ignited; liquid engines, in contrast, could be shut off should dangers develop.[40]

Although the solid-rocket versus liquid-rocket controversy would reappear in NASA history, the issue was moot in Apollo planning. The Nova, whatever its fuel, depended on missions to justify it and commitments to fund it. Von Braun argued that going ahead with Nova meant "giving up the race to put a man on the Moon in this decade even before we started." By late 1961, in contrast, preliminary research for the Saturn V was well underway. Thus once NASA decided that direct ascent could not meet its goal, the Agency stopped funding Nova, and Marshall's rocket designers quietly swept its plans from their drafting tables.[41]

By early 1962 mode options narrowed to a choice between EOR and LOR, short for lunar orbital rendezvous. The LOR mode called for two light, specialized spacecraft, a command spacecraft and a lunar lander-launcher. The two craft would travel to the Moon together. From lunar orbit, the lunar craft, more light in weight than its EOR counterpart, would descend to the Moon, blast off from the surface, rendezvous in lunar orbit with the command craft, and then be jettisoned. John Houbolt, an engineer from the Langley Research Center, was the great booster of LOR. Initially both Marshall and the MSC challenged his ideas, because his plan called for computer-controlled rocket firings behind the Moon and his estimates for the weight of the lunar craft were very low and optimistic. By January 1962, however, Houbolt had convinced the MSC of the utility of LOR.

At this point, the interpretation of the mode decision becomes controversial, and no definitive historical account exists. Participants and historians have offered conflicting accounts of the events leading up to the decision and of its implications. One reason for the lack of consensus has been the partisanship caused by disputes between the MSC and Marshall. The mode options would push the Agency in directions more favorable to one Center than the other. The MSC people favored LOR because developing two specialized spacecraft would be easier then developing a single multipurpose one, and because they could maintain control over human activities in space. Marshall favored EOR because its demands would help the Center grow from propulsion research into

Earth orbital engineering, and would require two Saturn launches per mission and thus generate more responsibility. In an interview in 1970, von Braun downplayed the rivalry. He contended that Headquarters had directed Marshall to study EOR and Houston to study LOR; Marshall never formally endorsed EOR but simply reported on it.[42]

Another reason for disagreements about the mode decision was the use of different engineering criteria. The MSC and most Headquarters officials evaluated any mode based primarily on whether it would technically simplify achievement of Kennedy's objective to land on the Moon by the end of the decade; by these criteria LOR was simplest.[43] Marshall and the PSAC evaluated modes based on the Apollo deadline, but also on ability to promote science and space exploration in the long term. EOR, they thought, would provide technology and experience in refueling, assembly and repair, and rescue in Earth orbit and better allow for a space station or lunar base.[44] The different criteria had created an impasse, but in March 1962, top NASA officials decided to choose the mode in June.

At this point, managers of the MSC resolved to sell LOR to NASA Headquarters and Marshall. They first went to Washington and convinced Dr. Joseph F. Shea, deputy director of Systems in the Office of Manned Space Flight (OMSF), and D. Brainerd Holmes, director, OMSF. Next representatives from Houston staged a day-long sales pitch in Huntsville in April 1962.

From that point until June, the behavior of Marshall Director von Braun is unclear. Stuhlinger, the chief of the Research Projects Lab, believed that von Braun preferred EOR but had become concerned that bureaucratic in-fighting would cause delays and could prevent meeting Kennedy's deadline. In the interest of promoting harmony in the Agency, Marshall's director therefore turned conciliator and favored LOR. When he announced his decision at a Center board meeting, Stuhlinger recalled, it caused a "storm" because many of his lab directors remained committed to EOR.[45]

Other evidence also suggests that von Braun was as much a wheeler-dealer as a diplomat. Headquarters officials Shea and Holmes held meetings with von Braun in May to discuss the mode options. They believed von Braun had questioned LOR mainly because he was concerned with its liabilities for Marshall. They reported later that von Braun kept asking what Marshall would gain if NASA selected LOR. Realizing that von Braun wanted his Center to branch beyond the propulsion business, Shea and Holmes offered Marshall a piece of the action on the lunar surface. Holmes later denied that a formal *quid pro quo* ever emerged, but Headquarters and von Braun discussed how Marshall could study lunar vehicles and base equipment.[46]

NASA made the mode decision on 7 June 1962 at a meeting attended by officials from the OMSF and the field centers. Formal presentations explained the modes, with Marshall engineers describing EOR. Following the presentations, von Braun said, "Gentlemen, it's been a very interesting day and I think the work we've done has been extremely good, but now I would like to tell you the position of the Center." Marshall, he then announced, supported the LOR mode. This was something of a shock to some Center personnel who had not known of his choice before the meeting.

Von Braun offered technical and political reasons for supporting LOR. Admitting that he had initially been "a bit skeptical" about the plan, he recognized its engineering simplicity. LOR's light spacecraft required only one Saturn V launch and thus eliminated the need for two successful launches. Moreover, a specialized lunar craft would simplify lunar landing and launching by eliminating the need for one heavy, multipurpose spacecraft. It would smooth construction by providing for the "cleanest managerial interfaces" between centers and contractors and by reducing the amount of technical coordination. At the same time that von Braun bowed to LOR's parsimony of engineering, he acknowledged schedule pressures. The mode controversy was delaying important design decisions and construction work; unless a mode decision was made "very soon," he said, "our chances of accomplishing the first lunar expedition in this decade will fade away rapidly." Von Braun concluded that, all things considered, LOR offered "the highest confidence factor of successful accomplishment within this decade."

At the same time, von Braun also recommended that Marshall develop a crewless, automated, lunar logistics vehicle to overcome the liabilities of LOR. Launched by a second Saturn V to accompany human missions, this vehicle would expand the duration and scientific benefits of lunar missions by providing supplies, equipment, and shelter.[47]

By agreeing to LOR, Marshall got credit for being a team player. Holmes and Shea felt that von Braun's decision helped stimulate inter-Center cooperation in the Apollo Program. Shea added that the Marshall director's endorsement of LOR was "a major element in the consolidation of NASA." With its top officials united, NASA formally selected the LOR mode using a Saturn V rocket and decided to study a lunar logistics vehicle.[48] Marshall immediately began studies of the craft, and although NASA never developed a flight model, the Center eventually oversaw construction of a moon car called the lunar roving vehicle.[49]

The choice of LOR mode shaped the Apollo Program, and debates about its merits continued long afterwards. Critics of the choice complained that NASA's narrow engineering mentality led the Agency to select the cheapest means in terms of money and time and to choose excessively specialized technologies; the mode meant brief lunar visits and restricted scientific research.[50] Long after the decision, many Marshall veterans continued to echo these sentiments. Von Tiesenhausen contended that LOR helped make Apollo essentially a "dead-end." Dannenberg also believed that rejecting EOR thwarted possibilities for constructing a space station and pursuing more open-ended missions in the 1960s. Others were less negative, believing that NASA expanded the scientific utility of Apollo technology by using the third stage of the Saturn V as the basis for the *Skylab* orbital workshop.[51]

The mode episode came to an ironic conclusion when von Braun publicly defended LOR before the national media. The issue came up on 11 September 1962 when President John Kennedy visited Marshall to look over Saturn development. The President brought with him Jerome Wiesner, the

Von Braun explains Saturn hardware to President Kennedy and Vice-President Johnson during their visit to Marshall on 11 September 1962.

chair of the PSAC, Vice-President Lyndon Johnson, and NASA Administrator Webb. While standing near a Saturn I stage and with the press listening, the group began discussing the merits of LOR. Wiesner argued fervently that LOR was neither as safe nor as scientifically useful as the other modes. An angry Webb and a calm von Braun contradicted Wiesner. Kennedy listened quietly, later telling Wiesner that he too doubted LOR and that they were alone in supporting the alternatives. The argument made national headlines but quickly passed from attention with the onset of the Cuban missile crisis.[52]

The choices of the lunar mission, the end-of-decade deadline, the Saturn V, and LOR all influenced Marshall's work. NASA had a clear mission, a definite schedule, and the necessary funds. Marshall would build the Saturn launch vehicles and have plenty of resources for the task. William Sneed, a manager on the Saturn project, recalled that Marshall had cash reserves to "accommodate the unknowns and unpredictables" and to fund more than one path of technological development. James Odom said that the parallel development of critical technologies allowed Center engineers to choose the most reliable option and to stay on schedule. Robert Marshall, a Center propulsion engineer in the 1960s, summarized the meaning of the decisions: "The schedule was fixed and the performance was fixed; money was a variable. We threw money at problems." After the halcyon decade of Apollo, no Center project would have such favorable conditions; in later efforts the money was fixed and the performance and schedule became variables.[53] The challenges and resources of the Apollo Program would also cause Marshall to grow bigger and develop new skills.

Growth and Change

To develop the Saturn stages, Marshall added more personnel and built new facilities. More significantly, the enormous technical and managerial challenges led Center personnel to change their organization and culture. Werner Dahm, an aerodynamic engineer, recalled how in the 1950s ABMA had been "a single-project outfit" that worked on one vehicle at a time with a couple of major contractors. The Apollo Program changed Marshall, making it a "multiproject organization" that developed many rocket stages and space technologies, managed multiple contracts, integrated diverse technologies, and coordinated far-flung organizations. The Center adapted to its new role by strengthening its capabilities in project management and systems engineering.[54]

Of all NASA's field centers, Marshall benefited most from the free-spending era of the early 1960s. Only the expenses incurred by the MSC rivaled those at Marshall. NASA allocated funds in three categories: Administrative Operations, Research and Development, and Construction of Facilities.[55] From 1961 through 1965, Marshall's accumulated Administrative Operations obligations (comprising principally salaries) were more than double those of any other Center.[56] Marshall's accumulated Research and Development obligations through June 1968 were larger than those of any other Center, five times those of every Center except Goddard and MSC. Only MSC came close to Marshall's figure.[57]

During the years in which Marshall built most of its Saturn test stands and assembly facilities, only the construction of the launch complex in Florida surpassed the Center's obligations for Construction of Facilities in Huntsville and at Michoud and the Mississippi Test Facility.[58]

Early 1960s test stand.

Marshall was also NASA's largest contract administrator. For six consecutive years (fiscal years 1961 through 1966), Marshall let contracts totaling more than any other Center, constituting more than 30 percent of NASA's contractual obligations. In mid-1968, Marshall held (either solely or jointly with other centers) six of NASA's eight largest contracts.[59] California, Louisiana, and Alabama, the major locations of Marshall business, ranked first, third, and fourth as recipients of NASA prime contracts from fiscal years 1961 through 1968.[60]

Other yardsticks measure Marshall's extraordinary growth in the early 1960s. The Kennedy goal of reaching the Moon by the end of the decade gave the Marshall Center a virtual *carte blanche*. When NASA established Marshall in

1960, it acquired land and facilities valued at $34,651,000. Within the next four and a half years, NASA funded new facilities worth more than $125,000,000. [61] Laboratories continued to operate in buildings inherited from the Army, but the Center expanded most of them and added new facilities. Test stands for the Saturn Project consumed much of the new facility money. In June 1963, 1,200 employees moved into a modern 10-story Headquarters building. Von Braun's office on the top floor overlooked a panorama of the Alabama countryside, rimmed by hills and sloping to the Tennessee River to the south, now punctuated by monolithic test stands. The government labeled the Headquarters Building 4200, but locals often called it the "Von Braun Hilton." Behind it, two smaller buildings in the same style completed a horseshoe-shaped Headquarters complex: the Engineering and Administration Building (4201) and the Project Engineer Office Building (4202).[62]

Other than the scale of the Saturn V, nothing demonstrated more dramatically the rapid growth of the American space program than Marshall's test complex at the southern end of the Center. Visible from the small Redstone Interim Test Stand were mammoth test stands used for Saturn development: Single engine test stands, static test stands, and the huge dynamic test stands.

Marshall Center's Test Area in 1978.

The construction of new facilities led to some conflicts between the Center and labor unions.[63] Beginning in August 1960 Marshall's arsenal system triggered jurisdictional disputes between the Center's Launch Operations Directorate (LOD) at Merritt Island, Florida, and building trades unions. The unions working on Launch Complex 34 (LC–34) were accustomed to Air Force practices. They expected to install ground support equipment with little direct supervision. LOD was accustomed to the arsenal system and thought that government scientists and engineers should install some equipment and closely inspect contractors. When LOD began introducing arsenal practices, the unions quickly complained that LOD personnel were doing too much construction and supervision. In a series of brief strikes, electricians, ironworkers, and carpenters walked away from LC–34, and the project lost 800 man-days of work from August to November. The disputes culminated in November when electricians went on strike to protest LOD civil servants installing cables and consoles in the launch control center.[64]

The Center justified applying its "army philosophy" to scientific projects by defining the launch complex as a "laboratory" intimately tied to the launch vehicle, which was itself a "flying laboratory." Logically NASA engineers and scientists should install some ground support equipment as part of "research and development."[65] Von Braun insisted that scientists with Ph.Ds sometimes had to use screwdrivers and wrenches; they had to get their hands dirty to make new machinery function and to maintain expertise. Von Braun promised that routine work would be contracted out, and this policy practically eliminated conflicts at the Cape after 1960.[66]

A labor dispute in Huntsville also occurred on a facility construction project but did not involve contractor-Civil Service issues. On 14 August 1962 a dispute between unionized and non-unionized contract workers led to a strike at Marshall's Saturn V Static Test Stand. Members of the International Brotherhood of Electrical Workers formed picket lines at Marshall's entrances and over 1,200 members of other building trades unions refused to cross. Work at the test stand and several other sites ceased.[67] With the strike continuing more than a week, construction delays and attention from the national media upset Marshall managers and the Huntsville elite. Von Braun argued that the dispute was costing $1 million a day and was causing the United States space program to fall further behind the Soviet Union. The *Huntsville Times* condemned the workers for causing the United States to lose "the competition between the free world and the forces of darkness which seek to engulf us."[68] A federal injunction ended the strike on 24 August and the National Labor Relations Board convinced the electrical union to refrain from strikes and secondary boycotts.[69]

The strikes in Huntsville and at the Cape taught Marshall a lesson, and in 1963 its managers sought to forestall strikes on other facility construction projects. With assistance from the Missile Sites Labor Commission, the Center held meetings with construction unions and contractors who would build the new test facility in Mississippi. The meetings sought to resolve potential problems and secure a union promise of three years without a strike. Marshall called it "the first such conference ever sponsored by the Federal Government in advance of the award of a construction contract."[70]

During the Saturn years, Marshall opened three new facilities in Louisiana and Mississippi. All three facilities helped NASA politically, helping the Agency garner support from federal legislators from those states. The sites also had technical advantages. The Michoud Assembly Facility in eastern New Orleans, selected in August 1961 by Marshall and NASA for the manufacture of Saturn lower stages, had once been a federal plant for manufacturing Liberty ships, cargo planes, and tank engines. It had a production building with 35-foot-high rafters and a 43-acre manufacturing floor, water access via the Gulf Intracoastal Waterway, closeness to skilled labor and industrial support in New Orleans, and proximity to sparsely inhabited land that could be used as a rocket test area.[71]

Two months after selecting Michoud, NASA chose a Saturn V test site on the Pearl River in Hancock County in southwestern Mississippi. The Mississippi Test Facility perfectly combined accessibility and remoteness. Only 45 miles from Michoud by water, and with few people to relocate, its surrounding swamps were large enough so that the tremendous sound waves created during rocket firings would not cause damage.[72] Constructing test stands, rail lines, and a canal took over four years and cost over $315 million.[73] The third site, the Computer Operations Office in Slidell, Louisiana, used an unoccupied building originally owned by the Federal Aviation Administration, and began activity in 1962. Located between the assembly and test facilities, Slidell's computers supported their work in engineering, checkout, and testing.[74]

Like other facets of Marshall's development in the 1960s, the Center's personnel numbers followed the curve of Saturn development: dramatic increases in the first half of the decade, reductions later. When it opened in July 1960, Marshall inherited 4,670 employees from the ABMA. By the end of the year, Civil Service employees numbered 5,367.[75] During its first six years, the Center experienced steady growth and by the summer of 1966, employment reached a peak of 7,740. Marshall was easily the largest NASA installation with 21.7 percent of the Agency workforce.[76] Marshall's combined workforce—contractor and Civil Service—peaked at over 22,000.[77]

The establishment of Marshall forced a reevaluation of NASA's allotment of excepted and supergrade positions above the grade of GS–15. Designed to make government management appointments competitive with the private sector, these positions were "among the most potent means by which the Administrator shaped the agency."[78] NASA received permission to increase its allotment from 260 to 290 to accommodate the so-called German positions inherited from the Army, and won increases to over 700 during the Apollo buildup.[79] Marshall held as many as 56 of these positions at the height of Saturn, after which its allotment quickly dropped by a third.[80]

Marshall's workforce was predominantly white, male, and well educated. Less than one percent of Marshall employees was black. The Center did not even begin to record statistics on the number of female employees until the 1970s, when the earliest figures showed that 16 percent were women.[81] Cutbacks in the late 1960s assured that there would be little change in the composition of the Marshall workforce, since reductions hit hardest in nonengineering classifications.

The greatest changes in the character of Marshall's workforce during the first several years were an increase in scientists and engineers, and a decline in wage board personnel. The number of engineers and scientists nearly doubled within the first four years and then remained relatively constant for the next four, an increase from 27.7 to 37.6 percent of Marshall's total employment. Wage board employees declined steadily during the same period from 1,925 (35.8 percent of the workforce) to only 835 (12.0 percent).[82] Von Braun explained the trends as a reflection of "the changing role of Marshall from an essentially in-house organization to one of program management."[83]

Von Braun's explanation highlighted the major change at the Center during the Apollo period. Although Marshall continued aspects of the Army arsenal system until the cutbacks at the close of the Apollo Program, Agency policy required that the Center adopt more of an Air Force system relying on private contractors. NASA Administrator Webb and other prominent officials criticized the arsenal approach. Federal employees, they charged, were more expensive than contractor workers. Reliance on civil servants led to fixed labor costs while contractors could be laid off at the end of projects. Federal experts unnecessarily duplicated skills in the private sector. In addition to its economic weaknesses, the arsenal system had political liabilities. It localized government spending and limited the number of regions participating in the space program. Besides, Webb, a corporate lawyer, former official in the Department of Defense, and former director of the Bureau of Budget, wanted to privatize federal research and development. The Agency Administrator was also a zealous champion of using public spending to stimulate private innovation and profit.[84]

Accordingly the Center and the rest of the Agency used the Apollo Program to expand the command economy in space hardware. Since the 19th century governments had created a command economy in military technology, becoming the sole buyer of weapons too expensive for private firms to develop on their own. After the Second World War, space hardware also became command technology.[85] Military methods provided much of the contracting apparatus for NASA, but the Apollo Program was so vast and complex that the Agency had to innovate. NASA created what its administrators called a "government-industry-university team," and Marshall and the rest of the agency improved methods for running R&D organizations, "managing large systems," and supervising business-government partnerships; their managerial methods became an "unexpected payoff" of the Apollo Project.[86]

For years as part of the military, the rocket veterans who formed the core of Marshall had worked with contractors. They had worked with business and university contractors at Peenemünde, White Sands, and in Huntsville. When ABMA employees transferred to NASA, armed services procurement personnel, procedures, and practices went along. Like the military, Marshall used technical specifications, drawings, performance requirements, and incentive fees to direct contractors. Marshall and NASA also often used military quality personnel to monitor contractors and inspect parts. The Center differed from military methods of monitoring contractors in the very detailed specifications its labs produced, the rigor of its testing, and the depth of its penetration of contractors.[87]

The increasing use of contractors and growing technical complexity of Apollo led Marshall to strengthen managerial and systems engineering groups so that all the parts and participants could be integrated. In the initial organization of 1960, the Center had no systems engineering group, and the laboratories, based on the practice of automatic responsibility, collectively resolved integration problems. A small Saturn Systems Office, with its three offices for the Saturn I/IB, Saturn V, and engines, handled project management of budgets and schedules. This organization differed little from those of Peenemünde and ABMA.

But by 1962, once complicated work began on both the Saturn I and Saturn V and once contracts were let across the country, the Center's traditional organization proved unwieldy. By the middle of 1963, Marshall's workload had increased more than four-fold in three years. The fiscal year budget had grown from $377 million in 1961 to $1.07 billion in 1963. Procurement had increased almost three-fold in three years, from $315.5 million to $949.7 million. The flood of responsibility swamped the Saturn Systems Office and the labs. Center officials worried that a lack of central controls could lead to excessive changes, cost overruns, and schedule slips.

By 1962 von Braun moved to forestall any problems. He told a management conference that his rocket team had changed from being a research and development organization to also being "a managerial group." To adapt, he oversaw a reorganization in 1962 that gave more authority to managers of a project (a "project" in NASA parlance was a discreet technology that was part of a larger "program"). Justifying the change in a three-page memo, "MSFC Management Policy Number 1," he explained that multiple projects necessitated stronger project offices. The labs would still be organized by technical discipline. Now, however, project offices would coordinate plans, assignments, and budgets for work involving more than one lab, and would oversee technical staff directly assigned to project work.

A major reorganization of the Center on 1 September 1963 formalized the new arrangements. One organizational branch called Research and Development Operations contained the labs, and another equal branch called Industrial Operations contained the project offices. In the Center hierarchy, lab directors and project managers were on an equal organizational rung for the first time. Within various projects, the project offices managed and the labs provided support. In addition, each lab had a Saturn Project Engineering Office to coordinate activities with the Saturn Project Office.[88]

Moreover, Marshall enhanced its abilities to handle integration problems. Pulling together the designs and hardware of the many pieces of a multistage vehicle was an enormously complex task. NASA had to help pioneer the relatively new field of systems engineering, and Marshall was in the forefront. In 1962 the Center established a Saturn/Apollo Systems Integration Office for working with other NASA centers. Marshall also enlisted a systems engineering contractor; Boeing, the contractor for the Saturn V first stage, became the

Saturn V Systems Engineering and Integration contractor. NASA and Marshall adopted similar practices for the Shuttle and later projects.[89]

After this reorganization, the project offices and labs acted as checks-and-balance on one another. Checks-and-balances were "built-in," Lucas recalled, because the labs and project offices had different interests. Scientists and engineers in the laboratories wanted to be thorough and inventive, and wanted the job done right with little concern about cost, schedule, or administrative nicety. In contrast, project offices were responsible for getting the job done on time and within budget. To meet deadlines and budgets, project managers sometimes had to limit technical innovations. Nonetheless the project offices, James remembered, did not make technical decisions based on managerial standards; they relied on "change boards" composed of lab experts who studied each proposed innovation and determined whether it was necessary. He also said that von Braun wanted to base hardware decisions on their technical merits rather than schedule or cost. Von Braun told James that "when you have an argument with the laboratories, I want you to know that I am on their side."[90]

As Saturn development progressed, Marshall hired more experienced project managers and pioneered new oversight methods. In 1964 the Center acquired on temporary assignment over a dozen Air Force officers who were veterans in running big, expensive, and complex aerospace projects; they had skills in budgets and schedules, and systems management. Also in 1964 Air Force General Edmund O'Connor became director of Industrial Operations, serving in that post throughout most of the Saturn years.[91]

The Saturn V Program Office, headed by Peenemünde veteran Arthur Rudolph, oversaw the crucial Apollo activities of the Center and its contractors. The office ensured that Saturn manufacturing stayed within budgetary and schedule guidelines and that all the contractors and components fit together in one system. This was an enormous problem because Marshall oversaw contracts with hundreds of companies in dozens of states. Rudolph thought his major problem was that "in a big program like the Saturn V you have many people involved and usually people want to go off on tangents," and so he tried to "get them all to sing from the same sheet of music." Saturn's self-styled "choir director" oversaw regular meetings in which Marshall and contractor officials reviewed and revised plans as the program evolved; sometimes the meetings would last until well after midnight.[92]

One novel feature of the Saturn V Program Office was a room called the Program Control Center. Rudolph's staff designed the room to enhance "visibility" and reveal problems. Three thousand square feet of visual aids and scheduling charts papered its walls. Based on systems developed for military missile programs, the charts graphed a path of progress for each part and showed crucial schedule checkpoints. Information for the charts passed up the Center-contractor organization, with each manager relaying data through superiors. Each chart directed attention to parts that were lagging so that managers could invest more resources on these critical parts.

Marshall officials were careful in how they used the charts. They sometimes regarded them as a "gigo" system—garbage in, garbage out—knowing that managers sometimes withheld information or exaggerated progress. James, Rudolph's successor as Saturn V manager, believed that Rudolph sometimes pretended that he could not understand the charts, using this pretext to question project managers about their progress. In remarks to Congress in 1967, Rudolph admitted schedules were often "soft" and could be set back. Nonetheless he thought the charts and schedule deadlines were useful managerial tools; in his words the "visibility" enforced "discipline" and got rid of "looseness." More importantly, the charts helped officials integrate the work of the Saturn team. NASA Administrator Webb loved the Program Control Center and its management charts. Webb brought dignitaries to Marshall just to parade them through the room which he said was "one of the most sophisticated forms of organized human effort" that he had "ever seen anywhere." When Webb looked at the charts, Saturn Program Control Manager Bill Sneed said, NASA's Administrator recognized that Marshall was doing more than building a lunar rocket; the Center was "innovating and developing management systems" that were "the best known to man."[93]

Marshall also worked with the rest of NASA to coordinate work on Apollo. Headquarters had an Apollo program office that made plans, allocated and monitored resources, set schedules, and maintained oversight of specifications and standards. A NASA Management Council, composed of top Headquarters officials and field Center directors, set broad policy. On technical issues, however, the centers had considerable autonomy. Experts from the centers staffed eight Inter-Center Coordination Panels on crew safety, instrumentation and communications, flight mechanics, flight evaluation, electrical systems, launch operations, mechanical design, and flight control operations. In this way experts assumed daily responsibility for coordination. Generally, these decentralized panels resolved disagreements, but difficult issues passed up the line to a Management Review Board composed of Headquarters officials Center directors, and program and project managers. The Centers and Headquarters also established a mirror organization, with functional offices matching each other to facilitate communication.[94]

Kurt Debus, Wernher von Braun, and Eberhard Rees watch the SA–8 launch in May 1965.

Headquarters also hired a systems engineering contractor to help it monitor the technical activities of the field centers. BellComm, a subsidiary of AT&T, helped review and define systems requirements, missions, tests, and quality programs. Both Marshall and the MSC complained about BellComm's role, questioning the legality of the company's access to proprietary information from other contractors and doubting the wisdom of duplicating expertise at the field centers. More importantly, both Marshall and Houston objected to micromanagement from Washington. Von Braun argued at a NASA Management Conference that there were "too many nuts and bolts engineers in Washington and too few managers" and that Headquarters wasted resources on "petty supervision" and efforts to "second guess" the centers. Nevertheless, Headquarters maintained a strong program office, and Shea, deputy director of Systems in the OMSF, defended the BellComm contract as "good insurance" that would proceed "regardless of Centers' wishes."[95]

Disagreements aside, the arrangements helped NASA smoothly coordinate Apollo activities. Such harmony contrasted with the planning controversies early in the program and on later projects. Technical and organizational factors also contributed to intercenter cooperation.

Marshall worked well with the MSC during the Apollo Program mainly due to technical factors. For Apollo, MSC and Marshall had a clear division of labor. Houston built the spacecraft and Huntsville built the launch vehicle, and one sat on top of the other. Interfaces between spacecraft and launch vehicle were clean and simple, mainly a matter of connecting wires and bolts. Disputes mainly resulted over weight; Marshall believed that Houston's spacecraft was too heavy while Houston thought Marshall's launch vehicle was too heavy. Von Braun credited the resolution of problems like this to mutual respect by the Centers and the unsung work of the intercenter panels.[96]

Social and technical factors helped Marshall work well with the Kennedy Space Center at Cape Canaveral. NASA's launch facility had originally been ABMA's Missile Firing Laboratory. When the Army rocketeers transferred to NASA, the lab remained under Marshall's organization as the Launch Operations Directorate. Kurt Debus, the launch team's director, had been von Braun's assistant at Peenemünde and Huntsville, and many members of the launch group continued to work in Huntsville. Alabama and Florida personnel worked closely together to ensure the compatibility of the assembly and launch facilities with the launch vehicles. Huntsville personnel helped design and construct some of the Cape's launch facilities.

By 1962, organizational problems emerged that led NASA to make the Launch Operations Directorate into an independent Center. Debus and von Braun worried about the managerial liabilities of having the launch team report to Marshall. Particularly problematic was the possibility that the launch team would have to arbitrate disputes between Marshall and another NASA Center. To solve these problems, NASA decided to make the launch team into an independent field center. Although Huntsville officials had lively debates about the merits of being a rocket "developer" or "operator," von Braun supported the change. On 1 July 1962 Marshall's launch laboratory became the Launch Operations Center, and, after President Kennedy's assassination, it became the Kennedy Space Center.[97]

The Apollo Program then led to changes at the Marshall Center in the 1960s. Apollo resources and challenges allowed Marshall to enhance its in-house research and development capabilities by adding new personnel and facilities. At the same time the Center modified its research organization and culture by adding new mechanisms and expertise in contractor management and systems engineering. Together the adaptations helped Marshall solve the enormous technical challenges of the Saturn launch vehicles.

---

1 Howard E. McCurdy, *Inside NASA: High Technology and Organizational Change in the U. S. Space Program* (Baltimore: Johns Hopkins University Press, 1993), pp. 36–37; Ernst Stuhlinger, Oral History Interview by Stephen P. Waring and Andrew J. Dunar (hereafter OHI by SPW and AJD), UAH, 24 April 1989, pp. 13–14.
2 William Lucas, OHI by SPW, UAH, 4 April 1994, pp. 17–18.
3 "Technical Facilities and Equipment Digest," Marshall Space Flight Center, January 1967, pp. 1–70.
4 "Marshall Directing Vehicle Development," *Aviation Week and Space Technology* (2 July 1962), p. 103.
5 Peter Broussard, OHI by SPW, 20 September 1990, Huntsville, pp. 5–9.
6 Lee B. James, OHI by David S. Akens, MSFC, 21 May 1971, pp. 4–5, James Bio. file, NASA Headquarters History Office (hereafter NHHO); Hans F. Wuenscher, "The Role of the NASA–MSFC Manufacturing Engineering Laboratory in the Development of Space Projects," 12 November 1964, pp. 7–8; Karl Heimburg, OHI by SPW and AJD, Huntsville, 2 May 1989.
7 Lucas OHI, 4 April 1994, pp. 36, 38.
8 James Odom, OHI by SPW and AJD, 26 August 1993, Huntsville, p. 9.
9 Wernher von Braun, "Management of the Space Programs at a Field Center," address, 9 May 1963, UAH Saturn Collection, pp. 17–19; Dieter Grau, "Saturn's Quality-Control Program," *Astronautics* (February 1962), pp. 43–62; John D. Beal, "History of MSFC Reliability Philosophy," address, 11 May 1967, UAH Saturn Collection.
10 Seamans and Ordway, pp. 286–87; Cropp, "MSFC Program Management," pp. 12–14; Bilstein, pp. 276–78; Tompkins, p. 13
11 Phillip K. Tompkins, *Organizational Communication Imperatives: Lessons of the Space Program* (Los Angeles: Roxbury, 1993), pp. 68–70; Bill Sneed, OHI by SPW, 15 August 1990, p. 9; McCurdy, pp. 40–42.
12 Tompkins, *Organizational Communication Imperatives*, p. 69; George McDonough, OHI by SPW, 20 August 1990, pp. 25–26.
13 Eberhard Rees, "MSFC Approach in Achieving High Reliability of the Saturn Class Vehicles," address, 28 July 1965; Lee B. James, *Management of NASA's Major Projects* (Tullahoma, TN: University of Tennessee Space Institute, 1973), pp. 92 ff., 97; Grau interview, pp. 11–12; Morris K. Dyer, "NASA Quality Requirements—Progress, Problems and Probabilities," address, 17 March 1967, UAH Saturn Collection, p. 4.
14 E. Rees, Comments on draft of "Statement of Work: Project Apollo," MSFC Archives, July 1961; Tompkins, pp. 7–8.
15 Walter Haeussermann, OHI by SPW, 18 July 1994; Heimburg OHI, 2 May 1989; K. Heimburg to Golovin Committee, "MSFC Large Launch Vehicle Testing Philosophy: An Answer to LLVPG Questions," 13 August 1961, Golovin Comm. folder, MSFC Archives; Lucas OHI, 4 April 1994, pp. 28–30. See also Bilstein, pp. 91, 97, 107–08, 183–84, 188, quotation, 108; Bob Ward, "Saturn 5 Program Boss Drives Toward '67

Goal," *Huntsville Times* (1 May 1966); Charles Murray and Catherine B. Cox, *Apollo: The Race to the Moon* (New York: Simon and Schuster, 1989), pp. 53, 155–56.

16 Konrad Dannenberg, OHI by Roger Bilstein, 30 July 1970, MSFC, p. 20, UAH Saturn Collection; Lee B. James, OHI by David S. Akens, 21 May 1971, MSFC, pp. 4–5, Lee James Bio file, NHHO.

17 Wernher von Braun, "Teamwork: Key to Success in Guided Missiles," *Missiles and Rockets* (October 1956), pp. 39–42; Wernher von Braun, "Management in Rocket Research," *Business Horizons* (December 1962), pp. 41–48.

18 Mathias Siebel, "Building the Moon Rocket," presentation to National Machine Tool Builders Association, 3 November 1965, UAH Saturn Collection; Mathias Siebel, "The Manufacturing Engineers in the Space Program," presentation to Society of Automotive Engineers, April 22–25, 1966, UAH Saturn Collection.

19 Haeussermann OHI, 18 July 1994; Belew OHI, 9 July 1990, pp. 8–9.

20 Wernher von Braun, "Management of the Space Programs at a Field Center," Address, 9 May 1963, UAH Saturn Collection; see Bilstein, pp. 69–74, 81, 262–63.

21 William J. Stubno, "The von Braun Rocket Team Viewed as a Product of German Romanticism," *British Interplanetary Society Journal* 35 (October 1982), pp. 445–49; Norman Mailer, *Of a Fire on the Moon* (Boston: Little, Brown, 1969), pp. 65–81, espec. 65, 75.

22 Ruth von Saurma, OHI by AJD, 21 July 1989, Huntsville, Alabama.

23 Konrad Dannenberg, OHI by SPW and AJD, 1 December 1988, Huntsville, Alabama; Georg von Tiesenhausen, OHI by SPW and AJD, 29 November 1988.

24 Dannenberg OHI (1988).

25 Dannenberg OHI, 30 July 1970, p. 40; Stuhlinger OHI, 24 April 1989, p. 23; McCurdy, *Inside NASA*, pp. 70–71.

26 Amos Crisp OHI, by Stephen P. Waring and Mike Wright, 1990, MSFC, pp. 1–8.

27 Mailer, p. 65.

28 Wernher von Braun, "The Curiosity which Impels Us into Space Pays off Here in $, Says von Braun," 19 July 1962, "Educating Engineers in the Space Age," 11 October 1962, MSFC History Office; untitled article, *Birmingham News* (February 1966), Alabama Space and Rocket Center, Drawer 12. See also, Marsha Freeman, *How We Got to the Moon: The Story of the German Space Pioneers* (Washington, DC: 21st Century Science Associates, 1993), chap. IX.

29 Von Tiesenhausen, OHI.

30 Wernher von Braun, "Teamwork: Key to Success in Guided Missiles," *Missiles and Rockets* (October 1956), pp. 39–42; Wernher von Braun, "Management in Rocket Research," *Business Horizons* (December 1962), pp. 41–48.

31 Heimburg OHI, 2 May 1989; Lee James, OHI by SPW, 1 December 1989, pp. 5–6; William Lucas, OHI by SPW and AJD, 19 June 1989, pp. 20; Georg von Tiesenhausen, OHI by SPW and AJD, Huntsville, 29 November 1988, quote p. 22; Bob Marshall, OHI by SPW, 29 August 1990, Huntsville, pp. 1–2; Bilstein, pp. 261–64, 269.

32 Bilstein, pp. 262–64; Phillip K. Tompkins, "Management Qua Communication in Rocket Research and Development," *Communication Monographs* 44 (March 1977), pp. 8–10; Tompkins, *Organizational Communications Imperatives*, pp. 63–66.

33 Lucas OHI, 4 April 1994, p. 44.

34 Robert A. Divine, *The Sputnik Challenge: Eisenhower's Response to the Soviet Satellite* (New York: Oxford, 1993), pp. 97–98; Bilstein, pp. 34–39; Frederick I. Ordway, Mitchell Sharpe, and Ronald Wakeford, "Project Horizon: An Early Study of a Lunar Outpost," *Acta Astronautica* 17 (1988), pp. 2–17; John Logsdon, *The Decision to Go to the Moon* (Cambridge, Mass.: MIT Press, 1970), pp. 48–53, 139.

35 David Akens, *Saturn Illustrated Chronology: Saturn's First Eleven Years* (MSFC History Office, 1971), pp. 1–3; Bilstein, *Stages to Saturn*, pp. 17–48.

36 Logsdon, pp. 141–42; Walter A. McDougall, *The Heavens and the Earth: A Political History of the Space Age* (New York: Basic, 1985), chap. 16; Dale Carter, *The Final Frontier: The Rise and Fall of the American Rocket State* (London: Verso, 1988), chap. 3. Logsdon and McDougall emphasize diplomatic motives; Carter emphasizes domestic motives; both need to be considered.

37 See Henry C. Dethloff, *Suddenly, Tomorrow Came...: A History of the Johnson Space Center* (Washington, DC: National Aeronautics and Space Administration Special Publication–4307, 1993), chap. 1–3, 6.

38 Don Neff, "How to Go to the Moon," manuscript, 19 February 1969, "Office of Manned Space Flight, Apollo Lunar Orbit Rendezvous" folder, NASA History Office; Murray and Cox, pp. 108–10, 118; "MSFC Unmanned Orbital Rendezvous R&D Program," 28 February 1962, Lunar Overall Program, March 1962–November 1963 folder, MSFC Archives; Ernst Geissler, "Project Apollo Vehicular Plans, Presentation to NASA Committee Meeting at Langley," April 1962, UAH Saturn Collection.

39 Murray and Cox, pp. 108–110; Wernher von Braun, "Concluding Remarks by Dr. Wernher von Braun about mode selection for the Lunar Landing Program given to Dr. Joseph F. Shea, Deputy Director (Systems), Office of Manned Space Flight (OMSF), 7 June 1962, "Office of Manned Space Flight, Apollo Lunar Orbit Rendezvous" folder, NASA History Office.

40 Maxime A. Faget, "Comments on 'Chariots for Apollo: A History of Lunar Spacecraft,'" 22 November 1976, "Launch Vehicles, Nova" file, NASA History Office; William Cohen to John Stone, "Recommendation for further work on solid-boosted Nova vehicles," 2 April 1963, "Launch Vehicles, Nova" file, NASA History Office; Dannenberg interview, by Roger Bilstein, 30 July 1975, UAH Saturn Collection, pp. 29–30.

41 H. H. Koelle, MSFC Future Projects Office, "Summary of Nova Studies," 1 May 1963, "Launch Vehicles, Nova file," NASA History Office; von Braun, "Concluding Remarks," p. 10.

42 W. von Braun, OHI by Robert Sherrod, 25 August 1970, p. 8–9, Sherrod Collection, NASA History Office.

43 See Robert C. Seamans, OHI by John Logsdon and Eugene Emme, 11 July 1969, pp. 9, 19–20, Apollo Oral History Interviews, JSC History Office.

44 Ernst Stuhlinger and Frederick I. Ordway, *Wernher von Braun: Crusader for Space* (Malabar, Florida: Krieger, 1994), pp. 172–79.

45 Stuhlinger and Ordway, pp. 172–79.

46 Neff, "How to Go to the Moon"; Courtney G. Brooks, James M. Grimwood, Loyd S. Swenson, *Chariots for Apollo: A History of Manned Lunar Spacecraft* (Washington, DC:

National Aeronautics and Space Administration Special Publication–4205, 1979), pp. 80–83.

47 Neff, "How to Go to the Moon"; von Braun, "Concluding Remarks," pp. 1–5; Murray and Cox, pp. 136–39, von Braun quoted, p. 139; Bilstein, pp. 57–67.

48 Von Braun, "Concluding Remarks"; E. Rees, "Conclusions Reached in the Management Council Meeting on 22 June 1962," 23 June 1962, Lunar Selection of Mode folder, MSFC Archives; W. E. Lilly, "Management Council Meeting Minutes," 22 June 1962, MSFC Archives; Brooks, Grimwood, and Swenson, pp. 80–83.

49 See documents in Lunar—Unmanned Lunar Logistics, 1962 folder and in Lunar—Unmanned Lunar Logistics, 1959–1964 folder, MSFC Archives. See also discussion of Lunar Roving Vehicle in Chapter 3 herein.

50 "LOR Narrows NASA Program," *Space Daily* (13 November 1962), p. 578; Harold L. Nieburg, *In the Name of Science* (Chicago: Quadrangle, 1966). The pundit Norman Mailer said of LOR that "for the first time in history a massive bureaucracy had committed itself to a surrealist adventure, which is to say that the meaning of the proposed act was palpable to everyone, yet nobody could explain its logic." He thought the decision symptomatic of NASA's strain of "the American disease": "Focus on one problem to the exclusion of every other." See Norman Mailer, *Of a Fire on the Moon* (Boston: Little, Brown, 1969), pp. 346, 386, 189.

51 Dannenberg OHI, pp. 7–8; Georg von Tiesenhausen, OHI by SPW and AJD, 29 November, 1988, p. 2.

52 Murray and Cox, pp. 139–43; E. W. Kenworthy, "Kennedy Asserts Nation Must Lead in Probing Space," *New York Times* (13 September 1962), p. 1 ff.

53 Sneed OHI, 15 August 1990, pp. 4, 11–12; Odom OHI, 26 August 1993, pp. 9, 14; Marshall OHI, 29 August 1990, pp. 8–10.

54 Werner Dahm, OHI by SPW, 21 July 1994, MSFC.

55 This threefold division of funding began in FY 1964. Three divisions had been used from FY 1959 to FY 1962, with Salaries and Expenses approximating the later Administrative Operations category. For FY 1963 only, S&E and R&D were combined into one category, Research, Development and Operations, complicating comparisons across time. Van Nimmen and Bruno with Rosholt, *NASA Historical Data Book, Volume I: NASA Resources, 1958–1968*, pp. 135–44.

56 Marshall, 530.4 million dollars; Lewis, 264.5 million dollars; Manned Spacecraft Center, 232.8 million dollars. MSC, however, was closing the gap by the end of this period. Van Nimmen and Bruno with Rosholt, *NASA Historical Data Book, Volume I: NASA Resources, 1958–1968*, p. 146.

57 Marshall, 8358.5 million dollars; MSC, 7990.8 million dollars; Goddard, 2457.5 million dollars. Van Nimmen and Bruno with Rosholt, *NASA Historical Data Book, Volume I: NASA Resources, 1958–1968*, pp. 164–65.

58 From 1963 through 1965, Kennedy Space Center's obligations for Construction of Facilities totaled 593.0 million dollars. The combined total for Marshall, Michoud, and Mississippi Test during the same period was 266.9 million dollars. Van Nimmen and Bruno with Rosholt, *NASA Historical Data Book, Volume I: NASA Resources, 1958–1968*, pp. 168–69.

59 Van Nimmen and Bruno with Rosholt, *NASA Historical Data Book, Volume I: NASA Resources, 1958–1968*, pp. 202–203.

60 Percentages for the first four states during this period were: California, 43.7%; New York, 10.0%; Louisiana, 7.5%; and Alabama, 6.0%. Ibid., p. 200.

61 Percentages for the first four states during this period were: California, 43.7%; New York, 10.0%; Louisiana, 7.5%; and Alabama, 6.0%. Ibid., p. 200.

62 MSFC Industrial Operations, *Facilities Directory by Location* (MSFC, May 1966), pp. 1–8; David Akens, "Historical Sketch of MSFC," (June 1966).

63 Our following discussion on labor unions in its draft manuscript form was used in the preparation of *Wernher von Braun: Crusader for Space* by Ernst Stuhlinger and Frederick I. Ordway III (Malabar, FL: Krieger Publishing Company, 1994), pp. 187–88.

64 Charles D. Benson and William B. Faherty, *Moonport: A History of the Apollo Launch Facilities and Operations* (Washington, DC: National Aeronautics and Space Administration Special Publication–4204, 1978), pp. 36–37.

65 Launch Operations Directorate, "Summary of Presentation to Labor Dispute Fact-Finding Committee," 30 November 1960, Legal Opinion Files, 1960–61, Labor-Misc. Opinions, MSFC Chief Counsel's Office.

66 "Report of Fact-Finding Committee Appointed by Sec. of Labor James P. Mitchell in Connection with Dispute of Saturn Missile Launching Sites at Cape Canaveral, FL," espec. pp. 9, 27–33, December 1960, Legal Opinion Files, 1960–61, Labor-Misc. Opinions, MSFC Chief Counsel's Office; Harry Gorman, Assoc. Deputy Director for Administration, MSFC, to Dr. Robert Seamans, NASA HQ, 5 December 1960, Legal Opinions Files, 1960–61, Labor-Misc. Opinions, MSFC Chief Counsel's Office; Benson and Faherty, p. 37.

67 "Space Center is Idle: Third Day in Strike," *New York Times* (16 August 1962).

68 See *Huntsville Times* for following articles and editorials: Bill Austin, "JFK Aides Demand End to Walkout," (15 August 1962); "NLRB Should Act at Once," (16 August 1962); "No Help to Our Space Role," (17 August 1962); "Thoughts on Happiness," (24 August 1962); "Let's Get on with the Job," (26 August 1962).

69 "Defy Goldberg Plea to Halt Missile Strike: Electricians Ignore Union Head, Too," *Chicago Tribune* (23 August 1962); "Moonshot Strike Halted," *Washington Evening Star* (24 August 1962); "Walkout Ended at Space Center: Electricians Halt Ten-Day Strike—Inquiry Slated" *New York Times* (24 August 1962); "Group Seeks End to Space Job Disputes," *Washington Star* (27 August 1962); Bob Ward, "Labor Dispute at Redstone Settled Here," *Huntsville Times* (27 September 1962).

70 Stuart H. Loory, "US Calls Business, Labor to Space Talks: Racing for the Moon—Strike Free?" *New York Herald Tribune* (28 June 1963).

71 Thomas P. Murphy, *Science, Geopolitics, and Federal Spending* (Lexington, MA: D. C. Heath), pp. 211–13; William Ziglar, *History of NASA MTF and Michoud. The Fertile Southern Crescent: Bayou Country and the American Race into Space* (NASA History Office, HHN–127, September 1972), pp. 13–15; Bilstein, pp. 71–72.

72 Van Nimmen and Bruno with Rosholt, *NASA Historical Data Book, Volume I: NASA Resources, 1958–1968*, pp. 433–35; Bilstein, pp. 73–74; Murphy, pp. 214–17; Public Affairs Office, MTF, "Mississippi Test Facility," no date, UAH Saturn Collection, pp. 3–5.

73 "Way Station to the Moon," *Business Week* (2 April 1966), quoted pp. 62, 63; "Mississippi Test Facility," pp. 2–3, 8; Public Affairs Office, MTF, "*NASA Facts," Coast Area Mississippi Monitor*, 1969, p. 54.

74 Bilstein, p. 72; Van Nimmen and Bruno with Rosholt, *NASA Historical Data Book, Volume I: NASA Resources, 1958–1968*, p. 432.

75 NASA filled 370 positions before official establishment of the Center, 81 of them from ABMA; 311 people transferred from the ABMA Technical Materials and Operations Division, a warehousing operation, on 1 July 1960; and 3,989 of the 4,179 employees of the ABMA Development Operations Division transferred to NASA on 1 July. Robert L. Rosholt, *An Administrative History of NASA, 1958–1963*, (Washington, DC: National Aeronautics and Space Administration Special Publication–4101, 1966), p. 122.

76 Marshall was the largest NASA center throughout this period. In comparison, in the summer of 1966 Lewis Research Center had the second highest number of NASA employees with 4,819, or 14.1%. Van Nimmen and Bruno with Rosholt, *NASA Historical Data Book, Volume I: NASA Resources, 1958–1968*, pp. 86–87, 91, 452; Akens, "Historical Sketch of MSFC," p. 12.

77 Contractor employment in Huntsville peaked before Civil Service employment at Marshall; at the end of 1965, MSFC contractor and Civil Service employees topped 22,000; by the summer of 1966, the number had dropped to 18,600. Akens, "Historical Sketch of MSFC," p. 28; MSF Field Center Development, "Economic and Social Benefits of the Manned Space Flight Program," 24 November 1969, p. 71, MSFC History Office.

78 Levine, p. 110.

79 NASA's allotment increased to 425 in 1962 in response to Apollo, and the agency increased its number of excepted and supergrade positions to over 700 by using nonquota positions by the beginning of FY 1967.

80 Levine, pp. 110–14; Rosholt, pp. 140–41; Van Nimmen and Bruno with Rosholt, *NASA Historical Data Book, Volume I: NASA Resources, 1958–1968*, pp. 94–97.

81 *NASA Pocket Statistics, 1981*, (Washington, D.C.: NASA, 1981).

82 Van Nimmen and Bruno with Rosholt, *NASA Historical Data Book, Volume I: NASA Resources, 1958–1968*, pp. 452–53.

83 Von Braun, "Status of Activities at the Marshall Space Flight Center," Appendix A of U.S., Congress, House, Subcommittee on Manned Space Flight of the Committee on Science and Astronautics, hearing, *1966 NASA Authorization*, Hearings pursuant to H. R. 3730, 89th Cong., 1st sess., March 1965, p. 422.

84 John N. Wilford, *We Reach the Moon* (New York: Norton, 1969), pp. 79–80; John Mecklin, "Jim Webb's Earthy Management of Space," *Fortune* (August 1967); Harold L. Nieburg, *In the Name of Science* (Chicago: Quadrangle, 1966), pp. 43–45 ff., 184–89, 233; McDougall, pp. 380–83.

85 William H. McNeill, *The Pursuit of Power: Technology, Armed Force, and Society Since AD 1000* (Chicago: University of Chicago, 1982), pp. 89–94, 278–80; Thomas Hughes, *American Genesis: A Century of Invention and Technological Enthusiasm* (New York: Viking, 1989), chap. 3.

86 See Leonard R. Sayles and Margaret K. Chandler, *Managing Large Systems: Organizations for the Future* (New York: Harper and Row, 1971), chap. 1; James E. Webb, "U.S.

Space Teamwork Comes of Age," *Missiles and Rockets* (29 November 1965), p. 37; Tom Alexander, "The Unexpected Payoff of Project Apollo," *Fortune* (July 1969); Robert C. Seamans and Frederick I. Ordway, "The Apollo Tradition: An Object Lesson for the Management of Large-scale Technological Endeavors," *Interdisciplinary Science Reviews* 2 (1977), pp. 270–304; George Mueller, "NASA Management Programs," address, 8 July 1968, UAH Saturn Collection, pp. 4–5.

87 Sneed interview, pp. 10–12; N. Lee Cropp, "Prepared Statement for Congressional Testimony," 1973, UAH Saturn Collection, p. 7; Wilbur Davis, Mike Hardee, and Jim Morrison interview, by John S. Beltz and David L. Christensen, 21 January 1970, UAH Saturn Collection, pp. 1–3; Dieter Grau interview, no date, UAH Saturn Collection, pp. 11–12.

88 Bilstein, pp. 264–69; Cropp, "Evolution of MSFC Program Management," pp. 6–13, 15–21, 30, 38–40; Bill Sneed interview, by Roger Bilstein, 26 July 1973, UAH Saturn Collection, pp. 7–8; Wernher von Braun, "MSFC Management Policy #1," 16 August 1962, UAH Saturn Collection. Draft review notes, Kim Jeffreys, MSFC comptroller, 1997.

89 Cropp, "Prepared Statement," pp. 14–18; MSFC review comments for Chapter 2.

90 Lucas OHI, 19 June 1989, paragraphs 47–50; James OHI, 1 December 1989, pp. 9–13; Tompkins, pp. 15–16; Bilstein, pp. 282–83.

91 Bilstein, p. 289.

92 Bilstein, pp. 271–74, 283–88; William Sheil, "Guidelines for Administrators," *Boeing Magazine* (January 1966), 6–7; Wilford, p. 163.

93 Bilstein, pp. 271–74, 283–88; William Sheil, "Guidelines for Administrators," *Boeing Magazine* (January 1966), 6–7; Wilford, p. 163.

94 OMSF, "Apollo Program Development Plan," 15 January 1965, Apollo Series, JSC History Office; Levine, pp. 5–6, 42–46, 61, 63; Mueller, pp. 3–5; Rocco Petrone interview, by BBC–TV, NASA History Office, Biographical file, 1979 folder, pp. 39, 44; "Marshall Directing Vehicle Development," p. 103; Bilstein, pp. 62, 266–81; Cropp, "Evolution of MSFC Program Management," pp. 17–18.

95 J. C. McCall, MSFC, "Systems Review Meeting, Houston, Texas," 10 January 1963; McCall, "Interim Report of the Ad-Hoc Committee on OMSF (Systems)/MSFC Interface," May 1963; John Hornbeck, Bell Comm, "Systems Engineering," 9 May 1963, Lunar-Systems Review Meetings folder, MSFC Archives; von Braun, "Speech to 7th NASA Management Conference," Langley Research Center, 5 October 1962, MSFC History Office.

96 Wernher von Braun, OHI, 28 August 1970, pp. 19–22, 26–28, UAH Saturn Collection; Lucas OHI, 19 June 1989, paragraph 62; James OHI, 1 December 1989, paragraphs 12, 48.

97 Benson and Faherty, *Moonport*, pp. 3–4, 14–15, 17, 34, 43–46, 135–36; von Braun, "Daily Journal," 11 and 22 December 1961, NASA History Office; von Braun, OHI by Robert Sherrod, 25 August 1970, p. 20, Sherrod Collection, NASA History Office.

Chapter III

Crafting Rockets and Rovers: Apollo Engineering Achievements

The most dramatic events at Marshall during the Apollo Program were the static firings of the enormous first stage of a Saturn V rocket. The five F–1 engines of the S–IC stage produced over 7.5 million pounds of thrust, enough to generate 119 million kilowatts, twice the power of all hydroelectric turbines on American rivers. The stage burned 4 million pounds of fuel in two-and-a-half minutes, and three trucks could park side by side in its fuel tank. The engines had valves as big as suitcases and pumps as big as refrigerators.[1]

First S–IC full five engines firing on 16 April 1965.

Test structures for the stage and its engine cluster were also gigantic. The S–IC Test Stand, first used by Marshall's Test Laboratory in April 1965, had a superstructure and derrick that rose 406 feet. Built massive to secure the huge rocket stage, it was anchored in bedrock 45 feet below ground and had as much concrete underground as above.[2] To dissipate heat and dampen sound, the stand's pumps fed 320,000 gallons of water per second from an adjacent reservoir into the flame bucket. Each test generated a white cloud of vapor and a thunderous roar that echoed (and even shook buildings) throughout Huntsville. Engineers claimed that as a noisemaker the S–IC was third only to atomic bomb blasts and the Great Siberian Meteor of 1883. One Marshall official recalled that before the first test people feared broken windows at the Center; unable to finish an important telephone call when the test began, he crawled under his desk and shouted in the receiver.[3] Von Braun liked to interrupt meetings so that everyone could witness the spectacle from the top floor of Marshall's administration building.

The sound and fury of such tests bore witness to Marshall's contributions to the space program in the 1960s. The Center's laboratories helped design, develop, and test crucial hardware for the Mercury and Apollo programs. Marshall's project offices oversaw dozens of contractors and forged individual efforts into a collective whole. The Center's step-by-step efforts on space vehicles helped NASA achieve a series of "firsts" in space flight: the Mercury-Redstone boosters lifted American astronauts on their first suborbital rocket flights, the Saturn rockets powered humans on their first trips to the Moon, and the lunar roving vehicle (LRV) first transported people across its surface.

Mercury-Redstone

Marshall's initial triumphs as a NASA Center came in Project Mercury, America's first entry in the manned "space race" with the Soviet Union. The Center contributed Redstone boosters for the early flights, helped the STG with integration of the booster and crew capsule, and oversaw the launch process. Involvement in the program began in October 1958, when NASA and the Army Ordnance Missile Command agreed that the ABMA would provide 10 Redstone and 3 Jupiter missiles for the space program. In the next year ABMA modified the Redstones to prolong the time of engine burn. Working with the Chrysler Corporation, the prime contractor, and the Rocketdyne Division of North American Aviation, the engine contractor, ABMA personnel elongated the propellant tanks.

Modifying the Redstone tanks was straightforward, but "man-rating" the rocket was not. Man-rating meant verifying the rocket's safety for human flight. Although the Redstone had many successful launches as a ballistic missile, man-rating led to technical disputes between Huntsville personnel and the STG. Huntsville's experience with missiles led them to consider the "payload" as a passive package. But members of the STG were "old NACA hands" who were experienced with airplanes and pilots.

The contrasting perspectives of Marshall and the STG led to quarrels over automatic flight abort procedures. According to Joachim P. Kuettner, ABMA's and later Marshall's manager for Project Mercury, Huntsville preferred "positive redundancy" which provided for automatic aborts whenever required; automation would ensure astronaut safety by restricting his role. Kuettner thought the STG wanted "negative redundancy" which avoided aborts unless necessary; with more control, astronauts would have more opportunities to finish missions. Panels of technical experts from Marshall and the STG worked out the differences, balancing pilot safety and mission success, machine automation and human control. Their contrasting perspectives improved the Mercury design and helped ensure success, but put the program behind schedule.[4]

Delays came from other sources. The STG often changed its designs, forcing Marshall to adapt its work on the Redstone. The McDonnell Company, contractor for the Mercury spacecraft, fell behind, slowing Marshall's ability to integrate the hardware of spacecraft and Redstone. But the Center's extensive hardware testing also took longer than expected and caused delays.[5]

Unfortunately more delays came from the failure of the first flight test of Mercury-Redstone. The crewless launch of Mercury-Redstone 1 (MR–1) on 21 November 1960 began with the rocket engine burning normally. After a flight of a few inches, however, the engine abruptly shut off. MR–1 fell back on its pad, resting upright and inert but for an escape parachute which released from the capsule and flopped limply in the breeze. An investigation traced the engine failure to the booster's tail-plug prongs, which connected the booster via an electrical cord to ground equipment. The prongs were too short to compensate for changes in the payload and thrust of the modified Redstone, and the tail plug pulled out, prematurely turning off the engines.[6]

After the failure, and a malfunction which caused the MR–2 engine to operate at higher than planned thrust level, von Braun wanted to avoid unnecessary risks. He therefore insisted on one flawless Mercury-Redstone flight before any manned mission and convinced NASA to insert an extra "booster development" mission. This mission with a boilerplate Mercury spacecraft (MR–BD) flew successfully on 24 March 1961. The extra mission, however, pushed back the schedule for America's first manned Mercury-Redstone flight (MR–3) and allowed the Soviet Union to capture prestige with Yuri Gagarin's first orbital flight on 12 April. This Soviet triumph overshadowed the success enjoyed by the United States, NASA, and Marshall on 5 May 1961 with the suborbital flight of astronaut Alan Shepard aboard MR–3. The final Mercury-Redstone mission occurred in July.[7]

During these first steps in human space flight, Marshall experienced some problems that would recur in later programs and learned important lessons. Kuettner noted several difficulties in relations with the STG. He observed that the group's control over funds "resulted in a tight technical control of the total vehicle by the payload people." The group tried to tell Marshall what to do even though they had less experience in managing complex projects. Rather than directives coming from one Center, Kuettner thought that "broad program control" should come from NASA Headquarters or negotiations between Center directors.

Kuettner also expressed chagrin at how the STG and NASA had handled publicity and had failed to promote Marshall's role. "Handling of Public Information affairs," he lamented, "has been considered unfair by most every participant in this program."[8] Eberhard Rees, Marshall's deputy director for research and development, thought that STG publicity for Shepard's flight merely mentioned Marshall's role without praise. Rees wrote to von Braun that "this is significant how STG thinks. Under these conditions we can not work in the 'Manned Lunar Program.'"Von Braun responded, "I agree."[9]

Although wounded pride had caused Center personnel to blame the STG, larger circumstances explain Marshall's lack of celebrity. The media and the public idolized the STG's astronauts, seeing them as heroic explorers, but largely took for granted the more prosaic contributions of engineers and managers; unfortunately for Marshall, the Center had no astronauts. NASA used this public fascination with the astronauts to bolster its image, attract political support, and justify big budgets for human space flight. Consequently press coverage of MR–3 mentioned the "Old Reliable" Redstone but seldom attributed it to Marshall. Even the *Huntsville Times* lionized Shepard with very little mention of local people.[10]

Regardless of such slights, Marshall personnel had contributed to the success of Project Mercury. Moreover they had learned about man-rating rockets and working with another NASA Center, lessons they applied to the Saturn project.

"Stages to Saturn"

Marshall's primary effort in the 1960s was the design, development, and testing of the Saturn launch vehicles.[11] The work helped lead to the extraordinary first human explorations of the Moon.

The three basic Saturn configurations fit into the Center's conservative "building block concept" in which less powerful and sophisticated launch vehicles preceded and tested designs of more advanced models. The Saturn I, originally called the Juno V and Saturn C–1, was a two-stage booster used to test multi-engine clusters, to qualify Apollo spacecraft, and to launch the Highwater and Pegasus experiments. The Saturn IB, also called the C–1B and Uprated Saturn, had more advanced upper-stage engines than the Saturn I. NASA used it to continue propulsion and spacecraft testing, and to launch the Earth orbital missions in the Apollo and *Skylab* programs. By far the most powerful was the Saturn V, also known as the Saturn C–5. It was NASA's largest launch system, and its three stages propelled the Apollo lunar missions and the *Skylab* workshop.[12]

Building the Saturns was a tremendous challenge for the Marshall team. During the less than 10-minute burn of launch, the engines had to generate tremendous thrust. The rocket structure, with all its seams and connections, had to withstand changing stresses. All the mechanical and electrical systems had to work to near perfection. Any breakdown could result in a fiery disaster.

To avoid this fate, the Center and its contractors drew from their experience in military rocketry. Ancestors of the Saturns included the von Braun team's V–2 and the liquid-fueled military rockets that North American Aviation's Rocketdyne Division developed for the Navaho cruise missile. Lessons from the Air Force's Thor and Atlas and the Army's Redstone and Jupiter contributed to the Saturn's engine, fuel, guidance, and launchpad checkout systems. The Saturns, like the Navy Vanguard, used gimballed, or swiveling, engines to control flight direction. The engine that powered the Saturn V's first stage, Rocketdyne's mighty F–1, began as an Air Force research project. Drawing on this military technology, Marshall and its contractors transcended it by increasing rocket size and thrust, reducing the weight of components, improving reliability, raising engine pressures, and developing faster fuel pumps.[13]

The military influence was especially strong on the first stage of the Saturn I (called the S–1) because the Center's rocket experts largely designed and developed the S–1 while still a part of ABMA. In April 1957 the Army began studies of a super-Jupiter. Recognizing the potential political liabilities and financial costs of a new booster, the goal was to maximize lift but build on current technology. The plan called for using the H–1 engine, an improved version of Rocketdyne's Thor-Jupiter S–3D engine, in a "cluster" configuration of eight engines to achieve 1.5 million pounds of thrust. Clustering engines was an untried concept; von Braun recalled that skeptics doubted that eight engines could fire simultaneously and called the S–1 a "plumber's nightmare" and "Cluster's Last Stand." The vehicle's structure also used existing technology, positioning eight Redstone tanks around one Jupiter tank. Not only would this save money, but multiple fuel tanks offered technical advantages; easy dismantling and reassembly would facilitate transportation, its RP–1 kerosene fuel and its oxidizer would reside in different tanks, and the number of interior fuel slosh baffles would diminish.

Saturn second stage acceptance test.

Following the August 1958 authorization to develop the Saturn I first stage, ABMA built the first eight vehicles in-house and then the Chrysler Corporation took over the work. With these measures, work on the S–1 proceeded quickly. Marshall began static firing of the first test booster on 28 March 1960, only three years after the project's conception and 19 months after its authorization. An improved, more powerful version of the S–I, designated the S–IB, provided the first stage of the Saturn IB.[14]

Because the S–IV and its more advanced progeny, the S–IVB, were the upper stages for the Saturn missions, they were the next boosters completed. In 1959 ABMA's initial designs for an upper stage called for using current military boosters with conventional rocket fuel. But the Jupiter, Altas, and Titan lacked the power needed for high altitude second stages. Using them with the S–1, observed Willie Mrazek, director of the Structures and Mechanics Lab, "was like considering the purchase of a 5-ton truck for hauling a heavy load and finally deciding to merely load a wheelbarrow full of dirt." Army and NASA planners began considering more powerful, innovative engines with liquid hydrogen fuel. This fuel was extremely volatile and flammable and had to be controlled with great caution, but it could boost heavier payloads.[15]

The rocket engineers at ABMA and Marshall drew on the work of others with liquid hydrogen engines. The United States Navy and Air Force, the Jet Propulsion Laboratory, Aerojet Corporation, and especially NACA's Lewis Research Center had developed the technology in the 1940s and early 1950s. In the late 1950s the military contractor General Dynamics worked on the Centaur upper stage with liquid-hydrogen engines developed by Pratt and Whitney. Marshall took over management of the Centaur contract in July 1960 and in August had Pratt and Whitney begin upgrading its propulsion for the Saturn project. After Marshall finished its designs, the S–IV had a cluster of six Pratt and Whitney RL–10 engines in a vehicle built by Douglas Aircraft. The Center made major contributions by conducting metallurgy studies to guide the selection of materials for the fuel tanks.

The S–IVB emerged from NASA's quest for even more powerful upper stages. A propulsion study committee headed by Abe Silverstein recommended a liquid-hydrogen engine of 200,000 pounds of thrust, far above the RL–10's 15,000 pounds of thrust. Marshall worked on the design and awarded a research contract to Rocketdyne in 1960. The final configuration awaited the outcome of NASA mission planning, and in 1962 the agency decided on one J–2 engine for the S–IVB. To increase tank capacity, Douglas Aircraft would widen the S–IV frame by a meter in diameter. A major challenge was developing technology for restarting the S–IVB in orbit for the reboost to the Moon. Since the liquid fuel would float freely in the microgravity, the Center and its contractors devised systems to position the fuel in the tanks, using pressurized mechanisms and small rockets to give the stage an initial boost.[16]

The largest of the Saturn boosters was the S–IC, the first stage of the Saturn V. Huntsville's propulsion experts began preliminary designs in the late 1950s, choosing RP–1 kerosene fuel because it would require less tank volume. Initial plans called for using four F–1 engines, but early in 1960 as the projected weight of the Apollo spacecraft continued to grow, NASA's engineers decided to add a fifth engine. Marshall's robust rocket structure with heavy cross-beams made addition of the fifth engine possible. The lifting capacity of five engines would prove invaluable when the weight of Apollo payloads increased.[17]

Installing S–IC–T stage in S–IC Test Stand in March 1965.

In December 1961 Marshall selected Boeing as the prime contractor for the S–IC, and for several reasons the two quickly formed an intimate relationship. Closeness was easier because, unlike other Saturn contractors, Boeing worked in Huntsville with offices at the center and in a converted textile mill called the HIC Building (Huntsville Industrial Center). Even when work moved to the Michoud Assembly Facility and Mississippi Test Facility, Boeing remained at Marshall sites. Moreover early design and development occurred in-house at Marshall. There the Center directly managed Boeing's work, integrating contractor personnel into Marshall teams and only gradually giving

Saturn S–II stage arrives at Marshall.

them independence. When manufacturing began in 1963, the Center used Boeing tooling to make the first three test models.[18]

Technical challenges also brought Marshall and Boeing together. The S–IC was so large, 33 feet in diameter and over 130 feet long, that its construction required new manufacturing methods. For example its bulkheads needed welds dozens of yards long to join the thin aluminum walls. To solve this problem Marshall helped its contractor devise new welding and inspection techniques. Center personnel invented an electromagnetic hammer to remove distortions in the bulkheads created by welding. The hammer functioned without physical contact, and technicians showed off its operation by inserting tissue paper between the electromagnetic coil and the metal part and removing the paper unscathed. Marshall also helped devise x-ray systems for inspecting the welds.[19]

Marshall's in-house activities for the Boeing contract sometimes led to problems. NASA Headquarters initially questioned the amount of arsenal work. During a visit to Marshall in 1962, one headquarters official "stated repeatedly that he believes Marshall should de-emphasize more the in-house operations in connection with S–IC development" and let Boeing handle the job. Marshall managers explained that the arsenal system saved money and time by allowing work to proceed while the contractor upgraded its skills and NASA constructed the facilities at Michoud and in Mississippi. Two years later the intimate relationship made it difficult for the Center to hold Boeing responsible for cost overruns. Marshall had so dominated the S–IC project that it was as responsible for the overruns as Boeing; one internal Center memo admitted that Marshall had "imposed our experience on their [Boeing's] minds to the point of their losing their identity as an independent contractor." Looking back after 30 years, Dr. William Lucas, then chief of the engineering materials branch in the Propulsion and Vehicle Engineering Lab, argued that the arsenal system provided Boeing with help it needed to solve the novel technical problems created by the Apollo mission; "there was not a contractor workforce out there willing and able to do the job."[20]

Marshall also had an especially close technical relationship with Rocketdyne for the F–1 engine. Saviero "Sonny" Morea, Marshall's manager for the F–1, recalled that the Center "used to drive them bananas with our technical prowess" and that "sometimes we penetrated more deeply than they desired us to penetrate" until Marshall was in Rocketdyne's "drawers quite deeply." Morea thought the Center and its contractor needed such a "team relationship" to solve technical problems and meet the end-of-the-decade deadline.[21]

Although the F–1 lacked the sophistication of the J–2, its size and thrust created new difficulties before 1965. To generate its 1.5 million pounds of thrust, its turbopumps and fuel lines had to deliver precise amounts of RP–1 kerosene fuel and liquid oxygen (LOX) to the combustion chamber. For each second of the two-and-a-half-minute burn, pumps provided 2 metric tons of LOX at minus-300 degrees Fahrenheit and 1 metric ton of RP–1 at 60 degrees. During operation the turbopumps warmed to 1,200 degrees and the combustion chamber reached 5,000 degrees.

One of the most severe problems addressed during the development of the Saturn V program was the issue of combustion instabilities in the F–1 engine. Combustion instability resulted from destructive pressure oscillations found in the engine's high-pressure, high-performance combustion chambers. The problem was so severe that some development engines were lost due to heat loads on chamber walls and damage to the injector; in several cases, instability caused catastrophic loss of entire engines.

Marshall formed an "ad hoc" committee to solve the F–1 problems. The committee was made up of engineers and scientists from government agencies, industry, and universities; this approach of pulling together the right people and resources to solve such problems was a strong point of Marshall's approach during Saturn development. The "ad hoc" committee analyzed the problems and developed a test program to study alternative designs. They ignited small bombs in the engine exhaust to induce instability, and tested prototypes until they failed. After considerable trial-and-error engineering, they reached a robust design that could compensate for combustion instability. The solution was a set of baffles in the combustion chamber which dampened the acoustic oscillations if they began. The process took some time, and Marshall did not certify the engine until January 1965.[22]

The S–II stage was the last completed, and Marshall's relationship with North American Aviation, the prime contractor, was its most troubled of the Saturn era. The story of the S–II reveals what Marshall expected from its contractors and how the Center responded to problems.

The design for the S–II began in late 1959 when NASA's Silverstein propulsion committee recommended upper stages with liquid-hydrogen engines. ABMA, and later Marshall, began preliminary studies and in 1961 selected North American Aviation for the contract. Unfortunately, however, NASA's choices about Apollo missions and escalating concerns about payload weight increases in 1962 led to changes in the S–II's technical requirements. NASA chose a cluster of five J–2 liquid-hydrogen engines, and wanted both to increase size to accommodate more fuel and to contain weight to allow for greater payloads.

To meet the S–II's complex requirements, Marshall and North American had to overcome many challenges. To save weight, their design used a single bulkhead between the LOX and liquid-hydrogen tanks rather than two separate tanks. The common bulkhead, however, needed insulation to prevent the liquid-hydrogen from boiling away. The material for the tanks had to be lightweight and compatible with the fluids in them. Marshall chose a pre-existing aluminum alloy for the tanks that its developer said was impossible to weld. Even worse, long welds were required to join the segments of a stage 10 meters wide and 24 meters high. Marshall and its contractors therefore had to develop new welding and inspection technologies.[23]

North American Aviation began manufacturing the S–II in the fall of 1963, but quickly encountered problems. Recognizing the technical complexity of the project, Marshall nonetheless concluded that the primary problems were managerial. Indeed for the next three years, reports of Center officials offered a litany of North American's management weaknesses. They complained that the company lacked a management system necessary for a complex research and development project and so it could not integrate budgeting, engineering, manufacturing, quality control, and testing. This led to unclear authority, piecemeal design, communications failures, unanticipated problems, crash efforts, rework, haphazard documentation, cost overruns, schedule slips, and unresolved technical weaknesses. In one case, Marshall project officials were stunned to find that North American had purchased the same vehicle checkout system from the same subcontractor as had Douglas Aircraft, but had paid 70 percent more. From the Center's perspective, excessive pride and optimism made the company reluctant to accept Marshall's directions. James Odom, Marshall's chief engineer for the S–II, recalled that Marshall had more experience in welding large structures than its contractor, but the experts at North American doubted the Center's technical advice. In addition, Center officials believed that NASA's

MSC contributed to the company's bad habits by lax management of North American's work on the Apollo Command Module.[24]

By spring 1965, the S–II had fallen so far behind that Marshall eliminated some test models so the contractor could work on flight stages. The structural failure of a stage during a load test in late September 1965, led General Edmund O'Connor, head of the Center's Industrial Operations, to warn von Braun that the project was "out of control" and "jeopardizing the Apollo Program." NASA Headquarters sent a team to investigate and advise. One Marshall engineer told the investigators that North American's "equipment is usually too complicated" and their work "is nearly always overpriced." "They accept direction readily if they agree with it. If they do not, they will stall, misunderstand, write, dither, and all the while continue along the same path until we are faced with a schedule impact if we force our position." Rees, the Center's technical deputy director, warned the company that failure to improve would result in transferal of the project to another contractor.[25]

Avoiding such a drastic step, Marshall sent managers and engineers to accelerate progress. North American changed project managers and reconfigured its managerial systems, but in May 1966 another stage was destroyed. Fortunately NASA's large Apollo budget and Marshall's arsenal system provided a wealth of money and expertise to throw at the problem. Even after 18 months of extensive assistance by Marshall, however, the S–II project remained in crisis. In December 1966 von Braun said the problems were "extremely urgent" and that Marshall would "apply whatever talent is necessary at whatever level, even at the expense of other Center programs." Finally, after the Apollo Command Module fire in January 1967, for which North American Aviation was the responsible contractor, NASA conducted another investigation and directed another project reorganization. The company added more talent to its NASA projects and another team from Marshall facilitated engineering changes and helped improve quality. During this time, Odom recalled, Marshall's Eberhard Rees told the team that "we will work 24 hours a day, 7 days a week, and if that is not sufficient, we will start working nights!" Although in August 1967 Center Director von Braun informed Headquarters that North American had "not yet demonstrated that it fully meets the standards expected of a NASA prime contractor," the first flight stages of the S–II were complete. By summer 1967 the stacking of the first Saturn V vehicles had begun in the assembly building at the Kennedy Space Center.[26]

In addition to working on the Saturn stages, Marshall people also labored over the vehicles' checkout and flight control systems. The checkout systems, which monitored the flight readiness of the vehicle on the launch pad, rested on military missile technology. The Center and its contractors advanced the state-of-the-art by automating more of the process with computers that read information from 5,000 data sensors on the vehicle.[27]

Marshall also helped design and develop the Instrument Unit (IU) that controlled the Saturn during launch. The Center, believing that an instrument unit provided redundancy, resisted efforts by the MSC for a single vehicle control system located in the Apollo spacecraft. Marshall's conservatism paid off when lightning struck AS–507 (Apollo 12) during launch; the spacecraft controls failed but the IU kept operating and NASA used its data to realign the guidance and control system in the command module. Located between the S–IVB and Apollo Service Module, the unit had systems for guidance and control, engine cutoff and stage separation, and data communication. Marshall began design and development as an in-house project, relying on German gyroscope technology, American electronics, and American military guidance systems like the Jupiter and Redstone. IBM became the contractor and manufactured the units at Huntsville's research park. The Center and its contractor improved guidance and control technology by using modular components, lightweight materials, microminature circuitry, and digital programming. When in 1965 the IU for the first Saturn IB launch (AS–201) fell behind schedule, Marshall and Boeing technicians jury-rigged a clean room on a barge, and continued work while chugging down the Tennessee and Mississippi Rivers.[28]

Before any Saturn stages reached Kennedy, Marshall and its contractors tested each one extensively in special facilities. Test stands stood in an irregular pattern around the East and West Test Areas. The largest was the S–IC Stand described earlier. The Static Test Tower had dual positions; it was constructed in 1951 to accommodate Redstones and Jupiters, modified in the 1960s for Saturn IB tests on one side, F–1 engine tests on the other, and reconfigured again in the 1970s for shuttle tests. A water-cooled bucket deflector absorbed the heat and sound of its exhaust. In one early test, however, enough acoustical energy bounced off low clouds to damage a Huntsville shopping mall, necessitating weather constraints on subsequent tests.[29]

Following successful static firing, the Saturn stages moved on to dynamic testing. Marshall engineers subjected each stage to a variety of stresses, such as the vibration induced by engine thrust and the sloshing of LOX fuel experienced during ascent. The Saturn V Dynamic Test Stand, a 360-foot tower topped by a 64-foot derrick, was the tallest structure in North Alabama. Marshall engineers assembled an entire 364-foot Saturn V with its Apollo capsule and enclosed it within the stand. Tests in 1966 and 1967 examined the effects of stress at 800 measuring points on the Saturn configuration.

Tests of the S–IC first stages and S–II second stages occurred not only in Huntsville but also at the Mississippi Test Facility (MTF) that Marshall managed. Built by the Army Corp of Engineers and operated mainly by contractors, the facility had a railway, a barge canal, laboratories, and three huge test stands.[30]

Transporting the huge Saturn stages led Marshall to develop its own ground, sea, and air fleet. Center engineers designed ground transporters; military trucks with aircraft tires carried the stages, which rested on assembly jigs that doubled as transport braces. In 1961 the Center began acquiring a fleet of barges, most of them converted World War II Navy ships, to ferry Saturn stages between Marshall, Michoud, Mississippi Test, and Cape Canaveral. The 3,500-kilometer barge trip from Huntsville to the Cape via the Tennessee and Mississippi Rivers and the Intracoastal Waterway took 10 days.[31] Marshall also used air transportation, contracting for a Boeing B–377 Stratocruiser with a lengthened and enlarged fuselage that could accommodate an S–IV stage. The "Pregnant Guppy," which separated in the middle for loading, carried its first Saturn stage late in 1963. This success and plans for larger stages prompted Marshall to contract for an even larger transport aircraft. The new "Super Guppy," large enough to hold the S–IVB stage, became operational in 1966. Both planes carried not only stages and engines, but other Apollo and *Skylab* cargoes.[32]

Flights and Fixes

More than an engineering development organization, Marshall assisted Kennedy Space Center with launch operations and the MSC with the first part of lunar flights. The Center helped oversee 32 successful Saturn launches, including 9 by Saturn Is, 10 by Saturn IBs, and 13 by Saturn Vs. No Saturn launch was a failure, a remarkable record for technology as complex as the Saturns and a stunning testimonial to the quality of engineering and management of the

Center, its contractors, and the whole Apollo team. Their expertise was especially evident after the second Saturn V flight when they rapidly corrected problems to clear the way for human exploration of the Moon.

Before launch Marshall and the Kennedy Space Center worked closely together, coordinating booster design with checkout and launch equipment, stacking the stages, and preparing for launch. During a launch, an elaborate communication system linked Marshall to Kennedy. For human missions, another network linked Huntsville to Mission Control at the MSC in Houston. This communication network relayed telemetry data to the Huntsville Operations Support Center, the Flight Evaluation and Operational Studies Division of the Aero-Astrodynamics Laboratory, and other units which monitored the Saturn stages.[33]

Marshall applied its "building-block" approach to the early Saturn flights, testing launch vehicles stage-by-stage, launching the first stage with dummy upper stages, and adding live upper stages only on later missions. The Block I flights, the first four missions beginning in October 1961, had dummy upper stages and primarily tested large rocket technology and clustered-engines. The missions validated Marshall's cluster concept and showed the Saturn's capability of launching with one engine out; the Center also learned that more baffles were needed to control fuel sloshing. The second and third launches also performed the engineering and atmospheric experiments called "Project Highwater."[34]

In 1964 NASA turned to Block II missions which tested fins on the lower stage and had the first flights of the S–IV upper stage. In January 1964, SA–5 successfully flew with live first and second stages successfully and boosted a heavier payload, albeit ballast sand, than the Soviet space program had. NASA press releases and media coverage described Marshall as closing "the missile gap." Representative headlines shouted "Out-Rocketing the Russians" and "We're No. 1 with Saturn I." Stories portrayed NASA as champion of the free world and the Saturn I as taller than the Statue of Liberty. From an engineering perspective, the Block II missions proved the liquid-hydrogen engines, verified the early versions of the IU, and carried the first Apollo spacecraft. In addition, the missions put in orbit three Project Pegasus satellites which detected micrometeoroid impacts to test spacecraft engineering concepts.[35]

Marshall's next building-blocks were the Saturn IB missions. Beginning in February 1966 the flights mainly tested the Instrument Unit and the S–IVB stage, which were nearly identical to Saturn V equipment. Especially successful were tests of the S–IVB which examined how liquid-hydrogen acted in orbit and proved that the engine could restart for the upcoming lunar missions. Later missions continued testing the Apollo Command Module. Launch vehicle SA–205 boosted the Apollo 7 capsule and the first crew into orbit in October 1968.[36]

Even as the Saturn I and IB flights were proceeding, NASA and Marshall abandoned the conservative, building-block method of flight testing for the Saturn V. George Mueller, who became NASA's associate administrator for Manned Space Flight in September 1963, argued that stage-by-stage tests were expensive and unnecessary. The test flights increased costs and delayed schedule without added assurance of safety or success. As an alternative Mueller proposed the "all-up" testing he had used as a systems engineer in the Air Force Titan II missile program. An all-up test launched an entire stack of live stages on the first flight. In a teletype of 1 November 1963, Mueller directed NASA Centers to prepare all live stage first flights for the Saturn IB and Saturn V; he further directed that the first Saturn V mission with a crew be the third rather than the seventh flight.[37]

Mueller's decision caused "shock and incredulity" among Marshall's engineers. All the lab chiefs and project managers initially opposed all-up testing, believing that it was an "impossible" and "dangerous idea." They particularly worried about problems from the liquid-hydrogen upper stages. Karl L. Heimburg, director of the Test Laboratory, expressed "immediate and strong opposition" and William A. Mrazek, director of the Structures and Propulsion Laboratory, thought Mueller had lost his mind. Lee James, project manager for the Saturn IB, said that "everybody explained [to Mueller] how complicated, how big this was, how the valves had never been used, how the engines had never been used."[38]

Nevertheless, Marshall quickly accepted the all-up approach. After some thought, Center engineers could neither refute the concept nor offer convincing technical justifications for stage-by-stage tests. Dr. Walter Haeussermann, director of the Guidance and Control Laboratory, and Dr. Ernst D. Geissler, director of the Aeroballistics Laboratory, concluded that the all-up concept could neither be proven right or wrong. Because of Marshall's conservative engineering and ground testing, there was "nothing to worry about."[39]

Von Braun and Rees sided with Mueller. Both initially had some doubts; later Rees said he "personally fought" the idea and von Braun said it "sounded reckless." After listening to the technical arguments, Marshall's director informed his people that all-up was the way to go. Von Braun and Rees decided that stage-by-stage launches would inhibit meeting the end-of-the-decade deadline, mainly because launch facilities would have to be reconfigured for each mission.[40]

First Saturn V launch, 9 November 1967.

Even so, many of Marshall's engineers felt uncomfortable with the policy and sometimes expressed doubts about all-up testing. James recalled that "I don't think anybody at Marshall believed it would work. I don't think anybody believed we would never have a failure in the Saturn program."[41]

Obviously the preparations for the all-up, first launch of the Saturn V booster AS–501 were very tense for Marshall and indeed the entire agency. The fact that checkout, prelaunch tests, and preparations took three weeks rather than one week only added anxiety. Consequently on 9 November 1967 everyone waited nervously. As the F–1 engines spitted flame and the Saturn V lifted off, von Braun could not contain his excitement and shouted, "Go, baby, go!" And after a flawless three-stage flight, he turned to Arthur Rudolph, the Saturn V project manager, and said that "he never would have believed it possible." Rudolph was just as surprised and even more pleased. The flight came on his 60th birthday, and he said the Saturn was "the best birthday candle" ever. The success made the whole Marshall team euphoric.[42]

Unfortunately, on 4 April 1968 the second Saturn V, booster SA–502, had many troubles that required emergency responses from Marshall. Each stage had problems. The S–IC first stage had severe vibrations from 125 to 135 seconds into the burn. Two of the five J–2 engines on the S–II second stage shut off prematurely and the stage required a new trajectory and longer burn. Once in orbit, the S–IVB third stage failed to reignite. If these problems had occurred on a lunar mission, NASA would have scrubbed it.[43] Unless Marshall could develop quick fixes, the agency could miss the end-of-decade deadline.

Marshall immediately assembled teams of experts from the Center and contractors. Following the discipline of "automatic responsibility," each lab checked flight and test data to investigate whether its specialty was involved. The Center worked primarily with Rocketdyne, the engine contractor, but the stage contractors also participated actively. To get independent perspectives, Marshall brought in consultants from the Air Force and academe and had other contractors investigate separately.

The experts determined that the S–IC had experienced the "pogo effect."[44] Pogo was longitudinal oscillation like the motion of a pogo stick in which the vehicle lengthened and shortened several times a second. The natural frequency of the stage structure of four cycles per second was very close to the operational frequency of the propulsion system (the fluid vibrations in the fuel lines and the hydraulic actions of the engines) of five cycles per second. As propellants drained, the structure's frequency increased until at 110 seconds into the flight it coincided with that of the propulsion system. The coupling of the frequencies amplified the up-and-down oscillations and caused tremors through the entire vehicle.

Pogo oscillations affected most large liquid-fuel rockets, but were not always severe. For example the Saturn I had no serious pogo problems. Even so Marshall had anticipated potential trouble and installed flight vibration detectors on the Saturn V. After AS–502 the Center's propulsion experts lacked proof that the S–IC's oscillations were dangerous. Nonetheless they worried that severe pogo could destabilize the propulsion systems, damage the command and lunar modules, or threaten the astronauts.

Two weeks after the flight and after identifying the pogo problem, Marshall formed a working group of about 125 engineers and 400 technicians from the Center, Rocketdyne, Boeing, and several other contractors. At Marshall, the Propulsion and Vehicle Engineering Lab performed the primary studies. Since the oscillations could not be duplicated on the ground, they relied on the Astrionics Lab and Computation Lab to create computer models of the phenomenon based on flight data and previous tests. The working group used a formal logic tree to assist their deliberations and identified several criteria to evaluate possible solutions; the optimal solution would prevent recurrence of pogo, would not adversely affect other systems, would be easily retrofitted, would not delay the Apollo schedule, and could be tested on the ground. Because of costs in development time and money, the team ruled out several proposals to change the vehicle structure or stiffen the fuel lines.

By 2 May, the team had decided to reduce the frequency of the propulsion system. Rocket engineers had already proven this approach; the Titan II had used a similar fix for the pogo effect, and in 1965 Marshall had applied that lesson to the S–IC fuel lines. Consequently the working group decided to test two alternative redesigns of the LOX intake system and divided tasks among the team. Marshall's Test Lab ran 9 of the 14 major types of tests which evaluated components, alternative LOX feed and pump subsystems, and the impact on the F–1 engines and S–IC stage. By July, static firings with the redesigns had produced data that the labs incorporated into computer models of flights; the tests and flight simulations verified that either design could suppress the oscillations.

Based on this information, the working group unanimously decided on 15 July that helium-charged accumulators in the LOX lines best met their criteria for a pogo fix. The solution took advantage of two preexisting parts of the S–IC. Helium gas was already on board to pressurize the fuel tanks, and the LOX ducts had a bulge called "a prevalve cavity" about 90 inches above the pump to detain oxidizer until ignition. The pogo fix would inject unpressurized helium in the cavity, and the redesign involved little more than adding a new helium line. The helium, which would not condense at the low LOX temperature, acted as a shock absorber to cushion the bottom of the LOX column. Ground tests confirmed that the helium accumulator reduced the operating frequency of the propulsion system from five cycles per second to two cycles. Later tests led the working group to conclude that an accumulator on the center engine could promote oscillations, so in the fall they decided to install the change only on the four outboard engines.[45]

While the pogo working group investigated the first stage, about a dozen engineers and 150 technicians from Marshall and Rocketdyne studied the problems with the J–2 engines on the second and third stages. Leading the way were experts from the Engine and Power Branch of the Center's Propulsion and

Vehicle Engineering Lab who gained clues from telemetry data from the Number 2 engine on the S–II stage. Temperature sensors inside the vehicle initially showed cold, evidence of a liquid hydrogen leak from the lines leading to the engine's igniter. Later the sensors read hot, signifying that the line had ruptured and the fuel burned inside the booster until another detector shut off the engine by closing a fuel valve. Unfortunately a mistake in electrical wiring had sent the shut-down signal from bad engine Number 2 to good engine Number 3 and turned off that engine as well. Exhibiting the same readings as the Number 2 engine, the J–2 engine on the S–IVB also had a rupture in the igniter line that prevented its restart.

In ground tests at Marshall the engineers subjected the igniter lines to greater pressure and vibrations than in flight conditions, but could not duplicate the failure. They then turned to vacuum tests of eight lines and found that all eight lines failed. They concluded that in ground tests the cold liquid hydrogen (minus 400 degrees Fahrenheit) had liquefied moisture in the air around a bellows section of the line; the ice then dampened the line's vibrations. In the rarefied upper atmosphere, there was no moisture to freeze and absorb the stress. Consequently fuel flow in the line caused vibrations of 15,000 cycles per second and led to ruptures. The engineers fixed the problem by eliminating the bellows section, reducing the diameter of the igniter line, and making the line more flexible by adding five bends. To be safe, they redesigned the LOX lines, even though these had experienced no problems. The engineers then performed vacuum tests and by the end of May had certified the reliability of the new configuration.[46]

The AS–502 investigations were so conclusive and solutions so reliable that the Marshall team convinced NASA that another test flight of the Saturn V was not needed. NASA decided to proceed with plans for a crew on the third Saturn V launch. On 21 December 1968, SA–503 (Apollo 8) sent people into orbit around the Moon for the first time.[47]

Of course the ultimate mission of the Apollo program was SA–506 (Apollo 11) which landed men on the Moon in July 1969. Norman Mailer observed lift-off from the observation site for the press located several miles away from the launch tower and lyrically described the sensations. He noted the eerie silence of watching the Saturn V rise before the sound reached his position; initially the liftoff, Mailer said, seemed "more of a miracle than a mechanical phenomenon, as if all of the huge Saturn itself had begun silently to levitate." The engine's bright blaze initially coursed along the ground in "brilliant yellow bloomings of flame," and after the Saturn rose above the launch tower, its "fire was white as a torch and as long as the rocket itself." When the sound reached Mailer, he heard "the thunderous murmur of Niagaras of flame roaring conceivably louder than the loudest thunders he had ever heard and the earth began to shake and would not stop." As the Saturn rose "like a ball of fire, like a new sun mounting the sky, a flame elevating itself," Mailer reflected that humans "now had something with which to speak to God."[48]

Celebration in downtown Huntsville of Apollo 11 landing.

Neil Armstrong's first footstep on the Moon completed Kennedy's challenge and accomplished an ancient human dream. Von Braun remarked that the lunar landing was the "culmination of many years of hard work, hopes and dreams." It was "as significant as when aquatic life first crawled on land" and "assured mankind of immortality." In a celebration in downtown Huntsville, crowds thronged around Marshall's engineers and managers, buoying them in a delirious outburst of happiness and hometown pride.

During the hoopla surrounding the mission, the media and public paid more attention to the Apollo 11 crew than to the Center responsible for the Saturn V booster. At a prelaunch news conference on 15 July, NASA officials fielded questions from the press about the upcoming flight. Lee James, Saturn program manager, represented Marshall, but the press did not ask him one question. The media, James reasoned, already believed that the Saturn V was "old stuff."[49]

For the Saturn V launches, the Center continued in its crucial, behind-the-scenes role. Marshall's engineers managed vehicle preparations, analyzed flight data, and corrected problems. As an example of this, the rocket engineers noticed a very small pogo effect that occurred on the S–II stage on the Apollo 8, 9, and 13 launch vehicles. Although the problem never endangered a mission, the experts took no chances and used computer simulations and static tests to isolate the phenomenon in the interaction of the center engine and the crossbeam on which the engine rested. Marshall added accumulators in center engine's LOX line and shut the engine down 90 seconds before the others, before vibrations in the propulsion and structural systems synchronized.[50]

Rees reiterated the Center's careful approach to space flight in a flight readiness review after Apollo 11. He encouraged his team to remain vigilant, saying, "this was the best launch vehicle we have ever had, but we should not be complacent over the success of this launch. We started calling these problems failures, then anomalies, now deviations. We should go into these deviations in detail and find out the causes. Then we should take corrective action where required."[51] This careful philosophy helped create the tremendous technical successes of the Saturn vehicles.

The Lunar Roving Vehicle

Marshall took its expertise in transportation in new directions by developing the LRV for the later Moon landings. The vehicle was the first human spacecraft built by the Center and was a harbinger of Marshall's diversification beyond its rocketry specialty. The lunar rover helped the Apollo astronauts explore the lunar surface and gather geological samples.

Von Braun and other engineers had proposed concepts for lunar cars from the 1950s.[52] Most Center planning for lunar vehicles, however, followed NASA's LOR decision of June 1962. In agreeing to the LOR mode, von Braun had proposed that Marshall build an Apollo Logistics Support System, a combined lunar taxi and shelter.[53] Immediately after this decision, Marshall initiated studies of lunar surface vehicles. For the next six years, the Center and contractors designed and developed various full-scale and subscale prototypes, investigating wheel design, drive systems, steering mechanisms, crew cabins and human factors problems, and navigation simulators.[54]

While Marshall engineers investigated designs, NASA clarified organizational assignments for the lunar missions. The division of labor between the Centers needed clarification because Marshall was entering Houston's domain in human space flight. Agreements of the Management Council of the Office of Manned Space Flight, which included the Center directors and Headquarters administrators, culminated in the August 1966 meeting at Lake Logan in North Carolina. There the Management Council assigned the MSC responsibility for lunar science, including planning for lunar traverses, lunar geology experiments, and biological and biomedical experiments. George Mueller said this gave MSC authority for the "overall management and direction" of the Apollo explorations and equipment. Marshall became responsible for what von Braun termed "devices of an engineering rather than a scientific nature." These included lunar vehicles like various types of surface rovers, a one-man flyer, or a remote controlled scientific surveyor.[55] Houston consented to Marshall's role in lunar engineering because of demands imposed by work on the Apollo spacecraft. As Joseph Loftus recalled, MSC had "an awful lot on our plate."[56]

Despite this division of labor, NASA as late as 1968 hesitated in its choice of a lunar transportation system. The choice of technologies was still open in November 1968 when Marshall requested proposals from aerospace companies to study a dual-mode rover that could carry one astronaut and undertake geological missions under remote control from Earth. But agency officials worried that the dual-mode vehicle would be too expensive and complicated.[57]

Houston's opposition delayed the decision on a lunar vehicle. The MSC stalled because of technical concerns rather than organizational jealousy of Marshall. MSC engineers, especially George Low, feared that a lunar vehicle would reduce lunar module (LM) fuel needed for safe landings; without surplus fuel as insurance, the LM could not hover and move to a suitable landing site. MSC's complaint, LRV Project Manager Sonny Morea remembered, was a "safety objection."[58]

NASA finally made a vehicle decision in late May 1969 by rejecting the flyer, and choosing a surface vehicle. By then the agency was confident that a landing could be done safely. Moreover a piloted Moon car would cost less than a remote-controlled unit and could do more science than a flyer. Indeed advocates of the LRV, especially the Marshall Center and George Mueller, overcame resistance by arguing for its scientific payoffs. On 27 May 1969, NASA authorized Marshall to develop the LRV.[59]

With the rover the Center faced imposing schedule constraints and technical challenges. The vehicle had to be ready by April 1971, making for a design and development schedule much shorter than the four-and-one-half to six years for other Apollo spacecraft and life support equipment. Marshall moved quickly, issuing requests for proposals on the same day as the first lunar landing in July 1969. Later in the month LRV work moved from Program Development to an LRV Project Office managed by Morea, who had previously supervised the F–1 engine program. The creation of a project office occurred before the normal initial steps of Program Development's phased project planning had been completed. In late October the Center chose Boeing as the prime contractor even though the company's bid of $19 million was far below Program Development's estimated cost of more than $30 million. Another unusual feature of the Boeing contract was how it sought to hasten the project and, in Morea's words, "cut out the bureaucracy." It specified performance requirements rather than any predetermined design and made the company responsible for systems integration; the company could authorize some hardware changes without formal NASA approval.[60]

The lunar module also affected vehicle design. MSC had authority over the LM and wanted to stabilize its design. Accordingly Houston refused to change the LM to accommodate the rover. In effect then Marshall was a contractor working for another Center and had to adjust to MSC's requirements; Morea lamented that Marshall "always seemed to get the short end of the string." The lunar car could not exceed a weight limit of 400 pounds but had to carry over 1,000 pounds of astronauts, equipment, and rocks. This meant that the LRV had to be built of light alloys and would collapse under a person's weight in Earth gravity. In addition, the vehicle had to fit in an LM storage bay about the size of a station wagon's, 66 inches wide, 60 inches high, and 49 inches deep.[61]

The lunar environment also shaped the rover. As Henry Kudish, Boeing's LRV project manager in Huntsville, observed, the vehicle was not a "lunar jeep" but rather "a very complex spacecraft." The vehicle had to operate in a vacuum and in temperature extremes of plus or minus 250 degrees Fahrenheit. It had to serve astronauts in cumbersome life support suits. The roving vehicle needed a navigation system to cope with the Moon's low sun angle and its effects on depth perception, lack of a magnetic north pole, and short horizon. It needed strength and stability to traverse rocks, crevasses, and steep slopes. Clinging lunar dust necessitated that everything be carefully sealed.[62]

Marshall and its contractors cooperatively designed and developed the LRV. As prime contractor Boeing used its expertise in aircraft structures to construct the folding aluminum chassis and to integrate the subsystems. GM and its Delco Electronics Division, Boeing's major subcontractors, drew from automotive experience to develop the wire mesh wheels with titanium chevrons as tread, torsion bar suspension, single stick control and all-wheel steering system, and harmonic drive assemblies. Other contractors built the silver-zinc batteries and communications system.[63]

Deployment testing of lunar roving vehicle in March 1971.

Other Centers, especially Houston, also helped. Marshall, MSC, and Kennedy established several intercenter panels to resolve problems on scientist-astronaut participation, crew systems and training, operational constraints, LM/LRV interface, prelaunch checkout, and communications with mission control. Astronauts from Houston helped with the crew station and suggested assists for getting in and out of the vehicle and upright seatbelts for sure visibility.[64]

Marshall, however, stamped its trademark on the LRV. The Center contributed to the vehicle's conservative engineering of several redundant systems, including two batteries which could individually power the vehicle, two independent steering systems on front and rear, a control stick that could be used from either seat, and separately powered wheels, each of which could be set to free-wheel should its drive assembly fail.[65]

Conservative engineering also showed in the number of rovers NASA purchased. The agency bought four one-sixth gravity flight models and seven test and training units. With enough funding for seven models, Marshall could require extensive tests. The test units included a rubber-wheeled Earth gravity trainer, a qualification unit for testing and troubleshooting during missions, a vibration test article, two one-sixth weight units used in deployment tests, a static mock-up for crew station reviews, and a test article known as "the glob" which Grumman used in early work with the lunar module. Marshall flew one test vehicle on parabolic flights in a KC–135 "Vomit Comet" allowing astronauts in space suits to investigate entry and exit in low gravity. So luxurious was rover's funding that NASA even wasted one flight model; when the Agency canceled an Apollo LRV mission, the LRV parts became spares.[66]

Most importantly the trademark of Marshall's arsenal system showed on the LRV. Marshall people worked on the project in functional teams organized in the Saturn system of matrix management.[67] The most significant contributions came from the Astrionics and Astronautics labs. The Engineering Division of the Astronautics Lab designed and developed a manual method to deploy the rover from the LM. Although designed as a backup to an automatic system, it became the sole deployment procedure. By pulling on two mylar tapes the astronauts unfolded the LRV from the storage bay and lowered it rear first to the lunar surface.[68]

NASA wanted a navigation system so that astronauts could travel widely to predetermined points and return safely to the lunar module. Engineers in the Astrionics Lab's Guidance and Control Division conceived the system because project managers feared that a disoriented navigation contractor had gotten lost with a costly, complicated mechanism. Center technicians constructed it mainly from components already available.

A team from the Sensors Branch developed a dead reckoning system. A processor used elementary trigonometry to make calculations based on a known starting point and measurements of vehicle attitude, direction, speed, and distance traveled. A console displayed distance traveled and distance from the LM, and heading and bearing to the LM. Three gyroscopes determined Lunar North, and a sun shadow compass, added by suggestion of MSC, checked the original heading and guarded against gyro drift.

Marshall worried that lunar soil might inhibit performance of the roving vehicle. Slippage on loose soil in the lunar vacuum could affect navigation and limit range. After considerable research, the Center decided to rely on odometer readings of the third fastest turning wheel to determine distance and speed.[69] In 1969 the Geotechnical Research Division in the Space Sciences Laboratory formed a Soil Mechanics Investigation Team that studied lunar soil samples, astronaut observations, photographs, and film. Marshall even conducted soil penetration and load bearing experiments on KC–135 flights. The research concluded that soil would not hamper a rover.[70]

The Center's technicians built the navigation system and performed tests in 1970 first in fields surrounding the Center and later in the lunar-like desert near Flagstaff, Arizona. Marshall's navigators imitated a rover by using a jeep with masked windows, a television camera on the hood, and the navigation system. The jeep driver found his way using a TV monitor, a map, navigation readouts, and a radio. A station wagon followed the jeep; the wagon's driver could see ahead but its passengers could not. Imitating mission control, the passengers used TV pictures and the navigation display and communicated advice to the LRV driver in the jeep. In this way the navigators tested both their mechanism and remote control methods. They found their way within two-percent error even on 19-mile trips.[71] The system was imprecise but cheap and simple, and team leader Peter Broussard said "we were being pragmatists" who just wanted to get the astronauts in sight of the LM. He recalled the "fun" of working a whole subsystem and seeing it from conception to operation, and remembered that nearly all the engineers who worked on the LRV said "that's the best project I ever worked on."[72]

In spite of Marshall's arsenal system, the rover contract fell behind schedule and went over budget. At one point the project was two months behind targets to meet the April 1971 deadline. Delays came partly because NASA was slow to select power, speed, and range requirements and partly because during vibration tests Boeing/GM found shorts in the electronic controls and broken gears in the harmonic drives.

NASA insisted that schedules be kept. Marshall Director Rees warned Boeing that "this project is simply too sensitive to allow further opportunity for embarrassment in either the technical or the cost area." Rocco Petrone, NASA's director of the lunar landing program, warned in January 1971 that he could only delay the summer launch of Apollo 15, the first rover mission, one month. If the vehicle was not ready after that, Petrone said, Apollo 15 would leave without it.[73]

Boeing made changes to catch up. It moved work from Huntsville to Kent, Washington, to get more skilled workers and to be closer to test equipment. The company conducted qualification testing and concurrently manufactured the first flight vehicle. But Boeing got back on schedule mainly by using more workers and paying them overtime. Most contract overruns went to pay overtime for skilled labor. As John Winch, Boeing LRV project executive said, "when we encountered problems something had to give. In this case it was cost." With the extra expenses, the company delivered the first flight roving vehicle in March 1971, three weeks ahead of the delivery order. The final cost of the project was $38.1 million, close to Marshall's projections but more than double Boeing's bid.[74]

Not surprisingly, critics blasted NASA for rover overruns. Columnist Jack Anderson charged that the agency had "goofed on the design" and compounded problems with a "head-in-the-clouds attitude toward Boeing's expenditures." He claimed that the cost of the project was $10 million more than the 1972 federal auto safety budget.[75] Much of the criticism rested on the assumption that the vehicle was merely an electric car.

But as a NASA official pointed out, the LRV followed "spacecraft rules, not automobile rules." H. Dale Grubb, NASA assistant administrator for Legislative Affairs, told one inquisitive senator that the vehicle was "in line with the cost of other equipment of similar novelty and complexity which NASA has developed and produced in the space program." And the lunar rover followed a 17-month schedule (and only 13 months from the contract award) that was far shorter than the 52 months for the Command Module, 62 months for the LM, 60 months for the astronaut suits, and 70 months for their portable life support systems. Given this rushed schedule and the gross overruns of later NASA projects, rover development seemed remarkably successful. Marshall project manager Morea believed that "unless we went into a mode of a crisis, a national emergency, we would not know how to do a program like that today. We could not do it today."[76]

Marshall assisted Houston on the LRV missions through the Huntsville Operations Support Center (HOSC) located in the Computation Laboratory. Marshall had used the HOSC to monitor earlier Saturn launches. On the LRV flights, however, Marshall extended its operations role and 45 vehicle specialists provided around-the-clock engineering advice to Mission Control in Houston. Center and contractor personnel checked vehicle performance, ensured proper operations, and responded to problems. To simulate any problems the astronauts might encounter, the Center also maintained an LRV qualification unit in a hangar.[77]

With the Center's help, the Apollo 15, 16, and 17 missions of 1971 and 1972 successfully used LRVs. On Apollo 15 the astronauts had some difficulty deploying the LRV, and on the first excursion an electrical short immobilized the front steering, allowing rear steering only. The next day, however, the front steering worked. Astronaut David Scott told Mission Control that "you know what I bet you did. . . . You let some of those Marshall Space Flight Center guys come up here and fix it."[78]

The success of the vehicle muted most criticism. The Apollo astronauts explored more territory and collected more geological samples with the LRV than ever before. On the three pre-rover missions, Apollo 11, 12, and 14, astronauts collected 215 pounds of samples and walked 10 miles in 36 hours. On Apollo 14, Alan Shepard and Edgar Mitchell tried to climb Cone Crater but had to give up after they got tired, disoriented, and began running out of time and air. But on a riding mission the astronauts could range farther, faster, safer. On Apollo 15 alone the astronauts traveled nearly four times farther than the three previous missions combined and collected 170 pounds. On all the LRV trips, NASA collected 635 pounds of samples, and traversed 134 miles in 122 hours.[79] Marshall's navigation system kept on track; the average position error at the end of a traverse was less than 200 meters on Apollo 15 and zero on Apollo 16 and 17.[80]

The rover won considerable praise from the astronauts. Scott said that the vehicle was "about as optimum as you can build." Gene Cernan and Jack Schmitt of Apollo 17 noted that they had "three good spacecraft," the CM, LM, and LRV, and believed "that thing couldn't perform better." They felt it was "a super performing vehicle. If you take a couple more batteries up there, that thing would just keep going."[81]

Meanings and Memories

In narrow terms, Marshall's work in the Apollo program offered many lessons and legacies. The Center contributed to technological progress, such as making advances in materials, metal bonding, and welding inspection, that proved useful in many areas. Marshall's engineering organization showed the value of comprehensive testing and multidisciplinary work teams that integrated specialists in design, manufacturing, testing, and inspection. The project management system successfully combined the efforts of dozens of businesses and government organizations. Marshall created rockets and rovers that allowed humans to explore space and the Moon.

In later years, Marshall's Saturn V cast a long shadow. In the 1980s, an era in which NASA's dreams of a space station were limited by the 25-ton capacity of the Space Shuttle, some longed for the 124-ton capacity of a Saturn V. Some aerospace companies and agency planners even sought blueprints of the Saturn V and its engines to gain inspiration for the next generation of rockets.[82]

In a wider sense, however, the Saturn V became something more than a powerful rocket. As decades passed, Americans reinvented the meaning of the Apollo Program and transformed it into a symbol of excellence. Why, they asked, could Americans not perform the way they had done during Apollo? In this context the Saturn V became a symbol of excellence in American society and government. In an era in which America seemed divided, anniversaries of the first landing on the Moon sometimes expressed nostalgia for the national commitment and unity of the Apollo Program in the 1960s. Looking back after 20 years, a Boeing engineer thought that the Saturn project was "the biggest single example I can think of getting the government-industrial complex together on a goal that had an established end and a monumental technical task before it."[83] That the Saturn V could become such a symbol in American culture was perhaps the most fitting tribute to the Marshall Center.

---

1 Statistics from R. Bilstein, *Stages to Saturn: A Technological History of the Apollo/Saturn Launch Vehicles* (Washington, DC: National Aeronautics and Space Administration Special Publication–4206, 1980), pp. 288, 354–55; Diana Woodin, "Saturn-Apollo: Boeing Engineers Reminisce about Early Days," *Huntsville Times* (16 July 1989), 1C.

2 The stand is an example of von Braun's engineering conservatism and his attempts to plan beyond current projects. Had Nova been built, this stand could have accommodated tests on its engines.

3 *Birmingham News*, 28 June 1969; William Lucas, OHI by SPW, 4 April 1994, pp. 34–35.

4 Loyd S. Swenson, James H. Grimwood, and Charles C. Alexander, *This New Ocean: A History of Project Mercury*, (Washington, DC: National Aeronautics and Space Administration Special Publication–4201, 1966), p. 172 ff., Kuettner quoted, p. 181.

5 *Ibid.*, pp. 185–87.

6 *Ibid.*, chap. X, espec. pp. 293–97.

7 *Ibid.*, pp. 323–24, 328–30, 332–65.

8 J. P. Kuettner, "Marshall–STG Management for Apollo, etc.," 14 July 1961, MSC/STG and MSFC folder, MSFC Archives.
9 E. Rees, routing-slip notes, 3 July 1961, MSC/STG and MSFC folder, MSFC Archives.
10 See James Kauffman, "NASA's PR Campaign on Behalf of Manned Space Flight, 1961–63," Public Relations Review 17 (Spring 1991), pp. 57–68; for coverage of MR–3, *New York Times* (May 5–10, 1961); *Huntsville Times* (May 5–10, 1961).
11 Roger Bilstein's *Stages to Saturn: A Technological History of the Apollo/Saturn Launch Vehicles* (Washington, DC: National Aeronautics and Space Administration Special Publication–4206, 1980) is the best history of Marshall's role in Saturn development. Many of the Marshall records used by Bilstein have subsequently been lost. Consequently this chapter relies heavily on his research.
12 Bilstein, p. 59; Linda Neuman Ezell, *NASA Historical Data Book, Volume II: Programs and Projects, 1958–1968*, (Washington, DC: National Aeronautics and Space Administration Special Publication–4012, 1988), pp. 54–61.
13 Bilstein, pp. 14–19, 89–96, 104–106, 235–41, 241–43.
14 Bilstein, pp. 25–31, 76–77, 81, 97–98; Wernher von Braun, OHI, 17 November 1971, UAH Saturn Collection, p. 5; Colonel Kaiser, OHI by John Beltz and David Christensen, 1970, UAH Saturn Collection, pp. 3–4.
15 Bilstein, pp. 36–45, quote, pp. 44–45; Ernst Stuhlinger and Frederick I. Ordway, *Wernher von Braun: Crusader for Space* (Malabar, Florida: Krieger, 1994), pp. 166–67.
16 Bilstein, pp. 129–43, 157–63; Stuhlinger and Ordway, pp. 186.
17 Bilstein, pp. 192–93; Wernher von Braun, OHI by Robert Sherrod, 25 August 1970, Sherrod Collection, NHO, p. 10; Karl Heimburg, OHI by AJD and SPW, 2 May 1989, Huntsville.
18 H. R. Palaoro, MSFC Chief, Vehicle Systems Integration Office, to O. H. Lange, Saturn Director, "Propulsion and Vehicle Engineering Contractor Manpower Requirements," 2 July 1962, Saturn V, (S–IC) 1962 folder, MSFC Archives, Bilstein, pp. 193–95.
19 Bilstein, pp. 191, 203; Stuhlinger and Ordway, p. 186.
20 M. W. Urlaub, S–IC Project Manager, and K. K. Dannenberg, Deputy Director of Saturn Systems, to von Braun, "Briefing for Mr. Tom Dixon, NASA HQ," 3 August 1962, Saturn V, (S–IC) 1962 folder, MSFC Archives; "Background Comments from the S–IC Stage Manager," October 1964, Saturn V (S–IC) 1964 folder, MSFC Archives; William Lucas, OHI by SPW, 4 April 1994, Huntsville, pp. 17–18.
21 Saverio Morea, OHI by SPW, 11 September 1990, MSFC, pp. 5–7.
22 Mark F. Fisher, MSFC Personal Communication to SPW, 29 August 1996; Bilstein, pp. 108–18; Charles Murray and Katherine Bly Cox, *Apollo: The Race to the Moon* (New York: Simon and Schuster, 1989), pp. 145–50; J. R. Thompson, OHI by AJD and SPW, 6 June 1994, Huntsville; L. W. Jones, M. F. Fisher, A. A. McCool, and J. P. McCarty, "Propulsion at the Marshall Space Flight Center: A Brief History," AIAA–91–2553, AIAA/SAE/ASME 27th Joint Propulsion Conference, June 24–26, 1991.
23 Bilstein, pp. 209–22; Lucas OHI, 4 April 1994, pp. 18–22.
24 NASA, "Contractor Performance Evaluation Report on NA Contract," 30 June 1964, Saturn V (S–II) folder, MSFC Archives; Bilstein, pp. 211–22; Lucas OHI, 4 April 1994, p. 23; James Odom, OHI by AJD and SPW, 26 August 1994, Huntsville, pp. 12–13.

25 See "S–II Stage Data for Dr. Rees [Binder]," 1965, Drawer 5, especially E. O'Connor to von Braun, "Background Data," 14 October 1965 and E. Rees to S. C. Phillips, "What is wrong with S&ID in general and on the S–II Program in particular?" 21 October 1965; D. Grau to A. Rudolph, "Excessive Program Costs," 12 January 1965, Saturn V (S–II) January–July 1965 folder, MSFC Archives; J. Bradley, "Comments to the special S–II Investigating Committee" in MSFC, "Saturn V S–II Survey Team: Report," 14 December 1965, Saturn V (S–II) Miscellaneous Reports 1965 folder, MSFC Archives.

26 Von Braun, "S–II Stage Project," 19 December 1966, Saturn–V (S–II), June–August 1967 folder; "S–II Special Task Team," 5 January 1967; E. Rees, MSFC to K. Debus, KSC, 13 January 1967, Saturn–V (S–II), January–February 1967 folder; James Odom, "Recollections from Apollo," NASA Alumni League Forum, MSFC, 20 July 1994; von Braun to G. Mueller, HQ, "S–II Stage Project," 11 August 1967, Saturn–V (S–II), June–August 1967 folder, MSFC Archives; Bilstein, pp. 226–33.

27 Bilstein, pp. 235–41.

28 Bilstein, pp. 241–54, 374–75.

29 "Technical Facilities and Equipment Digest," Marshall Space Flight Center, January 1967, pp. 71–83; MSFC Public Affairs Office Fact Sheet, 5 August 1966; Mike Wright, "Informal Notes for MSFC History Tour," 7 May 1990.

30 "Way Station to the Moon," *Business Week* (2 April 1966), quoted pp. 62, 63; "Mississippi Test Facility," pp. 2–3, 8; Public Affairs Office, MTF, "NASA Facts," *Coast Area Mississippi Monitor*, 1969, p. 54.

31 Bilstein, pp. 302–308. Of the seven barges used to transport Saturn stages, six were owned by NASA and operated by MSFC. The seventh, the USNS *Point Barrow*, was a Navy barge loaned to NASA which occasionally transported stages from California via the Panama Canal.

32 Bilstein, pp. 308–19.

33 Benson and Faherty, pp. 181–83; Murray and Cox, pp. 275, 309, 311–12; Bilstein, p. 323.

34 Bilstein, pp. 323–24.

35 Press clippings, January 1964, Launch Vehicles File, Saturn C–1 SA–5 Folder, NASA History Office; NASA, *Current News* 29 January 1964, SA–5 1–29–64 folder, MSFC Archives; Bilstein, 325–37.

36 Bilstein, pp. 337–45.

37 Bilstein, pp. 348–49; Murray and Cox, pp. 158–62.

38 R. B. Young to Mitchell R. Sharpe, 11 January 1974, UAH Saturn Collection; Lee B. James to Mitchell R. Sharpe, 10 January 1974, UAH Saturn Collection; Murray and Cox, pp. 161–62; Lee B. James, OHI by SPW, 1 December 1989, p. 17.

39 Walter Haeussermann Interview, by MSFC History Office, 14 December 1973, UAH Saturn Collection; Dieter Grau to Mitchell R. Sharpe, "Saturn History," 12 December 1973, UAH Saturn Collection.

40 Eberhard Rees, Director, MSFC, to Robert Sherrod, Secretariat Support Division, NASA, 4 March 1970, Launch Vehicles File, "Saturn 'all-up' testing concept" folder, NASA History Office; Bill Brown Interview, by Mitchell R. Sharpe, 13 December 1973, UAH Saturn Collection; Wernher von Braun Interview, 30 November 1970, UAH Saturn Collection, pp. 49–51; Murray and Cox, pp. 162–63; Arnold S. Levine, *Managing NASA*

in the Apollo Era, (Washington, DC: National Aeronautics and Space Administration Special Publication–4102, 1982), p. 6.

41 James OHI, 1 December 1989, p. 17.

42 James interview, paragraph 34; Murray and Cox, p. 249; Arthur Rudolph interview, 14 December 1973, UAH Saturn Collection; Bilstein, pp. 355–60.

43 Bilstein, pp. 360–61.

44 For the pogo story, see E. E. Goerner, MSFC, Chair of POGO Working Group, "Certification of POGO Solution for SA–503," 5 August 1968, SA–503 Technical folder; MSFC, "Summary and Assessment of AS–503 POGO Suppression," 14 December 1968, Misc. POGO 1968 folder, MSFC Archives; NASA, "Saturn V Tests Completed," NASA PR 68–128, 18 July 1968; W. von Braun, "The Detective Story Behind Our First Manned Saturn V Shot," *Popular Science* (November 1968), pp. 98, 100, 209.

45 See previous note.

46 Von Braun, "Detective Story," pp. 98, 99, 100; Lucian Bell and Jerry Thomson, MSFC, OHI by Robert Sherrod, 12 March 1970, Sherrod Collection, NHO; J. Thomson, MSFC to W. Lucas, MSFC, "AS–502 J–2 Engine Failure Test Report," 22 May 1968, 1968 Apollo 6 SA–502 folder; Rocketdyne, "AS–502 Report to NASA," 1968 Apollo 6 SA–502 folder.

47 W. von Braun, OHI by Robert Sherrod, 25 August 1970, pp. 12–16, Sherrod Collection, NHO; Murray and Cox, pp. 308–14.

48 N. Mailer, *Of a Fire on the Moon* (Boston: Little, Brown, 1969), pp. 95–96.

49 "Von Braun Says Feat Makes Man Immortal," *Huntsville Times* (22 July 1969), p. 1; "News Meetings Please James," *Huntsville Times* (15 July 1969), p. 49.

50 Rocco Petrone, Apollo Program Manager, "Minutes of the S–II Stage Low Frequency Oscillation Design Certification Review," 13 August 1970, AS–509 POGO—Apollo 14 1970 folder, MSFC Archives; "Apollo/Saturn V S–II Oscilllations," 4 March 1970, POGO file, Sherrod Collection, NHO; James Kingsbury, manuscript comments, 16 June 1996.

51 MSFC, "Minutes of Combined Staff and Board Meeting, 8 August 1969," 8 August 1969, AS–506 Technical folder, MSFC Archives.

52 Mitchell R. Sharpe, "Lunar Roving Vehicle: Surface Transportation on the Moon," XVth International Congress for the History of Science, Edinburgh, Scotland, (17 August 1977), NASA History Document Division (hereinafter NHDD), Lunar Exploration Vehicles folder; ABMA, "Project Horizon" Vol. I–IV, (Huntsville, 1959), Redstone Arsenal, Redstone Scientific Information Center.

53 Wernher von Braun, "Concluding Remarks by Dr. von Braun about mode selection for the Lunar Landing Program given to Dr. Joseph F. Shea, Deputy Director, OMSF," 7 June 1962, OMSF, Apollo Lunar Orbit Rendezvous folder, NHDD; William D. Compton, *Where No Man Has Gone Before: A History of Apollo Lunar Exploration Missions* (Washington, DC: National Aeronautics and Space Administration Special Publication – 4214, 1989), pp. 98–101. See also Wernher von Braun to Joseph F. Shea, 26 December 1963, cited in Roland W. Newkirk and Ivan D. Ertel with Courtney G. Brooks, *Skylab: A Chronology* (Washington, DC: National Aeronautics and Space Administration Special Publication –4011, 1977), p. 29 (hereinafter *Skylab Chronology*); Herbert Schaefer and Leonard S. Yarbrough, MSFC Systems Planning Office, Aero-Astrodynamics Lab, "Apollo Logistic Support System," 7 May 1964, UAH Saturn Collection.

54 Georg von Tiesenhausen, "Preliminary Presentation Notes [for] Lunar Trafficability Study" (30 November 1962) cited in Sharpe, pp. 11–12; MSFC Public Affairs Office, "Background on Lunar Surface Vehicles," 15 August 1969, MSFC History Office; *Astronautics and Aeronautics*, 1966, pp, 60, 91–92.

55 TWX, George Mueller to MSC, MSFC, and KSC, 13 September 1965, and Memorandum for record, E. J. Brazill, NASA HQ, "Meeting held on Monday, July 11, 1966 by Dr. Seamans, Dr. Mueller and Dr. Newell," 15 July 1966 both cited in *Skylab Chronology*, pp. 50, 82; Wernher von Braun to Robert Gilruth, 19 October 1966 and George Mueller to Robert Gilruth, 14 November 1966, Rice University, Center Series, Sullivan Management Series, Box 9, Management Council—Lake Logan Meetings Folder.

56 Joseph Loftus, OHI by SPW and AJD, Johnson Space Center, Houston, Texas, 13 July 1990, p. 6.

57 MSFC Press Releases 68–274, 69–110; *Marshall Star*, 4 December 1968, p. 1; "Lunar Vehicle Studies Begin," *Marshall Star* 16 April 1968, p. 1; Compton, pp. 98–101, 130–32.

58 Saverio F. Morea, "The Lunar Roving Vehicle: A Historical Perspective," NASA Paper LBS–88–203, Lunar Bases and Space Activities in the 21st Century, 5–7 April 1988, Houston, Texas, pp. 2–3; Saverio F. Morea, OHI by SPW, MSFC, 11 September 1990, p. 9.

59 Morea, "Lunar Roving Vehicle," pp. 2–3; Compton, pp. 212–13. William Compton's history of Apollo exploration offered no description of Marshall's management and role in the LRV project; see Compton, pp. 227–31.

60 David Akens, *An Illustrated Chronology of the NASA Marshall Center and MSFC Programs: 1960–1973* (MSFC, 1974), pp. 223, 231; Dr. William R. Lucas, OHI by SPW and AJD, 20 November 1990, Huntsville, Alabama, pp. 27–28; Morea interview, pp. 18–20.

61 Morea interview, pp. 7, 10.

62 Henry Kudish, "The Lunar Rover," *Space Flight* 12 (July 1970), p. 270; Morea interview, pp. 15–16; Morea, "Lunar Roving Vehicle," pp. 3–5; Saverio F. Morea and C. D. Arnett, "America's Lunar Roving Vehicle," no date, MSFC History Office, Folder 391.

63 Morea, "Lunar Roving Vehicle," pp. 8–9; Morea interview, p. 12; Richard G. O'Lone, "Lunar Rover Program Spurred," *Aviation Week and Space Technology* (15 February 1971), p. 50.

64 NASA Headquarters to MSC and MSFC, [telegram], 17 June 1969, Apollo Program Chronological Files, Box 071–43, JSC History Office; Morea interview, p. 11.

65 Kudish, pp. 270–74; Boeing Corporation, "The Lunar Roving Vehicle," 1972, MSFC History Office; Morea and Arnett, "America's Lunar Roving Vehicle"; Morea, "Lunar Roving Vehicle," pp. 9–19.

66 Joe Jones to Howard Allaway, "Notes on the LRV Contract," 17 August 1971, MSFC History Office, Folder 390; MSFC Public Affairs Office, Release 71–45, 16 March 1971, MSFC History Office; Morea, "Lunar Roving Vehicle," p. 13.

67 See Saverio F. Morea, "Organizational Study of the Lunar Roving Vehicle Program," MA Thesis, University of Oklahoma, 1974, MSFC History Office.

68 MSFC Public Affairs Office, Release 71–125, 21 July 1971.

69 Peter Broussard, OHI by SPW, 20 September 1990, Huntsville, Alabama, pp. 2–14; Morea, "Lunar Roving Vehicle," pp. 16–17; Kudish, pp. 273–74.
70 MSFC Public Affairs Office, Release No. 69–227, 15 October 1969, NHDD, Lunar Exploration Vehicles folder; MSFC Public Affairs Office, Release 70–169.
71 "Pinpoint Navigation System for LRV in Desert Testing," *Marshall Star*, 16 December 1970, pp. 1–2.
72 Broussard interview, 11, 14.
73 Bob Richey, "Mooncar is Behind Schedule," *Huntsville Times*, 8 January 1971, p. 1; "Boeing Admits Rover Work Lag," *Huntsville Times*, 9 January 1971; "Rover Costs Rise," *Aviation Week and Space Technology* (25 January 1971),p. 19; Eberhard Rees to George H. Stoner, Senior Vice-President Operations, Boeing Company, 3 October 1970, NHDD, Lunar Exploration Vehicles, Documents folder.
74 O'Lone, p. 49; Jones to Allaway; "Boeing Delivers Early: NASA Given Moon Buggy," *Washington Evening Star*, 11 March 1971.
75 Jack Anderson, "Washington Merry-Go-Round," *The Washington Post*, 11 August 1971, p. F11; *The Washington Post*, 30 January 1971, p. A3.
76 Jack Anderson, "Washington Merry-Go-Round," *The Washington Post*, 11 August 1971, p. F11; *The Washington Post*, 30 January 1971, p. A3.
77 MSFC Public Affairs Office, Release 71–128, 26 July 1971, MSFC History Office.
78 Zack Strickland, "Added Mobility Spurs Geologic Harvest," *Aviation Week and Space Technology* (8 September 1971), pp. 13–17.
79 Strickland, pp. 13–17; Morea, "Lunar Roving Vehicle," pp. 20–21; H. Dale Grubb to Miss Helen Sworn, 24 January 1973, NHDD, Lunar Exploration Vehicles folder; "Spacecraft with Wheels," MSFC film.
80 Earnest C. Smith and William C. Mastin, "Lunar Roving Vehicle Navigation System Performance Review," November 1973, MSFC History Office, Folder 391.
81 Strickland, pp. 13–17; Morea, "Lunar Roving Vehicle," pp. 20–21.
82 Daniel J. Connors, "Shuttle C/Saturn 5," *Aviation Week and Space Technology* (5 September 1988), p. 258; R. Jeffrey Smith, "Soviet Rocket Matches US '60s Model," *Washington Post* (20 May 1987); William J. Broad, "Hunt is On for Scattered Blueprints of Powerful Saturn Moon Rocket," *New York Times* (26 May 1987), p. C3.
83 Diane Woodin, "Saturn-Apollo: Boeing Engineers Reminisce about Early Days," *Huntsville Times* (16 July 1989), p. 1C.

Chapter IV

The Marshall Reconstruction

In launching the Apollo Program, NASA also launched a reconstruction of the South. In the Moon program's "Fertile Crescent" that stretched from Houston to Huntsville to the Cape and back to New Orleans, NASA helped reconstruct the region's economic, demographic, social, and educational landscape.[1] Agency administrators, as managers of the command economy of space, "planned" some of the changes, especially in the economy. Other changes were unanticipated; "spillover" effects could be seen in the space program's effects on civil rights and education. But the impact was pervasive, permanent, and driven by federal dollars. This "Second Reconstruction," one historian has suggested, "went beyond the pork barrel into the realm of social planning."[2]

In part the reconstruction resulted from Kennedy and Johnson administration promises concerning the lunar program. They promised that the Apollo Program, like other programs of the New Frontier and Great Society, would promote progress in terms of advances in material plenty and social equality for the entire nation.[3]

The reformist impulse, however, combined with regional promotion. The South benefited most from space spending; it controlled key committee chairmanships in Congress, and military and NASA installations already dotted the landscape. As one commentator observed, NASA's Centers in the South formed an "arch" through which federal money passed. Marshall was the "keystone of this arch."[4]

Civil Rights

In the early 1960s, the most dramatic story in Alabama came not from the test stands at Redstone Arsenal, but from the streets of Montgomery, Birmingham, and Selma.[5] The Heart of Dixie was the center of the civil rights struggle.

Alabama evoked images of the scorched skeleton of a bus abandoned by Freedom Riders in Anniston, the confrontation at the Edmund Pettus Bridge in Selma, Bull Connor's dogs and firehoses in Birmingham, and Governor Wallace standing in a doorway at the University of Alabama in Tuscaloosa.

Marshall Space Flight Center could not operate in a technological vacuum, isolated from events to the south. The Center's role in the unfolding civil rights story revealed the interplay between the Federal Government and the states over civil rights. The sizable federal presence in Huntsville helped civil rights progress in Madison County and facilitated desegregation. Concurrently, Alabama's culture of segregation slowed Marshall's progress in black recruitment in comparison to federal installations elsewhere.

NASA was vulnerable to the race issue, since its major installations resided in the South and Project Apollo was to showcase American virtues. More than any other federal Agency, NASA needed to avoid the stains of American racism and be a symbol, "clean, technically perfect, the bearer of a myth."[6]

Before 1963, Marshall was little touched by the civil rights maelstrom that swirled through Alabama. The Center avoided controversy in the early 1960s because Huntsville offered a less promising place for civil rights advocates to make a stand than cities to the south. Civil rights leaders learned early that nonviolent direct action was most successful in confrontations with recalcitrant segregationists, and Huntsville politicians and businessmen wanted to avoid controversy. Madison County's prosperity depended on the Federal Government, and few wanted to jeopardize that support. The Gospel of Wealth had more disciples in Huntsville than did the Gospel of White Supremacy.

Alabama Governor George Wallace and Dr. von Braun at Marshall.

Circumstances in North Alabama differed from those in the rest of the state. North Alabama developed differently from the Black Belt to the south; with smaller farms and fewer blacks, the north did not have the patterns of racial segregation that typified the southern plantation economy. Its politics had always been more liberal. In his successful races for governor in both 1962 and 1966, George Wallace received a smaller percentage of the vote in Madison County than in any other county in the state. Days before Wallace stood in the door of the University of Alabama in Tuscaloosa to bar the admission of a black student, the *Huntsville Times* said in an editorial, "One thing now is eminently clear—if U.S. troops are called to Tuscaloosa and to Huntsville, one man and one man alone bears the chief responsibility. That man is Governor George C. Wallace."[7]

Marshall contributed to the state's regional differences. "I never did feel that North Alabama should have been accused of some of the things that they were accused of," explained Art Sanderson, who worked in the Marshall Personnel Department in the 1960s. "We brought people into this area from all over the country. All cultures. They were not just Mississippians, Alabamians, Tennesseans. They were from all over, Boston, from the major big schools, from California, Florida. We brought people with all different cultures to make up the ABMA and later Marshall Space Flight Center. You have got all these cultures coming in here, and they weren't coming into Birmingham or Selma, they were coming here. I always felt that the people who came in here were quite a bit above the accusations about civil rights. It may have been true somewhere south of here. It was not true here. . . . I felt that everybody was here to do a job. We really didn't have time for that kind of business."[8]

If Huntsville was no Selma, neither was it a civil rights paradise. The Congress of Racial Equality (CORE) led sit-ins at Huntsville restaurants and lunch counters early in 1962. The protests led to several arrests and culminated in a visit by Martin Luther King in March.[9] Although not as violent as confrontations elsewhere in Alabama, these events showed that Huntsville shared in the state's culture of segregation.

"The fact of the matter," one of NASA Administrator James Webb's assistants observed, "is that Huntsville is in Alabama."[10] Public facilities and public schools were segregated, and African Americans struggled to find housing. Black per capita income in Huntsville was less than half that of whites.[11] Employment opportunities were limited; African Americans comprised 18 percent of Huntsville's population, but less than 1 percent of Marshall's workforce.[12] Clyde

Foster, one of the few blacks who worked at Marshall in the early 1960s, recalled that he was not able to participate in training sessions in Huntsville, where public accommodations were segregated. Accommodations on the Arsenal and at Marshall were no longer segregated, but blacks still encountered barriers. "Most definitely there was discrimination," Foster said. "There was this subtle kind of discrimination. Upward mobility just wasn't there."[13] In May 1962, two black Marshall employees filed complaints with the President's Committee on Equal Employment Opportunity. Joe D. Haynes charged discrimination barring promotion, and Joseph Ben Curry complained of assignments inappropriate to his job classification.[14]

Marshall nonetheless felt little pressure, mainly because the Kennedy administration did not promote civil rights in federal installations before the spring of 1963. The administration treated civil rights as a political issue, avoiding confrontations with southern politicians. Kennedy, who received overwhelming black support in his narrow victory in 1960, made gestures designed to appease civil rights advocates. He issued an executive order in April 1961 that established the President's Committee on Equal Employment Opportunity and mandated that executive agencies prohibit discrimination. Marshall replied that its activities conformed fully, and this was enough to satisfy the administration.[15] For the next two years Marshall focused on Saturn, and civil rights remained peripheral.

Events in the spring of 1963, many of them in Alabama, jolted the administration into action on civil rights. Marshall could not avoid repercussions of events transpiring a hundred miles to the south. Martin Luther King's crusade in Birmingham in May became a pivotal confrontation when Sheriff Bull Connor sent dogs to attack marchers and turned firehoses on children. A bomb in a church killed three black girls attending Sunday school classes.

The Birmingham campaign prompted new presidential activism on civil rights. For the first time President Kennedy proclaimed the issue a moral one and moved to initiate legislation. Attorney General Robert Kennedy, long a critic of Vice President Lyndon B. Johnson's leadership of the President's Committee on Equal Employment Opportunity, met with the committee on 18 June. Webb, a protégé of Johnson, represented NASA. Kennedy grilled Johnson, puncturing his vague claims of progress. After "making the Vice President look like a fraud," in the words of one observer, the Attorney General turned on Webb. "Mr. Webb,

I just raised a question of whether you can do this job and run a Center and administer its $3.9 billion worth of contracts and make sure that Negroes and nonwhites have jobs . . . I am trying to ask some questions. I don't think I am able to get the answers, to tell you the truth."[16]

As Webb reacted, Marshall moved from the shadows to the spotlight. Webb informed von Braun that "The Vice President has expressed considerable concern over the lack of equal employment opportunity for Negroes in Huntsville, Alabama." Johnson directed NASA, the Department of Defense, and the Civil Service Commission to formulate a plan to address the problem. The agencies met on 18 June, and decided to conduct surveys of housing and federal employment practices in Huntsville; to provide assistance to Alabama A&M College and Tuskegee Institute, historically black colleges in Alabama; to meet with Huntsville contractors to find out their plans to ensure equal employment opportunity; and to ensure that blacks be granted a fair proportion of summer jobs at Marshall. Webb directed von Braun to give personal attention to developing equal employment opportunity programs at Marshall.[17]

Marshall established an Affirmative Action Program in June, following recommendations offered by a Civil Service team from Atlanta. Dr. Frank R. Albert became the first Equal Employment Opportunity Coordinator. Albert hired Charlie Smoot as a professional staffing recruiter; Marshall claimed Smoot was "possibly the first Negro recruiter in government service."[18]

Federal pressure had an immediate impact in Huntsville. With nearly 90 percent of the city economy based on federal funds, Washington had more leverage in Huntsville than elsewhere in Alabama. Federal contractors, most of whom worked for the Army at Redstone Arsenal or NASA at Marshall, recognized that they could lose funding. They met on 5 July at Brown Engineering in Huntsville, formed the Association of Huntsville Area Contractors, or AHAC, and named as their spokesman Milton K. Cummings of Brown Engineering.[19] The committee agreed that contractors should take "immediate positive steps" to increase minority employment, to make "significant financial contributions" to aid black schools, to initiate immediate training programs for blacks, and to use their influence "to make our citizens more conscious of our responsibility in the area of housing, education, and the availability of private and public facilities."[20] AHAC agreed to keep NASA informed of its progress.[21] The group had an immediate impact. L.C. McMillan, a black man who had been a college administrator in Texas, arrived three months later to serve as executive director. "I was expecting the usual six months of preparatory meetings when I came here to start the program," he recalled, "but I was amazed to find that these people wanted to slice away the fat and get right down to the meat of a problem."[22]

The disappointing record of black recruitment at Marshall and its contractors stemmed from barriers that limited black access to scientific and technological education. Huntsville was a microcosm of a larger regional problem. The two colleges in the city divided along racial lines. Alabama A&M was a historically black college that conferred its first B.A. degrees in 1900. The University of Alabama established a Huntsville Center in 1950; like the main campus in Tuscaloosa, it was segregated. The curriculum at Alabama A&M centered on traditional programs at predominantly black colleges: teaching, social science, premedicine, and law.[23] The school had strong programs in the natural sciences and mathematics, but not in the modern engineering disciplines required by Marshall. A&M's regulations complicated its relations with the Center. As Clyde Foster explained, "Because of the system, we couldn't use available whites that were qualified to go out and teach at the Alabama A&M University." And it was difficult to recruit blacks from elsewhere to come to Alabama. Foster, one of the recruiters, remembered, "The image at that particular time was the George Wallace image and made it very difficult for people like myself to go out and to recruit other blacks who could qualify to move into Alabama."[24]

Steps toward alleviating inequities in higher education began in the summer of 1963. On 13 June, two days after Governor Wallace blocked for five hours the admission of the first blacks to the University of Alabama, Marshall mathematician David M. McGlathery became the first black to enroll at the university's Huntsville Center. Unlike the dramatic confrontation in Tuscaloosa, McGlathery's enrollment proceeded without incident.[25]

Marshall also began to improve its ties with Alabama A&M. Delegates from Marshall met with state officials to press for increased funding for A&M and for building a library at the school. Marshall representatives also met with A&M officials and officials from Huntsville's Oakwood College (a black sectarian college) to discuss grants-in-aid and internships. The Center reached beyond Madison County, sending representatives and surplus equipment to other black colleges, expanding recruitment, and inviting representatives from 12 black colleges to Marshall to discuss cooperative training programs. By the end of the summer, NASA Associate Administrator George Mueller called Marshall's equal opportunity program "imaginative and well rounded."[26]

Marshall came under fire again in August, when the hearing officer for the Haynes and Curry discrimination cases submitted his report. He found that both men had been victims of discrimination, and recommended that Haynes be promoted and Curry be reassigned to more appropriate duties.[27] The report noted that of 7,335 employees at Marshall, only 52 were black, and that blacks comprised only one-half of 1 percent of employees in GS–5 through GS–11 positions. It concluded that "a pattern of discrimination has and continues to exist at Marshall."[28]

Von Braun accepted the charge of discrimination, but objected to some of the charges in the report as "gratuitous and unwarranted under the circumstances." He contended that the report might damage efforts then underway at Marshall to ensure equal employment opportunities. "While the figures cited in the opinion may be accurate," he argued, "they fail to reflect Marshall's attempts to encourage Negroes and other minority groups to seek employment; that there are few qualified personnel in such minority groups who are located in the area, and that those employed elsewhere are reluctant to move here."[29]

After the Kennedy assassination and the accession of Lyndon Johnson to the presidency, Webb's advocacy of civil rights became more forceful. The Civil Rights Act of 1964 set new standards for federal agencies. Webb informed von Braun that the principal topic of discussion at a cabinet meeting he attended on 2 July had been the need for effective leadership to implement the Act, and suggested that Marshall's location in Huntsville made von Braun's support essential.[30]

Webb recognized that the difficulty in implementing equal employment opportunity at Marshall was larger than Huntsville.[31] On a speaking tour in Alabama in late October, he told civic leaders and businessmen in Montgomery that social conditions in the state made it difficult to recruit scientists, engineers, and managers, and suggested that leaders in Alabama should "address themselves in their own interests to the causes of these difficulties." Congressman Hale Boggs of Louisiana, after a conversation with NASA officials in Washington, announced that "hundreds" of Marshall's top personnel, including perhaps von Braun himself, might be transferred to Michoud.[32]

Alabamians reacted with consternation. "This is a big thing; this is a tragic thing; this is a terrible thing," railed former Congressman Frank Boykin, who termed the proposal a "dastardly deal." Boykin suggested that New Orleans, "down there in the marshes, . . . is a fine place to eat and drink, but there can be no better place on earth, if somebody wants to work and do some good for all mankind than Huntsville, Alabama."[33] Some feared that the state was being punished for political transgressions, since Democratic electors had been left off the state ballot for the upcoming presidential election, virtually conceding the state to Republican Barry Goldwater. One constituent urged Alabama Senator John Sparkman to retaliate for this "political blackmail" by doing something about "the Webb creature," and complained about "the Negroes having all the rights and the whites having none." Businessmen worried about the effect of the announcement on impending transactions.[34] Sparkman met with Webb and contacted the President, and received assurances that nothing would be done to move operations from Huntsville.[35]

Webb completed his Alabama tour with a stop in Huntsville. In a speech to Marshall employees and local businessmen, he assured them that NASA wanted to continue Apollo booster work at Marshall, and suggested that if people in Huntsville did their part, the number of employees at Marshall could increase over the next year or two. But he added a caveat: "If we cannot get the seasoned executives here that we need for the management function, then we will do more of this work at other locations." When questioned about the "apparent" image of Alabama, he replied, "There is an unfavorable image, and we feel it in our recruiting; and the problems we face right now are not as hard as the problems we're going to face a year from now."[36]

Reaction to Webb's visit was mixed. Civic leaders believed he had given insufficient consideration to the differences between Huntsville and the rest of the state, but at the same time they initiated reforms that made those contrasts more striking. Huntsville Mayor Glenn Hearn established a biracial Human Relations Committee to seek improvement in racial relations, particularly in housing and employment. He set up a civil rights complaint department. Marshall, too, continued to work with community leaders through AHAC, the Marshall Advisory Committee, and the Chamber of Commerce Committee for Marshall Space Flight Center.

Von Braun addressed the Huntsville Chamber of Commerce on 8 December. He reiterated Webb's argument, saying, "I think we should all admit this fact:

Alabama's image is marred by civil rights incidents and statements." He urged the businessmen to improve Huntsville's facilities for education, transportation, and recreation, but also challenged them to do more "for those less fortunate families who are bypassed by the big space and missile boom."[37]

In the months that followed, von Braun continued to urge attention to Alabama's racial problems. He lamented that Alabama ranked "near the bottom" in education, that barriers to voting formed "a Berlin Wall around the ballot box." He cautioned that resistance to federal desegregation orders could reduce NASA expenditures in the state. "Obstructionism and defiance . . . can hurt and are hurting Alabama," he warned. The national press referred to him as "one of the most outspoken and persistent spokesmen for moderation and racial reconciliation in the South."[38]

Other signs seemed to augur for constructive change. Alabama businessmen published a full-page ad in the Wall Street Journal and state newspapers calling for compliance with the Civil Rights Act. County school superintendents, in defiance of Governor Wallace, agreed to comply with provisions of federal law in order to continue to receive federal funds. Webb, taking note of these developments while preparing for a visit to Huntsville, conceded that "certain constructive forces in the state are endeavoring to move ahead to meet modern conditions and to get the past behind them."[39]

While von Braun and Webb pressed for resolution of Alabama's racial problems, Governor Wallace continued to proclaim "segregation forever." Neither Webb nor von Braun mentioned Wallace by name, but both criticized his policies. Wallace had already had other confrontations with federal officials, of course; another, with NASA, seemed likely. NASA debated protocol over Vice President Hubert Humphrey's planned visit to Marshall: "Governor Wallace has sent feelers about a visit to Marshall. Should he be invited for the V/P meeting? Can V/P and NASA ignore him in his state?"[40]

A confrontation came on 8 June, when Wallace, members of the state legislature, and 48 out-of-state newsmen visited Marshall for a Saturn test firing and addresses by von Braun and Webb. Von Braun urged his audience to "shed the shackles of the past," and suggested that Alabama might not achieve its promise of industrial growth under Wallace's policies. Webb added that "the size and importance of our operations in Alabama require us to add our support to the efforts of forward-looking and fair-minded leaders of the state."[41] When Webb and von Braun asked Wallace in a more informal setting if he would like to be the first person on the Moon, the governor replied, "Well, you fellows might not bring me back."[42]

By mid-1965, Huntsville's leadership—von Braun, businessmen in AHAC, civic leaders, and educators—had shown initiative in seeking to overcome the effects of racial discrimination. Webb's staff acknowledged that "the city of Huntsville is carrying out a very commendable effort on the local scene to improve matters," but cautioned that "the solution to the problem is not an impressive list of things that are being done in the Huntsville area. It is a statewide problem that will call for state-wide solutions."[43]

Despite a promising start, Marshall's equal opportunity program failed to alter the employment pattern at the Center. Marshall lagged behind other NASA Centers, consistently failing to meet minority hiring and promotion targets. By late 1969, Marshall had only eight blacks in grades above GS–11; the Manned Spacecraft Center in Houston had 21, and even Kennedy, with a much smaller workforce, had five.[44] A decade after Marshall initiated its affirmative action program, an internal NASA report singled out Marshall for its harshest criticism: "Most of the other Centers met their modest goals for the first year, with the exception of the Center which had the most extreme lack of proper staff and management support. This Center, located in Huntsville, Alabama, and in need of the <u>most</u> skilled compliance staff, had appointed only one totally inexperienced employee rather than the three highly qualified specialists required. The continuing failure of this Center to meet any of its goals has been repeatedly presented to NASA management which refused to take corrective action."[45]

Marshall's shortcomings represented a portion of a larger NASA failure. NASA lagged behind other federal agencies in implementing equal opportunity programs. NASA's minority employment rose only from 4.1 percent to 5.19 percent between 1966 and 1973, when overall federal minority employment reached 20 percent. Furthermore, most of its minority employees were clustered in lower grades. The Agency's own EEO staff concluded that "NASA has failed to progress because it has never made equal opportunity a priority."[46] Deputy Administrator George Low conceded that "Equal Opportunity is a sham in NASA," and derided the Agency's "total insensitivity to human rights and human beings."[47]

Marshall's achievements in fostering equal opportunity from 1963 to 1965 resulted from pressure from Washington. Webb, agencies charged to enforce the Civil Rights Act of 1964, and occasionally the White House pressured Marshall to change. This pressure declined in the late 1960s, even as the civil rights movement disintegrated into factions and lost popular support as riots charred the ghettos of northern cities. Establishment of a bureaucracy to further civil rights, the result of political pressure inside and outside government, undercut the political activism that had made civil rights progress possible.

Webb's message lost its sting. When asked again about Alabama's image on a visit to Huntsville in 1967, he responded that when he thought of Alabama he thought about the great job Marshall was doing, not about Wallace's opposition to desegregation. He reiterated that difficulties in hiring top managers persisted. Even these remarks, mild in comparison to early threats to move NASA business from Alabama, caused another furor; Huntsville businessmen contacted Senator Sparkman to see if he could do something about Webb.[48]

Institutional limitations also affected Marshall's ability to meet civil rights goals. The Equal Employment Opportunity Program started when Marshall employment was near its peak. In the late 1960s, the Marshall workforce declined in number. NASA continued to shift work to contractors, and imposed reductions- in-force on Marshall as work on Saturn for Project Apollo began to wind down. It was difficult to increase minority employment when overall manpower was declining. Federal regulations for reductions-in-force dictated that the last people hired should be dismissed first, leaving recently hired minorities vulnerable. For the relatively few black scientists and engineers seeking jobs, the uncertainties of NASA's future and the lure of higher salaries elsewhere made employment in the private sector more attractive. NASA argued that given the constraints under which it operated, it was not doing badly; 3.4 percent of NASA's scientists and engineers were black, not far below the national figure of 3.5 percent.[49] Finally, Alabama's image was slow to change; it continued to be difficult to attract blacks to the state who had the requisite technical training to take jobs at Marshall. Thus Marshall's greatest achievement in civil rights in the 1960s was not in its own record of minority hiring, but in its impact on the community.

Huntsville's Growth

NASA's reconstruction of Huntsville and Madison County extended beyond civil rights. The Space Center also helped change the area's economic structure, social patterns, and educational institutions. NASA decisions and the Saturn program led directly to demographic and material growth in the area.

The Saturn Project helped bring in thousands of "in-migrants" to Huntsville. Aerospace workers moved to the city and thousands of other people followed them, lured by opportunities in a boomtown. At the peak of its growth, local officials estimated that 36 new residents moved into the city each day.[50] Huntsville's population grew from 16,437 in 1950 to 72,365 in 1960 and to 143,700 in 1966.[51] The vast majority of the newcomers were white, young, urban, professional, and middle class. Huntsville's black population was relatively stable, meaning that the number of African Americans declined as a proportion of the total.[52]

As more and more people came in, the city faced incredible pressures. Mayor Hearn figured that with the addition of every 1,000 people, the city needed "92 acres of residential land, 23 acres of streets, 13 acres of public land, four acres of retail stores, 263 houses, 550 cars, three miles of paved streets, 150,000 gallons of water a day, two extra policemen, and two extra firemen."[53] But like any boomtown the city often could not keep up with its new problems. In the early sixties Huntsville suffered from an inadequate airport, nonexistent public transportation, overreliance on automobiles, traffic congestion, strip development and suburban sprawl, a stagnating downtown, and deficient educational and health institutions.[54]

The area addressed some of these problems relatively quickly. New facilities included a jet airport, three new hospitals, a four-lane "Parkway" to improve traffic flow, and a downtown redevelopment campaign that led to the construction of new civic buildings by the early seventies. Huntsville's public school system improved. School enrollments increased from 3,000 in 1950 to 15,500 in 1960 to 32,000 in 1967, and the city built an average of one new classroom per week between 1956 and 1968. Moreover educational standards and achievement improved. Such improvements came partly because Marshall-Redstone personnel had high expectations for their children and partly because their spouses often became teachers. By the end of the decade 80 to 95 percent of the city's high school graduates continued on to college as opposed to the Alabama average of only 20 percent.[55]

Marshall and its contractors also contributed to economic changes. NASA spending, combined with the aerospace spending on the Army's Redstone Arsenal, made Huntsville "virtually a one-economy city."[56] Economists estimated that 90 percent of the city economy in the 1960s was based on federal aerospace programs; Marshall accounted for 40 to 50 percent of the total. At the peak of Saturn work, Marshall and its contractors employed 29.4 percent of the total Huntsville workforce.[57] When city residents heard the sound of a Saturn test, then Mayor Robert B. Searcy said, "they heard the jingle of a cash register."[58]

The creation of a federal space industry made Huntsville-Madison County less like neighboring rural counties and more like other Southern metropolitan areas. Aerospace dethroned agriculture in the local economy and "King Space" took the seat of "King Cotton."[59] The overthrow took material form when Chrysler, IBM, and Boeing refurbished a textile factory in the old Lincoln mill district and used it for Saturn work.[60] But unlike agriculture or the textile industry, the space industry offered "good jobs." Research and development jobs were interesting and innovating, employed skilled professionals, managers, and technicians, and paid middle-class salaries. In the space economy most people worked for the Federal Government and big, prominent "core" firms like Boeing and Chrysler. These employers offered workers considerable financial benefits and career opportunities.[61]

Not surprisingly residents of Madison County during the early sixties were on average prosperous. The county had the highest per capita income of any county in the state.[62] The annual rate of growth of personal income in the city grew at more than twice the national rate between 1959 and 1966.[63] Huntsville, one visitor noted, was "an island of affluence afloat in agricultural Alabama."[64]

Despite overall gains, the Saturn program could not correct existing income inequalities in the area. Per capita income in Huntsville was 50 percent of the national average in 1960 and only 80 percent in 1967. Income was less equitably distributed than the national average; in comparison with the rest of the nation, more income in Huntsville went to the richest 20 percent of the population.[65] A wage gap existed between employees in the space sector and those in the county's service, agricultural, and industrial sectors.[66] In addition, since space jobs went primarily to qualified whites, space spending helped perpetuate racial inequities. So although professional and technical jobs constituted 60.5 percent of total employment by NASA's Huntsville contractors, only 30 percent of their black employees worked in professional and technical jobs.[67] Black average income continued to run far behind that of whites.[68]

The Saturn project changed the Huntsville economy. Local companies often blossomed with NASA contracts. For example, Brown Engineering, formerly Alabama Machine and Tool and currently Teledyne-Brown Engineering, grew from a small, local contractor to a prominent, national aerospace engineering firm. In other cases, Marshall helped firms use space hardware for commercial purposes. Technological "spinoffs" from Marshall's research and development in the 1960s included polyurethane insulation for construction and flat electrical cables and connectors.[69] Marshall also helped develop and disseminate to industry innovations in alloys, metal forming and bonding technology, welding techniques, metal grinding, and finishing machines. These improvements in metallurgy and machining were the Center's most important industrial innovations.[70] In other cases, the import of technical expertise encouraged the formation of new high technology companies that did not depend on government contracts. For instance, a computer specialist, who had originally come to Huntsville to work on Saturn's IU, formed Intergraph, a computer and software firm that by the 1980s would grow into a Fortune 500 company with worksites across the globe.

Despite these successes, in the late 1960s and early 1970s, Huntsville was as dependent on federal funding as a city could be. If NASA pulled the plug on space spending, half the city would go down the drain. In 1966 a Marshall study warned that Apollo budget cuts would result in mass exodus, "large numbers of home mortgage defaults, business failures, and a serious regression in the overall economy." Besides depression in the city, cutbacks would devastate the Center, "one of the world's finest technological institutions."[71] A NASA Headquarters report agreed, finding that the costs of allowing Huntsville's infrastructure to decline were "greater that the costs of sustaining it until it achieves a critical mass and diversification."[72] When NASA's spending on Apollo began to constrict in the late 1960s, both the city and the Center would face years of uncertainty and austerity.

Marshall officials foresaw some of the troubles and recognized that the Center and the city of Huntsville were interdependent. Von Braun worked with civic and commercial leaders to create a social and educational environment that could facilitate economic growth and diversity. He cooperated with Army, business, and civic leaders to establish Cummings Research Park. Research Park eventually became a center for businesses specializing in advanced technology research, manufacturing, and management.[73]

Von Braun also promoted education, especially university education. He recognized that Huntsville needed high quality academic and research institutions to attract and retain skilled people and to maintain NASA's investment. Therefore von Braun said his goal was to help Alabama get the nation's "Number 1 educational center for rocket and space technology" just as it had the "Number 1" football and rocket teams. He lobbied the state to upgrade the Huntsville Extension Center of the University of Alabama in Tuscaloosa. In 1961 von Braun successfully appealed to the Alabama legislature for a $3 million bond issue to create a research institute on the extension Center's campus. With Marshall's support, the Center extended its graduate offerings and in 1966 became the University of Alabama in Huntsville (UAH), an independent campus in the Alabama system. UAH specialized in science and engineering and soon had millions of dollars in NASA contracts.

By improving Huntsville's educational and research institutions and bringing in skilled people, von Braun and NASA helped create Apollo's most important spinoff. The schools and skilled workers created an "environment for growth" and planted the seeds that would, in the long term, produce economic diversification in Madison County.[74]

In addition, Marshall's Saturn rockets became the centerpiece in one of Huntsville's most visible concerns. The Space and Rocket Center opened in 1970 and housed an aerospace museum, theme park, and camp for children. The facility had a Saturn I and Saturn V on display and became the state's most popular tourist attraction. In becoming marketable as museum exhibits, the Saturns were a permanent spectacle that directed attention to the political and symbolic goals of the Apollo program.[75]

In sum, many of Marshall's important achievements in the 1960s were side effects of its main mission of space exploration and technological innovation.

Because of the Center, Huntsville and Madison County experienced a federal reconstruction of many social relationships and economic patterns. In race relations, the Center worked to open employment opportunities. In the economy, Marshall contributed to growth and diversity. In education, it helped improve public schools and form a new university and research center. The Marshall Center transformed Huntsville from the Watercress Capital of the South into Rocket City, U.S.A.

---

1 Loyd S. Swenson, "The Fertile Crescent: The South's Role in the Space Program," *Southwestern Historical Quarterly* 71 (1967–68), pp. 377–92.
2 Walter A. McDougall, *...the Heavens and the Earth: A Political History of the Space Age* (New York: Basic Books, Inc., 1985), p. 376.
3 Swenson, p. 376.
4 See Dale Carter, *The Final Frontier: The Rise and Fall of the American Rocket State* (London: Verso, 1988), esp. pp. 199–210.
5 Our following discussion on civil rights in its draft manuscript form was used in the preparation of *Wernher von Braun: Crusader for Space* by Ernst Stuhlinger and Frederick I. Ordway III (Krieger Publishing Company, 1994), pp. 187–88.
6 McDougall, p. 376.
7 Luther J. Carter, "Huntsville: Alabama Cotton Town Takes Off into the Space Age," *Science* 155(10 March 1967), p. 1228. "If Bayonets Come . . .," *Huntsville Times*, 9 June 1963.
8 Art Sanderson, Oral History Interview by Andrew J. Dunar (hereafter OHI by AJD), 20 April 1990, Huntsville, Alabama.
9 *Huntsville Times* clippings, 5 January to 24 March 1962; Richard Haley to Marvin (no last name given), 15 January 1962; and Richard Haley, "Case History of a Failure," 7 March 1962, Reel 17, Series V, Papers of the Congress of Racial Equality (CORE), Martin Luther King, Jr. Center, Atlanta, Georgia.
10 Ray Kline to Colonel L.W. Vogel, 11 May 1965, MSFC Directors Files, National Archives Depository Atlanta.
11 In 1962, black per capita income was $2,457; white per capita income was $5,386. Robert A. Myers, "Planning for Impact: A Case Study of the Impact of the Space Program on Huntsville, Alabama," M.A. Thesis, (New York University, 1967), p. 99.
12 One report claimed that "Negro housing, ghetto areas excepted, was virtually nonexistent." MSFC Manpower Office, "A Chronology of the Equal Employment Opportunity Program at MSFC," February 1971, pp. 2–3.
13 Clyde Foster, OHI by AJD, 23 April 1990, Huntsville, Alabama.
14 "Findings and Recommendations of the Hearing Officer, E. J. Spielman, in the Matter of the Complaints of Joseph Ben Curry and Joe D. Haynes," 5 August 1963, VI–M Discrimination Cases 1962 folder, Personnel—Legal Opinions File, Legal Office, MSFC.
15 Alfred S. Hodgson, NASA Headquarters, to Director, MSFC, telegram, 25 April 1961; von Braun to Hodgson, telegram, 27 April 1961, MSFC Directors Files, National Archives Depository Atlanta.

16 Arthur M. Schlesinger, Jr., *Robert Kennedy and His Times* (Boston: Houghton Mifflin Company, 1978), pp. 335–36.

17 Webb to von Braun, 24 June 1963, Minority Groups folder, NASA History Division Documents Collection, NASA Headquarters, Washington, DC.

18 MSFC Manpower Office, "A Chronology of the Equal Employment Opportunity Program at MSFC," February 1971, p. 5.

19 The organization was originally known as the Huntsville Contractors Equal Employment Opportunity Committee, but soon assumed the title AHAC. Cummings later founded the Huntsville-Madison County Community Action Committee, and became its first chairman.

20 Report of meeting of Huntsville Contractors, 5 July 1963, MSFC Directors Files, National Archives Depository Atlanta.

21 Report of meeting of Huntsville Contractors, 5 July 1963, MSFC Directors Files, National Archives Depository Atlanta.

22 *Huntsville Times*, 24 April 1966, 8 May 1966.

23 MSFC Manpower Office, "A Chronology of the Equal Employment Opportunity Program at MSFC," February 1971, p. 3.

24 Clyde Foster, OHI by AJD, 23 April 1990, Huntsville, Alabama.

25 *Huntsville Times*, 13 June 1963.

26 Von Braun to Webb, 15 July 1963; Mueller to Webb, 24 August 1964, MSFC Directors Files, National Archives Depository Atlanta. When the representatives of the 12 colleges visited Huntsville in July, they stayed at a local motel that had previously been segregated. MSFC Manpower Office, "A Chronology of the Equal Employment Opportunity Program at MSFC," February 1971, p. 6.

27 The recommendation regarding Curry was made moot by his resignation on 20 September. Haynes was promoted. Harry Gorman to von Braun, memorandum, 20 September 1963, MSFC Directors Files, National Archives Depository Atlanta.

28 "Findings and Recommendations of the Hearing Officer, E. J. Spielman, in the Matter of the Complaints of Joseph Ben Curry and Joe D. Haynes," 5 August 1963, VI–M Discrimination Cases 1963 folder, Personnel—Legal Opinions File, Legal Office, MSFC.

29 Von Braun to Webb, 24 September 1963, VI–M Discrimination Cases 1963 folder, Personnel—Legal Opinions File, Legal Office, MSFC.

30 Webb to von Braun, 13 July 1964, MSFC Directors Files, National Archives Depository Atlanta.

31 Webb had earlier acknowledged to von Braun that "The problems you face in this area [equal employment opportunity] are somewhat more difficult than they may be in other parts of the country." Webb to von Braun, 19 April 1963, MSFC Directors Files, National Archives Depository Atlanta.

32 *New York Times*, 24 October 1964. Speaking in Florence, Webb claimed that NASA had offered one vital post at MSFC to about 30 executives, all of whom turned it down. *Huntsville Times*, 19 January 1967.

33 Frank W. Boykin to Thomas W. Martin, 26 October 1964, Federal Government-NASA folder, Box 66A677, No. 7, John J. Sparkman Papers, Special Collections, University of Alabama, Tuscaloosa, Alabama. Boykin, who was from Mobile, retired from Congress in January 1963 after serving nearly 18 years.

34 Marian Cook to Sparkman, 28 October 1964; Hugh Morrow, Jr. to Sparkman, 27 October 1964; David Henderson Head to Sparkman, 26 October 1964, Federal Government-NASA folder, Box 66A677, No. 7, John J. Sparkman Papers, Special Collections, University of Alabama, Tuscaloosa, Alabama.

35 Sparkman to Boykin, 29 October 1964, Federal Government-NASA folder, Box 66A677, No. 7, John J. Sparkman Papers, Special Collections, University of Alabama, Tuscaloosa, Alabama.

36 *Marshall Star*, 4 November 1964; Webb, transcript of remarks to Huntsville Chamber of Commerce, October 1964, MSFC History Office. Webb to Congressman Robert R. Casey, 17 November 1964, NASA-Marshall Space Flight Center folder, box 170, Webb Papers, Harry S. Truman Library, Independence, Missouri.

37 Wernher von Braun, "Huntsville in the Space Age," Address to Huntsville-Madison County Chamber of Commerce, 8 December 1964, MSFC History Office.

38 Ben A. Franklin, "Von Braun's New Role—Racist Critic," *San Francisco Chronicle*, 16 June 1965. Franklin was a *New York Times* correspondent.

39 Webb to Drs. Bisplinghoff, Mueller and Newell, memorandum, 3 May 1965, Alabama State Legislature-MSFC folder, box 222, Webb Papers, Harry S. Truman Library, Independence, Missouri.

40 "JW" to Mr. Scheer and Mr. Callaghan, 10 March 1965, George C. Wallace Biography folder, NASA History Division Documents Collection, NASA Headquarters, Washington, DC.

41 *New York Times*, 9 June 1965. That evening, Webb addressed AHAC. He praised Milton Cummings, Arthur Pilling, and the organization for having "truly taken affirmative action to assure equal employment opportunity in Huntsville." "Remarks for Mr. Webb to Use at Meeting with Association of Huntsville Area Contractors," 8 June 1965, Alabama State Legislature-MSFC folder, box 222, Webb Papers, Harry S. Truman Library, Independence, Missouri.

42 Ben A. Franklin, "Von Braun's New Role—Racist Critic," *San Francisco Chronicle*, 16 June 1965.

43 Ray Kline to Colonel L. W. Vogel, 11 May 1965, MSFC Directors Files, National Archives Depository Atlanta.

44 George Low to Robert J. Brown, 24 June 1970, Minority Groups folder, NASA History Division Documents Collection, NASA Headquarters, Washington, DC.

45 "NASA Equal Opportunity Program," appended to Ruth Bates Harris, Joseph M. Hogan, and Samuel Lynn to James C. Fletcher, 20 September 1973, NASA History Division Documents Collection, NASA Headquarters, Washington, DC.

46 *Ibid*. The Headquarters staff further complained that "Appropriate statements have been issued, but when the commitment is tested, it is found lacking. A sound equal opportunity staff was permitted to be formed, but it has been continuously kept short of resources and under the control of insensitive middle management. Field installations have been required to establish equal opportunity offices, but in cases where they proposed to appoint unqualified, uncommitted persons to staff these programs, the objections of the Headquarters Equal Opportunity staff were overruled."

47 "Excerpt of Remarks by Dr. George M. Low before Conference on EEO," undated

(apparently 1973), NASA History Division Documents Collection, NASA Headquarters, Washington, DC.

48 Webb to Hubert H. Humphrey, 17 March 1967, Center file, MSFC Correspondence folder, NASA History Division Documents Collection, NASA Headquarters, Washington, DC.; *Huntsville Times*, 19 January 1967.

49 Levine, p. 121.

50 Roger W. Hough, "Some Major Impacts of the National Space Program, V. Economic Impacts," (Menlo Park, CA: Stanford Research Institute, June 1968), p. 23, MSFC History Office.

51 MSFC, "Apollo Phase Out Economic Impact Study," 15 September 1966, MSFC History Office.

52 Robert A. Myers, "Planning for Impact: A Case Study of the Impact of the Space Program on Huntsville, Alabama," (M.A. thesis, New York University, 1967), chap. V.

53 Dolan, p. 46.

54 Myers, chap. III. For a discussion of changes in Huntsville in the 25 years after the founding of the Redstone Arsenal, see a 16-page supplement to the *Huntsville Times*, 3 November 1974, hereinafter cited as Supplement.

55 Virgil Christianson, "Space Era's Impact on Community," *Huntsville Times* (1 January 1968); Don Eddins, "City Schools Faced Growth Unprecedented," Supplement, p. 11; Bilstein, pp. 390–94; Hough, pp. 20–21.

56 "'Rocket City' Lives on 'Space Economy,'" *New York Times* (28 June 1960).

57 Mary A. Holman, *The Political Economy of the Space Program* (Palo Alto, CA: Pacific Books), pp. 200, 203–205; Myers, chap. IV.

58 Paul O'Neill, "The Splendid Anachronism of Huntsville," *Fortune* 65 (June 1962), p. 238.

59 Patricia Dolan, "Huntsville—Space City, USA," *GE Challenge* (Spring 1968), p. 45.

60 O'Neill, p. 231.

61 See discussion of Standard Metropolitan Statistical Areas in William W. Falk and Thomas A. Lyson, *High Tech, Low Tech, No Tech: Recent Industrial and Occupational Change in the South* (Albany, NY: State University of New York, 1988), esp. pp. 11, 16, 37–49.

62 R.M. Nicholson, "Economic Impact Study, Apollo Program Supplementary Data: Comparison of Decatur, Alabama with Huntsville," 5 December 1966, MSFC History Office.

63 Holman, p. 206.

64 Rudy Abramson, "Huntsville—A City with a Case of Moon Madness," *Los Angeles Times* (5 June 1966).

65 Holman, pp. 207–09.

66 Myers, chap. IV, esp. p. 65.

67 MSF Field Center Development, "Economic and Social Benefits of the Manned Space Flight Program," no date, MSFC History Office, pp. 30, 32.

68 Myers, p. 99.

69 MSFC Manpower Office, Administration and Technical Services, "Impact of MSFC on Local Business and Government," 8 June 1971, MSFC History Office.

70 Bilstein, pp. 396–97, 399.

71 MSFC, "Apollo Phase Out Economic Impact Study," 15 September 1966, pp. 16–17, MSFC History Office.

72 MSF Field Center Development, "Economic and Social Benefits of the Manned Space Flight Program," no date, p. xxii, MSFC History Office.

73 Wernher von Braun, "Remarks to Alabama Legislature," 20 June 1961, "Ground Breaking Ceremony, UAH Research Institute," 20 December 1962, MSFC History Office; UAH, "Summary Report: Self-Study for Reaffirmation of SACS Accreditation," January 1985; Bilstein, pp. 392–93; "City's Research Park Among Nation's Finest," Supplement, p. 6; Don Eddins, "University of Alabama Spreads Wings," Supplement, p. 13.

74 Hough, pp. 8–15.

75 Michael L. Smith, "Selling the Moon: The U.S. Manned Space Program and the Triumph of Commodity Scientism," in Richard Wightman Fox and T.J. Lears, eds., *The Culture of Consumption: Critical Essays in American History, 1880–1980* (New York: Pantheon, 1983), pp. 175–209.

Chapter V

Between a Rocket and a Hard Place: Transformation in a Time of Austerity

"I'd like to see a little less 'crash' and a little more 'program.'"

—Wernher von Braun

Once the rockets are up, Who cares where they come down, "That's not my department!" Said Wernher von Braun.

—Tom Lehrer, 1965

On 9 November 1967 at seven o'clock in the morning, the first Saturn V launch lifted off from Cape Kennedy carrying the Apollo 4 mission into space. Wernher von Braun, who watched from the firing room, exclaimed at a news conference that "No single event since the formation of the Marshall Center in 1960 equals today's launch in significance."[1] Later in the day, von Braun learned that a reduction-in-force (RIF) would cut 700 people from the Center, some who had helped build the Saturn that had flown that morning. The juxtaposition of the two events on a single day dramatically showed the shift in Marshall's fortunes, for even at a peak of achievement, the Center faced an uncertain future.[2]

The irony symbolized by the concurrent success of the Apollo 4 mission and a budgetary crunch would recur through the next decade of Marshall's history. As television viewers throughout the world watched the powerful Saturn rockets roar into space and marveled at the spectacle of men on the Moon, Marshall engineers could take pride in their accomplishment of a national mission. Not only were they responsible for the rocketry that powered all of the lunar missions, they developed the roving vehicle used on the Moon's surface in the missions of the early 1970s. And the 1973–74 *Skylab* mission, the first American "Space Station," was a Marshall achievement. But people within the Center had little opportunity to revel in the triumphs of the space program, for in the midst of its success, Marshall confronted a protracted institutional crisis.

The causes of the crisis were many. Tom Lehrer's satiric song of the mid-sixties foreshadowed a shift in public opinion about space. As the Vietnam War and domestic divisions diverted attention from NASA, many Americans became bored with—in some cases antagonistic to—the Agency's programs. The national economy staggered under "guns and butter" budgets until hard realities mandated cuts that forced Marshall to move from the affluence of the early sixties to the austerity of the seventies. The politics of budgets increasingly defined the Center. Planning and decision-making shifted to Washington, where political priorities of the executive offices and Congress were more important than technological goals.

As the Center coped with external strains, it would be internally transformed. New leadership replaced many of the Germans and reshaped von Braun's organization. The arsenal system gradually gave way to the Air Force contracting system as in-house capabilities steadily declined. New, diversified scientific and technological responsibilities supplemented the Center's propulsion specialty. Management struggled with serious threats to the Center's well being, and even its survival, for NASA Headquarters considered closing Marshall. Funding cutbacks, RIFs, transfer of projects to other Centers, and changes in leadership were manifestations of a more fundamental question: What, if anything, was to be Marshall's role in the post-Apollo space program?

In the late 1960s, then, Marshall Space Flight Center slowly became the victim of its success, and the characteristics that made Marshall unique defined its crisis. Of all Apollo hardware, Marshall's Saturn launch vehicles had the longest lead time, the fastest buildup, and the largest workforce. The Saturn program peaked in the mid-sixties, however, and while other Centers were still building, Marshall began to retrench. Many of its facilities had been built for Saturn, rather than for long-term institutional needs, and had limited utility in NASA's post-Apollo plans.[3] In short, when the heady days of unlimited funding and ample manpower were over, Marshall faced the "crash" that inevitably follows any crash program.

The Perils of Post-Apollo Planning

NASA and Marshall were both slow to initiate planning for the post-Apollo space program, and planning was often encumbered by overly optimistic projections. In 1963 Marshall was still hiring, and expected to add 2,000 Civil

Service employees in two years before leveling off at 9,500.[4] The two years passed with only modest increases, but with 90 percent of his workforce devoted to Saturn work, von Braun expected Marshall manpower to remain constant through the remainder of Apollo. After all, contractors had already scheduled manpower reductions, and von Braun warned, "as the highly skilled engineering teams and contractor plants are disbanded, our in-house people must shoulder the burden to meet the unforeseen."[5] He compared Marshall's role to firefighters in a mid-size city—essential, but underutilized when there was no fire.[6] Initiation of the Apollo Applications Program late in 1965 raised rosy expectations of 1,500 to 2,000 new jobs at Marshall.[7] The Center's master plan was equally optimistic; it anticipated new construction and continued conversion of old Army facilities without consideration of financial constraints. Von Braun envisioned human planetary missions perhaps as early as the late 1970s, and he had established a Future Projects Office at the Center in the early 1960s.[8] But he had given less attention to short-range planning. When asked about the future of Marshall, his thoughts ran to NASA's vague plans for extensions of the Apollo Program and to possible work on post-Saturn launch vehicles.[9]

Nonetheless, critics who have chided NASA for its failure to plan for the aftermath of Apollo have been unduly harsh. Nobody anticipated a steep decline in the halcyon days of Saturn development, and NASA began to consider alternatives before the launch of the first Apollo mission. The budgetary cycle and the long lead-time on big science projects forced NASA to consider post-Apollo plans in the mid-1960s. NASA's worries that the Johnson administration's reluctance to commit to supporting space programs might precipitate the breakup of its team hastened Agency planning.[10] Contractors agreed in 1966 that "the erosion of the Apollo space team has already started."[11]

Marshall developed methods for long-range planning, but institutional constraints hampered the Center's efforts. Dr. Heinz Koelle directed an active Future Projects Office that had been formed in the fall of 1964 to draft plans for technical projects. Its tasks included launch systems, Saturn rockets, Nova, nuclear-thermal rockets, lunar stations, and Space Stations. It devised schemes for use of a spent-rocket stage as a manned orbiting laboratory that helped form foundations for *Skylab*. The Research Projects Laboratory conducted studies for science-oriented projects including High Energy Astronomy Observatories (HEAO), the Large Space Telescope, the Apollo Telescope Mount (ATM), early lunar rover studies, lunar science activities, and scientific projects for satellites.[12] But frequent changes in funding guidelines from Headquarters, uncertainties about the goals of the post-Apollo Program, and an increasingly bureaucratized procedure for task approval limited its ability to generate new projects.[13] Marshall executives knew that difficult years were ahead; as early as mid-1966, they discussed the impact that Vietnam and Lyndon Johnson's domestic programs would have on NASA budgets.[14]

As Saturn development crested, and long before the scale of the decline became evident, von Braun realized that funding limitations would force Marshall to broaden its mission beyond its traditional specialization in launch vehicles, the Center's "bread and butter." Marshall had a vast physical plant, proven engineering expertise, and demonstrated managerial ability. But how could those resources be applied? The Center was "a tremendous solution looking for a problem."[15]

Headquarters offered little guidance. George E. Mueller, NASA Associate Administrator for Manned Space Flight, told von Braun that Marshall should maintain its launch capability, but that NASA Administrator James Webb would ask, "Do they need 14,000 people to do that job?" Von Braun wanted Marshall to make the best pitch for all projects it could get, believing space science and operations looked promising.[16] "For us the essential thing is this," he told Headquarters. "We must be able to plant a new flag in Marshall in some new field."[17]

Unfortunately, internal NASA politics limited Marshall's flexibility to move into new areas. Each NASA Center had its own specialization and jealously guarded its prerogatives. Von Braun's diversification would encroach on Goddard's turf in space science and Houston's in operations. Huntsville had fewer options for expansion than other Centers. Any new field might compete with others, and even work on propulsion might meet challenges. As one veteran of intercenter competition observed, "There was nothing that Marshall had that was uniquely Marshall's."[18] No one rivaled Marshall's experience in large launch systems, but its expertise in launch vehicles was not unique: Lewis had rocket engine experience dating back to NACA, had built the Centaur, and had "staked out a role in advanced propulsion technology that Marshall could not expect to emulate."[19] Headquarters and Wallops Island managed LTV's development of the Scout, and Goddard managed McDonnell-Douglas's development of the Delta launch vehicle.[20]

Rivalry between Marshall and Houston's Manned Spacecraft Center (MSC) had been present since the days of ABMA and the Space Telescope Group in the late 1950s, and intensified as Apollo wound down. Apollo's neat division between Marshall's Saturn V and Houston's capsule separated authority into stages; plans for post-Apollo Programs made responsibilities in human space flight less distinct.

Marshall and Houston, described by one historian as "semiautonomous, almost baronies,"[21] guarded their realms fiercely. Houston challenged any proposal from Marshall that related to operations, astronauts, or manned systems. Competition with Houston was most pronounced in the Apollo Applications Program (of which *Skylab* was the centerpiece; see Chapter VI), but it touched all relations between the two Centers. "We had the perception that they weren't worrying about NASA or the space program, but they were worried about feathering their nest," recalled Houston's Chris Kraft.[22] The rivalry bothered von Braun, who told his staff that he was disturbed that a Marshall collision with MSC could jeopardize the lunar landing program.[23]

To clarify the post-Apollo division of labor Mueller summoned all three Manned Space Flight Centers to a three-day executive hideaway meeting at Lake Logan, North Carolina in August 1966. Marshall and Houston divided *Skylab* responsibilities, and worked out means to resolve future disputes. However, as one study observed, Lake Logan provided "a convenient formula, but did not eliminate the competition between Centers for post-Apollo work."[24]

Von Braun's designs for a Marshall role in astronomy met less resistance. In May 1966, he discussed future NASA missions with Mueller and Robert Gilruth, Center director in Houston. All three agreed that Marshall should get involved in astronomy, and Mueller suggested work on the Apollo Telescope Mount (ATM) might lead to Marshall becoming the lead Center in space astronomical observatories. When Homer Newell, head of space science at NASA, concurred, von Braun had secured one new niche for his Center. On some astronomy projects, Goddard would be considered a consultant to Marshall.[25]

The limits of space science as a new role for Marshall became clear with the Center's first venture into Big Science. The Center developed plans to support Voyager, an anticipated series of probes to Mars. Voyager work would place Marshall under the Office of Space Science and Applications (OSSA), and might open other opportunities outside of the Center's usual responsibilities under the Office of Manned Space Flight. Just as Marshall neared agreement on how to proceed on space science without jeopardizing Apollo, Congress postponed Voyager in August 1967. The projected cost had risen from $43 million to $71.5 million, and Congress suspected that the mission might lead to more costly human missions.[26]

Ernst Stuhlinger, Marshall's head of space science, worried that the Voyager postponement might divert the Center from expansion into space sciences. He considered development of projects under OSSA not merely good business, but essential to the Center's future. Supporters of manned programs and unmanned science programs had battled since NASA's formation, and scientists resented the dominance of Mueller's OMSF. Stuhlinger advised that Marshall's future would be most secure if the Center had a foot in both camps. Unless Marshall moved into space science, he cautioned, "our Center with its present one-project, one-HQ-boss orientation will give the image of an aging organization, unwilling to accept the challenge of broader responsibilities as the space program evolves."[27]

Marshall's Manpower Crisis

Even as Marshall struggled to diversify for the post-Apollo era, a manpower crisis transformed the Center. By the end of the decade, reassignments, RIFs, reductions-in-grade, and other personnel actions were stultifying its activities. Morale declined, and union action led to suits that challenged the Federal Government's reliance on support service contracts, which were used to supplement work done by civil servants. Young engineers left for more promising jobs elsewhere, and the average age at Marshall increased. Recruitment, already considered a Huntsville problem at Headquarters, became more difficult. "Marshall's mood became more and more defensive," remembered Bruce Murray of the Jet Propulsion Laboratory (JPL). "Relentless efforts to maintain employment levels replaced von Braun's dream of the stars."[28]

Marshall's dilemma first drew attention when it became clear that the Center had a larger workforce than was needed to complete its remaining Apollo tasks. Marshall transferred 200 people to Houston in 1965, and a year later much larger reductions seemed imminent.[29] Headquarters and other NASA Centers saw Huntsville as a source of manpower, and this "Marshall problem" became the major manpower management issue in NASA by the time the Agency's in-house workforce peaked in 1966.[30]

The issue prompted NASA Deputy Administrator Robert C. Seamans to request a review of Agency manpower policy. He directed a task force chaired by MSC Director for Administration Wesley Hjornevik to examine how "Center complements could be adjusted by management to meet the needs of changing roles and missions."[31] Hjornevik met with von Braun and his staff late in August 1966. Von Braun urged Headquarters to use its vacant floating manpower allocations (positions that Headquarters could assign at its discretion which usually totaled three percent of the NASA workforce) to obtain the flexibility needed for personnel adjustments, and to let Center directors work out manpower problems among themselves. Unfortunately, the problem was already larger than von Braun realized. NASA was already planning for 10-percent cuts, and needed an Agencywide policy. Marshall would feel the pinch first, but one of those listening to the discussion remarked that "It is apparent that the MSFC manpower problem of today is the NASA manpower problem of tomorrow."[32]

The Hjornevik group recommended that NASA adopt means to track personnel requirements, and suggested ways to match manpower to programs. Although the committee assumed that NASA manpower requirements would remain constant, its conclusions comprised "a warning that NASA would have to prepare for major changes within the near future."[33] The committee suggested RIFs, actually laying off people, might be necessary as a last resort: the final option of eight alternatives for restricting manpower.

NASA personnel policies were under attack from another quarter, and Marshall was at the center of the controversy. The General Accounting Office (GAO) reviewed support service contracts at Marshall and Goddard Space Flight Center, and concluded that both Centers could have saved money by relying on Civil Servants rather than support service contracts. Support service contracts are common throughout the Federal Government, so the investigations had potentially broad implications. The June 1967 report alleged that Marshall could have saved 19 percent on the three contracts examined. The GAO did not rule on the legality of the contracts, but submitted the Goddard cases to the Civil Service Commission (CSC) for further consideration.

Leo Pellerzi, CSC general counsel, ruled in October that the contracts were indeed illegal, since they involved on-site contractor work using government equipment in tasks expected to last longer than one year, established an employer-employee relationship, and had the effect of creating new government positions by using contract personnel to perform regular NASA work. Lacking any other guidance, NASA used these "Pellerzi Standards" to evaluate its support service contracts, and the courts used them to evaluate NASA's compliance with Civil Service regulations.[34]

Dire warnings became reality the next year. Congress slashed NASA's budget request for Fiscal Year 1968 Administrative Operations—the schedule from which salaries were drawn—by $23.1 million in August 1967, then cut another $20 million in October.[35] Headquarters warned that the budget cuts might require a personnel cutback (RIF) at Marshall. On 9 November Headquarters confirmed the need to cut 700 positions.[36]

On 29 November, von Braun delivered the bad news. He explained the circumstances leading to the RIF to Marshall employees sitting in Morris Auditorium and watching on television around the Center. He described Marshall's evolution from "a do-it-yourself, self-contained organization to a partner of industry," and explained the mandate to reduce Marshall's workforce to 6,386 by January 1968. Half the reduction was to come from wage board employees and technicians, half from among engineers. Attrition might reduce layoffs to 640. The personnel office expected further dislocations, with the RIF requiring 1,300 intracenter reassignments to adjust for those who would be separated. Support contractors would have to match Civil Service reductions on a one-to-one basis.[37]

Four weeks later, the Marshall local of the American Federation of Government Employees (AFGE) and six individual Marshall employees filed a complaint in the U.S. District Court for the District of Columbia requesting an injunction to stop the RIF. The complaint accused the Center director of unfair labor practices, and alleged that the RIF was illegal as long as contract support service personnel were engaged in the same work as Civil Service employees who were to be separated. The court issued a preliminary injunction halting the reduction on 11 January, just two days before the RIF was scheduled to go into effect.[38]

The court's order required NASA and the CSC to examine Marshall's personnel requirements and support service contracts in light of Civil Service law. The two sides reached agreement on 19 February 1968 and canceled all but 147 of the original 1,120 notices for termination, reduction, and reassignment. The court lifted its injunction on 12 March, and dismissed the complaint on 18 April. The plaintiffs appealed.[39] The case dragged on for years, and became a factor in negotiations between the union and Marshall in subsequent RIFs. The case was not settled until 1978, when Judge Joseph Waddy upheld NASA's use of support service contracts.[40]

By the time Marshall was able to proceed on 30 March 1968, attrition and other personnel actions reduced the number of employees who would be subject to RIF action to 147, of which only 57 were terminated—the others were reassigned or reduced in grade. This greatly understates the impact of the RIF, however. Marshall lost 787 employees by May, many of them through retirement or transfers, leading to "grave and serious imbalances in the MSFC workforce." No engineers or scientists left involuntarily, but more than twice the usual number during a comparable period departed during the four-month RIF period. The average age of scientists and engineers increased, since most of those who left—113 out of 145—were under age 40.

This trend raised questions about the future vitality of the Center, since college recruiting was made more difficult by rumors of another RIF and federal regulations that required that newly hired personnel be the first dismissed during reductions. Nor was Marshall given authority to do much recruiting; in FY 1968, the Center replaced only 1 of 14 people separated, by far the lowest replacement ratio of any NASA Center. Morale of both Civil Service and contractor personnel plunged, and post-RIF voluntary separations remained as high as they had been during RIF action.[41]

The RIF also had unanticipated ramifications. Many of those who received notices under the Center's original RIF plan were able to keep their jobs by the time Marshall implemented the RIF late in March, and voluntary departures and court action decreased management's ability to control the RIF. Marshall later estimated that it missed the planned post-RIF mix of skills by 47 percent. Management worried that its ability to deal with personnel issues might be impaired by the union's new image as a strong defender of employee rights.

The reductions also had an impact on the Huntsville economy. Approximately 480 people outside the Center lost their jobs as a result of the Marshall action. Local payrolls declined by $3.4 million a year, and retail sales declined by $1.6 million.[42] Prime contractor manpower in Huntsville dropped even more precipitously than Marshall's Civil Service employment, falling to less than a third of what it had been four years earlier.[43]

Reorganizing for the Post-Apollo Era

The dramatic personnel changes introduced a new dimension to the "Marshall problem" by the summer of 1968. Marshall's manpower continued to erode through attrition after the RIF, and NASA expected it to fall below 6,000 by the end of the year. Reductions at the lower levels had not been matched by corresponding adjustments in upper management. The Center was becoming top heavy, with an administration still geared to maximum workload. Headquarters worried that "the current Marshall structure does not recognize the program and operating situation under which Marshall activities will be conducted over at least the next several years."[44] Headquarters directed Marshall to cooperate with a NASA team in a review of the Center's organizational structure.[45]

The request raised fundamental questions about NASA planning, Marshall's future, and the relationship between Centers and NASA Headquarters. The idea originated in NASA's Organization and Management section rather than in the OMSF, Marshall's administrative superior. NASA seemed to be losing its sense of direction, with manpower and budget considerations driving program decisions. Von Braun questioned "the need for an analysis of the current organizational structure without even mentioning the requirement for an assessment of this Center's future tasks which must obviously be addressed first."[46]

The environment of the Apollo phasedown altered Center relations with Headquarters. Center autonomy had been the rule during Apollo, continuing a tradition that extended back to NACA. "The NACA figured that all Headquarters needed was somebody to go over to the Treasury to get the money," one veteran of the early space program recalled. "Wisdom is in the field, not in Washington."[47] Georg von Tiesenhausen described Marshall's attitude in the early years as "just give us the money, we were the boss."[48] Apollo, with its clear-cut division of authority, precise sense of mission, and end-of-the-decade timetable, perpetuated Center autonomy. Headquarters had "to interface with the

Congress, interface with the OMB, [and] set policy," Kraft conceded, but the Centers neither needed nor wanted direction from Headquarters.[49] One study of NASA management during the 1960s concluded that "Most planning, and almost all that mattered, was carried out by the Centers and program offices, not by Headquarters staff offices reporting to the Administrator."[50] Only occasionally—as in the case of Mueller's all-up testing decision—did Washington intrude. "Quite a few of us originally thought that all the directions from Washington should come through Dr. von Braun so that he is informed about what is going on," von Braun deputy Eberhard Rees explained. Marshall had "always thought that nobody from the outside should actually rule into our place here but through Dr. von Braun."[51]

As budgets, personnel limitations, and the uncertainties over future programs began to drive NASA decisions, authority shifted from the Centers to Washington. NASA began to set policy based on available resources rather than on program goals. With Great Society programs and Vietnam competing for funds, Congress began to challenge the Agency's budget. Moreover, post-Apollo Programs were vulnerable and unlike the lunar landing program were not blessed by any aura of national prestige. External pressures forced Headquarters to assume a new controlling role and make decisions that had been unnecessary in the boom years. NASA, despite Webb's reservations about the value of such an Agencywide enterprise, established a Planning Steering Group to review long-range plans, and OMSF established a Cost Reduction Task Force.[52] The burden fell on the Centers, and Marshall was the first to move into a less certain post-Apollo world.

Marshall's size, its manpower predicament, and the doubts about its future placed the Center at focus of a NASA end-of-the-decade self-examination. The Center's future had been under review for four years, and with uncertainty now an Agencywide phenomenon, Marshall's destiny was doubly in doubt. Von Braun's usual optimism could not withstand fear that he was presiding over the dismantling of his dream, and he occasionally lashed out. He described his mission as scrapping a vital industrial structure, and claimed that the goal seemed to be to ensure that there would be no capability left by 1972.[53] He decried the "rapidly deteriorating environment in our industrial complex," and feared that complacency about space research, scattering of subcontractors, and pressures to reduce costs were creating a "hazardous situation."[54]

The Headquarters requirement for a new Center organization typified the new NASA of scarcity and bureaucracy. Marshall had reorganized before, but the initiative had always come from within the Center. Now, von Braun reacted to circumstances beyond his control. He feared "irreparable damage to a working team that has been built up over a number of years," and asked Headquarters to grant him time to reconcile the Center's loss of manpower and change of mission before initiating precipitous changes.[55] He conceded that Marshall would have to realign its workforce in order to get future space projects.[56]

Within two months, Marshall developed a reorganization plan that responded to the Headquarters mandate and prepared the Center for changing times. Von Braun and some of his closest advisers worked out the basic plan on a hideaway at Jekyll Island in Georgia in the late fall of 1968. Particularly influential was William R. Lucas, Marshall's director of propulsion and engineering, who proposed a Program Development Organization to centralize planning at Marshall. Von Braun explained that the new organization would "help chart the course for this Center in the post-Apollo period," and he appointed Lucas as director.[57]

Program Development's planning process was unique in NASA. No other Center had Marshall's problems; no other Center needed something like Program Development. Marshall's managers reasoned that planning during the Apollo Program had suffered because laboratories and line personnel were too busy working on Saturn to attend to new projects. Maintaining line and lab attention was worsened by the long lead time between preliminary design and final development of a big science project. Therefore Marshall's managers separated planning from doing and new business from old. Program Development was, as Lucas recalled, "a new business organization," a central office to design and sell new projects and ensure that the organization would never run out of work.[58]

The staff of Program Development consciously acted as business people and quickly became Marshall's entrepreneurs. Indeed von Braun referred to Lucas as his "vice president for sales."[59] Like a business, Program Development studied the technical capabilities of the Center in order to find its marketable skills. They found that building rockets was so complex that Marshall had skills not only in propulsion but in general engineering, management of large systems, big structures, strong and lightweight materials, guidance and control, computing, power, and astrophysics. Next the office sold Marshall by seeking new customers in the scientific community. The selling was often difficult because many scientists doubted the Center's skills. Bob Marshall recalled that scientists often felt that "here is this group coming from the South, from Alabama with this funny talking language, trying to get into science."[60]

Even when customers were sold on Marshall, Program Development was not done. The office still had to assess feasibility, compare alternative proposals, develop preliminary designs, define support requirements, perform cost analyses, forecast NASA funding, and finally recommend the best projects to Center management. Marshall said that Program Development had to sell projects to outside groups ("We can do it") and to Center managers and engineers ("You can do it"). If management consented, the Center then solicited Headquarters for the final sale.[61]

At times the transition between Program Development and project offices encountered difficulties. Project offices found Program Development's oversight intrusive. "Some of our worst problems grew out of sending PD people who were not skilled managers over to a project office to lead a major project," recalled George McDonough, who saw several such instances during his work in project offices.[62] Program Development people sensed resistance in the project offices, and believed that project officers and laboratory personnel could lack understanding of and commitment to the new project; they could experience the "not-invented-here syndrome." To overcome this hand-off problem, Lucas and Program Development created pre-project teams. Headed by a pre-project manager, each team drew line personnel from the laboratories and worked on the first two parts of NASA's phased project planning, Phase A (preliminary analysis) and Phase B (definition). In the process, the pre-project team mediated between experts outside NASA and engineers in the Center. When the project got a "new start" and moved into Phase C (design) and Phase D (development/operations), the preliminary design team formed the nucleus of a formal project office.[63]

Program Development became an important source of projects at Marshall in the seventies and eighties. The office oversaw the Center's diversification from Saturn into Shuttles and satellites, solar energy and coal mining, telescopes and materials processing. When projects came out of individual efforts in the labs, Program Development often institutionalized them.[64] The resulting diversity created a new identity for the Center and would give it unique problems.

The other major change introduced during the 1968 reorganization was creation of the new post of associate director for science, acknowledging the importance space science would play in Marshall's future. Stuhlinger became the first to hold the position. Von Braun described him as the "scientific conscience of the Center," and directed him to work closely with the scientific community.[65] The new directorates fell directly below von Braun's two chief deputies, Rees and Harry Gorman.

Reorganization alone could not address all the Center's problems. The Center's appropriations were less than half of what they had been four years earlier.[66] Manpower continued to drop, pushed lower by hiring freezes, attrition, and low replacement ratios; by the end of 1968, Marshall's permanent Civil Service strength had fallen by more than a thousand positions since its peak four years earlier.[67]

Reductions eroded Marshall's historic strengths. Von Braun scrambled to find ways to maintain rudiments of the arsenal system. The Center reassigned wage board employees and technicians to replace support service contractor personnel for testing and quality surveillance, and retrained engineers who had been serving in management. Von Braun informed Mueller, "Our goal is to achieve a systems engineering capability in-house which will permit us to review in depth the design concepts of our stage contractors; and the technologies associated with the manufacture, test, quality maintenance, and reliability assurance employed by our current and future prime contractors."[68] These skills had been the foundation of Marshall's success in the 1960s; once lost, such skills would be difficult to regain in a time of retrenchment.

Charting a New Course

NASA's directive requesting Marshall to reorganize was but part of a larger Agency effort to chart a future course. NASA's prospects at the end of the 1960s were unclear. The Apollo 11 Moon landing in July 1969 culminated a national quest, and public interest in space waned. Ever-tightening budgets constricted vision, and changes at Headquarters brought in leaders with new goals.

Three changes at the top of NASA management had a substantial effect on Marshall. Administrator Webb resigned in the fall of 1968, and his deputy Thomas O. Paine took over as acting administrator. Webb's resignation would affect the Agency in countless ways over the years, but of more immediate impact on Marshall were two changes in the next echelon of NASA management. In November 1969, NASA announced that Mueller would retire as associate administrator for MSF, and that George Low, Apollo manager at MSC, would become deputy administrator.

Mueller, who left NASA to go into private industry, was best remembered at Marshall for his Saturn all-up testing decision, but as head of OMSF he had helped shape the Center in the late sixties. Presiding over NASA's two largest Centers—Huntsville and Houston—Mueller exploited their rivalry. "I think he played Johnson Spaceflight Center (JSC) and Marshall against each other," claimed Kraft. "He did that purposefully. I think he was Machiavellian in that respect." At a time when Marshall was declining, however, Mueller tried to prevent reductions from unduly crippling the Center, and emerged as something of an advocate. Houston sensed favoritism, and Kraft suspected that Mueller showed partiality because "he could tell Marshall what to do and they would do it."[69] Lucas agreed that Houston's intransigence influenced Mueller, and that as a result "Mueller did lean a little bit more to Marshall than to Houston, although I don't think that it was distorted."[70]

George Low's arrival in Washington signaled a change in environment, for if Mueller was in any sense Marshall's advocate, Low was Houston's. Low had served at Headquarters during NASA's first six years, and said later that during that period "I considered myself Bob Gilruth's representative in Washington." Like most of his colleagues in Houston, Low resented Mueller for his alleged Marshall bias. Just months before he became deputy administrator, Low claimed that MSC had always taken the lead on key Apollo decisions, and "as a Center it has generally prevailed, more often than not against Dr. Mueller's desires." He also shared the self-confidence that hallmarked Houston at the height of Apollo, and claimed "We have better people than will be found at the other Centers."[71] Marshall had a high regard for Low, but as the Center's problems deepened after 1969, Huntsville often saw him at the source.[72] Discussing the Marshall dilemma of the late sixties and early seventies, von Tiesenhausen recalled that "One Headquarters name pops up all the time in this context. George Low. He was von Braun's adversary."[73]

Paine inherited control of a NASA in transition. More committed to long-range planning than his predecessor, he announced an ambitious agenda for the Agency despite fiscal constraints. After the inauguration of Richard Nixon, Paine got little support from an administration less committed to space. A Democrat who always felt like an outsider in the administration, Paine nonetheless convinced the President to review national space policy.[74] Nixon appointed Vice President Spiro T. Agnew to chair a Presidential STG and develop a plan for America's next decade in space. The composition of the STG posed problems for the Agency. Its members included not only Paine and former NASA deputy administrator Robert Seamans, but the President's Science Advisor, Lee DuBridge; and placing planning for space in the hands of an external group decreased Agency leverage.

Formation of the STG enabled Paine to promote planning within NASA, for the Agency's suggestions would weigh heavily. Paine requested recommendations from field units, and at Marshall the new Program Development office headed by Lucas formulated the Center's response. The resulting Integrated Space Program showed how the Agency struggled to retain broad vision while recognizing budget limits: its "transcendent objective" was to "maximize space flight while minimizing funding requirements." Marshall's Program Development report acknowledged that "The dominating criteria in the development of new systems is to reduce the cost of space flight."[75]

Although the Centers contributed to the Integrated Space Program, Headquarters centralized the planning, and decision making again shifted away from the Centers. Mueller had been working on Agency plans long before Webb's resignation; a 1967 BellComm study under his direction had first targeted Mars as a post-Apollo goal for the manned space flight program.[76] "This integrated plan was pretty much Dr. Mueller's own activity," von Braun recalled. "It did not grow in the grass roots of the Centers, but it was something that he created with his Headquarters staff."[77] Both Marshall and Houston considered some of Mueller's cost projections unrealistically low.[78]

The Agnew STG September 1969 report was a "partial victory" for NASA administration. The report recommended both manned and unmanned missions, and a manned Mars mission before the end of the century. But the report did not commit the administration to anything, not even a specific target date for a Mars landing. Its suggested funding levels were merely alternatives, and within months the President endorsed the cheapest alternative and dropped mention of the Mars mission. In the end, NASA had discrete programs—scientific satellites and probes, *Skylab*, and a reusable Shuttle. But unlike the Apollo years, the Agency had no over-arching goal, "no post-Apollo space program."[79]

If the STG report did not commit the administration to an extravagant space program, neither did it forestall NASA's ambitious expectations. But all of NASA's plans were now constricted by the politics of budgets, and even the most visionary projections could not avoid the question of money. In the same month that the STG submitted its report, Mueller told von Braun of his hopes for manned space flight, including regular human visits to the Moon by the end of the 1970s at costs substantially below those of Saturn. He envisioned a Space Station and a reusable transportation system, programs that might lead to piloted trips to Mars and Venus in the 1980s. Mueller tempered his optimism with a caveat that was more predictive of the Agency's future: "Costs are of paramount importance. Unless we can substantially change our current way in doing business we will not be given the opportunity to demonstrate the unique capabilities that space provides."[80]

Cutbacks and the Huntsville Economy

NASA budget cutting burst Huntsville's space bubble. The city's Apollo boom became a post-Apollo bust. Signs of decline were already apparent by 1968. Restaurants were still busy at lunchtime, but dinner business was sparse. Sales were down. Unemployment rose. The real estate market suffered. Four motels had closed. Apartments had vacancies in a city that had waiting lists for motel rooms a few years earlier. People worried about whether the city could rebound. A laid-off engineer offered that "If they ever want to build it back up again it is going to take a lot of time and cost a lot of money."[81]

Amidst the gloom, some found grounds for optimism. Huntsville's economy was more diversified than it had been 10 years before. The Huntsville Industrial Expansion Committee, founded after World War II, had seen the city through previous cycles of boom and bust, and had promoted growth that was not solely dependent on the Federal Government.[82] In 1969, the committee could boast that it had just lured four major plants with no connection to the space industry to Huntsville.[83] A real estate salesman offered that "It may be that we profited from experiences of many years ago that have nothing to do with the space program." Paul Styles, in charge of manpower at Marshall, explained that "Von Braun helped to get Huntsville prepared years ago. He told the community leaders at every opportunity that they should broaden their economic base here, that they should get in more industry, that they should not be a one-industry town."[84]

Diversified or not, Huntsville's economy still rested on the town's two federal installations, Redstone Arsenal and Marshall. Marshall was not alone in feeling the uncertainties of federal funding in the late 1960s, for the 1,200 Redstone employees working on the Anti-Ballistic Missile defense system saw their jobs at stake in Congressional debates over limited ABM deployment. Civic leaders put their faith less in diversification than in a gushing federal spigot. One columnist observed that Huntsville had "an almost mystical faith" that Congress would not allow its considerable investment in steel and concrete go to waste, and that von Braun would not let the city down, but would "pound on desks in Washington until fresh money for more big programs is allotted."[85]

End of the Von Braun Era

Von Braun would indeed be in Washington, but not as a lobbyist for Marshall. Paine stunned Huntsville by announcing on 27 January 1970 that the man who had directed Marshall since its inception would move to NASA Headquarters on 1 March and become associate administrator for planning, the fourth-ranking position in the Agency. Paine wanted von Braun to help promote a Mars mission as NASA's next major goal, although von Braun had reservations about the Agency's ability to sell another large program to Congress.[86]

Speculation about why von Braun chose to accept Paine's offer abounded. The frustration of the post-Apollo phasedown, the hope that he might have a larger role in determining NASA's future in Washington, and his rapport with Paine were factors. At Headquarters he would be less pressured by daily crises. "I've spent ten years doing what was 'urgent,'" he explained, "and regrettably not doing what was 'essential.'"[87] Close associates believed that his wife may have influenced his decision.[88] That von Braun was on a seven-week vacation to the Caribbean when Paine announced the move increased consternation in Huntsville.

Von Braun appeared before Marshall executives on 2 February wearing a beard grown on his vacation, and told them, "I am leaving Marshall with nostalgia. I have my heart in Marshall. I love this place." He assured them that "the future of Marshall is the brightest of all NASA Centers."[89]

Huntsville declared "Wernher von Braun Day" on 24 February. Five thousand people turned out in cold, drizzly weather to bid farewell to him. A banner across the grandstand read "Dr. Wernher von Braun—Huntsville's First Citizen—On Loan to Washington." The city announced that its new $15 million civic center would be named for him, and unveiled a granite marker citing some of his achievements. Supporters established scholarship funds in his name at Alabama A&M University and the University of Alabama in Huntsville. *The Huntsville Times* lauded his contributions to the city's culture, education, and economy, and concluded, "Dr. von Braun leaves this community bigger and better than he found it."[90]

Von Braun's decade as Center director left an imprint on Marshall that is difficult to gauge, in part because he was a figure of legendary proportions. In the public imagination, his own role in the early years of America's space program overshadowed the Center. But Marshall took on a distinctive character under von Braun.

Von Braun's approach to management comprised an important part of his legacy to the Center. A blend of techniques applied at Peenemünde and the methods used by the American Army during the ABMA days, von Braun's organization was hierarchical, disciplined, conservative. Apollo veteran Bob Marshall described "a very conservative overview in management technique which went through the whole organization and even prevails today."[91] Not surprisingly, those who were part of von Braun's inner circle remembered it as a creative system. Many of the Germans who immigrated with him remembered teamwork as one of his most lasting legacies. "This team spirit that Wernher von Braun promulgated in his days still permeates the working laboratories at the Marshall Center," according to von Tiesenhausen.[92] Some of those who were lower in the hierarchy saw things differently. Von Braun's weekly notes brought forward "problems and bad things—very few good things got surfaced," according to Bob Marshall. "Nobody at the bottom really felt free to do anything unless he got it approved from the next level up, the next level up, the next level up."[93] One assessment criticized the notes as creating "an almost iron-like discipline of organizational communication."[94]

Whatever Marshall's acknowledged discipline and engineering skills, the Center's reputation for managerial excellence was not as high. Headquarters considered NASA's managerial expertise to rest at Houston. Bob Marshall recalled that Headquarters considered Marshall a "very good technical organization, but a poor management organization."[95] Von Braun's managerial technique contributed to this image. A 1968 study described von Braun as a model for the "reluctant supervisor" typical at Marshall—one who wanted to keep his hands dirty, and avoid red tape and committees.[96]

During the von Braun years, Marshall acquired a reputation for secrecy. "We were rather closed in regard to talking with reporters, journalists," von Tiesenhausen admitted. "That was a general policy then. It helped Von Braun to maintain his options."[97] Some of the younger engineers found this stifling, and one recalled that "People would not go outside the Center and say what they thought if they thought it was different than what management would want you to say. You were very careful. It was as if you did something wrong, you would be banished."[98]

Such caution was but a manifestation of the Center's defensiveness under von Braun. Marshall's defensive posture during the post-Apollo retrenchment was to be expected, but it had become a characteristic of the Center long before cutbacks began. Von Braun had always been an outspoken advocate for Marshall's position, but only to a point. He would back down rather than risk division, and did so several times in confrontations with Headquarters or other Centers. Marshall was a "good soldier," sometimes to its detriment. Key decisions, such as to make Huntsville's LOR in Florida an independent Center, to shift from the arsenal system to the Air Force contracting system, and to favor LOR over EOR cost Marshall. Mueller's "all-up testing" concept ran against the grain of Marshall's traditional engineering conservatism, but von Braun accepted it after voicing initial objections. Kraft noticed von Braun's unwillingness to go beyond a certain point in intercenter disagreements.[99] And Lucas noted the difference in relations between Washington and NASA's two largest Centers: "Headquarters would try to tell Houston what to do and they would ignore it. They just wouldn't do it. Marshall would argue until they were blue in the face, but then they would go ahead and do it."[100]

Von Braun's conciliatory attitude owed in part to the wartime origins of Marshall's German hierarchy. Seldom stated openly, it was from the start an unspoken presence in discussions with Headquarters. When the ABMA's Germans joined NASA, headquarters made clear that they could not bring their operating principles with them; Deputy Administrator T. Keith Glennan averred that those principles would not work in a democracy.[101] Charges regarding the Nazi past of Huntsville's Germans cropped up—with decreasing frequency—but enough to keep the issue alive, and enough to compel von Braun and his associates to maintain a "proper" humility. A film biography of von Braun in the early sixties entitled "He Aims for the Stars" inspired critics to add the subtitle "But Sometimes Hits London."[102] In the mid-sixties an East German publication accused von Braun of militaristic and bloodthirsty activities both in Germany and in the United States, and received some attention in the U.S.[103] Von Braun's relationship with Webb had always been proper but distant, and was tinged with the Nazi question. Paine claimed that Webb wanted to keep von Braun out of Washington: "I think Jim had the feeling that, well, the Jewish lobby would shoot him down or something. The feeling that basically you were dealing with the Nazi party here. And you could get away with it if he were a technician down in Huntsville building a rocket, but if you brought him up here. . . . "[104] Charles Sheldon, White House senior staff member of the National Aeronautics and Space Council in the early 1960s, remembered the resentment toward von Braun in Washington. People discounted rumors that von Braun might eventually head NASA, since "von Braun would never be given any political position. No one who had worked with Hitler and the Nazi government could be trusted."[105]

Webb could be patronizing, reminding von Braun that he was subordinate. During the civil rights crisis in the sixties, Webb lectured von Braun about the need to place a priority on progress in civil rights although it might divert attention from the Center's major task, even though von Braun had already taken action in advance of Headquarters interest. NASA executives resented von Braun's high profile. "When Von Braun appeared at certain occasions—symposiums, meetings at Headquarters—he, rather than the upper administrator, was the center of attention," von Tiesenhausen observed.[106] Webb once warned von Braun that his speeches contained overly optimistic projections of NASA capabilities, creating unrealistic expectations of what the Agency could achieve.[107] Later, Webb restricted the number of paid public appearances von Braun could make each year to four, and required that he submit a list of speaking engagements to Headquarters for approval. In each case, von Braun apologetically accepted direction. These were small matters, but they established subordination beyond what Marshall's principal rival in Houston would accept, and a perception in Houston and Washington of Marshall reticence.

Examination of the von Braun legacy invited comparison with Houston, the other major manned space flight center. Even in appearance, the two Centers revealed their contrasting origins. One Marshall veteran contrasted the difference between Marshall's "gun-metal gray, plain jane buildings" and Houston's "college campus atmosphere."[108] The looser, freer environment at Houston showed in differing approaches to NASA business. Bob Marshall remembered giving presentations in Washington with letter-perfect charts that had been dry-run at least three times, often before von Braun. "My counterparts from Houston or Kennedy would come in with charts that they made up on the way on the airplane," he recalled.[109]

Under New Management: The Rees Directorship

Dr. Eberhard Rees, Marshall Space Flight Center Director, 1970–1973.

Von Braun's departure left his deputy for technical and scientific matters, 62-year-old Rees, as Marshall's director. Rees was older than von Braun, and the two had anticipated that Rees would retire before von Braun would leave Marshall.[110] Von Braun's departure took everyone by surprise, however, and thrust Rees into command.

Rees had been at von Braun's side since Peenemünde, and provided continuity needed in a time of stress. He had the respect of von Braun's staff. "He knew us and we knew him," Stuhlinger recalled. "So that was a very easy transition for both parties."[111]
Rees's talents were very different from von Braun's. Von Braun was a visionary, a politician, a motivator. Rees had none of von Braun's charisma, but he was precise, practical, and a better disciplinarian than von Braun. Their collaboration had worked well. "The two complemented each other perfectly," according to von Tiesenhausen, who worked with them for more than two

decades.[112] Von Braun would originate ideas, Rees would carry them out. "Eberhard was the much more careful person," according to Konrad Dannenberg. Although he was seldom "looking as far ahead as Von Braun, . . . he was a really good man to do the detail planning, to find out what facilities do we need, what people do we need."[113]

Rees believed in centralized management. He reflected that one of the lessons of Apollo was the need to assign "all responsibility to single organizational management structures pyramiding into a single strong personality." Apollo had succeeded, he believed, because of "government-industry teams," but there remained a need for "contractor penetration" since industry's desire to work with only minor intervention by the government had led to "too many cases of severe program impact."[114]

The characteristics that made an ideal deputy did not necessarily correspond to those needed for a successful Center director, and Rees had the misfortune of assuming control of Marshall at the most difficult time in the Center's history. Succeeding a man of von Braun's stature added to the challenge, as Rees acknowledged when Paine introduced him as the new director to Marshall executives at Morris Auditorium. "Becoming the successor of Dr. von Braun is tough," Rees said, " and I'm convinced that anyone who would have got this position would have problems to live up to the standards of Dr. von Braun."[115]

Under Rees's leadership, Marshall followed the path charted by von Braun. The Center continued work on *Skylab*, and increased its involvement in space science. Astronomy became a Marshall specialty, as the Center began development of the Apollo Telescope Mount for *Skylab*, the Large Space Telescope, and the HEAO. Marshall developed life science and Earth resource experiments for *Skylab*. Rees was a top-flight engineer, and had the engineering problems associated with Apollo Applications and space science been his only challenge as director, his talents would have been suited to his responsibilities.

But Rees's administration would be consumed by the continuing phasedown that had confounded Marshall in the late 1960s. Rees soon confronted difficulties that even von Braun had been unable to master, for Marshall's retrenchment was not over. The Nixon budget for Fiscal Year 1971, announced just days after Paine presented Rees as the new Center director, seemed to offer

Marshall a respite. Marshall would only lose 60 positions, which could be absorbed by attrition. And overall Marshall funding would actually increase. But as Congress began to debate the budget, rumors of deeper cuts circulated. Rees tried to allay fears in an open letter to employees, but both House and Senate proposals threatened NASA with personnel reductions that could have affected as many as 1,300 employees.[116]

RIF Redux

On 15 July 1970, NASA Headquarters informed Marshall that it would have to institute another RIF to reduce its manpower to 5,804 Civil Service employees by 1 October. The Center issued RIF letters to 190 employees. Of the 190 employees separated, 99 left voluntarily. Eighty-five other employees were affected, either reduced in grade or reassigned. Headquarters concluded that the Marshall RIF had gone "fairly well," and that morale at the Center was "fair."[117] Unlike the 1968 reduction that singled out Marshall, that of 1970 was distributed among NASA Centers. Houston lost three more employees than Marshall, and four Centers and Headquarters had a higher percentage of employees affected.[118]

Nearly half of Marshall's Civil Service force belonged to the AFGE, and the union followed Center actions closely. However, unlike the 1967 RIF, the union did not initiate action against Marshall. RIF action enabled the union to grow and to organize more effectively.[119] But government unions cannot bargain for wages or strike, and except for their success in delaying the 1968 RIF, they could do little other than to monitor management, trying to ensure equitable treatment for employees who received notices.[120] As a result, the Center was able to execute the reduction under a "controlled environment."[121]

Marshall's handling of the RIF nonetheless raised legal issues. Without consulting Headquarters or the union, the Center had changed competitive designations of some employees in order to avoid the appearance of releasing personnel who were doing jobs performed by support service contractors. By increasing the number of job descriptions, Marshall could make it appear that employees who were doing similar work were performing different functions, and could then hand-pick those who were to be dismissed without fear of veteran or seniority protection.[122] Headquarters anticipated possible unfair labor practice charges from the AFGE, and in fact the issue would rise again as the 1967 RIF action found its way through the courts.[123]

The impact on employees who were released was greater than it had been during the 1968 RIF. The Huntsville economy was weaker, and fewer of those forced to leave were able to find new jobs in the local area. In four years, Huntsville had lost 11,000 space and defense related jobs, and unemployment was at its highest level in 10 years. Thirty-three of those affected filed appeals with the Civil Service Commission, and 10 percent wrote letters to congressmen.[124]

Among those affected by the RIF were a dozen German members of the von Braun team who had come to the United States immediately after the war. Seven of them lost their jobs, leaving only 38 still working at Marshall. Six of the seven were especially vulnerable, since they had chosen to remain in "excepted" status rather than become Civil Service employees at the time they became citizens, and none had the protection afforded by American armed service veteran status. All non-veterans were especially vulnerable at Marshall, since the Center had a higher percentage of veterans than its sister Centers. Given their ages and the depressed condition of the aerospace industry, prospects for jobs were slim, and they were bitter. "How would you feel?" asked Werner G. Tiller, one of the dismissed engineers.[125] Robert Paetz, one of the members of von Braun's team, had to accept reduction in rank from GS–15 to GS–12, and then lost his job in the next RIF. He filed an age discrimination suit against the Center that was not settled until 1988, when the court upheld the Center's RIF procedure.[126]

Marshall's ordeal continued. On 27 January 1971 the Center learned that it would have to undergo still another RIF. President Nixon's budget for Fiscal Year 1972 called for a reduction of another 1,500 NASA employees, of which Marshall's share was anticipated to be 297.[127] In July the Center proposed a plan to OMSF for the separation of 241 people, hoping to meet the remaining quota through attrition. Headquarters reduced Marshall's quota in an effort to minimize the impact on ongoing programs, and on 16 August, the Center issued notices to 183 employees. Before executing the RIF, the Center was able to salvage 42 positions of experienced technical personnel, promising to cover those reductions through anticipated attrition. The Center dismissed 141 permanent employees through RIF action on 2 October.[128]

The following year, Marshall had to endure another RIF, the fourth in five years. In June 1972, the Center lost 131 employees to RIF proceedings, and another 90 to other causes. Its Civil Service manpower fell to 5,377. The average age of its employees had risen by three years since the first RIF.[129]

Losses devastated the Center. Contractor strength declined even faster than Civil Service manpower. Marshall had lost 65 percent of its peak total manpower resources by early 1972. Rumors circulated, including one that 1,000 Marshall employees would be transferred to Houston, and morale plunged.[130] The Center expected further reductions, and the ability to use attrition to effect reductions declined each year; RIFs would have to be larger in the future. Prospects were so grim that the Center began to consider deeper RIFs as a means to restore vitality through hiring.[131]

Rees feared that continued losses would destroy whatever remained of the arsenal system. "I strongly believe that we have now reached the minimum acceptable level in Civil Service employment at MSFC," he told Headquarters in December 1972. "We absolutely need a period of no further strength reduction in order that we can better assess our situation and rebalance our skills from attrition." He argued that the Nixon administration's philosophy of reductions would lead to a situation in which industry, rather than NASA, would chart the nation's future in space. Without preserving the technical skills of its engineers and scientists, the Agency would no longer be able to evaluate and monitor contractors.[132]

Marshall had not been the only installation affected by reductions, and tension between the Centers and Headquarters increased. NASA conducted an internal survey of attitudes of the Centers and Headquarters toward one another at a meeting of Center directors in the fall of 1972, focusing on the impact of "institutional aging." Center personnel complained about growing Washington bureaucracy, strangling red tape, declining Center autonomy, and failing communications. Headquarters criticized the Centers for shortcomings that reflected the impact of reductions. By far the most frequent criticism of the Centers was the lack of new talent coming in, a problem that Marshall had been battling since the 1968 RIF. A complaint about obsolete organization ("structured for yesterday's program, not today's") also targeted Marshall's dilemma.[133]

The appointment of a new NASA administrator offered little hope that Marshall's problems might be alleviated. James C. Fletcher took command in 1971 following the resignation of Paine and a brief interlude in which George Low served as acting administrator. A Republican businessman, Fletcher lacked influence in the administration, and could not sell space to the White House.[134] Marshall could expect little relief from an administrator who considered Civil

Servants less efficient than contractors.[135] Although Fletcher fought hard to preserve funding for the Shuttle Program, he accepted reductions in other programs to preserve the Shuttle. Cost cutting became paramount, and overall operations at the Center suffered.

Budget battles with Washington proved wearing to Marshall Director Rees. On 17 November 1972, he spoke to Center employees in Morris Auditorium in an address that amounted to his valedictory, for he would announce his retirement the following month. "We have gone through some trying times together," he told them, "but we have survived these stern and sometimes anguishing ordeals without any great impairment of our performance." He announced another reorganization, one more suited to a scaled-down Center and diversified scientific missions. He tried to put Marshall's ordeal in the best possible light, claiming that "nothing in the basic intracenter relationships has changed," and that "our in-house capability remains." But he acknowledged budget pressures, and concluded that the NASA had to "either find low cost routes to our objectives or these objectives will dry up or be reduced in scope to the point where our proud space program will wither and America's significant space achievements will be just a memory."[136]

An Outsider Takes the Reins: Rocco Petrone as Center Director

Rees announced in December that he would retire in January 1973, three months before his 65th birthday. Headquarters selected Dr. Rocco Petrone, head of the Apollo lunar program, to succeed him. Although Petrone had served with ABMA, he was the choice of neither Rees nor von Braun. Von Braun had worked with him when Petrone had been launch operations director at the Cape during Apollo, and considered him too parochial, more concerned with Kennedy's independence than with the program. Von Braun and Rees both preferred Lucas, then Marshall's

Dr. Rocco A. Petrone, Marshall Space Flight Center Director, 1973–1974.

technical director. Von Braun had told Lucas in 1968 that he wanted Lucas to become Center director. Both Rees and von Braun had expected Rees's tenure as director to last only two or three years, and that Lucas would then move up.[137]

Petrone, the husky son of Italian immigrants, had played football at West Point. He had served with the Army Corps of Engineers after leaving ABMA, and supervised construction of launch facilities at the Cape. He became launch operations director at Kennedy Space Center after resigning from the Army in 1967, and had been the director of the last six Apollo flights. One of his colleagues at Kennedy described him as hard working and hard to get along with, explaining that "Nobody crosses him. I mean nobody."[138]

Why had Headquarters sent an outsider to Marshall? The Center's trials were not yet over, and Washington believed an outsider could preside over further retrenchment dispassionately. Deputy Administrator Low, the Agency's highest-ranking official with long NASA experience, saw the need for further tightening. Kraft believed that Low wanted "somebody strong and very virile. Somebody that could raise hell and cut throats and that sort of thing. He wanted somebody like that and saw it in Petrone."[139]

Marshall's remaining members of von Braun's German team bore much of the burden of reductions, and it is not surprising that some believed they had been singled out. They considered Petrone a "hatchetman," sent by Headquarters to clean house. "He literally threw out the whole von Braun team out the door," claimed von Tiesenhausen, whose own situation was one of many wrenching stories. "I was not eligible for retirement at that time, so I was demoted, which was one of the blackest days of my life. My whole pride was attacked, because I had always thought I had done a good job," he recalled. Others went through similar experiences, and he remembered some being reduced four or five grades.[140]

NASA's austerity program became even more stringent during Petrone's brief stint as Center director. Nixon's budgets continued to reduce funding for space. Even as Petrone prepared to assume control of Marshall, one observer described the Agencywide impact of new budget proposals, predicting "There's going to be some blood letting."[141]

RIFs became an annual exercise. Marshall lost another 199 employees in 1973, 97 of them terminated under RIF proceedings. While other manned space flight centers also experienced reductions, none bore as much of the burden as Marshall, which had absorbed 81 percent of the personnel reductions in manned space flight since the mid-sixties. Marshall's personnel ceiling dropped to 4,564 in Fiscal Year 1974 as the Center experienced its sixth RIF in seven years.[142]

In fact, NASA had been examining the impact of aging on the Agency for several years.[143] Marshall, with a higher average age than other Centers, was again the focus of attention. "Because we had some people who had been in rocketry longer than some others and we had a lot of people coming up for retirement," recalled Lucas, "the average-age situation made us stand out."[144] An independent study cited NASA's attempts to counter "age creep" and to hire younger personnel, but found that some of the methods employed had not worked. "Over-RIFing"—cutting personnel to open slots for recruits—failed when successive RIFs forced Centers to relinquish the new positions. The study worried that RIFs slowed promotions, forced young people of promise out, and shunted others to less challenging jobs.[145] Huntsville's Germans were victims of the desperate attempts of a besieged Agency to renew itself.

That the Germans thought they had been singled out, even purged, was understandable. Many fixed the start of the decline of the German team at the time of von Braun's departure for Washington, for it seemed that without his dominating presence in Huntsville, Headquarters could move against the Germans with impunity. Von Braun's own fate had been part of the tragedy, for his job at Headquarters was disappointing, and with NASA's reduced funding under Nixon, it became virtually meaningless. He retired from NASA in 1972 to accept a position at Fairchild Industries.

"The system forced us out," concluded Walter Jacobi, who had to accept reduction from a position as a mid-level branch chief to a designer in the structures division. RIF rules, with their protection for American armed service veterans, seemed stacked against the Germans. They dominated Marshall management; if the Center was to develop new leadership in a contracting market, it had to provide opportunities for advancement. Jacobi's fellow Germans attributed the break-up of the team to petty jealousies in Washington, reduced national interest in space, changes in Marshall's mission and philosophy. Marshall's characteristic reticence may have contributed. Karl Heimburg claimed that in the last years, "too much time was spent waiting for Washington to tell us what to do. I think we were too obedient. If you always wait for an order, that is stifling."[146]

But Petrone's assignment was not just a slash-and-burn operation. Retrenchment also involved reorganizing the Center for new responsibilities. "Rocco came to Marshall to reorganize Marshall," according to James Kingsbury, who helped implement Petrone's plan. Headquarters sent an outsider because reorganization "was going to have serious impact on the senior management at the Center, and unless an outsider did it, the senior managers of the Center would not make significant impact on themselves."[147]

Thus despite the furor over lost jobs and damaged careers, Petrone's most lasting impact on Marshall was not his administration of RIFs, but a May 1974 Center reorganization. The plan centralized the Science and Engineering Directorate and restructured its laboratories, eliminating duplication of functions characteristic of Marshall's labs since their inception. Kingsbury, part of a five-man team that had worked on the plan for a year, explained that before reorganization "every laboratory was by and large self-sufficient. It had a little of every other lab in it." The Center liked to describe itself as the "Marshall team," but because of autonomy in the laboratories, it had really been more of a "Marshall league."[148] The changes, McDonough remembered, "stripped all the administrative functions out of the laboratories."[149] By reforming the laboratories, the Petrone reorganization undercut part of the old German and ABMA engineering system.

Laboratory reorganization also reinvigorated Marshall's matrix management system. The use of ad hoc, problem-solving teams drawing specialists from various labs had been used in the 1960s. But the imperial laboratories of the Saturn years had provided an alternative to such functional teams. Experts from one lab could work full-time on one project. With lab reform, personnel cuts, and diversification, however, multilab teams were necessary. "Matrix management had been talked about in the Apollo Era," Bob Marshall said, "[but now] matrix management had to happen." The changes also reinforced the rise of project offices relative to the laboratories. The labs acted as contractors to the project offices, providing technical services and support. Lab directors, rather than being the leaders as they had been in ABMA days, shared authority with project officers.[150]

The Petrone reorganization also signified the formal end of another Marshall practice, the arsenal system. Petrone announced that "The in-house capability to manufacture, inspect and checkout major hardware projects has been eliminated."[151] Kingsbury believed that the arsenal system had been a luxury of the Saturn boom and that the post-Apollo bust forced NASA to end it. The change, especially the loss of support contractors, he thought, forced Marshall's engineers to become less complacent and more self-reliant.[152] But most "old hands" lamented the loss and worried that the Center was less able to monitor contractors and achieve technical excellence. McDonough said that "we couldn't do anything anymore. Our shops went, our technicians went."[153]

Petrone implemented reorganization "parallel with the necessary reduction-in-force."[154] The Center mailed a thousand letters to notify employees of changes in position.[155] Simultaneous implementation of reduction and reorganization eased the turmoil of the most dramatic internal change in Center history. "The lab directors, by and large, were all new," Kingsbury explained. Since the older former lab directors had retired, "we didn't have a lot of trouble putting it into place."[156]

The Threat to Close Marshall

As reductions continued at Marshall, people inevitably began to wonder if the Center would survive. The question had arisen informally in earlier Headquarters discussions about the post-Apollo phasedown, and in the mid-1970s NASA reopened the issue for serious consideration. "There was a good, strong possibility that the Center could have been closed before the end of the seventies," recalled Lucas. "We came very near to it, nearer than most people know."[157] NASA twice conducted studies that considered closing Marshall: in 1975, under Fletcher; and again in 1977 when the Carter administration cut space funding during Robert A. Frosch's tenure as NASA administrator.

The challenge to Marshall's survival resulted from further threats to NASA manpower. By 1975, the Agency recognized that even if its budget remained constant, it would have to reduce Civil Service strength by 5,000 by 1979. In April 1975, Fletcher met with his staff to discuss realignment of the Centers in the face of new reductions. They concluded that "the reduction in Civil Service positions could be reached by closing a Center." Fletcher assigned E. S. Groo, associate administrator for Center operations, to develop a plan for reducing people, saving money, and realigning the Centers.[158]

For the next several months, Headquarters studied options for Center realignment. Groo and his staff, along with representatives of the Centers, debated the reassignment of tasks, reduction of personnel, and the feasibility of closing a Center. Ames, Lewis, Wallops, and JPL received scrutiny, but most attention focused on Marshall. The group developed a scenario for closing Marshall that anticipated phasing out space science, applications, and nuclear technology by 1978, and closing the Center in 1982. Marshall's Shuttle and Spacelab development would have transferred to Johnson and Kennedy, its space science research to Ames and Goddard, its smaller projects distributed throughout the Agency.[159]

Position papers formulated for the discussion of closing Marshall considered the Center's strengths and weaknesses, and showed insight into Marshall's problems. The committee wondered whether Marshall's "skill mismatches," old facilities, and its competition with Johnson for new programs met NASA's long-term needs. Reductions in resources for piloted vehicle development seemed likely, and without a major new program, the Center would likely have to be reduced even if it remained open. Constant reductions had inhibited the Center's future planning, but its "typically innovative" approaches were likely to benefit Shuttle development.

Closing Marshall would have serious implications for NASA's future. It would have been a "clear signal" that the Agency was not about to undertake ambitious missions such as space industrialization, sending men to Mars, or colonizing the Moon. NASA would have lost Marshall's capacity to develop large space systems.[160]

Groo decided that closing Marshall was neither practical nor feasible. Closure would have disrupted the Shuttle program. A required two-year phasedown was unworkable, particularly since Marshall facilities were needed for ongoing NASA programs. Too many programs required Marshall's capabilities; not only large lift vehicles, but the Space Station, space industrialization, and future piloted planetary exploration drew on the Center's talents. Marshall gave the Agency flexibility; with Goddard's workload near saturation, Marshall could absorb the overflow. Marshall would remain open.[161]

Marshall's respite was short-lived. When the Carter administration instituted more cuts to NASA's budget, the issue rose again, for as one Headquarters assessment noted: "Agency internal reactions are always aimed at closing MSFC whenever an institutional crisis occurs. They have few advocates."[162] Lucas, Center director at the time, recalled that "we set up what we called a 'mole-hole operation.' We had a few key people doing strategic planning in the basement determining how we could posture ourselves to move on. As a matter of fact, we had made the decision early in the 70s to diversify. . . . Had we not we would have been closed."[163] Again, the Center survived.

The Impact of Retrenchment

The decade from the mid-sixties to the mid-seventies had been extraordinarily difficult. Marshall descended from a major role in one of mankind's great scientific achievements to a fight for survival. In 1975, Marshall had 4,100 Civil Service employees. By 1978, the figure dropped to 3,760, less than half what it had been at peak a dozen years earlier. Other Centers were still growing when Marshall began to retrench, then experienced smaller cutbacks. In 1965, Houston's workforce was 57 percent as large as Marshall's; in 1975, 89 percent. Kennedy was 32 percent as large as Marshall in 1965, 55 percent in 1975.[164]

Dr. William R. Lucas, Marshall Space Flight Center Director, 1974–1986.

Retrenchment destroyed Marshall's attempts to increase minority employment. Compounding the recruiting impediments imposed by Alabama's negative image in civil rights was the fact that new employees were more vulnerable to RIFs. In 1975, only 2.6 percent of Marshall's personnel were minorities, the lowest of all NASA installations, at a time when NASA had increased minority employment to 6.8 percent. Marshall's minority employees were clustered at low-level positions. Fifty-five percent of the Center's minority employees did not have a college degree, compared to 41 percent of all employees.[165]

Morale at Marshall was low not only because of the constant threat of RIFs. Marshall ranked lowest of all NASA installations in 1975 in promotions and quality-within-grade increases. In 1974 and 1975, the Center still had the largest Civil Service workforce in NASA, yet its employees received fewer promotions than any other installation. In 1974, only eight-tenths of one percent of Marshall employees received promotions, compared to the NASA average of 11.2 percent. Marshall's workforce was equal to the oldest in NASA, but ranked below the NASA average in grade, and below the other two manned space flight centers in percentage of salary increases.[166]

NASA underwent a painful transition after Apollo, and Marshall felt the impact disproportionately. The politics of budgets drove NASA's agenda. The contrast with the 1960s was telling. As Lucas explained, during Apollo, the performance (landing on the Moon) and the timetable (by the end of the decade) "were both fixed items. The variable was funds. The schedule and performance were fixed. They were not variables. In the seventies, the funds were the only things that were fixed. The schedule and the performance were the variables. That is the best way to waste money that I know of, to stretch out the schedules."[167]

Dr. Lucas (center) in conference.

The nature of the Center had changed by the mid-seventies. The arsenal system, the heart of the von Braun approach to development, fell victim to small budgets and demands from the private sector aeronautics industry. "The in-house capability of building things was given up with great reluctance. In retrospect, that weakened the Center," Lucas remembered. The arsenal system "is no longer practiced and industry doesn't want it to be practiced because they want to do all the work. There is merit in that argument. I don't knock it. But it does say that an agency of the government is more nearly a captive of industry than they might have otherwise been."[168]

Marshall's employees became monitors of contractors, rather than "dirty hands" engineers. "There was paperwork to do rather than technical work," according to Walter Jacobi. Bernard Tessmann, former deputy director of the Astronautics Laboratory, retired in 1972 because he did not "want to be a paperboy and push paper."[169] The transition affected the entire Agency. NASA became more centralized, more bureaucratic. One historian observed that "Increasing centralization, contracting out and the natural forces of aging have tipped the balance within NASA in favor of the forces of organization as opposed to the forces supporting the original NASA culture."[170]

The Center nonetheless had reasons for optimism that transcended its mere survival as an institution. Marshall's diversification had done more than allow the Center to survive; the Huntsville Center was in the forefront of new NASA work in space science, and continued to be one of the two largest installations for development of piloted space projects. Even during the most arduous period of retrenchment, individuals at Marshall made major contributions to the nation's space program. In 1975, only Houston exceeded the Center in the percentage of employees receiving sustained superior performance awards.[171] Marshall emerged from its transition a very different organization than it had been a decade earlier, but it was still at the center of the American space effort.

Nonetheless, Marshall's transition had affected the Center in ways that would not become apparent for years. One engineer reflected that cuts went deeper than the fat and were "so austere that I think we went into the red meat."[172] At the time, attention focused on space spectaculars to which Marshall contributed: lunar landings, *Skylab*, Apollo-Soyuz—triumphs that eclipsed institutional developments. Decreasing budgets, pressure from aerospace firms to increase contracts, and the centralization of NASA decision-making precipitated traumatic changes that transformed the Center. When NASA encountered problems in major programs in the eighties, people looked for technological explanations and individuals to blame. The agony and the austerity of Marshall's transition had faded from public memory. But these institutional changes were the foundation of Marshall's future.

1 Roger E. Bilstein, *Stages to Saturn: A Technological History of the Apollo/Saturn Launch Vehicles* (Washington, DC: National Aeronautics and Space Administration Special Publication–4206, 1980), pp. 355–57.
2 "Washington Roundup" clipping, MSFC General 1959–1969 folder, NASA HQ History Archive.
3 Paul E. Cotton, "An Analysis of NASA In-House Manpower Management: Manpower Analysis Comment Copy," 10 February 1976, p. B–1–7, Management Studies—1970s Notebooks, NASA Headquarters History Archive; Arnold S. Levine, *Managing NASA in the Apollo Era* (Washington, DC: National Aeronautics and Space Administration Special Publication–4102, 1982), p. 134.
4 "MSFC Center to Employ 9,500 Within Two Years," 22 May 1963, Marshall SFC folder, NASA HQ History Archive.
5 "Von Braun Briefing Book, Teague Committee," February 1967, Fiche No. 317, MSFC History Office.
6 Cited in U.S., House, 90th Cong., 1st sess., "Apollo Program Pace and Progress: Staff Study for the Subcommittee on NASA Oversight of the Committee on Science and Astronautics," 10 March 1967, pp. 1130–31.
7 "AAP Means New Jobs Here," *Huntsville Times*, 19 December 1965.
8 MSFC Press Release No. 66–73, 5 April 1966, Post-Apollo/Manned Interplanetary Flight folder, NASA HQ History Archive; Jonathan Spivak, "NASA Sees A Cutback if Post-Apollo Projects Aren't Approved Soon," *Wall Street Journal* (18 July 1966), p. 10.
9 Von Braun referred to the Apollo Extensions Systems (AES) program, later the Apollo Applications Program (AAP). Von Braun to Teague, 26 April 1965, Box 005–22, Shuttle Series Chronological File, JSC History Office.
10 Spivak, "NASA Sees A Cutback," *Wall Street Journal* (18 July 1966), pp. 1, 10.
11 Roy H. Beaton, Missile and Space Division, General Electric, to George E. Mueller, 12 October 1966, Fiche No. 317, MSFC History Office.
12 Ernst Stuhlinger to Mike Wright, 3 January 1991, comments on chapter draft, MSFC History Office; Levine, p. 245; W. David Compton and Charles D. Benson, *Living and Working in Space: A History of Skylab* (Washington, DC: National Aeronautics and Space Administration Special Publication–4208, 1983), pp. 22–23.
13 Ernst Stuhlinger, "A Proposed Long-Range Supporting Development Program for MSFC," 17 December 1964, Record File, Stuhlinger Papers, Alabama Space and Rocket Center Archival Library, Huntsville, Alabama.
14 Ray Kline, "Minutes of Combined Staff and Board Meeting," 16 August 1966, Von Braun Daily Journal, NASA HQ History Archive.
15 Compton and Benson, p. 5.
16 Von Braun Daily Journal, 11 May and 26 May 1966, NASA HQ History Archive.
17 Von Braun Daily Journal, 14 July 1966, NASA HQ History Archive.
18 Joe Loftus, OHI by AJD and SPW, 13 July 1990, Johnson Space Center, Houston, Texas, p. 10.
19 Levine, p. 134.

20 Stuhlinger to Mike Wright, 3 January 1991, comments on chapter draft, MSFC history office.

21 Levine, p. 172.

22 Chris Kraft, OHI by AJD and SPW, 11 July 1990, Johnson Space Center, Houston, Texas, p. 33.

23 Von Braun Daily Journal, 11 May 1966, NASA HQ History Archive.

24 Compton and Benson, *Living and Working in Space*, p. 52; Ray Kline, "Minutes of Combined Staff and Board Meeting," 16 August 1966, Von Braun Daily Journal, NASA HQ History Archive.

25 Von Braun Daily Journal, 18 May and 15 July 1966, NASA HQ History Archive.

26 Robert W. Smith, *The Space Telescope: A Study of NASA, Science, Technology, and Politics* (Cambridge: Cambridge University Press, 1989), pp. 116–17; Eberhard Rees, "Briefing on MSFC Organization and Management," House Oversight Committee Staff Study on Apollo Program Management, 18 October 1968, Huntsville Alabama, p. 8, Fiche No. 1724, MSFC History Office; Linda Neuman Ezell, *NASA Historical Data Book, Volume II: Programs and Projects 1958–1968* (Washington, DC: National Aeronautics and Space Administration Special Publication–4012), p. 339.

27 Stuhlinger to J. Shepherd, 29 February 1968, Historical (Miscellaneous) folder, Boxes 12 and 13, Stuhlinger Papers, Alabama Space and Rocket Center Archival Library, Huntsville, Alabama.

28 Bruce Murray, *Journey into Space: The First Three Decades of Space Exploration* (New York: W. W. Norton & Company, 1989), p. 213.

29 Von Braun, memo to MSFC employees, 13 August 1965, Box 065–65, Apollo Program Chronological Files, JSC History Office.

30 Cotton, p. B–1–6; Levine, pp. 127–28.

31 Cited in Levine, p. 127. Harry Gorman represented Marshall on the committee.

32 Ray Kline, "Comments at Session of NASA Manpower Committee on 30 August 1966," 31 August 1966, Von Braun Daily Journal, NASA HQ History Archive.

33 Quote from Levine, p. 129. The task force submitted its report, entitled "Considerations in the Management of Manpower in NASA," on 8 September 1967. Cotton, pp. B–1–7 to B–2–6.

34 Levine, pp. 127–30; Lodge 1858, American Federation of Government Employees, et al., Plaintiffs vs Administrator, National Aeronautics and Space Administration, et al., Defendants, Civil Action No. 3261–67, United States District Court for the District of Columbia, Final Summary and Judgment Order, 12 August 1976 (hereinafter Waddy Case, Final Summary Judgment and Order), pp. 4–5, Fiche No. 1620, MSFC History Office.

35 "Marshall Space Flight Center Reduction," 23 April 1968, Teague (1968 NASA Oversight Committee) folder, NASA HQ History Archive.

36 Von Braun, memo to MSFC employees, 23 October 1967; "Washington Roundup" clipping, MSFC General 1959–1969 folder, NASA HQ History Archive.

37 Von Braun, Presentation to MSFC Employees, 29 November 1967, Fiche No. 1626, MSFC History Office.

38 Judge Alexander Holtzoff rendered the decision. Waddy Case, Final Summary and Judgment Order, pp. 1–3; "FY–68 RIF at MSFC," Fiche No. 868, MSFC History Office.
39 "Marshall Space Flight Center Reduction."
40 Waddy Case, Final Summary Judgment and Order, Fiche No. 1620, MSFC History Office.
41 *Ibid.*; "FY–68 RIF at MSFC;" "Adverse Effects of RIF (Based on MSFC 1967–68 RIF)," Reduction-in-Force folder, George M. Low Papers, NASA HQ History Archive; Levine, p. 136.
42 "Adverse Effects of RIF (Based on MSFC 1967–68 RIF)."
43 Huntsville area in-house prime contractor equivalent manpower reached a high of 9,785 in Fiscal Year 1965, and fell to only 3,165 in Fiscal Year 1969. Four years later, it was reduced nearly by half again, to 1,660, and then dropped to only 920 in Fiscal Year 1974: less than ten percent of what it had been less than a decade earlier. "Marshall Space Flight Center Huntsville Area Prime Contractor Manpower," Fiche No. 836, MSFC History Office. Direct prime contractor personnel dropped from a high of 42,000 to 19,600 in Fiscal Year 1969. Rees, "Briefing on MSFC Organization and Management," p. 15.
44 Harold B. Finger, Associate Administrator for Organization and Management, to George E. Mueller, memo, 9 August 1968, MSFC Organization 1968 folder, Box 14, MSFC Director's Files, Federal Records Center, Atlanta.
45 Finger to Mueller, 9 August 1968; Mueller to von Braun, 30 August 1968, MSFC Organization 1968 folder, Box 14, MSFC Director's Files, Federal Records Center, Atlanta.
46 Von Braun to Mueller, 17 September 1968, MSFC Organization 1968 folder, Box 14, MSFC Director's Files, Federal Records Center, Atlanta.
47 Joe Loftus, OHI by AJD and SPW, 13 July 1990, Johnson Space Center, Houston, Texas, p. 10.
48 George von Tiesenhausen, OHI by AJD and SPW, 29 November 1988, Huntsville, Alabama, p. 38.
49 Kraft, OHI, p. 21.
50 Levine, p. 261.
51 Rees, "Briefing on MSFC Organization and Management," p. 5.
52 Webb, never an advocate of agencywide planning, established the Planning Steering Group (PSG) under the direction of Homer Newell early in 1968, apparently in response to outside pressure. It was meant to coordinate planning between OSSA, OMSF, and the centers. but had a mixed record. Von Braun's assessment of PSG was that it "succeeded in identifying the interesting areas in applications and science where the main thrust ought to be, but since it dealt with people of many different disciplines—astronomers, meteorologists and math majors and people like that—the result was about as incoherent as the science community that was involved in getting the story together." Arthur L. Levine, *The Future of the U.S. Space Program* (New York: Praeger Publishers, 1975), pp. 120–22; Von Braun, OHI by John Logsdon, 25 August 1970, p. 7, NASA HQ History Archive; Mueller to Robert R. Gilruth, 7 August 1968, Box 070–24, Apollo Program Chronological Files, JSC History Office.

53 Cited in transcript, "ABC's Issues and Answers," 15 March 1970, Box 005–65, JSC Shuttle Program Files, JSC History Office.

54 Von Braun, discussion with House Oversight Subcommittee on "Future Manned Space Flight Effort and Launch Production Rate," 20 October 1969, Fiche No. 1738, MSFC History Office.

55 Von Braun to Mueller, 17 September 1968, MSFC Organization 1968 folder, Box 14, MSFC Director's Files, Federal Records Center, Atlanta.

56 "MSFC Must Realign Work Force, Von Braun Says," *Huntsville Times*, 3 November 1968.

57 "Major Reorganization Underway at NASA-Marshall," *Space Daily* (10 December 1968), p. 171.

58 W. R. Lucas, OHI by AJD and SPW, Huntsville, Alabama, 20 November 1990, p. 1.

59 Lucas, OHI, 20 November 1990, p. 1.

60 Bob Marshall, OHI by SPW, Huntsville, AL, 8 September 1990, p. 23.

61 Marshall, OHI, pp. 16–23; Bill Sneed, OHI by SPW, Madison, AL, pp. 18–20; W. R. Lucas, "Initial Documentation for Establishment of Program Development," 20 December 1968, Fiche No. 221, MSFC History Office.

62 George McDonough, undated comments to authors on chapter draft, MSFC history office.

63 William Snoddy, Terry Sharpe, Herman Gierow, OHI, by AJD and SPW, Program Development Office, MSFC, 30 October 1990.

64 George McDonough, undated comments to authors on chapter draft, MSFC history office (for comment on projects developing in the labs).

65 Von Braun, "Adjustment to Marshall Organization, Announcement #4, 16 January 1969, Apollo Program Chronological File, Box 70–62, JSC History Office. Other changes included the appointment of David Newby to the new position of director of Administration and Technical Services; renaming of Industrial Operations as Program Management under Major General Edmund O'Connor; and renaming Research and Development Operations as Science and Engineering, under Hermann Weidner.

66 Marshall's total appropriations were $1,618.7 million in Fiscal Year 1965, and $800.3 million in Fiscal Year 1969. Von Braun, Preliminary Draft, Introductory Remarks before the Subcommittee on Manned Space Flight of the Committee on Science and Astronautics, U.S. House of Representatives, 6 March 1970 (prepared for delivery, but visit was canceled).

67 "MSFC Civil Service Manpower Summary," in "Backup Book: 1973 Hearings before the Subcommittee on Manned Space Flight," 2 March 1972, Fiche No. 1772, MSFC History Office.

68 Von Braun to Mueller, 21 January 1969, Box 070–63, Apollo Program Chronological Files, JSC History Office.

69 Chris Kraft, OHI by AJD and SPW, 11 July 1990, Johnson Space Center, Houston, Texas, pp. 16, 17.

70 William R. Lucas, OHI, 20 November 1990, pp. 12–13.

71 George M. Low, Interviews, 9 January, 14 January, and 4 February 1969, "From STG to Apollo," Center History Discussions File, JSC History Office.

72 Ernst Stuhlinger to Mike Wright, 3 January 1991, comments on chapter draft, MSFC history office.

73 George von Tiesenhausen, OHI by AJD and SPW, 29 November 1988, Huntsville, Alabama, p. 12.

74 Joseph J. Trento, *Prescription for Disaster* (New York: Crown Publishers, 1987), p. 94.

75 Terry H. Sharpe and Georg von Tiesenhausen, "Internal Note: Integrated Space Program, 1970–1990," 10 December 1969, MSFC.

76 J. P. Downs, W. B. Thompson, and J. E. Volonte, "Rationale for an Integrated Manned Space Flight Program," 31 July 1967, UAH History Files, File #169.

77 Von Braun, OHI by John Logsdon, 25 August 1970, p. 8, NASA HQ History Archive.

78 Ernst Stuhlinger to Mike Wright, 3 January 1991, comments on chapter draft, MSFC history office.

79 Levine, *Managing NASA in the Apollo Era*, pp. 260–61.

80 Mueller to von Braun, 11 September 1969, Drawer 1, General Files, General Missions and Future Trends File, Ordway Collection, Alabama Space and Rocket Center Archive, Huntsville, Alabama.

81 Lawrence C. Falk, "Huntsville Boom May Be Ending with Apollo," *Birmingham Post-Herald*, 24 October 1968; "Slowdown in Space Programs: Its Impact on the Southeast," Monthly Review 54(May 1969), 61.

82 Elise Hopkins Stephens, *Huntsville: A City of New Beginnings* (Woodland Hills, California: Windsor Publications, 1984), pp. 113, 120.

83 Bruce Biossat, "Rocketville," *Washington News*, 25 March 1969.

84 Jim Maloney, "Huntsville Takes Cut, Turns to Light Industry," *The Houston Post*, 30 March 1970.

85 Bruce Biossat, "Rocketville," *Washington News*, 25 March 1969.

86 Von Braun wondered if Congress would consider any large programs, saying "The question is how bullish can you get in a bear market." Von Braun, OHI by John Logsdon, 25 August 1970, pp. 13–14, NASA HQ History Archive.

87 Cited in Jack Hartsfield, "About Leaving: Von Braun's Feelings are Mixed," *Huntsville Times*, 3 February 1970.

88 Lucas, OHI, 20 November 1990, p. 21.

89 Jack Hartsfield, "About Leaving: Von Braun's Feelings are Mixed"; "MSFC, MSC to Share Shuttle Work," *Huntsville Times*, 3 February 1970.

90 Erik Bergaust, *Wernher von Braun* (Washington, DC: National Space Institute, 1976), pp. 445–46; Carolyn Maddux, "Patience in the Rain," *Huntsville Times*, 25 February 1970; "Wernher von Braun: A Tribute," *Huntsville Times*, 22 February 1970.

91 Bob Marshall, OHI by SPW, 29 August 1990, Huntsville, Alabama, p. 1.

92 Georg von Tiesenhausen, OHI by AJD and SPW, 29 November 1988, Huntsville, Alabama, p. 9.

93 *Ibid.*

94 Phillip K. Tompkins, "Organization Metamorphosis in Space Research and Development," in *Communication Monographs* 45(June 1978), p. 116.

95 Bob Marshall, OHI, p. 4.

96 Tompkins, p. 116.

97 Georg von Tiesenhausen, OHI by AJD and SPW, 29 November 1988, Huntsville, Alabama, p. 38.
98 Bob Marshall, OHI, p. 2.
99 Chris Kraft, OHI by AJD and SPW, 11 July 1990, Johnson Space Center, Houston, Texas, p. 34.
100 Lucas, OHI, 20 November 1990, p. 12.
101 Memorandum, Walter T. Bonney to Glennan, 30 September 1958, NASA-Army (ABMA) folder, NASA HQ History Archive; Dryden to Lieutenant General Arthur G. Trudeau, 25 February 1959, Krafft Ehricke—The Peenemünde Rocket Center folder, Vertical file, JSC History Office.
102 Another example from this era was an article entitled "Wernher von Braun: The Ex-Nazi Who Runs Our Space Program," which had a subtitle reading: "His rockets may be aimed at the moon now, but once they destroyed London—and he was trying for New York!" *Confidential* 8 (October 1965), 16–17, 42–43.
103 Webb to von Braun, 20 December 1963, Von Braun Miscellaneous Documents folder, NASA HQ History Archive.
104 Cited in Trento, pp. 89–90.
105 Cited in Peter Cobun, "A Footnote is Enough," *Huntsville Times*, 8 August 1976.
106 Georg von Tiesenhausen, OHI by AJD and SPW, 29 November 1988, Huntsville, Alabama, p. 12.
107 Webb to von Braun, memo, 17 December 1966; Von Braun to Webb, 18 January 1967 (note attached: "Apparently not sent. vB met with Webb on January 18, '67 and presumably discussed the matter. Mitch Sharpe."), Box 067–62, Apollo Program Chronological Files, JSC History Office. At the bottom of the memo, Ray Kline wrote: "Von Braun is worried about contractors relaxing if too much pessimism gets to the press." But the tone of von Braun's response to Webb was entirely conciliatory.
108 Bob Marshall, OHI, p. 1.
109 *Ibid.*, pp. 1–2.
110 Lucas, OHI, 20 November 1990, p. 24.
111 Ernst Stuhlinger, OHI by AJD and SPW, 24 April 1989, Huntsville, Alabama, p. 23.
112 Georg von Tiesenhausen, OHI by AJD and SPW, 29 November 1988, Huntsville, Alabama, p. 19.
113 Konrad Dannenberg, OHI by AJD and SPW, 1 December 1988, Huntsville, Alabama, pp. 19–20.
114 Rees, "Project and Systems Management," presented at XVI CIOS World Management Congress, Munich, Germany, 25 October 1972, Speeches and Articles box, Rees Papers, Alabama Space and Rocket Center Archival Library, Huntsville, Alabama.
115 "Rees Looking to Tough Task, *Huntsville Times*, 28 January 1970.
116 James Fre, "Space Budget Slash Won't Hurt Huntsville," *Birmingham News*, 2 February 1970; "Rees' Letter Kills Rumors," *Huntsville Times*, 26 May 1970.
117 "FY–71 RIF at MSFC," Fiche No. 868, MSFC History Office; Boyd C. Myers, II to George M. Low, 9 October 1970, Reduction-in-Force folder, Low Papers, NASA HQ History Archive.

118 4.6% of Marshall's employees were affected, compared to 12.4% of those at Headquarters, 47.1% of the small NASA office in Pasadena, 8.3% at MSC, 5.3% at Kennedy, and 4.7% at Ames. Only MSC had more personnel (356 to 275 for Marshall and 247 at Headquarters) affected, however. Myers to Low, 9 October 1970.

119 Levine, *Managing NASA in the Apollo Era*, p. 137.

120 Art Sanderson, OHI by AJD, 20 April 1990, Huntsville, Alabama, p. 10; Lucas, OHI, 20 November 1990, p. 14.

121 "FY–71 RIF at MSFC," Fiche No. 868, MSFC History Office.

122 Marshall was not the only Center to use the technique to control RIFs. Kennedy was the worst abuser, and had created 46 competitive levels for 46 GS–6 secretaries, making it possible to select individuals for dismissal without fear of reprisal. Marshall had 2,800 competitive levels for its 5,600 employees. John Cramer, "12 Employees Lose Gamble Against NASA Rules," *Washington Evening Star and Daily News*, 28 August 1972; John Cramer, "NASA's 'Evil Invention' May Be on Way Out," *Washington Evening Star and Daily News*, 22 September 1972.

123 Myers to Low, 9 October 1970.

124 "FY–71 RIF at MSFC," Fiche No. 868, MSFC History Office; Myers to Low, 9 October 1970; "Over Four Years: Space, Defense Job Losses Total 11,000 in Huntsville," *Birmingham News*, 10 December 1970.

125 John Noble Wilford, "NASA Layoffs Hit von Braun Team," *New York Times*, 3 September 1970; National Academy of Public Administration Foundation, "Report of the Ad Hoc Panel on Attracting New Staff and Retaining Capability during a Period of Declining Manpower Ceilings," June 1973, p. 19, Management Studies—1970s Notebooks, NASA Headquarters History Archive; Pat Houtz, "At Huntsville: Emotions Run High at NASA Cutback," *Birmingham News*, 23 August 1970.

126 "Robert Paetz v. United States of America," Case No. CV 84–HM–5231–NE, in the District Court of the Northern District of Alabama, Northeastern Division, May 1988, Fiche No. 324, MSFC History Office.

127 Jack Hartsfield, "Reduction in Force Anticipated by Rees," *Huntsville Times*, 29 January 1971.

128 "FY–72 RIF at MSFC," Fiche No. 868, MSFC History Office.

129 "Dr. Petrone's Backup Information, MSF Subcommittee Visit," 1 February 1974, Fiche No. 874, MSFC History Office. The average age of Marshall employees rose from 40.4 in June 1968 to 43.4 in June 1972.

130 The story appeared in the *Houston Post*, attributed to a MSC source, but was denied by Houston, Marshall, and Headquarters. Rees to Marshall employees, 13 March 1972, Fiche No. 1183, MSFC History Office.

131 "Manpower Situation" in "Backup Book, 1973 Hearings before the Subcommittee on Manned Space Flight," 2 March 1972, MSFC History Office.

132 Rees to Low, memo, 2 December 1972, Management Development, 1972–73 folder, NASA HQ History Archive.

133 George M. Low to All Attendees at Center Directors' Meeting, 11–12 September 1972, memo re: Bruce Lundin's Questionnaire, 22 September 1972, Personnel (1970–75)

folder, NASA HQ History Archive. Headquarters also criticized the centers for having a relationship with other agencies that was more adversarial than cooperative, which may have reflected Headquarters sensitivity to center rivalry, although in context it is not clear whether this referred only to outside agencies.

134 Trento, p. 105.

135 See, for example, Harvey W. Herring, "Meeting Record," 24 March 1975, Management Studies, 1970s Notebooks, NASA Headquarters History Archive.

136 "Dr. Rees' Text for Director's Briefing," 17 November 1972, Director's Briefing 11–17–72 folder, Speeches and Articles 1959–1972 box, Rees Papers, Alabama Space and Rocket Center, Huntsville, Alabama.

137 Barry Casebolt, "Rees Planning to Stay in Huntsville," *Huntsville Times*, 24 December 1972; Von Braun, telephone conversation with Kurt Debus, Von Braun Daily Journal, NASA HQ History Archive; Lucas, OHI, 20 November 1990, p. 1.

138 Barry Casebolt, "Petrone Selected New Chief at MSFC," *Huntsville Times*, 21 December 1972.

139 Kraft, OHI, p. 27.

140 Von Tiesenhausen, OHI, pp. 10–11.

141 Thomas O'Toole, "White House Said Reducing Space Budget," *Huntsville Times*, 4 January 1973.

142 "Dr. Petrone's Backup Information, MSF Subcommittee Visit," 1 February 1974, Fiche No.874, MSFC History Office.

143 See, for example, Boyd C. Myers, II to Low, memo re: "Aging of the Work Force," 7 July 1972; and Low to All Attendees at Center Directors' Meeting, 11–12 September 1972, memo re: Bruce Lundin's Questionnaire, 22 September 1972, Personnel (1970–75) folder, NASA HQ History Archive.

144 William R. Lucas, OHI, AJD and SPW, 19 June 1989, Huntsville, Alabama, p. 36.

145 National Academy of Public Administration Foundation, "Report of the Ad Hoc Panel on Attracting New Staff and Retaining Capability during a Period of Declining Manpower Ceilings," pp. 18–25.

146 Cited in Peter Cobun, "A Footnote is Enough," *Huntsville Times*, 8 August 1976.

147 James Kingsbury, OHI by SPW, 22 August 1990, Madison, Alabama, p. 28.

148 Kingsbury, OHI, 22 August 1990, pp. 28–29.

149 George McDonough, OHI by SPW, MSFC, 20 August 1990, p. 7.

150 McDonough, OHI, pp. 14–15; Marshall, OHI, pp. 24–25, 32; Kingsbury, OHI, pp. 20–22.

151 Petrone, "Memorandum to All Employees," MSFC, 5 March 1974.

152 Kingsbury, OHI, pp. 28–30.

153 McDonough, OHI, pp. 18.

154 *Ibid.*

155 "Marshall Reorganization Plan Now Set for Implementation," *Huntsville Times*, 6 March 1974.

156 Kingsbury, OHI, 22 August 1990, p. 31.

157 Lucas, OHI, 19 June 1989, p. 34.

158 Henry E. Clements, "Meeting Record," 22 April 1975, Management Studies—1970s Notebooks, NASA Headquarters History Archive. Attendees included Fletcher, Low, and

Petrone, who had moved to the Headquarters staff after William Lucas assumed office as center director at MSFC.

159 "MSFC," Reduction Plans, 1975, Management Studies—1970s Notebooks, NASA Headquarters History Archive.

160 "MSFC," Reduction Plans, 1975; "Second Institutional Assessment Report to Dr. Fletcher, Dr. Low," 11 September 1975, Management Studies—1970s Notebooks, NASA Headquarters History Archive.

161 Dick Wisniewski to Ray Kline (with attached position paper), 3 May 1976, Marshall SFC Correspondence folder, NASA Headquarters History Archive.

162 "Assessment: Marshall Space Flight Center," in "Assessment of NASA Centers," 13 September 1977, p. M–3; "Second Institutional Assessment Report to Dr. Fletcher, Dr. Low," 11 September 1975, Management Studies—1970s Notebooks, NASA Headquarters History Archive.

163 William R. Lucas, OHI, 20 November 1990, p. 3.

164 See Appendices.

165 Marshall Space Flight Center, "Strength Distribution Report, FY 1973 to FY 1988," Fiche No. 1290, MSFC History Office; NASA Office of Personnel, "NASA, The In-House Work Force: A Report to Management," 1975, Management Studies—1970s Notebooks, NASA Headquarters History Archive.

166 In 1974 and 1975 combined, Marshall had 141 promotions; even Wallops, with one-third the number of employees as Marshall, had 142 promotions. NASA Office of Personnel, "NASA, The In-House Work Force: A Report to Management," 1975.

167 Lucas, OHI, 20 November 1990, p. 10.

168 Lucas, OHI, 20 November 1990, pp. 28–29.

169 Cited in Peter Cobun, "A Footnote is Enough," *Huntsville Times*, 8 August 1976.

170 Howard E. McCurdy, "The Decay of NASA's Technical Culture," *Space Policy* (November 1989), 309.

171 NASA Office of Personnel, "NASA, The In-House Work Force: A Report to Management," 1975.

172 Bill Sneed, OHI by SPW, 15 August 1990, Huntsville, Alabama, p. 7.

Chapter VI

Skylab: Competition and Cooperation in Human Systems

Like many Marshall people, Wernher von Braun had dreamed of building spacecraft for human flight to the planets since his youth. The dream was so strong that as director of Marshall he sought adventures analogous to space conditions. Funded by a National Science Foundation grant in 1966, von Braun and Ernst Stuhlinger, chief of Marshall's Space Science lab, took Robert Gilruth and Maxime Faget of the Manned Spacecraft Center (MSC) on an expedition to Antarctica. The four space officials experienced the hostile environment, toured scientific installations, and examined equipment, learning lessons that could help NASA. Mixing research and pleasure, NASA's top officials walked around the South Pole, orbiting the earth every five seconds.[1]

The expedition symbolized new directions for Marshall in the late sixties and early seventies, revealing its diversification from rocketry into human spacecraft and its new intimacy with Houston's Manned Space Center. The diversification emerged because Marshall had started work on the Saturn rockets long before NASA had settled Apollo plans and so had a headstart on its part of the lunar landing mission. By the late sixties Marshall needed new challenges. As von Braun told Congress, the Saturns had closed the "missile gap" but now NASA suffered from a "mission gap."[2]

NASA recognized that Marshall needed new work and that Houston was still busy with Apollo. The Apollo fire had delayed Houston's work on the Apollo spacecraft; lunar mission planning and operations continued to be major tasks. Accordingly NASA Headquarters officials, especially George Mueller, head of the Office of Manned Spacecraft Flight, encouraged Marshall to develop America's first Space Station.

Marshall's diversification into human spacecraft engineering, however, led to competition with the MSC. Houston officials worried that in an era of diminishing resources Marshall's gains in new projects would mean Houston's losses. Consequently, *Skylab* planning and preliminary design activities led to considerable controversy and in-fighting. NASA sought an effective division of labor and eventually found beneficial forms of competition and cooperation that helped make *Skylab* a scientific and engineering success. Dramatic accomplishments came when Center personnel helped solve problems with *Skylab*'s defective micrometeoroid shield and effectively managed the workshop's orbital decay.

Diversifying into Human Spacecraft

Skylab emerged from the Marshall Center's quest for post-Apollo work. The Center was, as the official *Skylab* history has suggested, "a tremendous solution looking for a problem."[3] Marshall's search for new business would lead not only to *Skylab* but also to new, sometimes competitive, relationships between the NASA Centers.

Building a Space Station had been an old dream for many at NASA, and Marshall people had envisioned various concepts. Von Braun presented designs for Space Stations in the 1940s and in his *Collier's* articles in 1952. Hermann H. Koelle in 1951 also sketched plans, and in 1959 with Frank Williams helped draft ABMA's Project Horizon report which suggested using a "spent stage" as an orbiting workshop.

The idea of outfitting a spent rocket stage as a Space Station had charmed the Germans since Peenemünde because on an orbital mission, the final rocket stage went into orbit with the payload. From the beginning of the Saturn project, Ernst Stuhlinger recalled, von Braun had talked of the spent stage concept as a preliminary step to a sophisticated Space Station. And of course von Braun and the Center's laboratory chiefs had initially favored the earth orbital rendezvous mode for Apollo in order to develop an "orbital facility" and ensure the race to the Moon led to advanced missions.[4]

The Douglas Aircraft Company, a contractor building the Saturn S–IV stage under Marshall's supervision, shared enthusiasm for a spent-stage station. The company wanted to get into the manned spacecraft business and had built a mock-up spent stage station for the London *Daily Mail* Home Show in 1960. In November 1962 Douglas presented Marshall with an unsolicited plan for such a craft. The Center's Future Projects Office, managed by Koelle and Williams, researched the idea, and a study contract with North American Aviation continued the work. By March 1965 Marshall had begun detailed studies of an empty S–IVB stage workshop.[5]

NASA Headquarters in the early 1960s developed the Apollo Extensions Support Study to investigate how Apollo technology could be used for other purposes. The study incorporated various Space Station concepts proposed by the military and other NASA Centers, including the Langley Research Center's work on the Manned Orbiting Research Laboratory.[6]

But for several reasons NASA's post-Apollo planning was, as one historian has said, "pedestrian, even timid." External problems constrained the Agency. Unlike the Apollo program, no presidential directive defined a follow-up mission. By the mid-sixties, presidents and congressional leaders were preoccupied with war and welfare rather than space. NASA administrators worried that beginning an expensive new project while Apollo was still underway could lead to underfunding of both efforts.[7] Constricted support restrained Agency ambitions for a new project like a Space Station.

Agency politics also inhibited planning. Without an external directive, the Agency had to choose post-Apollo goals. In NASA's decentralized structure, the field Centers had different specialties and interests, but had to agree for plans to proceed. Marshall's plans, however, would realign Center roles. If Marshall converted a spent rocket stage into a manned station, it would encroach on the MSC's turf in manned spacecraft.[8] Marshall managers explicitly recognized that their plans required their entering competition with Houston in this territory.[9] Not surprisingly Houston resented Marshall's intrusion. As Chris Kraft recalled, Houston believed that being "in charge of manned space flight" was their "birthright" and so "whenever Marshall Space Flight Center tried to penetrate that part of manned space flight, I think it was felt as a competitive move." Faget thought they were "always trying to get into our business from the very start."[10]

To overcome Houston's qualms, Marshall needed an influential sponsor in NASA Headquarters and found one in Mueller. As chief of Manned Space Flight, Mueller had several reasons for becoming Marshall's ally. He wanted to use Apollo technology and teams to promote space science, maintain public attention on space flight, and provide a transition between the lunar landings and later missions. He also hoped to help Marshall avoid crippling losses in personnel and keep the Agency's team together through the end of Apollo.[11]

In August 1965 Mueller established the Apollo Applications Program (AAP) Office in the Office of Manned Space Flight (OMSF). The centerpiece of Apollo Applications was Marshall's spent stage. In a classic case of what political scientist Howard McCurdy called "incremental politics," Mueller hoped to use old technology for a new mission and thus avoid controversy and possible rejection in Congress. Leland Belew, manager of the Center's AAP Office after March 1966, said that Mueller wanted a station but knew "it had to be cheap, it had to be salable and such that it didn't impose on the Apollo Program itself." Planners sold the program as an "orbital workshop" or a "spent stage laboratory" because, Belew explained, "you didn't dare call anything a Space Station. It had to be framed right, because there was no way to get a new start." Asking Congress for approval would have been "no-go."[12] As an example of the AAP sales pitch, Stanley Reinartz, Belew's deputy, reassured Congress in 1966 that the spent stage was "not really a program" because it would exploit surplus Saturn IBs. The spent stage thus became the camel's nose under the flap of the Apollo tent. Based on incremental politics, the workshop became, Reinartz later recalled, "an awful lot George Mueller's program. . . . George was a very patient, continuing, ongoing, very bright but patient individual, who would just keep pushing and working and finding a way to keep things moving forward."[13]

After August 1965, planning accelerated on the spent stage workshop. All OMSF Centers, including Houston, participated. Marshall, however, did most of the planning. In December, Mueller made Marshall responsible for development plans and in February gave the Center responsibility for workshop design and integration. The Center's Apollo Applications Office quickly became an auxiliary planning staff for Mueller. Reinartz remembered that one week he and Ludie Richards worked in Mueller's office at Headquarters and phoned changes suggested by Mueller back to Huntsville.[14]

In Apollo Applications planning throughout 1966, NASA concurrently decided technical and managerial issues. Technically, AAP orbital workshops would have several major parts with Marshall overseeing the S–IVB spent stage and Houston an airlock module. Because of the entangled responsibilities, the two Centers were feuding by spring 1966. Kraft complained to Headquarters that Houston was losing its responsibility over manned systems.[15]

To resolve Center disputes and put the AAP Humpty-Dumpty together, the Manned Spacecraft Flight Management Council met in August 1966 at Lake Logan in North Carolina. The agreement reached at Lake Logan, historians have argued, was "perhaps the most fundamental statement of intra-NASA jurisdictional responsibilities since the Marshall Center first became a part of the agency and MSC emerged as a separate field Center."[16] The council confirmed Marshall's role in developing manned spacecraft and proposed handling the new division of labor among Centers with two guiding ideas, the "module concept" and the "lead Center/support Center concept."

The module concept assumed that any spacecraft had several parts or modules. Clean hardware interfaces between modules would allow the Centers to divide labor yet easily integrate the pieces. The Lake Logan agreement established a clear division of labor in some areas, especially by continuing the Apollo pattern with Marshall in charge of propulsion and Houston the "command post" including communication and control systems.

But the dividing lines between some modules were very fuzzy because Marshall took over some of MSC's traditional responsibilities for manned systems and space science. Marshall and MSC divided responsibility for the "mission module" and "experiment modules." Marshall was in charge of large structures, quarters, laboratories, some power and environmental systems, and the astronomy experiments; the Center was also responsible for workshop and experiment integration. Houston had life support and some power systems on the airlock module, medical research, earth experiments, astronaut activities, and flight operations. But living quarters mingled with medical research, astronomy equipment with crew management, and so on. As Belew recalled, "*Skylab* had no clean interfaces." The fuzzy division of labor produced technical disputes that the Centers could resolve only with careful negotiations.[17]

The Lake Logan agreement proposed the lead/support Center concept as a managerial formula for resolving problems. A lead Center would have overall managerial responsibility and set hardware requirements for the support Center which directly oversaw module development. For Apollo Applications, Marshall would be lead Center for workshop development and MSC lead Center for mission operations. Having two lead Centers was supposed to correspond to the two stages of development and operations, but the two stages were seldom distinct. A mixing of development and operations was natural because the developer would customize hardware to the demands of the operator. In effect this meant that Marshall became a contractor to MSC. As Marshall's Belew said "we structured to meet the requirements of the customer. They were our customer."[18]

After the Lake Logan meeting, Marshall's preliminary planning on what would become *Skylab* would be affected by the interplay of several factors. A design emerged from NASA's quest for a follow-up to Apollo that could get political acceptance, and from technical debates within the agency, especially discussions between Houston and Marshall.

Negotiating a Design

Interchanges among NASA Headquarters and the field Centers shaped the orbital workshop's mission, configuration, and launch system. Marshall contributed to changes in *Skylab*'s design even as the Center and its contractors began development of hardware.

Initial planning for Apollo Applications outlined two missions, the spent stage workshop and the solar science of the Apollo Telescope Mount (ATM). The first Apollo Applications schedule of March 1966 called for three workshops and three ATM missions. The first orbital workshop missions would be very simple, with basic mobility and biomedical experiments, amounting to little more than zero-gravity calisthenics in a pressurized S–IVB tank. The ATM missions were more sophisticated, fulfilling NASA plans dating to the early 1960s to put manually operated solar telescopes in a storage bay of the Apollo service module. In March 1966 the Goddard Space Flight Center, the agency's astronomy specialist, became lead Center for the ATM. By the end of the year, however, the two Earth-orbit missions converged, and NASA decided to reassign the ATM to Marshall and make it part of the workshop.[19]

Politics shaped the decisions. Mueller worked at "selling" the Office of Space Science and Applications on the idea of moving the ATM to Huntsville. Marshall's leaders, especially von Braun and his chief scientist Stuhlinger, also petitioned the agency, pointing out that Marshall had developed scientific payloads for the Explorer and Pegasus satellites. At the same time, NASA Associate Director for Space Sciences John L. Naugle, NASA chief astronomer Nancy Roman, and Mueller began questioning the utility of ATM-service module missions. By the summer of 1966 they realized that mating the ATM to a modified lunar module (LM) would allow for larger instruments and use more Apollo hardware, justifying transfer of the ATM–LM to Marshall because the Center had more experience with complex systems and manned missions than Goddard.[20] A desire to hold the Marshall team together also motivated Mueller.

When a Houston official challenged him for assigning the solar observatory to Marshall partly for political reasons, Mueller replied that his motives "were not partly political but completely political."[21]

Technical factors also influenced the telescope mount decisions. NASA officials realized that ATM–LM missions restricted instrument size, limited observation time, and wasted Saturn lifting capacity. And of course an ATM–LM mission would still be brief. So by the fall of 1966 NASA realized that mating the solar observatory in some way to the orbital workshop would allow for longer missions and larger instruments.[22] Such a configuration also justified giving the telescope mount to Marshall, the lead Center for workshop development, and legitimized the workshop by giving it an important scientific mission.

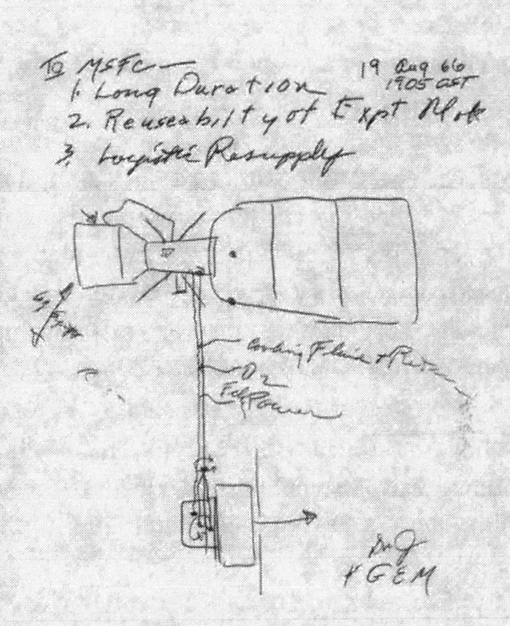

George Mueller's initial sketch of orbital workshop.

These decisions culminated in the fall of 1966 with the "cluster concept." On a visit to Huntsville in August, Mueller sketched a configuration that had an ATM–LM tethered to the workshop by a power cable. The design looked so bad, Reinartz remembered, that "nobody could figure out what it was, so it got the name of "the kluge." Mueller did not like that name so "in more polite terms it was called "the cluster."[23] Within a few weeks the tether gave way to a new cluster concept in which the ATM would be launched separately. A Marshall-built chamber called the multiple docking adapter (MDA) would anchor the telescope mount and the command module to the workshop.[24]

The observatory decisions proved controversial. Some questioned whether Marshall should build the telescope mount rather than have a contractor do so.[25] Abe Silverstein believed that mating the mount to a lunar module created "a monstrosity" and felt that jury-rigging Apollo hardware for new purposes wasted money. Some on the President's Scientific Advisory Committee wondered whether astronauts could contribute much to space astronomy. Since the ATM would be remotely controlled and not built for repair, astronauts on board the spacecraft could contribute no more than operators on the ground. Moreover, human contamination and motion could impair observations.

Center managers, worrying about the criticism, reminded their personnel that Marshall needed to succeed with scientific payloads. Von Braun declared in October 1966 that the telescope mount was "of particular significance to our Center, as our successful performance in this endeavor will determine MSFC's participation in similar projects."[26] Moreover Center officials defended the ATM choices. They admitted that repairable instruments would be more expensive and were really unnecessary since unmanned satellites had proven reliable, but pointed out that fitting the mount to the workshop allowed for larger, more complex instruments than an unmanned satellite and for photographic film which offered better resolution than electronic telemetry. Astronauts could change film canisters and return them to Earth.[27]

Such discussions were mild compared to quarrels over the spent stage or "wet workshop" idea. The Mueller-Marshall plan called for the first workshops to be launched by a Saturn I–B with a live S–IVB rocket stage. The plan initially assumed that all Saturn Vs would be used for the lunar program, and so a live upper stage was needed to achieve orbit with a I–B. Before reaching orbit, the workshop interior—the inside of the S–IVB fuel tank—would be "wet" with liquid oxygen and hydrogen. Once in orbit, suited astronauts would go on extravehicular activity (EVA), purge leftover fuel, move in the shop, outfit it, pressurize the cabin, and make it habitable.[28]

Marshall's engineers acknowledged problems with the wet workshop. As Eberhard Rees said, problems with habitability and EVA would make it "primitive," but the exercise would be enormously educational in learning about space. Moreover, the use of surplus Apollo hardware would minimize costs and give the wet workshop political advantages. NASA could not move openly for a Space Station because the Apollo Program was expensive and unfinished so expediency dictated "no new starts."[29] "The wet workshop was for us and for von Braun," Stuhlinger recalled, "always only an intermediary step."

Like the Center's preferred step-by-step method of testing rockets, Apollo Application plans called for several increasingly sophisticated wet workshop flights. The long-term goal, however, was a real Space Station, some sort of "dry workshop" that would be fully equipped on the ground. Dating from the first Apollo Applications schedule in March 1966, plans called for a mission with an S–IVB dry workshop launched with a Saturn V. Nevertheless the program from 1966 to 1969 only had enough money for Marshall to develop a wet workshop. The Center's policy until 1969, Stuhlinger said, was that the wet workshop "would be limited, but it could be done" and was worth doing.[30]

As early as 1966 Marshall had begun bending metal for a spent stage station. When engineers discovered structural weaknesses in the dome of the S–IVB, von Braun found money to install a quick-opening hatch large enough to support the dome and accommodate a suited astronaut. Later the laboratories tested interior materials for stress, corrosion, toxicity, and odor. They particularly checked the S–IVB's insulation on the inside of the fuel tank for flammability and outgassing of dangerous fumes. When high-velocity penetration tests showed that a puncture by a micro-meteoroid could cause the insulation to ignite, the Center sealed the insulation with aluminum foil. The labs studied ways of fastening equipment to the thin walls of the rocket. They installed two grid floors to allow for liquid hydrogen flow. The Center also began designing the telescope mount and EVA equipment for activating the workshop.[31]

The laboratories performed most of the EVA research in the Neutral Buoyancy Simulator where the wet workshop really was wet. One of Marshall's unique facilities, the simulator had a 1.5 million gallon water tank that was 75 feet in diameter and 40 feet deep to provide an environment that approximated zero gravity for testing hardware. After being denied Cost of Facilities money, Marshall called the simulator a "tool" and built it using $1 million appropriated for Research and Development. This creative financing led to a GAO audit and reprimand, but became a legendary example of Center resourcefulness.[32]

For workshop efforts, divers submerged mock-ups of the workshop in the simulator. To simulate the weightlessness of space, astronauts had suits and tools weighted to attain "neutral buoyancy," neither rising nor sinking. A team of engineers, psychologists, and human factors specialists monitored the astronauts through windows, television, and physiological displays. By early 1969, the team began to test hardware and devise methods for performing tasks, using tools, installing lights, sealing meteoroid penetrations, and changing ATM film canisters.[33]

The simulator aroused some friction with Houston. The Lake Logan agreement had confirmed MSC's responsibility for the astronauts and their equipment on spacewalks. But Marshall's responsibility for "large structures" and for studies of "EVA equipment and procedures which may be used to carry out . . . operations on large space structures" created ambiguities. Houston's managers resented this crossing into their territorial waters. MSC Director Gilruth believed that Marshall's tank needlessly duplicated Houston's capabilities in order to become "a manned space center." Despite this early jealousy, Marshall's Neutral Buoyancy Simulator immediately became a marvelous agency resource.[34]

Houston officials also objected to the wet workshop concept. No dispute since the lunar mode decision was so controversial. Robert F. Thompson, manager of Houston's Apollo Application's office, said that for the first time two Centers were competing for future work; until the wet workshop idea was abandoned in 1969, Apollo Applications was "not a program" but "a dogfight." Marshall's George McDonough recalled that one intercenter discussion of the wet workshop got so tense that Thompson wanted to take him out and fistfight.[35]

Houston's engineers doubted the technical merit of making a Space Station from a spent stage. They questioned whether suited astronauts in zero gravity could outfit an effective workshop. Because the Mueller-Marshall cluster conglomerated disparate hardware for a new purpose for which it had not been designed, MSC called it a "kluge," or more commonly, a "goddamn kluge." They believed that the wet workshop would waste money, risk failure, and, by perpetuating Apollo technology, prevent progress.[36]

As an alternative, Houston proposed an experiment carrier that would substitute for the lunar module on a Saturn I–B. Kraft recalled that Houston thought this would be "a Space Station, not a kluge." Less than half the size of the S–IVB, the experiment carrier would be "dry," constructed on the ground, and outfitted each time for progressively complex orbital missions. Houston thought it would be superior to a spent-stage station for about the same cost. Marshall Center engineers saw no technical advantages in Houston's carrier, which they derisively called "Max's can" (after Max Faget). They thought Houston was "extremely unrealistic" in expecting Congress to approve new hardware.[37] Most importantly Marshall worried that the experiment carrier could threaten its survival as a major Center. In a July 1966 message, Belew reminded von Braun that unless NASA built an S–IVB station "our allotted funds will be extremely small since our only other orbital station involvement is in the area of experiments." Approval of Houston's cans would mean that "the dollar split . . . [between MSC and Marshall] would tend toward 75%–25% rather than today's 50%–50% split." An S–IVB station, Belew wrote, was necessary "in order to fully utilize the skills that Marshall wants to retain and would insure a substantially more stable resource level for both Marshall internal and contractor operations."[38]

Luckily for Marshall, the rest of NASA also questioned Houston's experiment carrier. Most agency officials felt the S–IVB workshop was feasible, worried about wasting the money and effort already spent on the workshop, and feared delay in turning to new hardware. So in November 1968 NASA rejected the carrier idea.[39] So Houston in the spring of 1969 changed tactics by proposing to launch the S–IVB with a Saturn V rocket as a fully equipped dry workshop.

Although only a recapitulation of the original Marshall plan for an AAP mission, Houston has always claimed full credit for the dry workshop idea. Robert Thompson said, "unquestionably the thrust for the dry workshop came out of this center [Houston]." Kraft argued that by sponsoring a new means to achieve the goals of the Apollo Applications Program, Houston "saved the damn thing."[40]

Marshall engineers resented the implication that the spent stage idea had been bad from the beginning. They responded to MSC's criticism by laboring hard to improve the spent stage and prove that it would succeed. But, Belew said, the Center had all along believed that the wet concept "was never the best notion of doing something if you had an option different." And NASA's original options were limited; since all the Saturn Vs were committed to the lunar mission, a live second I–B stage was needed to achieve orbit.[41]

Moreover, Belew thought Houston's claim to be the inventor of the dry workshop was "only half true." Marshall had formulated the plans to use an S–IVB as a Space Station and helped draft the original AAP plans which had, in the long run, called for Saturn V dry workshops. Stan Reinartz believed Houston could not take full credit for the dry workshop because their preferred alternative was the can; by proposing the experiment carrier, "they tried to kill" the S–IVB station. Houston only warmed to an S–IVB workshop as a last resort.[42]

Marshall's engineers credited Houston, however, with forcing NASA to consider alternatives. Houston's position, Belew recalled, "drove you to a real hard decision of what we really ought to do." In addition, circumstances changed dramatically by the fall of 1968. Declining budgets forced a reconsideration of Apollo Applications, and the agency realized that it lacked resources for several wet and dry workshop missions. Marshall's work on the wet workshop was already behind schedule, with officials complaining they were getting only two-thirds of the money needed to meet deadlines. Moreover, after the success of Apollo 8 in December 1968, NASA concluded that a Saturn V could be used for an Apollo Applications mission. So from the fall of 1968 to the spring of 1969, the agency conducted an exhaustive study of its options.[43]

Marshall had studied the dry workshop before but now Mueller directed a small group at the Center to reassess the concept. Because they were regarded as "pariahs" in Huntsville, McDonough recalled, the dry group operated discreetly and even held a secret poolside meeting with Mueller in a motel at the Cape. After hearing the group's report in early 1969 and recognizing the changed circumstances, von Braun concluded that the wet workshop was no longer the best option.[44]

In May 1969 the Management Council met in Houston and Mueller gave them several options, all of which drastically reduced the number of AAP workshops. Basically the council had a choice of missions involving one wet or one dry workshop. A dry option emerged as their favorite. Von Braun then convinced some of his reluctant lab directors that a ground-outfitted configuration improved the design. In a letter to Mueller on 23 May, he acknowledged that although the wet workshop could meet AAP's scientific objectives on time and on budget, this would "take substantial hard-nosed scrubbing down of some of the current methods." Von Braun thought a dry workshop offered "real and solid advantages over the present program." With the greater lift of the Saturn V, reliability could be improved by using sturdy and redundant hardware and by installing and checking equipment on the ground, and habitability could be improved by eliminating liquid hydrogen.[45]

Gilruth of Houston seconded von Braun, and on 18 July 1969 NASA Acting Administrator Thomas Paine used the success of Apollo 11 as an opportunity to announce plans for the dry workshop. The Apollo Telescope Mount would be launched with the workshop rather than on a separate flight, eliminating the makeshift ATM–LM and a complicated rendezvous with the workshop. The telescope system could be simplified by attaching the instruments to a heavier, specially designed rack and by creating a deployment system; upon reaching orbit, the mount would swing out perpendicular to the workshop. The solar observatory could also duplicate the power, communication, and control systems of the workshop. In addition, by the fall NASA decided to avoid putting all its eggs in one basket by building an identical qualification workshop and equipment that would be used in tests and refurbished to back up the flight model. The competition between the Centers had helped improve the design.[46]

In February 1970 the workshop got a new name. In mid-1968 NASA had held a contest to name the project and an Air Force officer assigned to the agency proposed "*Skylab*," short for laboratory in the sky. NASA people were initially nonplused by "*Skylab*," Reinartz remembered, but still avoided calling the project a Space Station. Wanting to build a more elaborate station later and fearing that identifying an expensive new project would offend Congress, the agency waited two years to sanction the name officially. *Skylab* became the only NASA project never to get formal congressional approval of a "new start" through the phased planning process.[47] The incremental strategy of Mueller and Marshall was successful and the Center could develop something more than a spent stage station.

Building the Workshop

As Lead Center for *Skylab*, Marshall oversaw diverse, complex development problems. Marshall used ideas from Space Station studies conducted by NASA contractors and Centers, especially the Langley Research Center. During the development phase, Marshall would again work closely with the Manned Spacecraft Center, and their complementary expertise helped solve the technical challenges of the project.

The technical challenges were formidable. No American manned spacecraft had used solar energy to generate all of its electrical power. No manned spacecraft had needed precise pointing control for a solar observatory. No previous manned mission had required equipment and life support systems for nine months. Crew systems had to be not only functional but habitable in order to maintain productivity and morale for long-duration missions.

Other design problems were less novel but still challenging. Onboard and Earth-bound communication and control systems were necessary. The space laboratory and its scientific equipment had to survive a harsh and dynamic environment. The workshop had to withstand changes in inertial loads during launch acceleration, bending forces caused by engine thrust and gimballing, temperature, vibration, and atmospheric and acoustic pressure. In orbit it had to endure vacuum, micrometeoroids, radiation, and docking impacts equivalent to earthquake shocks.[48]

Skylab's designers overcame these complex challenges with a series of systems and structures. The new dry configuration meant that engines and flight hardware could be removed and experiments, life support equipment, and storage units added. For launch the workshop was pressurized with dry nitrogen to maintain rigidity and was vented during ascent to equalize atmospheric loads. Because the orbital configuration could not withstand the pressures of launch, diverse mechanisms deployed the payload shroud, antenna booms, solar observatory, workshop micrometeoroid shield, and solar arrays on the ATM and workshop. Thermal control came from passive systems using insulation and exterior surface coatings and active systems using heaters, coolant pumps, heat exchangers, and radiators. The oxygen and nitrogen laboratory atmosphere required methods for purification, humidity regulation, circulation, and odor removal. Pressure tests guarded against leaks.

Skylab also had systems for power, communications, and attitude control. Electrical power came from solar cells that provided power during sunlit phases of the orbit and from batteries that discharged during shaded phases. Communications systems could transmit data, hardware commands, video, and voices. The workshop had over 2,000 data sensors and could receive more than 1,000 digital commands. Attitude and pointing control for the 100-ton *Skylab* came from three control moment gyroscopes. The gyros were the first used on a manned spacecraft and were chosen because a gas reaction system would have required too much propellant for the long mission; cold gas thrusters served only as an auxiliary. The control system employed a computer, Sun sensors, a star tracker, and rate gyroscopes to determine position and angular rate.[49]

Marshall divided work on these systems between itself and contractors. As Lead Center for development, the Center was responsible for systems engineering, contractor management, and cluster integration. Boeing helped with systems engineering. McDonnell Douglas modified the S–IVB into a space station in Huntington Beach, California, and built the airlock module that contained power and life support systems in St. Louis. Houston initially monitored the airlock contract, but Marshall soon took it over to simplify project management. TRW built the solar arrays for the workshop and the ATM. Martin Marietta of Denver was responsible for payload and experiment integration; Marshall also assigned the corporation the MDA.[50]

For *Skylab* development, the Center drew on technology and organizational methods from the Saturn era. Its approach to monitoring contractors was essentially the Saturn method. Belew's *Skylab* Program Office established a project office for each major hardware component and for experiments, set up resident manager offices to penetrate contractors, and designated "tiger teams" of specialists to solve crises. The biggest contractor problem came when McDonnell Douglas fell behind schedule in mid-1971 during the enormously complicated final integration of the workshop. The Center's William K. Simmons, project manager of the orbital workshop, organized a 10- to 15-member tiger team that stayed in California until mid-1972. McDonnell Douglas's problem, Simmons believed, was that its management system for manufacturing airplanes was "geared to quantity" and "a lot of their practices weren't compatible with building one-of-a-kind." Particularly, the company managers were isolated from development problems and had not established an integrated schedule for incoming components. The Marshall team imposed order by drawing a master schedule, working alongside McDonnell Douglas's managers, and getting the company president to act as program manager.[51]

Skylab also drew from the remnants of the arsenal system at Marshall. The Center maintained a mock-up *Skylab* in Huntsville to test alternatives and monitor contractor performance. Marshall built two shells of the multiple docking adapter and turned them over to Martin Marietta for final development. Marshall also tested hundreds of components and helped build hardware for many *Skylab* experiments.[52]

The greatest scientific instrument produced by Marshall's arsenal system was the Apollo Telescope Mount. None of the Center's previous scientific payloads had been as sophisticated as the solar observatory. Marshall's experience with vehicle engineering, however, prepared it for payloads. ATM Project Manager Rein Ise said, "once you have applied structures to large vehicles, there is essentially no conversion involved in taking knowledge and designing the structure for a solar telescope."

Teams from the Astrionics, Space Sciences, and Manufacturing Engineering laboratories took on the challenge of the telescope mount. They used components from contractors; Bendix provided the control moment gyroscopes, Perkin-Elmer the pointing system, IBM the computer, and experimenters the instruments. But the Center designed and developed the solar observatory system. To mount the eight solar telescopes, engineers built an octagonal spar 11 feet in diameter and 12 feet long. Their design had subsystems for orbital deployment, communication, electrical power from four solar cell arrays, and attitude and pointing control.

The requirements for the pointing control system were very complex. The telescope needed accuracy within two arc-seconds, which meant an error of no more than the width of a dime at a distance of two kilometers. Yet the accuracy and stability of the telescope system could be affected by the movements of the *Skylab* spacecraft and the astronauts. Moreover large bundles of stiff electrical wires connecting the telescope tub and spacecraft could limit the telescope's pointing motion and accuracy. To solve the wiring problem, an engineering team led by Wilhelm Angele from Marshall's Astrionics Lab developed flat electrical cables that were so flexible that they allowed the telescope mount to move with very little mechanical resistance.

For the pointing system, Marshall chose a design using three control moment gyroscopes, actuators, a computer, photoelectric sun sensors, and a star tracker. The Center tested the system on specially built engineering simulators that used analog devices and computer models. The engineers struggled to simulate the performance of the control moment gyroscopes in microgravity; they compensated for gravity distortion by floating an ATM simulator in a mercury bath. But still ground tests could only prove the accuracy of the pointing system within six arc-seconds. Marshall engineers waited until *Skylab* was in orbit to learn that the system worked well and that astronomers could not measure pointing errors.

Marshall helped solve other technical problems for the solar observatory. When scientists became concerned that the South Atlantic Anomaly, a high radiation area that *Skylab* crossed in orbit, could expose film used in the observatory,

Skylab's Apollo Telescope Mount is prepared for Thermal Vacuum Test–1970.

Marshall engineers worked with Eastman Kodak to develop special films that could survive in the radiation environment. They devised computer programs that duplicated the anomaly and so could predict the fogging on film. Center personnel also developed crew trainers and operating procedures for the solar observatory. Marshall constructed an ATM checkout facility for final integration and equipped it with automatic monitors and air control equipment that made the whole building a clean room.[53]

The Center engineers and scientists who worked on the ATM believed that in-house manufacturing accounted for the success of the telescope mount. Dr. Walter Haeussermann, director of the Astrionics Lab and later head of Central Systems Engineering, claimed that the arsenal system allowed for "tremendous flexibility" in inventing new technology. Technicians could build models, allowing designers to execute modifications without making elaborate drawings and wasting time and money. Dr. Tony DeLoach, an experiment scientist for one of the ATM instruments, believed the system centralized management and engineering. When work was done in-house rather than by contractors spread across the country, teams of experts could quickly confer to solve complex problems.[54]

Since the lives of astronauts depended on *Skylab*, Marshall's design incorporated conservative engineering ideas and redundant systems. Marshall set high quality standards and sought to achieve them with heavy structures, existing technology, and extensive testing. Launching *Skylab* with a Saturn V reduced weight problems, allowing for heavy hardware and backup systems. Moreover, using tested ideas and mature technology reduced development time and saved money. The Center, according to Robert G. Eudy, deputy chief of the Structures Division, "relied heavily upon existing technology, available hardware, and hardware concepts" for *Skylab*. Marshall engineering teams used hundreds of components from the Gemini program; recognizing that using proven components could save money and time, the teams tested Gemini technology for its suitability for the longer *Skylab* mission, for example, adopting Gemini hatch latches for the airlock module hatch. Other systems adapted for *Skylab* included a separation system for the payload shroud from the Titan IIIC and a scientific airlock originally designed for the Apollo Command Module hatch. The workshop itself was a modified S–IVB rocket stage with its liquid oxygen tank used for waste disposal, its liquid hydrogen tank used for habitation, and interior structures attached to cylinder rib intersections.[55]

In addition, the workshop had redundant batteries, chargers, electrical circuits, and solar arrays. The ATM controls, Ise said, used "a belt-and-suspenders approach in that we designed redundancy throughout the system" and had three rather than two control moment gyroscopes to change attitude. The gyros were new technology for a manned spacecraft, but Marshall stayed conservative by choosing big, heavy wheels that spun relatively slowly. Moreover, the Center carefully tested equipment; the ATM, for instance, went through functional, vacuum, and vibration tests. And because NASA built prototypes for qualification tests and then refurbished them as spares, the agency had a backup *Skylab*.[56]

Perhaps the greatest Saturn legacy to *Skylab* was relatively liberal funding. To be sure, Marshall experienced budget cuts throughout the late sixties and early seventies and laid off hundreds of Civil Servants. And as the only surviving AAP mission, *Skylab* became the first major NASA program in which budgetary shortfalls caused schedule delays. (*Skylab* was launched in 1973, six years after AAP's target for the first wet workshop.) Nonetheless, compared to later programs, Skylab's budgets allowed for backup hardware and extensive testing. Looking back after almost 20 years, ATM manager Ise saw few funding pressures on *Skylab*. "I am sure that the *Skylab* manager didn't get everything he wanted, but he got almost everything he wanted," he said, "*Skylab* had the money when it needed it."[57]

Marshall's internal management during *Skylab* also continued the same pattern as the Saturn program. During *Skylab* the Center distributed management authority between the project offices, which oversaw budgets, schedules, and contracts, and the laboratories in Science and Engineering, which handled design, development, and testing. Also like the Saturn era, Center managers struggled to find the best division of labor between centralized offices and specialized labs. Their balancing act became more difficult as Marshall diversified from a propulsion specialty and took on more projects. The balance can be seen in relations between the "lead laboratory" system, the project offices, and the Central Systems Engineering Office.

The lead lab system originated in the Center's practice of automatic responsibility. The goal was to empower the technical experts, fuse planning and doing, and keep engineers' hands dirty. Research and Development Operations, the laboratory side of Marshall, assigned technical responsibility for a component or subsystem to one laboratory. For example, the Astrionics Laboratory had responsibility for the telescope mount and the Propulsion and Vehicle Engineering Laboratory had the Multiple Docking Adapter. Each lead lab developed hardware specifications and managed interfaces. Initially project offices for hardware components were decentralized in the laboratories, rather than being centralized under Belew's *Skylab* Program Office.[58]

One of the lead lab's major tasks was soliciting support from other labs. This often meant time-consuming negotiations with other specialists to resolve differences in engineering methods or technical requirements. Indeed von Braun expected the lead lab system to encourage cooperation, Haeussermann recalled, and the lead lab never commanded others. When the system worked well, the lead lab organized a team of experts drawn from other labs that collectively overcame problems in design and development.[59]

Sometimes, however, the system could be frustrating. Decentralized labs often struggled to solve complex problems with multiple specialists and components. Especially troublesome was establishing requirements for a whole system, getting the labs to cooperate, and forming multi-lab teams. For example the Astrionics Lab moved so quickly that ATM design became fixed and not easily changed to meet the needs of labs working on other parts. Ise remembered that the German laboratory directors "had a little bit of this fiefdom philosophy where each one ran their own little kingdom. One laboratory was not very effective in being able to manage other laboratories that also had to participate in a very key way on the whole project." McDonough thought that the boundaries between labs sometimes became "war zones" and to get the support of other labs specialists had to go "up, over, and down" the chain of command. William Lucas, then chief of the Propulsion and Vehicle Engineering Lab, remembered how he struggled to get other labs to commit resources to his tasks. He believed the limitations of the lead lab approach proved the "old Chinese proverb that says, 'If two guys are going to ride on a horse, one has to ride in front.'"[60]

To put somebody in front, Marshall managers sought ways to centralize managerial and engineering authority. Some early centralization for *Skylab* was makeshift and accommodated the labs. James Kingsbury, deputy director of the Astronautics Lab, often worked as ad hoc chief engineer for *Skylab* and helped resolve problems.[61]

Formal mechanisms also existed. A Technical Systems Office in Research and Development Operations, renamed the Systems Engineering Office in July 1967, controlled design requirements, and helped specialists in the labs integrate the many pieces of a scientific space station. Systems engineers became another layer in the Center's hardware hierarchy of lab specialists, chief laboratory engineers, and project managers. Von Braun, recognizing that the Center now had too many projects for him to oversee, strengthened the office in late 1968 and early 1969.[62]

The systems engineering office had its limitations too. Laboratory personnel worried that centralized design and integration, whether in a staff office or a systems engineering contractor, would be ineffective without engineers keeping their hands dirty and maintaining skills. Moreover excessive centralization would weaken the labs. Lucas, answering von Braun's questions about systems engineering and lead labs in November 1968, argued that giving labs responsibility for systems engineering would foster "an entrepreneurial climate" and "let the workers be the master of their own fate." Robert Schwinghamer, head of the lab's Biomedical Experiment Task Team, agreed, worrying that centralized systems engineering would convert technical decisions into financial ones and thereby weaken "the in-depth technical capability of Marshall laboratories." Technical deterioration, he thought, would call into question the need for the Marshall Center because "a purely management function not supported by a strong technical institution could as well be performed in Washington."[63]

Finding the right balance between the labs and project offices was sometimes controversial as well. As *Skylab* progressed, the project office sought more programmatic control over the project engineers in the labs. Chief engineers colocated in both project offices and the labs and answered two bosses—the project manager and the director of Science and Engineering. This change, for instance, meant that the telescope mount project manager more directly supervised the budgets and schedules of the Astrionics Lab.[64]

This quest for greater programmatic control by project managers sometimes annoyed laboratory personnel who feared a loss of technical autonomy. When Belew's project office sought programmatic control over the Propulsion and Vehicle Engineering Lab's development of the biomedical experiments, Schwinghamer resisted. He claimed that greater control by the project office would sabotage the "quick response, economy, and flexibility" necessary to get the experiments done on schedule.[65] Nevertheless, the culture of the labs and their relationship with staff offices remained essentially the same during *Skylab* as during Saturn. New programs started after *Skylab* tended to rely less on the Center's labs and more on contractors.[66]

Because *Skylab* had more complicated technical interfaces and more interaction between Houston and Marshall, its design and development was more controversial than Saturn. The Centers worked together using intercenter panels of lower and middle level officials. Hardware interface and systems panels met regularly to coordinate technical plans in areas of divided authority. Unsolved problems passed on to periodic, face-to-face meetings of upper administrators. Unsolved disputes between Houston and Marshall were passed up to Headquarters.[67] J.R. Thompson, who headed Marshall's Man/Systems Integration Branch and oversaw the Center's interaction with Houston's astronauts and human factors specialists, remembered that the disputes were "good, honest differences of opinion" about "the best technical solution." He explained that usually "Marshall had a stronger engineering solution and Houston had a stronger operational solution. So you tried to find the best of both of them." Marshall, for example, wanted a fireman's pole to extend through the workshop; but Thompson recalled that Houston's astronauts believed this was superfluous and they never deployed the pole.[68] Such technical disputes between Centers became most intense over ATM controls, workshop habitability, and biomedical systems and experiments.

Marshall built the telescope mount controls, but Houston's astronauts would use them. Feuds erupted in 1967 and 1968 when Houston complained that Marshall lacked understanding of crew instrumentation, that the astronauts would have little control over the mechanisms, and that some toggle switches flipped up in the off position and some flipped down. Marshall accepted many of

Houston's recommendations, but Center engineers, who often judged the cost of equipment in terms of luxury cars, joked that the redesign cost "umpteen more Cadillacs." The controls were "probably the most complicated ever flown in a spacecraft" yet worked well during the *Skylab* missions.[69]

Marshall and Houston also struggled to improve "habitability" and make the S–IVB an efficient, comfortable, and pleasant place to live and work for long missions. Center interactions were complicated because NASA never formally defined which one was really in charge of the workshop interior. Headquarters merely divided a list of hardware items, so Marshall had *Skylab* structure while Houston had the habitability experiment—which affected the entire structure.[70]

Making the workshop habitable had been a low priority in Marshall's original planning. Wet workshop designs had been necessarily austere. Center engineers had been mainly concerned about workability, ensuring that equipment functioned properly. Moreover, William Simmons, the workshop project manager, pointed out that Marshall lacked experience with manned systems. "Man-rating a vehicle is one thing," he said, but "making it livable or adaptable for a man is really something else." Reinartz said that emphasis on workability over habitability came because "our guys had been building rockets. We hadn't had people around." He admitted that there was a certain amount of "lack of appreciation by the Marshall people of the concerns for being in these tin cans for up to ninety days." To learn about the problem, Marshall engineers studied designs of ships, submarines, and railway cars and consulted with astronauts.[71]

By 1968 Houston's spacecraft designers, transferring from Apollo spacecraft work, began criticizing Marshall for its lack of concern with workshop habitability. The criticism intensified after the mid-1969 dry workshop decision when Marshall was slow to recognize the new priority for habitability. Recalling an inspection of Marshall's workshop in 1969, Mueller said that "nobody could have lived in that thing for more than two months. They'd have gone stir-crazy." Mueller helped bring in two industrial designers, Caldwell Johnson of MSC's spacecraft division and Raymond Loewy, an internationally renowned industrial design consultant.

Johnson and Loewy thought Marshall's designs lacked creature comforts and aesthetic qualities. They complained that sleeping chambers were too big and living quarters and storage compartments too small. Lighting was random and cold. Loewy said that the color of the workshop was "Sing-Sing green," the same as the death cell at Sing-Sing prison, and the grid floors cast "cage-like" shadows. The interior pattern of cylindrical walls, rectangular equipment, and triangular grid floors was confusing. The workshop lacked a wardroom and a window. Accordingly they recommended changes and received support from NASA Headquarters. Marshall responded by improving the lighting, layout, color scheme, and by adding a window.[72] In 1969 Marshall continued habitability research in "space station analogs," sending an engineer on the Gulf Stream Drift Mission in which a six-person submarine traveled from Florida to Nova Scotia. Marshall also sent personnel to the Tektite II underwater habitat in the Virgin Islands.[73]

Despite the improvements, Houston again proposed major changes in the spring of 1970. After a tour of *Skylab* at the Douglas plant, Houston's Kraft argued that workshop habitability was still inadequate, especially in terms of hygiene and waste management. Acknowledging that the contractors and Centers were "all partially to blame," he thought that Marshall and its contractors had relied too much on astronauts who accepted "a make-shift situation on the basis of 'that's the way things have been done in the past.'" But for prolonged *Skylab* missions, a comfortable spacecraft was necessary to maintain crew productivity. Proposed changes included better environmental control, storage, lighting, sleep restraints, and housekeeping devices as well as the addition of an entertainment center and an alternate waste-disposal system.[74]

Full-scale mock-up of Skylab at Marshall in April 1973.

Kraft's rhetoric prompted Rees, who became Marshall's director on 1 March 1970, to ask his *Skylab* program office to make the changes. Rees remembered how during a research trip to Antarctica "without a shower for six days we really felt rotten." The Center director, however, reversed course after Belew explained that additional changes would put *Skylab* over budget and behind schedule. Moreover Marshall had already improved habitability by expanding the wardroom, rearranging the waste management area, and adding individual sleeping compartments, a window, a food freezer and warming oven, and a trash airlock.[75]

To stay within *Skylab*'s limited resources, Rees decided to oppose Houston's proposals. He argued that for more than three years MSC had gone along with Marshall's designs and then began constantly changing requirements. Houston's habitability proposals had changed *Skylab* from an "experimental astronomy program" to "a very sophisticated and unprecedented medical experiment." By changing the ground rules and upgrading hardware, Rees thought, MSC was threatening the whole program. J. R. Thompson acknowledged that "amenities" were necessary, but contended that Houston wanted to spend money on "interior decorating" rather than on improving equipment. If equipment like *Skylab*'s toilet failed, then he doubted that "any color scheme recommended by any committee would make much difference in improving the habitability of the Waste Management Compartment."[76]

Houston got Headquarters to overcome Marshall's resistance. In July 1970 Charles W. Mathews explained to Rees that the changes were necessary because "*Skylab* may be the only manned missions flown for an uncomfortable number of years between Apollo and early shuttle missions. It is critical that we make the most of this opportunity consistent with our resources." Mathews acknowledged, however, that budgets and schedules had to be kept. Such constraints led the Centers to stabilize habitability designs after the fall of 1970.[77]

Marshall and Houston also cooperated on biomedical experiments that would monitor the effects of microgravity on physiology. Marshall would develop a waste management unit that disposed of urine and feces and preserved samples for return to Earth. In addition, in a meeting at the Cape in 1968, Dr. Charles A. Berry, Houston's chief medical researcher, told von Braun that he was having difficulty getting medical hardware built. Von Braun offered his Center's services for in-house development of an ergonometer with a physiological monitor and a lower-body negative pressure device. Marshall engineers believed biomedical projects would "firmly pave the way for future Marshall missions" and "establish a capability essential to future activities." Houston, while desiring Huntsville's help, also wanted to maintain control over biomedical research and operations.[78]

Consequently by December 1968, the Centers negotiated an agreement that "followed the same general mode of operation as any other contract that MSC has where a contractor is providing flight hardware for medical experiments." Von Braun accepted this agreement as "the best we can get." Nonetheless he worried that the contract's technical requirements would deny Marshall the "leeway" to assist Houston "not only with our hands, but also with our imagination and inventiveness." Von Braun's worries were well grounded because the contract did not prevent the Centers from arguing over the biomedical equipment; the official history of *Skylab* has described the design of the urine collector as "probably the most vigorously contested point in the entire workshop program."[79]

Throughout the multi-year project, Houston's doctors and Marshall's engineers had difficulty communicating. When the doctors "started talking medicine," recalled Henry B. Floyd, head of Marshall's experiment office, "it was just traumatic; it was a whole new language." The doctors were "as much in the dark about engineering language." Schwinghamer, who directed the medical work, said the engineers and doctors acted like "two dogs sniffing at each other" and that "Houston was worried about us getting into their britches."[80]

Marshall's people approached the biomedical equipment as just another engineering problem. To test the fecal management system, the Center installed prototypes in a KC–135 airplane and collected "data points" by having specimens defecate in the half-minute of zero gravity. For the urine collector, Schwinghamer had his people urinate into beakers to determine the appropriate vessel volume, but during tests astronauts sometimes found that their cups runneth over. Schwinghamer expanded its volume to meet conservative engineering standards.

The engineering approach to the urine collector peeved the doctors. Houston pointed out that all medical labs preserved urine by freezing it. Nonetheless Marshall questioned the utility of freezing "urisicles" and believed drying the samples would be simpler, cheaper, and lighter. The stream of invective over the urine collector continued for months with Houston recommending freezing and Marshall drying. Houston's Dr. Berry said "you could not get through to them." Eventually the doctors convinced the Headquarters program director to choose freezing. But Marshall's Simmons insisted "until my dying day I'll always say . . . we should have dried the urine instead of freezing it."[81]

After *Skylab*'s success, participants downplayed design and development controversies and believed that the disagreements had improved the program. Gilruth praised Marshall's engineers, saying "they're a bunch of craftsmen . . . and the stuff turned out well." The chief of MSC's Bio-engineering Systems Division praised "the outstanding performance of the medical experiments hardware" that met its requirements even through extended missions. Caldwell Johnson's final habitability report, while critical of storage and restraint problems, praised many parts of the workshop, including its up-down architecture and ergonometer. Kenneth S. Kleinknecht, MSC's *Skylab* program manager after February 1970, thought the habitability complaints improved the workshop and felt Marshall "welcomed the strong positions we took [because] that helped them with their money."[82]

Marshall's Belew also believed the competition had been healthy and that *Skylab* habitability compared favorably with that eventually built into the Shuttle. The workshop's features were "not slouchy looking things even some twenty years after." Astronaut Jack Lousma went further, saying the waste management hardware was a "no fuss, no muss, no smell system" and the Shuttle system was a "step backwards." Center conflict, Marshall's Haeussermann argued, was mainly restricted to a project's early phases of task division and hardware design; in these periods quarrels arose mainly because of disputes about resources and responsibilities and because working level people had different ideas about what was the best possible system. Disputes were usually set aside during development and operations when the Centers closely collaborated.[83]

An example of this pattern was the planning for *Skylab* operations. As early as 1967 Marshall sought some role in mission operations. No longer just a propulsion specialist, the Center was building a spacecraft and believed the engineers who built it could best operate it. Houston refused to give up its operations monopoly and wanted to use Marshall personnel only if they were subordinate to MSC's managers and part of its organization. Houston should "operate spacecraft developed by MSFC," Gilruth argued, "in the same way that SAC [the Strategic Air Command] flies bombers designed by several contractors."

After heated discussions, the Centers in May 1970 established a flight planning team with a Houston majority and Marshall representatives. Houston would manage daily mission operations and respond to immediate problems but would consult with Huntsville on hardware matters and long-term problems. Sophisticated communication systems linked Houston's Mission Control Center and Huntsville's Operations Support Center (HOSC). Marshall assigned over 400 engineers to 10 mission teams, providing mission support for the systems and experiments it developed. The teams helped with problem analysis and crew training, staffing simulators such as the neutral buoyancy facility and the solar observatory backup unit, as well as developing computer programs for thermal and environmental control, attitude and pointing control, and electrical power. The agreements enabled the Centers to function as one team during *Skylab* missions.[84]

Rescuing *Skylab*

The NASA Centers showed their shared commitment to mission success during *Skylab* operations. Marshall helped rescue, repair, and run the orbital workshop in its three long missions.

Perhaps the most dramatic episode in Marshall's history occurred as it helped to salvage *Skylab* 1, the unmanned orbital workshop, from the damage incurred during launch on 14 May 1973. The Saturn V rocket fired normally, and the launch seemed successful. But 63 seconds into the flight, controllers in the HOSC read telemetry signals showing early deployment of one solar array and the micrometeoroid shield, a thin protective cylinder surrounding the workshop. Designed to provide thermal protection with a pattern of black and white paint, it was supposed to fit the workshop snugly during ascent and then extend five inches in orbit. Although the workshop attained orbit, its solar wings failed to provide electrical current, and temperature readings on its Sun side were off the scale at 200 degrees F. Later investigations determined that the meteoroid shield had ripped away during the launch, taking with it one array and jamming the other.[85]

Skylab was in a crisis. Heating could spoil food and film and cause the S–IVB's insulation to give off poisonous fumes. Lack of electricity would cripple the workshop. Acting quickly, NASA postponed launch of *Skylab* 2, the first crew for the workshop, from 15 May until 25 May. NASA Centers and contractors had 10 days to develop remedies. For Marshall these days were so eventful that Center Director Rocco Petrone said, "We lived through 'ten years in May,' not ten days in May."[86]

Within an hour of the *Skylab* 1 launch, the Center had shifted to a crisis footing. H. Fletcher Kurtz, head of the HOSC's Mission Operations Office, remembered that he "quickly became a landlord with about a hundred very unhappy guests. The chain of command went out the window as senior managers increasingly moved into key positions in the HOSC, working directly with those most concerned with the rescue." Petrone appointed a special team headed by Kingsbury of the Astronautics Lab and William Horton of the Astrionics Lab to coordinate trouble-shooting. The director told the team to "keep the vehicle in a mode where we can inhabit it and find out a way to fix it. Whatever you need at the center is yours. This is the one thing we are going to do at the moment." The team complied and Kingsbury said "we turned on everything and everybody we had who could do anything."[87]

Contractors, support teams, project offices, and laboratories acted with selfless dedication and spontaneous teamwork. Schwinghamer, who had driven with his wife to the Cape to watch the launch, recalled driving back to Huntsville all night so he could help. People worked long hours, sometimes sleeping in their offices or going for days without sleep. Sometimes their dedication was dangerous since tired people made mistakes. Ludie Richards would walk up to people, hold up a few fingers, and ask "how many?" He sent home those who could not count. "It was long hours," James Ehl, an engineer in the Manufacturing Engineering lab, said, "but everybody seemed to enjoy it. It was a challenge." Kingsbury said "we could not drive people away. . . They just did not want to leave. It was their baby and it was in trouble, and they were here to fix it." And the remarkable thing was "it came right in the middle of a small . . . reduction in force and an announced sizable reduction in force in the coming months. Nobody said, 'I don't care. I'm not going to be here next year.' It was, 'Let's get it fixed.'"[88]

Top administrators who had kept their hands clean for years showed up in the labs. Belew remembered that "everyone that had a role was apt to be any place, any time of day or night." Petrone "was running it. . . . He was there all the time." Reinartz said that the director, who "was like a bull in a china shop normally," was even more excited. Petrone worried that the teams were disorganized and would ask "who was in charge?" and when nobody knew "he would just hit the ceiling." To keep the chain of command clear, some began wearing signs saying "I am in charge!"[89]

Marshall's first priority was lowering the temperature and ensuring electrical power in the workshop. George Hopson and Dr. J. Wayne Littles, co-leaders of the HOSC Thermal, Environmental Control and Life Support Team, began changing the workshop's attitude. They performed a delicate balancing act: reducing temperatures required shading the workshop by pointing the MDA end at the Sun and cutting off solar power; increasing power required pointing the ATM's solar arrays at the Sun and heating the workshop. In balancing these goals, Marshall's close involvement with operations paid off because the Center could direct the spacecraft. Hopson said that "one of the things that has been most gratifying to me was the close cooperation between Marshall and JSC. They have been more than helpful, with everybody trying to help the other fellow solve his problems."

Optimizing temperature and electrical power was trying because attitude changes would freeze one side and scorch the other. The craft had to be maneuvered continually and judging angle and position was difficult because pointing control instruments had not been set. Slight changes brought tremendous joy or despair. The team worked around the clock and Littles said "that first 'day' for many of us was forty-four hours long." Within 10 days the maneuvering had used almost half of the entire mission supply of nitrogen gas in the control thrusters. Petrone told the team, "you're pouring out liquid gold you know!" Eventually the Center pointed *Skylab* so that its sidewalls were at a 45-degree angle to the Sun which reduced interior temperatures to 122 degrees F but still generated some electrical current.[90]

Center engineers test methods for freeing Skylab's *solar array in the Neutral Buoyancy Simulator in June 1973.*

Meanwhile a group managed by James Splawn and Charles R. Cooper constructed a mock-up of the damaged spacecraft in the Neutral Buoyancy Simulator. They had a good picture of the jammed array from radar images and photographs from Air Force spy satellites. By 19 May, Marshall engineers and Navy divers rigged an underwater model they called "the junk pile." NASA air freighted a mock-up of the Apollo Command Module that the team immersed in the tank to test whether an astronaut could stand in an open hatch and pry the solar wing open with a 10-foot pole. Throughout the crisis, the simulator group tested tools, repair procedures, and workshop shields.[91]

Beginning 17 May, Marshall engineers tested cutting tools for opening the wing, restricting themselves to existing tools to save development time. They even tested the surgical bone saw included in the *Skylab* medical kit. NASA Centers and businesses around the country sent devices. Eventually Section Chief A. P. Warren of the Auxiliary Equipment office got an idea from tree-trimmer shears purchased from a Huntsville hardware store. Working with a manufacturer of electrical cable tools, Marshall helped develop pulley-driven cable cutter and shears and a two-prong universal tool. Each had attachments so five-foot sections of aluminum pole could be added.[92]

Other teams throughout NASA were designing systems to protect the workshop from the Sun. In 10 days the Agency tested hundreds of combinations of designs and materials. In Huntsville, Schwinghamer experimented with spray painting and tried it in a vacuum chamber; he determined that spraying would lower the temperature but could coat ATM lenses.[93]

The solution evolved in discussions between Marshall engineers and the crew of *Skylab* 2. Because the astronauts were in preflight quarantine, Center personnel wore surgical masks, giving the meetings a macabre atmosphere. A 75-person shade team conferred through the night of 16 May, sketching designs on a chalkboard. By the early morning of 17 May, they decided on a method in which astronauts on EVA would attach two telescoping poles to the telescope mount. Then, using lines and pulleys, they could stretch a protective cloth between the poles in much the same way as they would run out a clothesline.[94]

Developing the twin-pole shield was hectic. Henry Ehl found the aluminum sections to make the 55-foot-long booms by calling vice-presidents of aerospace companies in the middle of the night. Marshall flew in two seamstresses from NASA's spacesuit contractor in New Jersey to make the sail by sewing together three-foot-wide strips of cloth. There was even some humor. While Petrone and Thompson watched the sewing, one of the seamstresses pushed the material ahead with her foot. "It just isn't right," Petrone muttered, "You're not supposed to kick flight hardware."[95]

Center Director Rocco Petrone (seated second from left) and Deputy Center Director William R. Lucas (standing) are briefed about twin-pole sunshade.

But considering the circumstances, a clear division of labor existed with Schwinghamer and his Materials Division working on sail development, Gustave Krull's Engineering Division designing flight hardware, and J. R. Thompson's Human Factors Branch handling 1-g deployment tests. These engineers tried to remain conservative by using simple materials, testing everything, and following standard development procedures. They made the sail from the same ripstop nylon used for spacesuits and performed 37 tests on the system in seven days. These included tests on its latex coating to ensure it would not deteriorate in ultraviolet light, and on the deployment system on a *Skylab* mock-up in Building 4619, and on the "junk pile" in the Neutral Buoyancy Simulator. The engineers conducted the normal hardware reviews, although at a

Seamstresses sew Skylab's *solar shield at Marshall.*

Testing the twin-pole sunshade at the Skylab *mock-up in Building 4619.*

rushed pace. Marshall made the first sail on 19 May and tested a mesh mock-up in the simulator on 22 May. At four o'clock on the morning of 23 May, development teams were still working. After final review at six o'clock, Marshall sent the flight article to the Cape. The 112-pound folded sail was vacuum sealed in a breadbox-size container and launched on 25 May.[96]

Skylab *in orbit with Marshall's twin-pole sunshade.*

Marshall's efforts paid off and helped rescue *Skylab*. The Huntsville Operations Support Center changed the workshop atmosphere four times to purge it of any dangerous gases before the astronauts entered. Using the cutting tools and repair procedures developed in the Neutral Buoyancy Simulator, the *Skylab* 2 crew freed the jammed array. Although the astronauts had initially deployed Houston's parasol sunshade, it had not been treated to resist ultraviolet light and began to deteriorate. When temperatures in the workshop began to rise again, the *Skylab* 3 crew deployed the twin-pole shade in a six-hour EVA on 6 August. The workshop temperature quickly dropped to near nominal levels, and *Skylab* became a very successful program. The rescue of the workshop, J. R. Thompson thought, showed that "NASA functions best when it's flat on its back."[97]

The Agency established a board headed by Bruce T. Lundin, director of NASA's Lewis Research Center, to investigate the sources of *Skylab*'s problems on 22 May. The board visited the major *Skylab* Centers and contractors and quickly determined that the workshop's meteoroid shield had been poorly designed. Marshall and McDonnell Douglas had selected a deployable shield because it was lighter than a fixed shield.[98] But design engineers did not provide enough vents to allow air trapped underneath to escape, and development engineers did not cinch it close enough to the workshop to eliminate air. As *Skylab* gained altitude, the trapped air rose in pressure and eventually peeled off the shield.

Lundin's board decided that the "design deficiencies" had not been caused by improper procedures, limited funding, rushed schedules, or poor workmanship. The fault had been "an absence of sound engineering judgment" at McDonnell Douglas and Marshall. *Skylab* engineers had assumed that the shield was "structurally integral" with the S–IVB hull. Thus the Center and its contractor had failed to assign a systems engineer to the shield and project reviews had failed to discuss aerodynamic stress on the shield during launch. This led to "a serious failure of communications among aerodynamics, structures, manufacturing and assembly personnel, and a breakdown of a systems engineering approach to the shield."

To prevent such failures from recurring, the Lundin report offered two recommendations. First, each hardware project and subsystem should have a chief engineer responsible for "all aspects of analysis, design, fabrication, test and assembly." Second, NASA should encourage direct, hands-on examination of technology and avoid formal, abstract, ivory-tower engineering. Marshall and the rest of NASA implemented the first recommendation, and a chief engineer became a normal part of hardware development.[99] Ironically, however, other NASA policies undercut Marshall's ability to perform dirty hands engineering. Reductions in force and destruction of the arsenal system would increasingly make Center engineers into monitors of contractors rather than builders of hardware and would pressure them to rely on abstract information. Not surprisingly, problems like *Skylab*'s meteoroid shield would happen again.

Looking back on the shield problem, Center personnel had mixed feelings. Belew believed the Lundin report had been wrong; the design was efficacious. The problem, he thought, had been improper cinching of the shield to the spacecraft. But most Marshall engineers agreed the design was flawed. Kingsbury wondered how the Center had overlooked the flaw. Stuhlinger recalled that the Aeronautics Lab had warned that trapped air had to be vented, but this advice had not been heeded.[100]

Ironically the shield had been unnecessary. Marshall's engineers had incorrectly employed data from the Center's own Pegasus meteoroid detection satellites.[101] Marshall's Space Science Lab had analyzed information from the three Pegasus satellites and had determined that the potential danger of meteoroid hits to spacecraft was negligible. If *Skylab*'s designers had used Pegasus information, they could have deleted the shield because it improved penetration protection only marginally. A coat of paint could have provided thermal protection.[102]

After the rescue, the Marshall Center helped Houston operate *Skylab*'s power, control, and environmental systems and solar instruments.[103] Marshall personnel also provided engineering support for *Skylab* systems. While much of this was routine, Center engineers helped Houston and the astronauts conduct repairs. During the first mission, for example, a solar observatory power conditioner failed and a Marshall team decided that a physical blow to the switch might correct the problem. Working with backup equipment, they determined the location that the astronauts should strike with a hammer. The astronauts carried out Marshall's instructions and the power conditioner resumed functioning, thanks to the big hit. A more complex problem arose with the rate-gyroscope processors used to control the workshop. Several gyros overheated and had drift rates much higher than expected. A Marshall team studied the problem, detecting design flaws which could be corrected. Using the Neutral Buoyancy Simulator, the second *Skylab* crew learned how to make the repairs. They took replacement rate-gyros into orbit and successfully fixed the workshop control system. Repairs like these proved the necessity of linking development teams with operational teams.[104]

Skylab offered many lessons like this and Marshall's *Skylab* Program Office, at the request of Headquarters, compiled a list of "lessons learned." The primary lesson, the program office argued, was that management and engineering must be integrated and all parts of a program should be seen as one system. When many organizations develop "a single hardware entity" from many components, careful attention must be paid to systems engineering and integration. Clear design requirements should be established early in the program, interfaces should be carefully controlled, all changes must be tracked, and many different levels of review should be held. Among the many technical lessons was the necessity of designing hardware for in-flight repair.[105]

With the completion of the manned missions, NASA shut off the workshop's systems and closed down the *Skylab* program offices in March 1974. The next year Marshall helped write the denouement of the Apollo program when the Center provided the Saturn I–B launch vehicle and materials processing experiments for the American and Soviet Apollo-Soyuz Test Project.[106] Apollo was over, but a final chapter remained in Marshall's relationship with *Skylab*.

Managing Reentry

The Center was a principal actor in the story of *Skylab*'s fall to Earth. Marshall studies made during the mission assumed that *Skylab* would remain in orbit long enough for the Agency to complete the Shuttle. The fifth shuttle flight could then carry in its cargo bay a Marshall-built teleoperator retrieval system and propulsion module that could boost *Skylab* into a higher orbit for later reactivation.

The Center miscalculated, however, because solar activity was more intense than the predictive models anticipated. The hotter Sun was heating the Earth's upper atmosphere and increasing drag on *Skylab*. Indeed the Center's predictions were so much more optimistic than Houston's or the National Oceanic and Atmospheric Administration, some journalists and scientists charged that Marshall deliberately ignored the early decay of *Skylab* in order to justify funding for the teleoperator system. Dr. Charles Lundquist, head of the Center's Space Sciences Lab, denied the charges and argued that the different predictions were innocent products of different scientific models. In any event, budget crunches and technical problems were delaying Shuttle development and a possible reboost.

To keep *Skylab* from falling down before the Shuttle could fly up, NASA decided in January 1978 to reactivate the workshop and change its attitude to reduce drag.[107] At the end of February a team of eight—four from Houston and four from Huntsville—went to Bermuda to the only tracking station that could communicate with *Skylab*'s now archaic equipment. Heading the team was Marshall's Herman Thomason, who worked in the Systems Engineering Lab. Dr. Thomason had written his doctoral dissertation in 1969 on *Skylab* control methods. He later joked that he got the job because he "had been griping to management that something had to be done about *Skylab*. I guess I talked too long." His work was made difficult by the fact that many old *Skylab* hands had retired or joined contractors and *Skylab*'s technical documentation was lost or gathering dust.

With radar support from the North American Air Defense Command (NORAD), the Bermuda team made radio contact with *Skylab* on March 6. Initially communication was sporadic because the workshop was tumbling and could only transmit when its solar panels pointed at the Sun. Thomason's team tried switching to *Skylab*'s ATM batteries, but these kept shutting off because of low voltage readings. By April the team recharged the batteries by sending signals every 1.5 milliseconds, ordering the batteries to remain on and receive power from the arrays. Days passed before the batteries recharged. Meanwhile NASA trained more operators and activated four other tracking stations so that *Skylab* could be monitored continuously. Finally on 8 June, *Skylab* had sufficient power to operate the telescope mount's control moment gyros, and Thomason thought to himself, "we are in Fat City." The next day the team turned the workshop about and began a complicated balancing act; for a year they tried to maintain an attitude that minimized drag and fuel expenditure and maximized solar power.[108]

As the work continued through the summer and fall, NASA changed its policy. In December 1978, the Agency decided that the Shuttle would not be ready in time to reboost *Skylab*. Rather than trying to keep the workshop aloft, NASA would manage its reentry. The goal was to reduce risk of damage and avoid anything like the scare caused by the reentry over Canada of a Soviet satellite containing radioactive materials. NASA studies argued that the risk was minimal; *Skylab* was passing over a path that was 75 percent water and where 98 percent of the land had less than one person per acre. A person in the "footprint" had only slightly more chance of being hit by a piece of *Skylab* than by a meteorite. Even more than might have been the case otherwise, in an era of limited funding NASA wanted to avoid any blemishes.[109]

Thomason's team at Marshall played a major role in managing reentry. Officially the same division of labor existed between the Centers, with Houston controlling flight operations and Marshall providing engineering support. But Marshall's 40-person team worked in shifts around the clock in a *Skylab* Control Center and wrote computer programs to adjust attitude. They improvised computer and communication equipment because the original *Skylab* equipment had been scrapped or transferred to other projects. The team continually updated the programs to adjust for increased drag as the workshop fell and sent the programs to Houston where they were relayed to *Skylab*. It was a Marshall program, issued on orders from Headquarters, that on 11 July 1979 caused the workshop to enter its final tumble and end its flight. As a result *Skylab* passed over the east coast of North America and fell harmlessly over the Australian outback and the Indian Ocean.[110]

The Center, however, got little credit. Virtually all the credit went to Houston or to Headquarters. Newspaper reports were datelined from Houston and the official history of the *Skylab* program praised "the Houston team." This slight irritated some at Marshall. One engineer complained that "sure we are part of a team, but even in football the starting line-up has their name announced." Kingsbury, head of the Center's Science and Engineering directorate, said "I guess this is something like the guard or key tackle on a football team. No matter what they do, the camera points at the quarterback."[111] Ironically the Center that had played the largest role initiating *Skylab* got the least mention at its end.

Veterans of *Skylab* remembered the program fondly. Ise, the ATM manager, summarized the views of many by saying that *Skylab* was "the highlight of anybody's career that was associated with it." The project lasted only eight years from beginning to end, and in-house manufacturing created pride in workmanship. "The whole thing was just wrapped up in a nice, neat package with a bow on it. Then you can go back and look at it and say, 'That was it and I was a part of that.' It is something that is not so easy to do today." The difference between *Skylab* and later payload projects, Ise felt, was "the difference between building an Empire State Building and building a bunch of houses."[112]

Skylab indeed closed the Apollo era and helped open the way to a new period in the history of Marshall and NASA. As part of the Apollo era, *Skylab* benefited from arsenal practices, the Saturn V's heavy lift capability, and budgets and schedules which allowed adequate spares and testing. Later programs evolved with more restricted testing, fewer spares, and greater risks. *Skylab* also opened a new era in which Marshall diversified from propulsion to multipurpose engineering. Organizationally, the diversification contributed to a new NASA politics in which Centers competed for control of projects and technical designs.[113] Technically and scientifically, Marshall's diversification helped create a space station of a kind that made splendid contributions to space engineering, Earth observations, astronomy, medicine, and physics.[114]

Unfortunately NASA did not follow up the successes of *Skylab*. As former Houston *Skylab* Program Manager Robert F. Thompson observed at a *Skylab* reunion in 1988, *Skylab* was a "beautiful tactical program" that had "numerous shortcomings" as a "strategic program." *Skylab*, he said, had not been designed for in-flight repair, resupply with air and water, refurbishment with improved technology, revisitation for reboost to a higher orbit, or restructuring as part of a larger station. Consequently it could not, and did not, lead to a strategic, sustained human presence in space. Alternatively, as Marshall's Stuhlinger argued, NASA failed to establish such a long-term presence less because of the workshop's design and more because of the Agency decision not to launch the second *Skylab*.[115]

Even so from the perspective of the design and funding crises over a space station in the 1990s, the success of *Skylab* loomed very large. Many in the Agency wished that *Skylab* was still in orbit, and others, with only a little whimsy, wanted to take the backup workshop on display in the National Air and Space Museum and launch it.[116] Indeed in retrospect *Skylab* came to represent how Marshall and NASA had achieved important successes by imaginative use of existing hardware and pragmatic adaptation to budgetary realities.

---

1 Wernher von Braun, "A Space Man's Look at Antarctica," Popular Science, May 1967, pp. 114–16, 200; Tillman Durdin, "Antarctic Studies Said to Aid US Space Program Research," *New York Times* (15 January 1967).

2 Subcommittee on NASA Oversight of the Committee on Science and Astronautics, U.S. House of Representatives, 90th Cong., 1st sess., *Apollo Pace and Progress* (Washington, DC: U.S. GPO, 1967), pp. 1130–32.

3 W. David Compton and Charles D. Benson, *Living and Working in Space: A History of Skylab* (Washington, DC: National Aeronautics and Space Administration Special Publication–4208, 1983), p. 5. Compton and Benson's study of *Skylab*, while excellent, has a definite Houston-centered perspective.

4 Ernst Stuhlinger, notes, no date, Stuhlinger Collection, Box 6, Papers by Dr. Stuhlinger, 1956–62, Unfiled, Space and Rocket Center (hereinafter SRC); Roland W. Newkirk and Ivan D. Ertel with Courtney Brooks, *Skylab: A Chronology* (Washington, DC: National Aeronautics and Space Administration Special Publication–4011, 1977), pp. 4–6; Ernst Stuhlinger, Oral History Interview by Stephen P. Waring (hereafter OHI by SPW), Huntsville, AL, 8 January 1991, pp. 6–7; MSFC Committee for Orbital Operations, Orbital Operations Preliminary Project Development Plan, 15 September 1961, MSFC Archives; Compton and Benson, pp. 5–9, 20–27. For more on EOR, see Chapter 2 herein.

5 *Skylab Chronology*, pp. 10–11; David S. Akens, *Skylab Illustrated Chronology, 1962–1973* (MSFC, 1 May 1973), p. 3.

6 Leland F. Belew and Ernst Stuhlinger, *Skylab: A Guidebook* NASA EP–107 (Washington, DC, 1973), p. 6; *Skylab Chronology*, pp. 16, 23–25, 38.

7 Arnold S. Levine, *Managing NASA in the Apollo Era* (Washington, DC: National Aeronautics and Space Administration Special Publication–4102, 1982), pp. 148, 176, 204–05, 239–43.

8 Levine, pp. 275, 261–73; Compton and Benson, pp. 33, 52.

9 See E. Rees to H. Gorman, "Comments to your draft of AAP Plan date January 26, 1966," 27 January 1966, AAP—January–February 1966 folder, MSFC Archives.

10 Chris Kraft, OHI by AJD and SPW, JSC, 11 July 1990, p. 4; Maxime Faget, OHI by AJD and SPW, Webster, TX, 13 July 1990, p. 3.

11 Compton and Benson, p. 22; *Skylab Chronology*, pp. 47, 63; MSF Management Council Planning Session [summary], 23–26 May 1968, Apollo Applications 1968 folder, MSFC Archives; Faget OHI (1990), p. 4.

12 Leland Belew, OHI by SPW, Huntsville AL, 7 September 1990, p. 11; For incrementalism, see Howard E. McCurdy, *The Space Station Decision: Incremental Politics and Technological Choice* (Baltimore: Johns Hopkins University Press, 1990), pp. viii–ix, 66.

13 Stan Reinartz, OHI by SPW, Huntsville AL, 10 January 1991, pp. 3–4, 12.

14 *Skylab Chronology*, pp. 43, 47, 52; Akens, *Skylab Illustrated Chronology*, pp. 3, 7; Reinartz, pp. 11–12.

15 *Skylab Chronology*, p. 89.

16 *Skylab Chronology*, pp. 89–90.

17 Wernher von Braun and Robert R. Gilruth to George F. Mueller, "Post-Apollo Manned Spacecraft Center and Marshall Space Flight Center Roles and Missions in Manned Spacecraft Flight," 24 August 1966; NASA HQ to Gilruth, 7 October 1966, JSC, Center Series—Sullivan Management Series, Box 9, Management Council—Lake Logan Meetings Folder; Belew OHI, pp. 16–17.

18 Belew OHI (1990), p. 18.

19 Akens, *Skylab Illustrated Chronology*, pp. 7–11; Belew and Stuhlinger, p. 8; *Skylab Chronology*, pp. 69, 82, 83, 87; G. Mueller to Deputy Administrator, "Apollo Telescope Mount Installation," 19 July 1966, ATM 1969 folder, MSFC Archives.

20 Stuhlinger OHI, pp. 7–9; S/Associate Administrator for Space Science and Applications to AD/Deputy Administrator, "Establishment of Apollo Telescope Mount Project," 17 March 1966, Documentation ATM Folder, *Skylab* files, NHDD, NASA Release

66–185, "ATM Assigned to MSFC," 13 July 1966; L. Belew, "Weekly Notes—AAP," 6 June 1966, 13 June 1966, Fondren Library, Rice University.

21 Rein Ise, OHI, by SPW, MSFC, 18 December 1990, pp. 4–5, 9–10; Stuhlinger OHI, pp. 7–9; Compton and Benson, pp. 69–73, quote p. 73.

22 Ise OHI, pp. 6–10.

23 Reinartz OHI, p. 7.

24 Compton and Benson, pp. 69–73.

25 Dixon L. Forsythe, Program Manager Solar Observatories to Dr. John E. Naugle, "ATM—Proposed In-House Development Plan," 20 April 1966, Documentation ATM folder, *Skylab* files, NHDD; "Wide Ripples from ATM In-House Ruling," *Technology Week* (25 July 1966), p. 4; Henry Simmons, "AAP—Wednesday's Child," *Astronautics and Aeronautics* (February 1966), pp. 5–9.

26 See W. von Braun, " MSFC Responsibilities for the ATM Project," 6 October 1966, W. von Braun to H. K. Weidner and E. F. O'Connor, "AAP Management," 15 September 1966, H. K. Weidner and E. F. O'Connor, "MSFC Responsibilities for the ATM Experiments," 23 November 1966, ATM January 1966 folder, MSFC Archives.

27 Leland Belew, "Note to Dr. von Braun," 29 January 1968, *Skylab* Chronological File, Box 550, Rice University; MLP–1/Edward J. Lievens to MLP/Director, SAA Program Control, "Potential Trouble Spots on the ATM," 9 December 1966, Documentation ATM folder, *Skylab* files, NHDD; E. Stuhlinger to H. Weidner, "Discussions during Social Gathering with PSAC on April 11," 14 April 1967, Stuhlinger Collection, Box 2, January 1967, Record File—R&D Administration; PSAC, Joint Space Panels, "The Space Program in the Post-Apollo Period," February 1967, ATM folder, *Skylab* files, NHDD; Henry J. Smith, Chairman, Solar Physics Subcommittee to Dr. William Erickson, Astronomy Department, University of Maryland, 5 May 1966, Documentation ATM folder, *Skylab* files, NHDD.

28 Compton and Benson, p. 28.

29 Eberhard Rees, OHI by W. D. Compton, MSFC, 28 January 1976, pp. 2, 4, 7–8, Rice University.

30 Stuhlinger OHI, pp. 6–7, 14–15; E. Stuhlinger to M. Wright, 3 September 1993, MSFC History Office; Leland F. Belew OHI, by W. D. Compton, MSFC, 6 November 1974, p. 5, Rice University. See MSFC PAO, "Saturn I Workshop," 15 March 1968, MSFC History Office. For AAP schedules calling for Saturn V dry workshops, see Akens, *Skylab Illustrated Chronology*, pp. 7, 10–11, 16, 23, 24; J. Disher, Deputy Director Saturn/Apollo Applications, "Saturn/Apollo Program Summary Description," 13 June 1966 and "S–IVB Stage Space Station Concepts," 13 July 1966, *Skylab* Series, Box 538, ML–AAP, NASA HQ, January–June 1966, Rice University, Fondren Library.

31 Belew OHI (1990), p. 12; James Kingsbury, OHI by SPW, Madison, AL, 22 August 1990, pp. 9–11; Robert Schwinghamer, OHI by W. D. Compton, MSFC, 7 October 1975, p. 2, Rice University.

32 Levine, p. 197; Robert Schwinghamer, OHI by SPW, MSFC, 16 August 1990, pp. 12, 21.

33 NASA Release P69–10004, "NASA Underwater Tests," 15 January 1969, MSFC History Office.

34 NASA HQ to R. R. Gilruth, 7 October 1966, JSC Center Series—Sullivan Management Series, Box 9, Management Council—Lake Logan Meetings folder, Fondren Library, Rice University; Dr. R. R. Gilruth, OHI, by W. D. Compton and C. D. Benson, JSC, 6 August 1975, p. 13, Rice University.

35 Robert F. Thompson, OHI, by W. D. Compton, 18 December 1975, JSC, Rice University, pp. 4, 13; George McDonough, OHI, by SPW, MSFC, 20 August 1990, p. 4.

36 Faget OHI (1990), p. 11; Max Faget, OHI by W. D. Compton and C. D. Benson, 11 December, 1975, JSC, Rice University, pp. 8, 14; *Skylab Chronology*, pp. 71–73; Kraft OHI, pp. 5, 14; Gilruth OHI, p. 66; Compton and Benson, p. 92.

37 Belew OHI (1990), pp. 12–13; Reinartz OHI, pp. 12–14; Stuhlinger OHI, pp. 14–15.

38 Belew to von Braun, "Endorsement of Saturn/Apollo Applications Program Concepts," Stuhlinger Collection, Box 1, July 1966 Record file, R&D Administrative, ASRC.

39 Kraft OHI, p. 15; Compton and Benson, pp. 92–96.

40 Faget to Gilruth, "A Study of Apollo Applications Program using Saturn V Launch Vehicles," 23 April 1969, Rice University; Julian M. West to Gilruth, "Extended AAP Flight Program," 25 April 1969, Rice University; Thompson OHI, p. 11; Kraft OHI, p. 5.

41 Belew OHI (1990), pp. 12–15.

42 Reinartz OHI, pp. 13–14.

43 Belew OHI (1990), pp. 12–15; Compton and Benson, pp. 105–110; *Skylab Chronology*, p. 160.

44 McDonough OHI, pp. 3–6.

45 Haeussermann OHI, pp. 19–21; von Braun to Mueller, 23 May 1969, Fondren Library, Rice University.

46 *Skylab Chronology*, pp. 161, 167–68, 176–77; Akens, *Skylab Illustrated Chronology*, p. 32; Haeussermann OHI, pp. 19–20; MSFC *Skylab* Program Office, *MSFC Skylab Mission Report* (Washington, DC: National Aeronautics and Space Administration Technical Manual–X–64814, October, 1967), p. 2.1.

47 Draft of *Origins of NASA Names* SP–4402, *Skylab* (Apollo Applications) 1970–71 folder, Disher Correspondence file, NHDD; Reinartz OHI, p. 10.

48 MSFC, L. Belew, ed., *Skylab*, Our First Space Station (Washington, DC: National Aeronautics and Space Administration Special Publication–400, 1977), pp. 10–11, 18–21: MSFC *Skylab* Mission Report, p. 4.1.

49 The best technical summaries of *Skylab* systems are L. F. Belew and E. Stuhlinger, *Skylab*: A Guidebook NASA EP–107 (Washington, DC, 1973), chap. IV; MSFC, L. Belew, ed., *Skylab, Our First Space Station* (Washington, DC: National Aeronautics and Space Administration Special Publication–400, 1977), pp. 18–25, 44–50, 64–70, 95–101; MSFC *Skylab* Program Office, *MSFC Skylab Mission Report—Saturn Workshop* (Washington, DC: National Aeronautics and Space Administration Technical Manual—X–64814, 1967), chap. 4–10.

50 Belew and Stuhlinger, p. 10; Reinartz OHI, p. 19.

51 E. Rees to J. T. Shepherd and L. Belew, "*Skylab* Overruns," 14 July 1971, AAP 1971 folder, MSFC Archives; William K. Simmons, OHI, by W. D. Compton, MSFC, 30 October 1974, Rice University, pp. 13–14; Compton and Benson, pp. 207–11.

52 Reinartz OHI, pp. 19–21.

53 E. Stuhlinger to M. Wright, 3 September 1993, MSFC History Office; Ise OHI, pp. 10, 12, 14–15; Belew and Stuhlinger, pp. 73–80; MSFC Release 70–87, "Apollo Telescope Mount," 8 May 1970, *Skylab* files, NHDD; John A. Eddy, ed. by Rein Ise, prepared by MSFC, *A New Sun: The Solar Results of Skylab* (Washington, DC: National Aeronautics and Space Administration Special Publication–402, 1979), pp. 47–55; Compton and Benson, pp. 166–74; Haeussermann OHI, pp. 15–18.

54 Haeussermann OHI, pp. 23–24; Dr. Anthony DeLoach, OHI by SPW, 11 May 1992, pp. 14–15.

55 William C. Schneider, Director, *Skylab* Program to Associate Administrator, OMSF, "Examples of *Skylab* Cost Savings at the Expense of Payload Weight," 30 November 1972, *Skylab* Costs file, NHDD; Robert G. Eudy, "MSFC's Utilization of Existing Design Technology in the *Skylab* Program," Paper presented to American Astronautical Society 20th Annual Meeting, Los Angeles, 20–22 September 1974, pp. 1–9; Roger Chassay, OHI by SPW, MSFC, 24 July 1992, pp. 5–7.

56 Reinartz OHI, pp. 21–23; Ise OHI, pp. 16–17.

57 Ise OHI, p. 17.

58 R–1210, "R&DO Organizational Issuance, Lead Laboratory Assignment for Specific Tasks," 16 October 1967, R&DO Biomed Interface Data folder, *Skylab* Biomedical Experiments File, MSFC History Office; Phillip K. Tompkins, "Organization Metamorphosis in Space Research and Development," *Communication Monographs* 45 (June 1978), pp. 110–12; Ernst Stuhlinger, "Establishment of Lead Laboratories," 10 August 1964, Stuhlinger Collection, Box 1 Record File July–August 1964, ASRC.

59 Haeussermann OHI, pp. 24–27.

60 Tompkins, pp. 114–15; Ise OHI, p, 12; McDonough OHI, pp. 12–15; Lucas OHI (1989), p. 24.

61 Kingsbury OHI, pp. 9–11; Reinartz OHI, pp. 24–25.

62 Tompkins, pp. 112–14; *Skylab Chronology*, p. 117; McDonough OHI, pp. 10–11, 13, 16–17; Haeussermann OHI, pp. 11–12; "Central Systems Engineering Charter," Number 35, 18 December 1969, MSFC History Office.

63 W. R. Lucas, "Answers to Questions related to the Organization and Implementation of the Systems Engineering Effort and Related AAP Management Areas at MSFC," September 1968, R. Schwinghamer, informal notes, Action Items from Dr. Lucas file, W. R. Lucas to H. Weidner, "Lead Laboratory Concept," 8 November 1968, Biomed Management Chronological Master folder, *Skylab* Biomedical Experiments File, MSFC History Office.

64 Tompkins, p. 117; Ise OHI, p. 12.

65 L. Belew to H. Weidner, "AAP Activities Relating to MSFC Biomedical Support and Experiment Integration," 11 February 1969, R. Schwinghamer to R. Cook, 29 January 1969, R&DO Biomed Interface Data folder, *Skylab* Biomedical Experiments File, MSFC History Office.

66 See Chapter 4 herein.

67 *Skylab Chronology*, p. 105; Compton and Benson, pp. 123–25; Thompson OHI, pp. 11–12.

68 J. R. Thompson, OHI by AJD and SPW, 6 June 1994, pp. 16–17, 22.

69 Ise OHI, pp. 20–21; Reinartz OHI, p. 18; Compton and Benson, pp. 173–74.

70 Compton and Benson, pp. 132–33.

71 Simmons OHI, p. 8; Reinartz OHI, p. 16; Belew, *Skylab, Our First Space Station*, p. 18.

72 Raymond Loewy/William Saith, Inc., "Habitability Study: AAP Program," February 1968, MSFC Archives, pp. 4–7; Compton and Benson, pp. 130–40; Reinartz OHI, pp. 16–18.

73 "MSFC Man Will Take Part in Scientific Undersea Trip," *Marshall Star*, 9 April 1969; Chester B. May, OHI, by Jo Cummings, 20 December 1990, Huntsville, AL; Chester B. May, "The Man-related Activities of the Gulf Stream Drift Mission," (Washington, DC: National Aeronautics and Space Administration Technical Manual–X–64814, 1967); Jacques Piccard, *The Sun Beneath the Sea*, tr. by Denver Lindley, (New York: Charles Scribner's Sons, 1971); D. P. Nowlis, "Tektite 2 Habitability Research Program," NASA–CR–130034 (30 March 1972).

74 C. Kraft to R. Gilruth, MSC, "Habitability of *Skylab*" 6 April 1970, R. Gilruth to E. Rees, 10 April 1970, AAP 1970 folder, MSFC Archives; Compton and Benson, pp. 132–33, 139.

75 E. Rees to L. Belew, "*Skylab* Habitability," 15 May 1970, L. Belew to E. Rees, "*Skylab* Habitability," 7 May 1970, AAP 1970 folder, MSFC Archives.

76 Eberhard Rees to Robert Gilruth, 7 May 1970; Rees to Gilruth, 27 May 1970; Rees to Gilruth, 16 June 1970; Rees to Dale D. Myers, Associate Administrator, OMSF, 16 June 1970; Rees to Charles W. Mathews, Deputy Administrator, OMSF, 21 July 1970; Rees to Gilruth, 19 August 1970; J. R. Thompson, Chief, Man/Systems Integration Branch, to W. Simmons, "Habitability Provisions of the Orbital Workshop," 27 May 1970, all *Skylab* Chronological File, Boxes 500 and 551, Rice University.

77 Mathews to Rees, 30 July 1970, *Skylab* Chronological File, Box 551, Rice University; Compton and Benson, pp. 144–48.

78 Dr. Charles A. Berry, OHI, by W. D. Compton, Houston, 10 April 1975, p. 6, Rice University; J. Kingsbury, Chief Materials Division to R. Schwinghamer, "Biomedical Experiments," 22 September 1968, Correspondence from Team Members folder; W. Lucas, Directors Propulsion and Vehicle Engineering, "Bioastronautics Task Team Appointments," 13 September 1968, Extra Copies folder; MSFC Bioastronautics Task Team, "Preliminary Research and Development Plan for Biomedical Experiments," 11 April 1969, Experiment Support Systems folder, *Skylab* Biomedical Experiments File, MSFC History Office.

79 von Braun handwritten notes on R. Gilruth to von Braun, 16 December 1968; "Management Agreement for AAP Medical Experiment Hardware Items," 16 December 1968, p. 1, Extra Copies folder, *Skylab* Biomedical Experiments File, MSFC History Office; Compton and Benson, pp. 149–65, quote p. 155.

80 Floyd OHI, p. 5; Robert Schwinghamer, OHI by SPW, MSFC, 16 August 1990, p. 8; Berry OHI, p. 8.

81 Floyd OHI, p. 10; "Belew Notes," 5 May 1969, 14 July 1969, 24 November 1969, 16 March 1970, Fondren Library, Rice University; Schwinghamer OHI (1990), pp. 8–9; Berry OHI, pp. 8–10; Simmons OHI, pp. 14–16; Compton and Benson, pp. 155–65.

82 Gilruth OHI, p. 13; John C. Stonesifer, MSC to R. Schwinghamer, "*Skylab* Medical Experiment Hardware, Outstanding Performance During Flight," 21 February 1974, Lessons Learned on *Skylab* folder, MSFC Directors Files; C. C. Johnson, "*Skylab* Experiment M487: Habitability/Crew Quarters," 20–22 August 1974, MSFC History Office, folders file; Kenneth S. Kleinknecht, OHI, by W. D. Compton, JSC, 10 September 1976, pp. 5, 16, Rice University.

83 Belew OHI (1990), pp. 19–20; Haeussermann OHI, pp. 31–32; Jack Lousma, Habitability and Food Systems Panel, "*Skylab* Revisit: A Conference in Recognition of the Fifteenth Anniversary," 11 May 1988, Huntsville, videotape, MSFC History Office; for examples of cooperation on the medical experiments, see R. Schwinghamer, "Bioastronautics Task Team, Biomedical Experiment Development Weekly Notes," 18 November 1968, *Skylab* Biomedical Experiments File, MSFC History Office, and "Heimburg Notes," 3 February 1969, *Skylab* Series, Chronological Files, MSFC, Box 550 Fondren Library, Rice University.

84 R. Gilruth to C. W. Mathews, "Proposed Management Responsibilities—AAP," 29 March 1968, AAP January–March 1968 folder, MSFC Archives; MSC, Flight Control Division, "*Skylab* Program FOMR Operations Plan," 1 June 1972, *Skylab* 1972 folder, MSFC Archives; Compton and Benson, pp. 213–15, 217–18; *MSFC Skylab Mission Report*, pp. 3.32–35.

85 *Skylab Chronology*, pp. 305–306.

86 Paul Houtz, "Space Exec Speaks—Worked for 'Ten Years in May' Saving *Skylab*," *Birmingham News* (10 October 1973). Petrone became MSFC Director when Eberhard Rees retired on 26 February 1973.

87 Bob Lessels, "The *Skylab* Missions: Problems, Triumphs 15 Years Ago," *Marshall Star* (11 May 1988); James E. Kingsbury, OHI, 12 June 1973, pp. 19, 20, NHDD, Biographical file.

88 Schwinghamer OHI (1990), p. 10; James H. Ehl, OHI, 12 June 1973, p. 5, NHDD Biographical file; George Hopson, OHI, 12 June 1973, p. 13, NHDD Biographical file; Kingsbury OHI (1973), pp. 20, 50.

89 Belew OHI (1990), p. 24; Reinartz OHI.

90 Hopson OHI, pp. 7–13; "*Skylab* Missions"; Barry Casebolt, "The Saga of Salvaging *Skylab* at Marshall," *Huntsville Times* (29 July 1973), pp. 13–14; Jesco von Puttkamer, "*Skylab*: Its Anguish and Triumph—A Memoir," *Journal of the British Interplanetary Society* 35 (1982), pp. 541–42, 545, 547; Reinartz OHI, pp. 30–32; MSFC Public Affairs Office, "A Narrative Account of the Role Played by the NASA Marshall Space Flight Center in the *Skylab* SL–1 and 2 Emergency Operations: May 14–June 22, 1973," 24 June 1973, pp. 3–5, 11, 14, 57, 63–64, 67, quote p. 13, MSFC History Office.

91 von Puttkamer, pp. 542–43, 545, 547.

92 Eudy, pp. 9–12; "Narrative Account," pp. 39–41; Thomas O'Toole, "Astronauts to Try Again to Free Solar Panel," *Washington Post* (2 June 1973); von Puttkamer, p. 545; Casebolt, p. 15.

93 Schwinghamer OHI (1990), pp. 11–12; Robert J. Schwinghamer, "Performance of Solar Shields," 20–22 August 1974, pp. 3–4, MSFC History Office.

94 von Puttkamer, pp. 543–46; "*Skylab* Missions." See U. S. Senate, Committee on Aeronautical and Space Sciences, Hearing on the Status of *Skylab* Mission," 23 May 1973 (U. S. GPO, 1973).

95 Ehl OHI, pp. 2–4; "*Skylab* Missions."

96 "Narrative Account," 18–37; Schwinghamer, pp. 3–30; Casebolt, pp. 14–15; von Puttkamer, pp. 545–48; Linda Cornett, "From Parachute to Sunshade," *Huntsville News* (9 August 1973); Ehl OHI, pp. 6–7, 11–13.

97 "*Skylab* Missions;" Compton and Benson, chap. 14, pp. 300–302. See Wernher von Braun, "The Rescue of *Skylab*," *Popular Science* 203 (October 1973), pp. 110, 170–72.

98 Compton and Benson, pp. 88–89, 243–44, 276–77.

99 "Statement of Mr. Bruce T. Lundin, Director NASA LRC before the Committee on Aeronautical and Space Sciences, US Senate," 30 July 1973, Lundin Report folder, Disher Correspondence file, NHDD; George Low to James Fletcher, "Statement for *Skylab* Failure Investigation Hearings," 25 July 1973, Low Papers, NHDD; Rocco Petrone to OSS/NASA HQ, "NASA Investigation Board Report," 11 August 1973, *Skylab* Investigation folder, MSFC Archives. John P. Donnelly, NASA's Deputy Administrator for Public Affairs, thought the Lundin report was too mild and believed the report said, in effect, "Sorry old boy. We goofed on that one, but no one's perfect." He found it "incredible that an agency that so prides itself on redundancy . . . would presume no vibration, acoustic, flutter or aerodynamic testing of the shield to be necessary." See "Dirty Questions from John Donnelly," n.d., Lundin Report Folder, Disher Correspondence file, NHDD.

100 Belew OHI (1990), pp. 25–30; Kingsbury OHI (1990), pp. 16–17, 23–25; E. Stuhlinger to M. Wright, 3 September 1992, MSFC History Office. See also Reinartz OHI, p. 28.

101 See Chapter 7 herein.

102 Simmons OHI, pp. 5, 6, 12; E. Stuhlinger to M. Wright, 3 September 1992, MSFC History Office.

103 DeLoach OHI, pp. 1–8; Eddy, pp. 55–57.

104 Eddy, pp. 89–91, 115–17; "MSFC *Skylab* Mission Report," pp. 6.24, 7.42–43.

105 MSFC *Skylab* Program Office, "Lessons Learned on the *Skylab* Program," 22 February 1974, Lessons Learned on *Skylab* folder, MSFC Archives.

106 Edward C. Ezell and Linda N. Ezell, *The Partnership: A History of the Apollo-Soyuz Test Project* (Washington, DC: National Aeronautics and Space Administration Special Publication–4209, 1978); "Saturn Mission Implementation Plan for the Apollo-Soyuz Test Project, Prepared by Mission Planning and Operations Branch, System Requirements and Assurance Office," 7 May 1975, MSFC History Office, Fiche 1654; "MSFC to Manage Experiments for Apollo-Soyuz Project," *Huntsville Times* (17 October 1973).

107 John J. Fialka, "*Skylab* Could Fall Too Soon for Targeting," Washington Star (8 May 1978); R. Jeffrey Smith, "The *Skylab* is Falling and Sunspots are Behind it All," *Science* 200 (4 July 1978), pp. 28–33; Compton and Benson, pp. 361–64.

108 H. E. Thomason, "*Skylab* Reactivation," 31 March 1978, *Skylab* 1978–1979 folder, MSFC Archives; MSFC Release, "Engineering Tests Awaken *Skylab*," March 1978, MSFC History Office, Folder 489; Herman Thomason, "*Skylab* Mission Report: The

Final Days," August 1979, MSFC History Office, Folder 490; Garrett Epps, "The Last Days of *Skylab*," *Washington Post Magazine* (8 May 1979), pp. 13–14; "Those U. S. '*Skylab* Pieces': You Can Still Get Burned," *Huntsville Times* (13 July 1979); NASA Press Release, "*Skylab* Maneuver Planned to Extend its Orbital Life," 5 June 1978, MSFC History Office, Folder 485; Everly Driscoll, "*Skylab* Nears Its Reentry from Orbit in a Race with Time and the Sun," *Smithsonian* (n.d.), pp. 80–89, MSFC History Office, Folder 485.

109 Robert A. Frosch, NASA Administrator, "Statement to Subcommittee on Government Activities and Transportation, Committee on Government Operations, House of Representatives," 4 June 1979, MSFC History Office, Folder 490; "Questions/Answers on *Skylab* Reentry," Spring 1979, MSFC History Office, Folder 480; NASA, "The *Skylab* Reentry," no date, MSFC History Office, Folder 480; Compton and Benson, pp. 364–67.

110 MSFC Release 78–16, "*Skylab* Doing Well Under 24-Hour Surveillance," 20 October 1978; MSFC Release 78–120, "*Skylab* Orbital Attitude to be Reversed," 31 October 1978; Christine Duncan, "*Skylab* Reentry Draft," 18 July 1979, MSFC History Office, Folder 59; Thomason, pp. 30–31.

111 NASA Current News, "The *Skylab* Re-entry Special, Part 1: Post-Impact News Stories," August 1979, MSFC History Office, Folder 485; Compton and Benson, pp. 367–72; "*Skylab*, Pride of NASA, Now Pieces of Metal," *Decatur Daily* (12 July 1979); "Marshall Center: Skylab's Real Hero," (12 July 1979); "*Skylab* Chief Relieved at Splashdown," *The Birmingham News* (12 July 1979). As an example of Houston-centered coverage, see William Stevens, "Shift in Predictions Push Skylab's Controllers into a Final Gamble," *New York Times* (13 July 1979).

112 Ise OHI (1990), pp. 19–20.

113 See Howard E. McCurdy, *Inside NASA: High Technology and Organizational Change in the U. S. Space Program* (Baltimore and London: The Johns Hopkins University Press, 1993), pp. 126–27.

114 See Chapter 7 herein.

115 Robert F. Thompson, Space Vehicle Systems Panel, "*Skylab* Revisit: A Conference in Recognition of the Fifteenth Anniversary," 11 May 1988, Huntsville, videotape, MSFC History Office; E. Stuhlinger to M. Wright, 3 September 1992, MSFC History Office.

116 James E. Oberg, "Skylab's Untimely End," *Air & Space* (February/March 1992), pp. 74–79.

Chapter VII

Beyond "The Gate of Heaven": Marshall Diversifies

"Open the gate of heaven." With these words, recalled Ernst Stuhlinger, Wernher von Braun defined the Center's mission during the early Saturn years. Marshall would develop rockets for scientists and astronauts to use.[1] But cuts in NASA's funding in the late 1960s led the Center to redefine its role. As Saturn development wound down in the mid-1960s, however, Marshall had a head start in dealing with hard times. Consequently Von Braun reorganized his Center to compete with other NASA Centers for scarce resources. In 1968 von Braun designated Dr. William Lucas as "his vice president for new business" and head of the new Program Development office.[2] Diversification continued under the leadership of Eberhard Rees and Rocco Petrone, reaching fruition after Lucas became director in 1974.

By the mid-seventies Petrone's wish that Marshall become "a scientific bounty hunter" had come true.[3] The Center made major contributions to *Skylab*, scientific instruments, satellites, applied engineering, and a wide range of space sciences. Diversification would culminate when Marshall became Lead Center for NASA's two major scientific projects for the 1980s, Spacelab and the Hubble Space Telescope.[4] Such a variety of projects involving piloted and scientific spacecraft, and both engineering and scientific research were unmatched by other NASA Centers. Praising Lucas for making Marshall "a very diversified Center," Andrew J. Stofan, director of NASA's Lewis Research Center from 1982 to 1986, said, "Bill diversified that Center beautifully. That's one thing he really did well."[5]

When Marshall diversified, Center personnel confronted new technical and managerial challenges. Their solutions changed Marshall's culture and relationships with other organizations. Internally Marshall enhanced its scientific sophistication by adding researchers with doctoral degrees and expanding cooperation between engineers and scientists. Externally the Center extended circles of cooperation with academic scientists, other NASA Centers, commercial interests, and other government agencies. Such growth, not surprisingly, was accompanied by struggles to control new territory. Marshall's success in many struggles propelled the Center beyond the "gate of heaven."

From Specialty to Diversity

Throughout most of the 1960s, Marshall personnel worked primarily on one very big engineering project—the Saturn launch system. The technical and managerial challenges of developing the mammoth boosters and supporting the lunar landing mission necessarily led to specialization in engineering rather than scientific research. Center strengths were in areas related to propulsion technology such as metallurgy and fluid dynamics. German and American engineers avoided intimacy with science and scientists unrelated to rocketry, making the popular term "rocket scientist" a misnomer. A kind of polarization developed between scientists and engineers; Stuhlinger recalled that engineers often argued that "we will build a spacecraft, and when it is all said and done and we have the lock-and-key job completed, then the scientists may come in and hang their pictures on the wall."[6]

In part this narrowness was a legacy of Army practices. At White Sands, V–2 rockets had launched the instruments of American scientists.[7] But the real task of the Army Ballistic Missile Agency had been to develop launch vehicles. ABMA rockets nonetheless continued to offer opportunities for scientific research in the upper atmosphere.

Accordingly the German and American rocket engineers worked with outside scientists in a relatively clear division of labor. The Army provided launch vehicles and the scientists provided instrument packages. In 1958 on Explorer I, the first American satellite, the Jet Propulsion Laboratory and James Van Allen of the State University of Iowa developed instruments; ABMA supplied the Jupiter C booster and integrated the instruments into an ABMA satellite. The teamwork paid off when Explorer I discovered radiation belts in the Earth's magnetosphere. Even after ABMA's group became the Marshall Space Flight Center in 1960, outside scientists and the rocket engineers continued this relationship in the Explorer and Pioneer programs. Relying on scientists from universities and research institutes, of course, was nothing new for NASA, but Marshall never had hundreds of experimenters like the Goddard Space Flight Center (GSFC) or Jet Propulsion Laboratory (JPL).[8]

Marshall's few scientists primarily supported engineers. Personnel with science training worked in all of the Center's laboratories. Scientists in the Aero-Astrodynamics Lab studied wind loads during launch and others in the Test Lab investigated the acoustic-seismic effects of engine tests. Most scientists, however, worked in the Research Projects Lab headed by Stuhlinger. Research Projects had the fewest permanent personnel of any of the Center's eight labs; it had only 87 permanent onboard slots while five other labs had over 600 each.[9]

ABMA created the Research Projects Lab in 1956 and teams working on the Explorer and Pioneer projects formed its nucleus until 1962. Personnel supported the satellite programs with management and design studies, devising scientific requirements for engineering development. While still part of the Army, the lab designed and built spacecraft for Explorers I, III, IV, and VI, and later in NASA did the same for Explorer VIII and XI. This was a major task since so little was known about the thermal, radiation, and meteoroid environment of space. By 1962 the lab widened and deepened its experimental research. Experts worked on spacecraft thermal control, radiation environment and shielding analysis, meteoroid protection, electric (plasma and ion) propulsion, materials research, and lunar soil and terrain studies.[10]

Despite the utility of their research, the team struggled to get respect in an engineering-centered organization. Both German and American engineers expressed patronizing attitudes for payload work, referring to Research Projects as "Stuhlinger's hobby shop." Von Braun contributed to this attitude, Stuhlinger remembered, because the Center director preferred providing services for outside scientists to specializing in science. ABMA originally designated Stuhlinger's group as the Research Projects Office, rather than as a laboratory, signifying their inferiority to the engineering labs.

The scientists also lacked resources for research. Even in the early days in NASA, Research Projects had no budget allotment for scientific equipment. Bill Snoddy, then a young American scientist in the lab, recalled how his colleagues in 1961 and 1962 had to bootleg hardware using procurement lines from other labs. Von Braun, though reluctant to support science at Marshall, was delighted when finally shown the fully equipped scientific laboratory.[11] Within its engineering mandate, the Research Projects Lab still did useful science and played the leading role in two science projects, High Water and Pegasus. Project High Water was an experiment in atmospheric physics that emerged partly in response to criticism of the absence of science in stage-by-stage testing. The Block I Saturn I test flights lacked scientific instruments and had a dummy upper stage filled with tons of ballast sand. To add science to the tests, Marshall developed the simple High Water experiment, which NASA Headquarters publicized as "the first purely scientific large-scale experiment concerned with space environments" and as a "bonus" project that took advantage of Saturn's wasted lifting capacity.

On the second and third Saturn I flights in 1962, Marshall replaced the ballast sand with 86,000 kilograms of water and used explosive charges to release the water into the upper atmosphere. When exposed to the low pressure of the ionosphere, the water boiled violently, then quickly evaporated and became a frozen mist. Within three seconds an ice crystal cloud expanded to 10 kilometers in diameter and produced electrical discharges much like a thundercloud. Scientists on Earth, in planes and on ships, studied the events using cameras, radar, and radio receivers. From High Water they not only learned about clouds, but also about the effects of fluid and gaseous discharges on telemetry.[12]

A more sophisticated mix of scientific and engineering research came in the three meteorite sensing satellites of Project Pegasus. The idea for Pegasus came in 1961 when the earth-orbital-rendezvous mode was still under discussion and Marshall engineers were worrying about meteoroid impacts on orbiting vessels. To maintain conservative standards and check designs of spacecraft and fuel tanks, they wanted more information about meteoroid size and frequency. Accordingly the Research Projects Lab conceived detection satellites, and Center personnel and the Fairchild Corporation built them.

The Pegasus satellites were mounted on an S–IVB second stage, and each had detection panels with a wingspan of 15 meters, electronic sensors and communicators, and solar power panels. Making use of Saturn's lifting capacity, they were NASA's largest satellites to date and were easily seen from Earth, having a surface area 80 times larger than Explorer meteoroid detectors. NASA launched the satellites in the spring and summer of 1965. Although Pegasus I had a flawed communications system, the second and third missions worked perfectly with Marshall's improvements. Marshall personnel monitored the missions from a Satellite Control Center at Kennedy Space Center and quickly analyzed the data so NASA engineers could use them immediately.[13]

One newspaper columnist criticized the program, writing that Pegasus set "a record for futility even in the annals of the National Aeronautics and Space

Administration." He thought that the program's hidden purpose was to justify the cost of Saturn I, Wernher von Braun's "$900 million dead-end kid," which, before Pegasus, had launched nothing "more glorious than a few tons of over-priced, 'space-rate' ballasting sand."

Such criticism unfairly ignored the achievements of Pegasus and how the satellites had yielded valuable information about meteoroid size and frequency. Before Pegasus, data had been highly uncertain and had indicated that spacecraft would be vulnerable to meteoroid damage. Pegasus data showed that the danger was minimal and that protective standards could be greatly relaxed. NASA engineers used Pegasus to create criteria for spacecraft design and ensure the success of the Apollo Program. Von Braun believed that the "Pegasus data have really become the main criteria . . . for all manned and unmanned spacecraft."[14]

Pegasus notwithstanding, in comparison to later years, Marshall personnel in the 1960s worked on science projects that were limited in number and range. Stuhlinger grumbled in a Weekly Note in 1969 that science at NASA remained just "a stepchild," and in 1966 one of his lab's division chiefs lamented that scientists had "the lowest priority in the budget."[15] As Saturn development came to a close in the late 1960s, however, Marshall personnel found opportunities to diversify.

By 1969 the *Skylab* Program and the Program Development Office sponsored multiple, sophisticated scientific projects. *Skylab* was significant not only because it represented the first big project outside of propulsion, but also because it combined manned flight and space science.[16]

At the same time Marshall was coping with the new technical challenges of *Skylab*, von Braun and his top assistants worked out a new Center strategy and organization.[17] Faced with declining budgets, manpower limitations, and Headquarters' pressure, Center managers decided in late 1968 that Marshall's survival depended on winning new projects, especially big science projects. Consequently the organizational changes sought to make science more prominent. Von Braun appointed Stuhlinger to the new post of Associate Director for Science. Von Braun created the position reluctantly, Stuhlinger remembered, and only after NASA Administrator James Webb urged him to improve the "image" of Marshall among scientists. Von Braun also created the Program Development Office and chose Dr. Lucas as its head. The broadening of mission also showed in new organizational names. The Research and Development Operations Directorate, the name for the laboratory side of the Center, became Science and Engineering. In 1969 the Research Projects Lab became the Space Sciences Lab.[18]

Following these changes, the Center diversified into new areas. By the mid-seventies, new research and development work included multiple *Skylab* and shuttle projects on solar astronomy, Earth resources, biophysics and materials processing, the HEAO series of satellites for high-energy astronomy, and the Hubble Space Telescope for planetary, stellar, and galactic astronomy.

To attract and support such scientific projects, Marshall began hiring scientists with doctoral degrees. This was necessarily a slow process given NASA's hiring limits and Marshall's personnel gaps. Stuhlinger, when asked in 1991 to describe Marshall's strengths in the space sciences during the late sixties, replied, "Sorry, almost none. There was practically no support for scientific work from Center management, and consequently not much from Headquarters." Although this was an overstatement, it was clear that to build strength, Marshall managers needed the support of Headquarters. In 1971 Center Director Rees complained to Harry Gorman, NASA Deputy Administrator for Management, that Marshall was working on a wider variety of important science projects than any other Center, but with fewer scientists. The Center, Rees said, had "an urgent need to continue to strengthen our in-house capability in space-related sciences."[19]

Marshall's Space Sciences Lab did become stronger. Finally protected by Center leaders from the reductions-in-force that decimated the rest of Marshall, the lab maintained about 150 personnel and gradually added Ph.D. scientists. While Center personnel was declining by one-third overall, the number of people holding scientific doctoral degrees increased.[20] By 1980 the Center had specialists in atmospheric science, solar physics, magnetospheric physics, high-energy astronomy, X-ray physics, superconductivity, cosmic rays, infrared physics, and microgravity science. Nevertheless Marshall never became a dominant NASA research center. The Center's managers had accepted the role of a development center, but had argued for the latitude to propose science projects. They laid out their position to Headquarters in 1968, just after the peak of the Apollo-Saturn program:

"Roles and missions [for field centers] are desirable only in a way which makes the best possible utilization of the Center's capability, experience, and motivation. The Centers should be encouraged to maintain a competitive position with other Centers within reasonable bounds. There is a danger in setting irrevocable roles and missions. We need to foster the Headquarters/Center relationship in much the manner that a customer/contractor relationship exists. The Centers should be free to submit competitive bids for projects for which they have the capability and capacity. The competition must not go to the point where the inter-Center relationships and mutual trust are damaged. For example, research Centers probably should not get heavily involved in development. Nor should development Centers get heavily involved in research. It would be equally wrong to legislate against research Centers doing any development or development Centers doing any research."[21]

Marshall's diversification and enhanced scientific sophistication did not bring a revolutionary change in culture. Dr. Charles R. Chappell, a physicist who came to Marshall in 1974 and later became associate director for science, observed that "S&E," the Science and Engineering Directorate, was "mostly E." The Center still had hundreds fewer scientists than Goddard or JPL. Moreover Marshall's scientists continued to play a role in engineering support as they conducted space science research. They sometimes believed that they lacked the autonomy experienced by their NASA peers and the resources needed to conduct research and maintain expertise.[22]

Most resources went to propulsion projects like the Space Shuttle. But in addition to being a propulsion Center, Marshall became an engineering organization for big science projects. As in the Saturn era, the Center's mission remained providing spacecraft and instruments for science rather than conducting all of the scientific research. Much of the experiment conception and analysis came from external scientists. Stuhlinger, in view of the strong orientation of the Center's top management toward engineering rather than science—but determined to maintain a high standard for Marshall's scientific projects—set forth the philosophy in 1966, arguing that Marshall should avoid the Goddard Space Flight Center's "authoritative way" of in-house science. Marshall should only help define the mission, provide cost and schedule constraints, and select competent project managers. The experimenter should define the goals of research, and NASA should provide assistance in producing a "flyable package that does not compromise the experimenters' objectives."[23]

By becoming an engineering center for space science Marshall diversified and survived. Essentially, the new strategy evolved out of rocket engineers' alliance with external scientists. Marshall took advantage of its strength in engineering and avoided confinement to particular scientific fields like NASA's research Centers. With this strategy, almost any field was open, and over time the Center's space scientists became known and respected by colleagues at other Centers and universities. The first big step through the "gates of heaven" was *Skylab*.

Skylab Science

For *Skylab*, the first American space station, Marshall was Lead Center, designing and developing the workshop and a substantial portion of its scientific hardware. The Center also led NASA efforts to solicit experiment proposals from external scientists, managed experiment integration, and ensured that scientific hardware mated with the workshop. Moreover, Marshall helped with engineering, operations, and research support for *Skylab* science. Particularly significant were the contributions to the Apollo Telescope Mount (ATM).

For *Skylab* experiments, NASA relied primarily on scientists from universities and research institutes. The complexity and quantity of experiments on board the workshop, however, led Marshall to develop a more formal organization for managing science and coordinating its activities with other Centers and outside scientists. A new Experiment Development and Payload Evaluation Project Office supported NASA's system for selecting experimenters and helped scientists build hardware.

Marshall managed 51 of 94 experiments flown on *Skylab*, including experiments in astronomy and solar science, engineering and technology, materials processing, student experiments, and science demonstrations. Engineering studies gained insights into thermal controls, habitability, crew vehicle disturbances, and spacecraft environment. The processing experiments examined metallurgy, fluid dynamics, and crystal growth (which are discussed later in this chapter).[24] NASA initiated the student experiments in 1971 in order to attract interest in *Skylab*. NASA and the National Science Teachers Association held a competition among high school students and Marshall helped select the winners from 3,409 entries. The Center also developed hardware for the 11 studies which ranged from fluid mechanics to spider web formation to earth orbital neutron analysis.[25]

During the *Skylab* missions the Space Sciences Lab also conceived several demonstration experiments. Astronauts on the first two missions asked for simple experiments to perform during their free time. In addition to their scientific and educational results, the demonstrations gave the third *Skylab* crew a change of pace. Since some experiments had clear objectives but offered limited guidance, the astronauts could choose the best method in orbit. The Space Sciences Lab devised demonstrations involving minimal equipment and studying such microgravity phenomena as the slow diffusion of liquids and the stability of a toy gyroscope.[26]

The variety and complexity of *Skylab* science forced Center engineers to adjust. Dr. Stuhlinger recalled that working on a project that included such a large program of purely scientific investigations was a new situation for Marshall. In the past, engineers at the Marshall Center had been working with other engineers, with engineering contractors, and with project and program managers from Headquarters. Much of the scientists' thinking, their way of planning and rationalizing, even their language was unfamiliar to them. During the early phases of the *Skylab* project, *Skylab* engineers and *Skylab* scientists lived in two different worlds. The engineers complained that the scientists "didn't really know what they wanted," and that they "changed their minds all the time"; and the scientists complained that the engineers "didn't even try to understand their viewpoints, and the needs of a scientific experiment."[27]

The Space Sciences Laboratory tried to bridge the gap between engineers and scientists. A team of Center engineers and scientists serviced each scientific specialty. An engineer worked full time on one or two experiments, helping in design, development, and qualification. Integration engineers worked on a group of experiments to maintain compatibility with *Skylab* systems. "What was new" for engineers, observed Rein Ise, project manager of the Apollo Telescope Mount, "was the appreciation of the science itself, that is the understanding of what the scientists were trying to achieve and the system [that] could best support them." Experiment scientists from the Space Sciences Lab acted as "representatives" for the principal investigators and helped engineers resolve development problems, thereby winning new prestige with their engineer colleagues and also with outside scientists.

Not until *Skylab*, when Marshall engineers became dependent on in-house scientists, Snoddy recalled, did they stop making references to the Space Sciences

Lab as a "hobby shop." "All of a sudden they had all these experiments from throughout the world that were flying on that thing . . . and suddenly they found it kind of handy to have some people here at the Center who understood this stuff and could interface with the scientists." Stuhlinger said "improvements came slowly [but] during the later phases of Project *Skylab*, cooperation between engineers and scientists worked well; MSFC had learned a few good and very useful lessons."[28]

Achieving cooperation between scientists interested in particular experiments and engineers involved with the whole workshop was not always easy. The ATM system, Marshall's first experience in developing and managing a science payload for a manned mission, was especially troublesome. The Center had to coordinate ATM operations with other experiments and resolve conflicts with the earth resources or medical experiments. The problems in planning operations were compounded by the lack of a chief project scientist at Marshall and at NASA Headquarters. One ATM investigator, Dr. Richard Tousey of the Naval Research Laboratory, complained to Stuhlinger in 1968 that "most of the problems which have plagued us in the ATM project are caused by the lack of a science-oriented person within the ATM project structure." Acting as liaison for the scientists, Stuhlinger warned Frank Williams, director of the Center's Advanced Systems Office, that "workshop planning" and "astronomy planning" were not "on a converging course" and that "if we lose the astronomers as customers. . . it will be most difficult to maintain a workshop development program."[29]

Conflicts over mission planning culminated in meetings in late 1970 and early 1971. The ATM principal investigators rejected the operations plan of Martin Marietta, Marshall's experiment integration contractor. Without informing NASA, the scientists developed their own plan. After the shock of this rebellion subsided, NASA accepted most of the scientists' program.[30]

Marshall's Space Sciences Lab managed the scientists' joint observing program. Lab personnel and the principal investigators established a team of scientists and technicians for each ATM instrument. Before *Skylab's* launch, the teams developed plans for maximizing research, making routine observations, and tracking dynamic solar events. Also before the mission, they practiced coordination with mission controllers and ground-based observatories. Cooperation with ground-based researchers around the world allowed for synergistic study of solar events and took advantage of ATM's ability to make simultaneous photographs in multiple wavelengths.

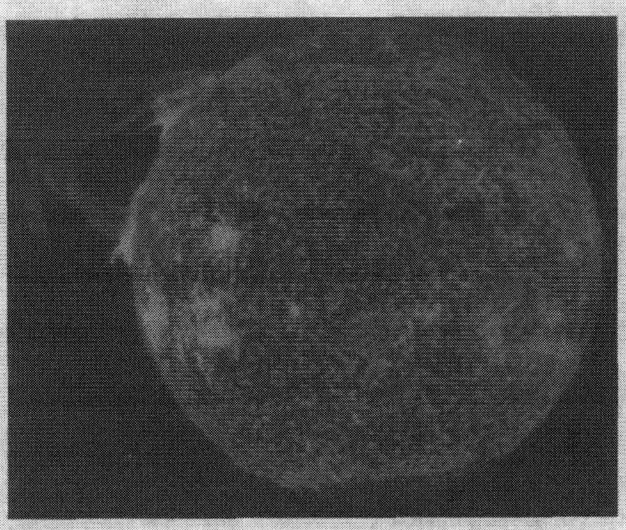
The dynamic Sun, photo from Skylab's *ATM.*

During the 13 months of *Skylab* missions, the Space Sciences Lab's assistant director and over 20 specialists moved to Houston and helped run the joint observation program. The NASA teams met daily with the investigators to plan observations and coordinate work with ground-based observatories and 300 solar scientists around the world. While operating, an "ATM czar" from Marshall oversaw a console in mission control and sent digital commands to *Skylab* and written instructions to the astronauts via a teleprinter.[31]

Marshall used similar procedures to help study Comet Kohoutek. Discovered in March 1973, astronomers expected it to be very bright. NASA developed a rush observation program using ATM instruments, and the four *Skylab* astronauts took into orbit the electronographic far-ultraviolet camera designed as backup for Apollo 16. Marshall managed the *Skylab* observations from November 1973 through February 1974, and Goddard coordinated NASA's work with other institutions. Marshall Center scientists contributed to studies of the comet's anti-tail and brightness. If the public was disappointed because Kohoutek proved dimmer than the media predicted for "the comet of the century," *Skylab*'s surveillance was a scientific success and showed the flexibility of a piloted orbital observatory. Kohoutek became "the best observed and studied comet in history," and the ATM instruments proved sensitive enough even though designed to view the Sun. Spectral evidence supported current theories that comets were composed of ice and primordial materials.[32]

After the missions ended, Marshall helped scientists interpret the data. The most elaborate support went to the astronomy experiments. With $15 million from NASA, Marshall managed an ATM Data Analysis Program that funded data archives, analysis, reports, and conferences. Teams of scientists from around the world met in three solar workshops to discuss and report findings. The wide spectrum of ATM instruments, the scientists found, revealed new information about the transition region between the cooler chromosphere and the hotter corona, coronal holes at the solar poles, magnetic fields around the Sun and their effects on the earth's upper atmosphere, and the dynamics of solar change. Scientists analyzed these discoveries for more than a decade and the ATM became, according to Leo Goldberg of the Kitt Peak National Observatory, "one of the most important milestones in the history of solar astrophysics."[33]

The success of *Skylab* and its science programs left a long legacy for Marshall. The contributions of Center scientists to *Skylab* made them "mainstream" and laid a foundation for cooperation with engineers on later projects.[34] Moreover *Skylab* formed the basis for later growth. Development of the workshop and the integration of its experiments helped Marshall become Lead Center for Spacelab and get a large role in Space Station efforts. Operations support during the missions set a precedent for Huntsville's science operations control facility for Spacelab. And the Center's work on *Skylab*'s scientific payloads, especially in solar astrophysics and materials processing, helped establish credibility among scientists and enabled diversification to continue.

The Satellite Business

Even during research and development for *Skylab*, the Center was already working on several satellites and scientific probes. These payloads were automated, unlike *Skylab's* ATM, and as a result Marshall had to work closely with other NASA Centers. The Center led successful efforts in high-energy astronomy, geophysics, and astrophysics.

One of the most elegant spacecraft was Marshall's Laser Geodynamic Satellite (LAGEOS). In 1964 geophysicists at the Smithsonian Astrophysical Observatory speculated that lasers aimed at a reflective satellite could help analyze the exact shape of the Earth and movements in its crust. They described their ideas in August 1969 at a NASA conference in Williamstown, Massachusetts, and later received support from Marshall. Since even the very thin atmosphere at orbital altitude would disturb the satellite, the experts recognized that the mass-to-surface ratio of the satellite should be as large as possible. The more massive the satellite was, the more stable it would be. Therefore the scientists proposed Project Cannon Ball, a four-ton sphere to be launched by a Saturn I–B. The designers had thought too big, however, and NASA Headquarters rejected the proposal because of the configuration's high cost.[35]

Marshall and the investigators returned to the drawing board, and in 1973 Headquarters approved a scaled-down satellite designated as LAGEOS. The new design carefully optimized weight and diameter. The passive satellite weighed over 900 pounds and had no moving parts or instruments. Its aluminum shell and solid brass core optimized the mass-to-surface ratio. Brass alone would have been too heavy to launch cheaply, and aluminum alone too light to orbit stably. The designers of LAGEOS also had to choose a diameter large enough to maximize the number of mirrors and small enough to minimize drag. They chose a 24-inch diameter which allowed 426 fused silica retroreflectors, making the completed LAGEOS look like a "cosmic golf ball." Because the sphere would stay in orbit for more than eight million years, NASA decided to mount a plaque inside to show its geologic mission.[36]

Although LAGEOS development was a team effort, Marshall did most of the work in-house. Perkin-Elmer made the laser retroreflectors. Originally the Center intended to contract for a full-scale prototype and a flight model, but since machine shops in the Test Lab and the Quality Lab were working 30 percent below capacity, Center management decided to build the prototype in-house. Technicians machined

Assembly of LAGEOS at MSFC.

the holes and mounting rings for the retroreflectors and assembled the sphere even as a RIF was under way to lay them off.[37]

Manufacturing LAGEOS was "a very precise high-tech job," Marshall engineer Lowell Zoller noted, which "benefited from the very specialized manufacturing capabilities that we developed during the Saturn Program." The Center's prototype was so finely crafted and performed so well in tests that NASA made it the flight model. "The guys did such a great job with the first one," said James Kingsbury, head of the Center's Science and Engineering Directorate, "that we never built the second one. I think that it is the only program in the history of NASA that came in under fifty percent of cost and on schedule."[38] Throughout the design and development phase, Marshall received scientific and technical support from the Smithsonian, Goddard Space Flight Center, and Bendix Corporation. Goddard also tested the mirrors, leading Marshall to alter the retroreflectors because six did not conform to design specifications. A Delta rocket launched LAGEOS in May 1976 and put it in a nearly perfect circular orbit.[39]

Thereafter Goddard coordinated research using LAGEOS, which had an operational lifetime of 50 years. Laser ranging stations around the world projected lasers at the satellite and its mirrors reflected the beams back to Earth. By timing the round trip of the beams, geophysicists could compute a location on Earth within two inches of accuracy. This enabled measurement of shifts in polar ice, tectonic plates, and fault lines. In addition to improving knowledge about changes of the Earth's crust, scientists hoped LAGEOS would help predict earthquakes.[40]

In the early 1970s, Marshall also managed Gravity Probe–A (GP–A), which had science as elegant as LAGEOS and more exasperating engineering challenges. In the late 1960s, scientists—again from the Smithsonian Astrophysical Observatory—proposed a redshift experiment to explore the structure of space-time and test one of Einstein's thought experiments in his theory of relativity. According to his "equivalence principle," the effects of gravity and constantly applied acceleration could not be distinguished, a fact that would cause "warping" of cosmic space-time. Consequently, two clocks located at two different places with different gravity levels would tick at different rates. A higher gravity level would cause a slower rate. Thus by comparing the two clocks, one stationary on the surface of the Earth, and the other moving in weightlessness onboard a free-coasting spacecraft, the earthbound clock would lag behind the spaceborne clock. The two-clock experiment would measure how several simultaneous effects contributed to the time difference: first, the classical Doppler effect between a stationary observer and a moving source; second, the relativistic Doppler effect between observer and source, described by one of the Lorentz equations in the special relativity theory; and third, the relativistic equivalence effect described in Einstein's general relativity theory. Effects one and two were experimentally well proven and accurately known. Gravity Probe–A would allow measurement of the third and thereby test Einstein's general theory. In late 1969, NASA Headquarters asked Marshall to help define this experiment. After rejecting a proposed satellite in an eccentric orbit for excessive cost, in 1971 NASA accepted Marshall's proposal for a suborbital flight, Gravity Probe–A.[41]

GP–A was a joint project of the Smithsonian and the Center. The experiment required two super-accurate clocks, which the Smithsonian developed using atomic hydrogen technology. The clocks lost less than two seconds every one hundred million years and functioned within five thousandths of one percent of prediction. In addition to supporting the Smithsonian's work, Marshall designed and built in-house the payload container and its power and communication systems. The Center also integrated the container with the clocks and instruments, tested the communications systems and the entire package. The finished probe was 45 inches long and 38 inches in diameter, weighed 225 pounds, and would spin during its hour-long flight.[42]

Perhaps not surprising given the sensitivity and complexity of the equipment, the development of the probe was difficult. The Center and its partners encountered problems with its very stringent thermal-control system, electronic parts, the clock and leaks in its pressure vessel, and the probe's spin dynamics and communication systems. The technical challenges, however, were exacerbated by people problems.

Initially Marshall blamed the Smithsonian for managerial failures which led to technical breakdowns. But Center managers admitted in August 1974 that "MSFC had underestimated the difficulty and complexity of the project" and failed to penetrate its contractor and provide enough resources. Therefore the Center had added more people and assigned a resident manager to the Smithsonian. It also required that the Smithsonian assign more people and improve its quality practices. Nevertheless, by late November, NASA's Office of Space Science and Applications informed Marshall that it was "considering cancellation of the GP–A Project in view of the long series of incidents."[43]

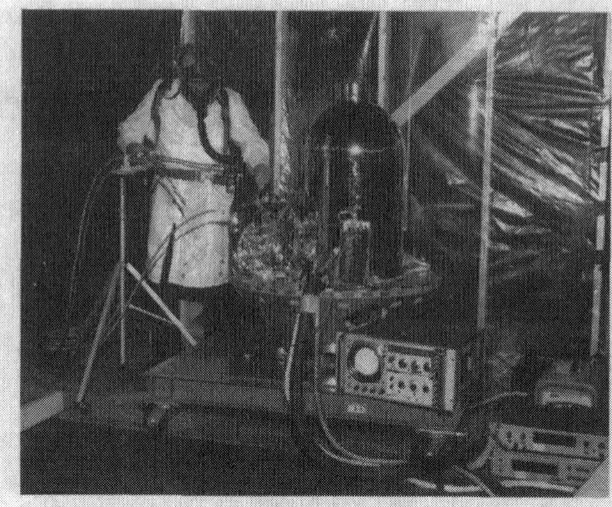

Final check-out of Gravitational Redshift Probe–A at MSFC.

Problems continued, culminating in a test failure. In December the Systems Dynamics Laboratory ran a vibration test on the entire probe payload, unaware of its limited capacity to withstand lateral axis shock. The test was too strenuous and damaged parts of the probe, a serious error since Marshall was using a "protoflight" concept in which the qualification model used for testing would be refurbished and used in flight. An internal investigation revealed a "breakdown in communication" between the development and test organizations "similar to the problems that caused the loss of the meteoroid shield in *Skylab*." Center managers took technical responsibility from the project office and assigned it to the labs. Although communication problems did not recur, technical glitches slowed development. Gravity Probe–A went two million dollars over budget to cost nine million dollars and its schedule slipped over one year.[44]

In June 1976 a four-stage Scout D rocket launched the probe from Wallops Island on a two-hour elliptical flight over the Atlantic. The probe attained a peak altitude of 6,200 miles and scientists compared readings from its clock with another at Cape Kennedy. The experiment was a full success and demonstrated the validity of this part of the General Relativity Theory to an accuracy never before attained. After the flight the principal investigators thanked the Center for helping "benefit the science of the experiment." They stated that Gravity Probe–A was "the first direct, high-accuracy test of the . . . [equivalence principle] and a beginning in the use of high accuracy clocks in space to measure relativistic phenomena."[45]

The biggest satellite project Marshall managed between *Skylab* and Hubble was HEAO, the High Energy Astronomy Observatories. The new discipline of high-energy astronomy studied X and gamma radiation and cosmic-ray particles. To detect these forms of radiation, which have shorter wavelengths and higher frequencies than visible light, astronomers in the discipline depended on access to space. Initially they used instruments flown in sounding rockets or balloons, but recognized that satellites would be better. To get a satellite program, they formed a coalition in the late 1960s, drawing help mainly from the Harvard-Smithsonian Center for Astrophysics, Naval Research Laboratory, MIT, American Science and Engineering Corporation, and the Space Science Board of the National Academy of Sciences. Attracting support from NASA scientists, the coalition needed the backing of a field center.[46]

Meanwhile, before the formation of the Program Development Directorate, Marshall's Research Projects Lab was looking for new work. Stuhlinger met with the astronomers and wanted the project. Although the lab had no X-ray and gamma-ray astronomers, its Special Projects Division had radiation experts who had worked on NASA's defunct nuclear propulsion program. Stuhlinger organized these people, under the leadership of Jim Downey, into an Electromagnetic Radiation (EMR) Project team.

The EMR team was less an instigator for the project and more an integrator, helping the scientists to conceive instruments and define technology for HEAO. Initial plans, similar to early concepts for the ATM, called for reconfiguring a lunar module to support X-ray instruments and using a Saturn V launch vehicle. From the beginning, the EMR team, Downey recalled, had many obstacles to overcome. First, since it had been put together on an ad hoc basis, the group lacked the sophistication and standing to build a coalition behind high-energy astronomy. "We were just trying to get some ideas so that we would have a respectable proposition" to present to Headquarters. The team "bootlegged the work" for more than a year, he said, on a strictly "catch-as-catch-can" basis. Even though the EMR team was moving the Center into a new area, support from von Braun and lack of bureaucracy created "an environment of innovation and creativity." "We just didn't know what we were supposed to be able to do," Downey thought. "Maybe we were just too young to be as easily constrained to a system. I don't know, but I don't think we could do it today" because a project has to become "kind of official before you can start working on it now. To me in those earlier days, we would create the project."[47]

Even though Goddard had more experience in astronomy, Marshall got NASA's formal support for the HEAO proposals in late 1969. Several reasons accounted for this. Goddard was busy with other projects, and during austere times NASA could not provide more personnel. In contrast Marshall had a personnel surplus. Moreover Goddard supported the project because of its scientific merits and because Marshall's role would not threaten its dominance in astronomy. As Lead Center for development, Marshall would "provide assurance that the HEAO Project is technically sound, remains on schedule, and is accomplished within available resources." This mated Huntsville to another Center, with Marshall managing the design and development of launch vehicles, spacecraft, experiments, support facilities, and vehicle operations. Goddard would be Lead Center for science, having charge of mission planning and data analysis; project scientists for the first HEAO missions would be Goddard experts.[48]

Based on this division of labor, Marshall established a project team in Program Development to make detailed experiment plans and vehicle designs. In March 1970, NASA released an Announcement of Opportunity for four HEAO missions and by November had already selected experiments for the first two satellites. TRW became the prime contractor for the spacecraft. HEAO plans called for "the largest payloads ever considered for an automatically operated US spacecraft," weighing 21,000 pounds and stretching 40 feet. Downey believed that Marshall encouraged the astronomers to "think bigger than they had been thinking" because the Center "had the big rockets" and "we thought big." Unfortunately Marshall's plans may have been too big, because NASA suspended HEAO in January 1973.[49]

Budget cuts by the Nixon administration led the Agency to slash funding for automated projects and to "descope" (NASA's term for downgrade) the HEAO series. HEAO would have to be redesigned to cost one-third to one-half as much. In dealing with monetary constraints, Marshall faced management challenges far different from the lush funding of Saturn or even *Skylab*. Survival of HEAO, observed Dr. Fred A. Speer, Marshall's program manager, "depends upon our success here at Marshall in outlining a lower cost program which will obtain a major part of the scientific results sought in the original HEAO plan."[50] In the first months of 1973, Marshall's project office planned the reductions with the HEAO astronomers and contractors. They decided to postpone the beginning of the missions, setting back the launch of the first satellite from 1975 to 1977, and to economize by shortening the mission. Three small satellites replaced the original four large ones. At three tons each, they were one-third the weight of the originals, but nonetheless very heavy scientific satellites. Atlas-Centaur boosters, rather than Titan IIIs, would launch the spacecraft. More than half the original experiments stayed; the X-ray instruments were light enough, but the cosmic- and gamma-ray instruments were too heavy and had to be redesigned. To keep costs low, Marshall also decided to use as much off-the-shelf hardware as possible.[51]

Using old hardware led to some awkwardness when the Grumman Corporation claimed that it could readily make HEAOs using hardware from Orbiting Astronomical Observatory satellites, a NASA program dating from the 1960s. As a result, Marshall decided to retain TRW as prime contractor. Although the Center justified its decision on legal and technical grounds, it also worried that the Grumman alternative would cost more money; moreover, building a new satellite would provide Marshall with more work than merely adapting an old one. At one point Center Director Petrone kept Grumman executives at bay by claiming his calendar was full for an entire month.[52]

Meanwhile Marshall tried to maintain support behind the HEAO program. The Center stuck with HEAO, Speer recalled, because getting work "was always a consideration after Apollo." To maintain support, Speer at the time counseled the investigators in "the need to act quickly and in keeping criticism on actions taken under control." Although at least one scientist referred to the descoping as NASA's "massive insult to science," most contained their resentment. Realizing that their specialty bound them to the Agency, the scientists launched a campaign for HEAO in NASA and Congress. Success of UHURU, the first X-ray satellite, made their lobbying easier. In October 1974, NASA Administrator James C. Fletcher promised that HEAO would be the Agency's "Number One priority" between Apollo-Soyuz and Shuttle, and in July 1974 development funding for HEAO resumed.[53]

In the restructured program, each HEAO satellite had a specialized mission. HEAO–A and HEAO–C scanned the heavens to make maps of the whole sky. Each rotated end-over-end every half-hour but kept its solar arrays pointed at the Sun for power. HEAO–A scanned for X-ray sources and low-energy gamma flux and HEAO–C for gamma-ray emissions and cosmic-ray particles. HEAO–B pointed at sources identified by HEAO–A and had the first pointed X-ray telescope ever built. Its instruments, 1,000 times more sensitive than any before, turned on a "lazy Susan," rotating in the focal plane of the telescope mirror.[54]

To achieve HEAO's scientific goals, Marshall realized that budget and schedule constraints had to be maintained. If the Center and its partners did not use their resources wisely, the scientific instruments would never reach orbit. "The cost and schedule" of HEAO, Speer said in 1980, were "tightly controlled right from the beginning."[55] Accordingly time and money determined many technical decisions during the development phase of the mid-seventies.

To minimize testing, Marshall and TRW used off-the-shelf space hardware like gyroscopes and star sensors. Almost 80 percent of the components for the HEAO–A spacecraft came from Pioneer, OSO, GEOS, and other satellites. The Center and its partners also standardized the three HEAO spacecraft with common computers, solar arrays, and equipment modules to support instruments.[56] Another way of saving development money was substituting "protoflight" for "prototype" testing. The traditional Marshall engineering approach was to build a prototype, or qualification model, for testing, and then use the lessons learned to build an improved flight article. Protoflight used a single piece of hardware for tests and flight. Richard E. Halpern, director of high-energy astrophysics at NASA Headquarters, told the Center to take a protoflight approach because the project lacked the money to build both a prototype and a flight model. HEAO's budget shortfall, Speer remembered, led his team "to rethink some of these Marshall traditions. One of the first campaigns I took was to persuade my lab directors and my Center Director to give up on this prototype concept." Marshall accepted protoflight partly because Goddard had used it successfully, but mainly because it helped "bring the price tag down." In the end protoflight reduced costs 30 percent below original cost estimates of prototype-based development.[57] Marshall's efforts to maintain budgets and schedules sometimes triggered conflicts with the scientists and their contractors. Protoflight reduced costs only if Marshall minimized hardware changes. The astronomers, however, often worried that resistance to change could prevent improvements and ultimately jeopardize research. Dan Schwartz of the Center for Astrophysics argued that "if you don't do it with a certain quality, you get nothing. I felt that NASA was always pushing that threshold." Another investigator believed that Marshall thought like a "bridge builder." "It would be a disaster to build a bridge an inch too short, it would be silly to make it a foot too long. They very much stuck to the minimum requirements, when a little extra might have yielded a substantial gain in quality." In one case the scientists resorted to subterfuge to get improvement. When Marshall turned down a telemetry system checker that monitored data errors, the scientists resubmitted the same device as a "block encoder" and Marshall allowed it.[58]

The Center's close management of contractors and insistence on proper records also caused conflict. To the scientists, government record keeping was oppressive red tape. They later griped that if they had done all the paperwork, "the thing would still not be in orbit." Disagreements culminated on HEAO–B experiments in 1975, when the Center blamed the scientists and American Science and Engineering for being lax and raising costs. This charge incensed Dr. Riccardo Giacconi, the pioneer in high energy astronomy, who protested to Headquarters that the scientists had more "carefully husbanded" resources than Marshall, and that "the level of visibility was neither sufficient for MSFC to closely monitor expenditures, nor adequate to foresee difficulties before they occurred."[59]

When the issue resurfaced in 1976, the president of American Science and Engineering complained to Speer about Marshall's excessive oversight of the project. Marshall's management, he argued, had "deteriorated to the point where it is not useful and is, in fact, detrimental to the program." He believed that Marshall was making so many requests for so many kinds of information from so many people that responses "often require the expenditure of effort in conflict with our internal priorities." The controls, he said, prevented his firm from "meeting our contractual requirements on schedule and with minimum costs." Speer agreed that the goal should be "more efficient communication, not less" and that Marshall would change its practices and seek only meaningful information through as few channels as possible.[60]

Generally, however, Marshall people defended the way they managed HEAO, pointing out the differences between the approaches of scientists and engineers. Astronomers, Speer observed, "didn't particularly enjoy being X-rayed on their design project. . . . The PI (principal investigator) felt that he was in control of his experiment and he knew better than anyone in the world what it should do and how it should be built. He minded somebody from Marshall whom he considered not on par with his scientific capabilities to start questioning him on some things." But Speer thought that success of HEAO caused the scientists to admit that, "Yeah, we didn't particularly like it, but we agree now that it probably was not a bad idea to go through this sort of scrutiny."[61]

Dr. Thomas Parnell, a Marshall employee and project scientist for HEAO–C, said that Marshall's penetration was "a real shock to people who haven't been through it. We must prethink everything in nagging detail, everything that could go wrong and prepare for it. Also we have to worry about cost. The paperwork raises the cost, but it guarantees that, when we launch, everything we can do to ensure success is done." By the same token, working on scientific instruments led Center engineers to change their attitudes. Parnell believed that in the beginning, engineers thought a scientist was "esoteric and not very practical" and "should write his requirements out on paper initially and then get out of the way." But whenever problems emerged in development, the engineers had to abandon preconceived paper requirements and seek the advice of scientists. Thus HEAO's technical challenges, Parnell concluded, forced Center engineers to become more flexible in how they managed scientific projects.[62]

Marshall also tried to save money by performing some tasks in-house. Arsenal capabilities, however, were mostly gone by the middle 1970s and the Center built no HEAO components. Marshall contributed to development more as designer and manager than as manufacturer. The Center's labs helped with spacecraft design, especially with troublesome gyroscopes. The Quality Laboratory ran a control center for electronics parts, and other labs helped with systems engineering and testing. The most lofty tests occurred aboard high altitude balloons. Marshall coordinated tests of cosmic- and gamma-ray detectors conducted aboard five balloons between September 1974 and May 1977.[63]

Marshall employees recover Stratoscope II telescope after balloon flight near Bald Knob, Arkansas, in September 1971.

HEAO–B telescope in MSFC's X-Ray Calibration Facility.

A more lowly but lengthy test of HEAO instruments occurred in the summer of 1977 in Marshall's X-ray Calibration Facility. Marshall built the facility in 1975–1976 to simulate X-rays from distant celestial objects and thus test an American Science and Engineering telescope for HEAO–B. The Center estimated that construction would cost $7.5 million but used surplus equipment from previous programs and cut costs to $3.9 million. The facility consisted of a variable X-ray source connected by a pipe 1,000 feet long and 3 feet in diameter to a chamber that housed the telescope. The source, pipe, and chamber had to be evacuated together. The long distance was needed to test the telescope focus and produce an X-ray beam of very small angular divergence, approximating the parallel X-rays arriving from celestial sources. Original planning called for a six-month test period, but a lag in the construction schedule forced Marshall to condense the tests into one month. Marshall technicians and the principal investigators worked 24 hours a day in two overlapping 13-hour shifts. They conducted nearly 1,400 tests and found problems that led to reworking the telescope hardware. The computer software developed for data retrieval during testing was later used for the same purpose during flight.[64]

Marshall's management of HEAO costs was very successful. During a time in which the consumer price index rose more than 50 percent, the high technology program finished within 20 percent of the original cost projection. Center Director Lucas told a HEAO Science Symposium in 1979 that HEAO–A had been built "at a lower cost per pound than any other NASA automated spacecraft."[65]

The Center co-managed operations for the three HEAO satellites launched from 1977 to 1979. Marshall established an HEAO operations office at Goddard and the two Centers divided authority. GSFC's role was mainly scientific, supervising mission planning, scientific observation, and data analysis. Marshall's role was primarily engineering. Although MSFC personnel helped plan observations with Goddard and the investigators, they primarily directed spacecraft communication and control.[66] The partnership played to the strengths of both Centers.

The operations role lasted longer than expected because NASA extended the lifetime of the HEAO missions. NASA had anticipated that the lifetime of the satellites would be limited by the amount of thruster gas needed for attitude control. But the earth's upper atmosphere proved less dense and the satellite control systems more flexible than expected. Marshall and contractor technicians developed techniques to maximize scientific observations while minimizing attitude changes, and to use computer programs and spacecraft gyroscopes to economize on thruster gas. These methods allowed for dramatic mission extensions; HEAO–2, expected to last only 15 months before its fuel ran out, kept going for nearly 30 months.[67]

With the help of Marshall managers and technicians, the HEAO program became a great scientific success. For the first time, astronomers had clear images of high-energy radiation sources. HEAO–A found more than 1,200 new celestial X-ray sources. The focusing telescope on HEAO–B found thousands more sources and made detailed studies of the brightest ones. The first X-ray image of Cygnus X–1 from HEAO–B, one scientist said, was "almost like a religious experience." By providing new insights on supernovas, cosmic rays and heavy elements, superbubbles, flare stars and stellar coronas, neutron stars, black holes, pulsars, degenerate dwarfs, and quasars, the satellites showed the limitations of optical astronomy and the significance

First picture of the X-ray star Cygnus X–1, by HEAO–B, also known as the Einstein Observatory, November 1978.

of studying high-energy emissions. Thus, according to Wallace Tucker, an astrophysicist and a historian of the project, HEAO "not only changed our knowledge of the astronomical universe, it has changed the way that astronomy is done."[68]

The satellites of the seventies not only produced important scientific results, but also contributed to Marshall's growing reputation as a multiproject Center. The projects created opportunities such as the astronomy facility Astro–1 on Spacelab, and the relativity experiment Gravity Probe–B. Especially the HEAO series, Fred Speer observed, "opened the door to a new dimension of our business," establishing the Center as "a member of the Space Science club." In part because of the project, Marshall would become Lead Center for AXAF, the Advanced X-ray Astrophysics Facility, with instruments 1,000 times more powerful than HEAO.[69]

Space Materials and Microgravity Research

When Marshall began diversifying, its arsenal system engineering culture and propulsion specialty made the materials studies of microgravity science and applications a fertile field. Developing space hardware meant that Center engineers had to be experts on the properties of materials in space and allies of physical scientists studying the effects of microgravity. This collaboration pushed back the frontiers of a new science and would draw the Center into national debates about NASA's mission and the commercialization of space.

Under ABMA and in the early NASA years, the rocket engineers contributed to materials research, because developing boosters required producing new materials and knowledge about the effects of the space environment. For the Explorer satellites, the Research Projects Lab discovered how to protect spacecraft from large temperature swings with thermal control coatings. The rocket engineers, especially in the Materials Laboratory, certified that materials met requirements. The Center's labs developed Redstone graphite jet vanes, ablative nose cones, aluminum alloys for liquid oxygen and liquid hydrogen engines, and methods for welding and inspecting aluminum. They used the Pegasus satellites to gather information on the effects of striking particles on spacecraft. They learned how to manage liquids in low gravity and control liquid fuel floating in partially filled Saturn tanks. For *Skylab*'s crew waste and shower systems, Center technicians experimented on liquid dynamics in space.[70]

During the late sixties, furthermore, the Center helped NASA make a transition from materials engineering into the new field of microgravity research. Inevitably early research in a new discipline was exploratory and involved trial and error. In 1965 Marshall personnel established a drop tower in the Saturn V Dynamic Test Stand in which they could release containers for several seconds of weightless freefall. Although initially used to study the effects of low gravity on fuel in rocket tanks, Marshall also used the drop tower for scientific experiments in microgravity research. During and after the *Skylab* program it helped test procedures and develop equipment.[71]

For Apollo 14 and 16 Marshall also helped devise "suitcase" experiments which studied how low gravity lessened convection, causing materials to mix and heat in other ways than on Earth. The investigations, recalled Dr. Robert Naumann, one of Marshall's leading materials scientists, were "try-and-see" experiments that lacked the controls necessary for solid science. Nonetheless the Apollo experiments showed clearly that spacecraft did not experience real zero-gravity; gravity gradients, thruster firings, atmospheric drag, and crew motion created sources of small acceleration vectors which disturbed fluid motion and caused other small, but perceptible effects in materials processes.[72] These discoveries caused scientists to change the designation from "zero-gravity research" to "microgravity research."

The real breakthrough for microgravity science, however, came with *Skylab*. NASA added materials studies late in *Skylab* planning, largely because Dr. Mathias Siebel, director of the MSFC Manufacturing Engineering Lab, persuaded Headquarters to include them. For these experiments the Center also designed and developed a materials processing facility with a work chamber that included an electron heating gun and a Westinghouse-developed electric furnace. The late addition of this research program, Naumann remembered, meant that "We had something like eighteen months from the time that it was decided to add these experiments to the *Skylab* until the hardware was actually delivered. Given what it takes in time to do things today, that's a pretty remarkable feat!"[73]

Marshall personnel acted either as managers or principal investigators for three general types of materials experiments on *Skylab*. They examined construction methods in space and tested welding and brazing as means of joining structures. Demonstration experiments studied various effects of microgravity, such as the melting of ice or the mixing of oil and water. Finally Marshall helped investigate metallurgical, chemical, and biological processes in microgravity and the potential of manufacturing novel materials in space, for example producing homogeneous alloys and growing pure crystals for electronics.

The experiments showed how gravity affected materials through convection, buoyancy, sedimentation, and hydrostatic pressure. Since materials processing in space was such a new field, results from *Skylab* were often isolated and unpredictable, yielding more questions than facts. Nonetheless Siebel observed that "the longest journey begins with a single step. This first step has been successful. We're all ecstatic."[74]

Unfortunately after *Skylab*'s first big step, Marshall and NASA were forced to take only little ones because of funding constraints. Progress in microgravity research slowed because no regular, sustained access to space for the scientists existed until the shuttle. Moreover the Agency gave the research low priority. A General Accounting Office study in 1979 showed that annual funding for microgravity studies amounted to one-half of one percent of the terrestrial applications spending which itself was only eight percent of the total NASA budget.[75]

Some progress came in the only manned orbital mission between *Skylab* and shuttle, the Apollo-Soyuz Test Project. For the mission in 1975, Westinghouse and Marshall improved *Skylab*'s processing furnace and the Center managed eight materials experiments that followed those done on *Skylab*. An electrophoresis experiment was particularly successful, separating biological cells by type and function and demonstrating the utility of microgravity research for medicine. Nevertheless, Naumann believed that Apollo-Soyuz was "about a level of sophistication lower" than *Skylab*. Not only did Apollo-Soyuz have less power, stability, and longevity than *Skylab*, but the short two-year interval between missions meant that NASA and materials scientists had little time to learn lessons from *Skylab* and introduce changes.[76]

Through the late 1970s, materials specialists at Marshall searched for creative ways to continue their research. They conducted experiments in NASA's KC-135 aircraft, the Center's labs, and in the drop tower. Struggling against restricted budgets, the Center created a new facility, a drop tube for containerless experiments. Lew Lacy of the Space Sciences Lab scrounged materials for the tube, finding in a warehouse one-foot diameter liquid oxygen pipes from Saturn rockets that had failed to meet specifications. Still facilities on the ground or in KC–135 airplanes were at best poor man's microgravity, offering only seconds of free fall in which to do research.[77] Accordingly, Marshall proposed and managed a sounding rocket program. The Space Processing Applications Rockets (SPAR) Program had 10 flights from 1975 to 1983, each with five minutes of research time as the rocket returned to Earth on a suborbital flight. Marshall ran SPAR on tight budgets with each flight costing about one million dollars. To save money, the Center worked with Goddard, which already had a sounding rocket program at White Sands Missile Range. Goddard supplied the Black Brant VC vehicles and directed launch and payload recovery. Marshall also saved money with in-house development of some investigators' hardware; the Center's labs designed, manufactured, integrated, and checked out about half of the experiment payloads. Roger Chassay, SPAR project manager, recalled that he was a "one-person project office" who chose the project name, wrote its plan, and in the first year wore through the soles of two pairs of shoes walking from lab to lab.[78]

Managing a low cost program like SPAR forced NASA to tolerate higher than customary technical risks. Chassay said he had to convince lab personnel to use different technical standards because SPAR could not afford to follow the Center's traditional quality standards for manned missions. "That was always difficult for me," he said, "to have our management and our engineers relax their standards, their technical standards, to allow them to be compatible with the tight schedule and the tight budget of SPAR."[79]

Headquarters had to be convinced as well. When all four experiments on SPAR IV failed, John Carruthers, Headquarters' director of materials science, acknowledged that the scientists were responsible for their hardware, but nonetheless recommended that Marshall increase its testing and penetration. In response Marshall objected to Carruthers "overstepping his bounds and telling us how to do our job" and thought "returning to the 'Apollo mode' of integration and penetration" would be "a big mistake." Marshall Director Lucas appointed a chief scientist to improve communication between external scientists and Center engineers and promised the Center would use more testing and simpler technology to avoid failure and "unnecessary criticism." But he also thought Headquarters should lower its ambitions for an experimental program and recognize that "scientific objectives can best be achieved after the apparatus has been proved in flight."[80]

Material processing experiments for a SPAR flight.

The SPAR flights had scientific, technical, and organizational payoffs. Microgravity specialists continued their research and improved their instruments. They developed containerless processors that suspended materials in an acoustic or magnetic field. SPAR also tested equipment for the Shuttle and Spacelab. Moreover, scheduling and integrating scientific experiments for a succession of flights taught Marshall payload managers lessons that proved useful for the Shuttle program. Project Manager Chassay remembered that Center Director Lucas enjoyed SPAR briefings, probably because the reports were "a pleasant diversion from some of the Shuttle problems for our Center management. They could see something positive going on. We would fly anywhere from four to nine experiments on a single flight and do that successfully."[81]

Despite impressive early achievements, microgravity research and applications suffered the growing pains of an immature field. To grow, the field needed scientific credibility, a political constituency, and lots of money. NASA needed these things too in the lean years after Apollo. The Agency sought programs that could yield beneficial results and bring political support for space exploration and corporate backing for the Shuttle. By the mid-seventies, the Agency decided to fund microgravity materials research in major corporations. Consequently NASA defined microgravity research as an applications program and promoted it as investment in "the industrialization of space." The common title for the field, "materials processing in space," emphasized its practicality.[82]

By the late 1970s, Marshall had assumed the leading role in promoting the commercialization of space processing. Press releases held out the promise that the research would eventually produce new materials, improvements in tools, electronics, and medicine, and ultimately "space manufacturing." The goal of the publicity, according to Marshall's Director of Program Development, was to create "a broad-based interest, and the climate and structure needed to sustain it within the context of our economic and political system." Then the field could become commercial, and the NASA-business partnership could "add material benefits to man's life style, satisfaction, and enjoyment, as well as make a positive economic contribution."[83]

Technological progress and material benefits, of course, had been a justification for the space program since its inception.[84] But NASA's claims about materials processing in space would later become very controversial because NASA claimed it might also become commercially viable. The merits of this claim became part of discussions about the utility of the Shuttle, Spacelab, and a proposed Space Station. Thus Marshall's efforts to commercialize materials processing in space helped provoke debates about the mission of NASA and the role of the government in the economy. What were the proper relations between business and government? Should government fund commercial R&D projects that had little business support? Could government officials anticipate the marketplace and pick commercially viable areas for research?

Whether the Agency was financing a boondoggle or a bonanza was unclear, and even optimistic Center engineers predicted a payoff only years in the future. But even as Center engineers envisioned commercial ventures in space, others worked on Marshall's down-to-earth energy enterprise.

The Energy Business

By the early 1970s, a national economic slump deepened post-Apollo cutbacks. NASA's plight became more serious when the 1973 Arab oil embargo touched off an energy crisis and a severe recession. Americans questioned the value of the space program. With the first Shuttle flight years away and the Apollo Applications program nearing an end, the Agency had few ways to capture public attention, and had to compete for scarce resources with other federal agencies. The new environment led NASA Centers to compete for the first time in space spinoff projects. NASA had worked with the Defense Department since its inception, but in previous contacts with other agencies NASA had always taken the lead. Now for the first time NASA would become subordinate to the Departments of Energy, Interior, and Housing and Urban Development. "We were used to doing things where we were the customers," according to Bill Sneed of Program Development. But now the Center was developing technology for commercial companies, homeowners, or other government agencies, and "we had difficulty acclimating to that."[85] Marshall and other Centers struggled to define new relationships with each other, with Headquarters, with other federal agencies, and with contractors in unfamiliar industries.

Diversification reached its limit when Marshall helped develop new coal mining technology. In 1974 a coincidence of interests between NASA and the Department of the Interior led Marshall to turn from the heavens to the earth's interior. NASA sought ways to keep its name before the public during the flightless years of shuttle development, and Interior's Bureau of Mines wanted fresh ideas to stimulate a flagging industry. New safety regulations and outmoded equipment had reduced mining productivity by 25 percent over a five-year period, and miners hoped new technology might stimulate the industry. Secretary of the Interior Rogers C. B. Morton challenged NASA Administrator Fletcher to apply NASA's engineering talent to develop automated mining technology that would increase mining safety, minimize environmental damage, and increase productivity.[86]

Notwithstanding the irony of the Space Agency setting its sights below the Earth's surface, the proposal had merit. NASA hoped to justify more generous appropriations by demonstrating that it could deliver more than space spectaculars. Coal mining offered a unique opportunity for NASA to help solve the national energy emergency.

That Fletcher selected Marshall as the Lead Center for NASA's coal mining work was not surprising. The Center's diversification plans had already led to active involvement in Earth resources programs in the Southeast. In the early 1970s Marshall had worked with state governments to develop a land classification system, to provide remote sensing for land surveys, to detect trees infested with the Southern Pine Beetle, and to develop a satellite-assisted system for the management of information on resources.[87] In January 1975, the Department of the Interior and NASA announced an interagency agreement for coal extraction. Marshall's Program Development organized a task team to coordinate work with contractors, Interior, and NASA Support Centers.[88]

Down-to-earth application of space technology for mining industry.

Marshall identified the automated extraction of coal from deep mines as the area most likely to benefit from NASA's expertise. Automated mining techniques were replacing the traditional room-and-pillar method, which required that much quality coal be left behind for roof support. The new longwall shear system allowed miners to carve out entire seams, making mining both faster and more efficient. The greatest obstacle to automated longwall mining was the lack of an effective means of adjusting shearing equipment. Cutters needed to take as much coal as possible without penetrating into the roof or floor beyond the seam, and thereby diluting the quality of the coal or leaving too little coal for support. An improved system thus needed both sensors and a control system to guide cutting drums. Preliminary studies indicated that such equipment could extract as much as 95 percent of the coal from a seam while reducing the rock collected from five to one percent.[89]

The task team found parallels to their customary work. Like space, mines were a hostile environment. "Everything about it is hostile. There's dust, shock, vibration," remembered Peter Broussard. "In space it's really in a way more benign." This meant that aerospace engineers had to adapt to the way miners worked. "A lot of it is sledgehammer stuff," explained Broussard. "You have to be able to make things so they will withstand the thousand natural shocks they're going to get either from the environment or the miners."[90]

Marshall's fresh perspective produced profits. Using space-derived technology, the task team demonstrated that devices using gamma rays, radar beams, impact devices, or reflected light could improve performance of longwall shearing equipment. A Wyoming mining company used a Marshall depth-measuring device to save an estimated $250,000 a month. Industry praised the Center's achievements. The Department of Energy, created since initiation of the project and now responsible for its administration, hoped to see it continue.[91]

At the same time as it assisted the coal industry, Marshall broadened its energy research to include solar heating and cooling for residential and commercial use. Marshall and Lewis Research Center initiated Earth-based solar studies before other Centers and won the backing of Headquarters for their efforts. When NASA made a bid to gain the lead government role in solar energy research in the fall of 1972, Marshall was already planning solar energy prototypes. NASA won only a supporting role, but early involvement ensured that Marshall and Lewis would be the focus of the Agency's solar energy activity.[92]

Test of a solar collector in simulated sunshine at the Marshall Center, 1978.

As its first solar energy project, Marshall proposed developing a demonstration building heated and cooled by solar energy.[93] Headquarters approved plans in October 1973, and by December engineers had constructed a prototype solar collector, the "heart of the test article," mounted at a 45-degree angle to simulate a roof. Nearby they positioned three surplus house trailers with 2,500 square feet of floor space to serve as the model solar house. The demonstration project went into operation in June 1974. Marshall's *Skylab* experience helped advance the state of the art: a solar absorptive coating replaced black paint on the collector panels and absorbed 93 percent of the available solar heat, and computer simulations aided design and performance predictions.[94]

Federal agencies jockeyed for energy funds with the advent of the energy crisis. Marshall's position became clearer in the fall. In September Congress passed the Solar Energy Heating and Cooling Demonstration Act, which established the Energy Research and Development Administration (ERDA). NASA named Marshall as the Lead Center for the Agency's responsibilities under the act, but advised that other Centers must be encouraged to participate. As Lead center, Marshall would develop solar heating and cooling equipment and manage the ERDA Commercial Demonstration Program.[95]

The Lead Center assignment in solar energy research testified to the dynamism of Marshall's diversification and the energy of Program Development. Marshall led the Agency into applied fields and charted a new entrepreneurial course for NASA. With Headquarters discussing the possible closing of Centers, however, Center Director Lucas knew that Marshall remained in a perilous position. Moreover, the Lead Center assignment in solar energy differed from one in development of space technology where the Lead Center could draw on other Centers to produce hardware for NASA. In applied fields, interagency contacts and institutional commitments constantly shifted, making the entrepreneurial environment even more competitive than normal Center relations, which were combative enough.

Consequently Lucas vigilantly guarded Marshall's lead in solar energy. When Langley Research Center asked for Marshall participation in an "Energy Conservation House" project, Lucas worried about "what appears to be our lack of initiative and resourcefulness in maintaining our apparent lead in developing ways of utilizing solar energy in residential and commercial activities."[96] Program Development offered participation to other Centers, but promised Lucas that "we will be very selective in our acceptance of their proposals."[97] Lucas offered participation to Johnson Space Center and Lewis Research Center only after Headquarters exerted considerable pressure.[98] Other agencies exploited the rivalry between NASA Centers; when Marshall complained that a Department of Housing and Urban Development procurement plan would make NASA technically responsible without management authority, HUD replied that the decision would be made at NASA Headquarters, not Marshall, and in any case Johnson could support them if Marshall would not.[99]

Indeed much of Lucas's concern stemmed from his belief that Headquarters had retained more control over the solar energy program than space programs. The organization chart placed two management control levels above the Lead Center program manager while other NASA programs had only one. Harrison Schmitt, who administered NASA's energy programs at Headquarters, acknowledged a new environment in which "traditional words of management may have to be applied in new ways." Schmitt confirmed Marshall's lead on technical matters, but insisted that Headquarters would lead in contacts with other federal agencies.[100]

Marshall's contributions to the nation's solar energy program grew, and during the second half of the decade the Center seemed destined to fulfill von Braun's promise: "Huntsville helped give you the moon and I don't see why Huntsville can't also help give you the sun."[101] After NASA negotiated an agreement with the Energy Research and Development Administration (ERDA) in March 1975, the Center helped select and manage ERDA commercial demonstration projects.[102] Marshall assumed technical management for a Department of Energy project to introduce solar energy into federal buildings.[103] By 1980, Marshall had responsibility for 106 of the 285 commercial solar energy projects selected by ERDA and the Department of Energy. Center personnel assisted the solar industry with over 150 system design reviews.[104] The Center developed a system to record sunfall for solar energy programs. Marshall engineers developed a solar collector that used air instead of water for heat transfer.[105]

The energy projects helped the Center grow beyond propulsion and apply its space expertise to Earth uses. It also protected personnel slots. In April 1979, NASA Administrator Robert Frosch agreed to allow Marshall to increase its manpower commitment to energy programs from 135 to 235 over the next three years if the Center's civil service manpower allotment could accommodate the increase.[106] Six months later, Frosch suggested that NASA might increase its commitment to energy from 3 percent of its manpower to 10 percent.[107] A GAO survey in 1980 found "diversification into expanded energy work a positive force in maintaining Center vitality."[108]

Despite Marshall's success, by 1981 NASA began reconsidering its energy programs. Opposition came both from within the Center and from Washington. Kingsbury, director of the Science and Engineering Laboratory, had never warmed to the idea of the Center devoting efforts to mining, an activity so removed from NASA's central mission. Center Director Lucas believed NASA should have a role in energy programs, but it should have its own mission rather than be responsible to other agencies.[109]

Political winds in Washington had also shifted. In spite of the Carter administration's limited support for NASA, energy seemed to be one area in which growth was assured. The Reagan administration, however, disapproved of technological development projects by federal agencies that could be conducted as well by private industry. Soon after Reagan's inauguration, Budget Director David Stockman announced plans to trim the Carter solar energy budget by 23 percent in 1981 and 62 percent the following year. Both solar energy and coal would be limited to long-term studies with the potential for large returns.[110]

With NASA manpower undergoing another reduction, energy programs became expendable. A budget amendment in May 1981 slashed NASA's direct energy research and development appropriation in half. NASA Headquarters directed Marshall to transfer its energy project to the Department of Energy by the end of the year.[111] The Center received permission to continue its coal research until February 1982 to complete work already underway, but then the Center's eight-year entrepreneurial energy ventures came to an end.[112]

However unlikely, Marshall's contributions to the earth-bound energy business were successful. Rather than waiting for private industry to apply ideas from the space program, Marshall directly sought space spinoffs. The mining inventions profited an old industry, and solar innovations yielded useful knowledge in a new field.

When Marshall's energy work was complete, its commercial undertakings were not. The experience influenced the way Marshall did business. Zoller recalled that the energy projects "certainly influenced how we dealt with the scientific community," and led the Center to involve industry and the scientific community in decision making. "We developed a working relationship first of all with industry in the solar business, then through commercialization, then through the scientific community to make them more part of the engineering management team," Zoller explained.[113]

Conclusion

Marshall's diversification took the Center far from propulsion and created problems as well as possibilities. The greatest problems of diversification were managerial. The Center had to manage, in addition to the science projects described here, the Shuttle, Spacelab, and the Hubble Space Telescope. At the same time that projects were increasing, personnel lines were decreasing.

Thus a flexible organization using ad hoc teams of specialists became a necessity. "Matrix management had been talked about in the Apollo Era," Bob Marshall said, but now it "had to happen." Rather than many engineers from one lab specializing on a problem, a handful of people worked full time and received support from dozens of part-timers who were working on several other projects. Critical staff shortages in some key technical specialties compounded the problems. Funding limitations and personnel caps prevented the Center from hiring experts for all its new fields.[114]

Naturally Center managers worried about having too few people with too little experience on too many projects. George McDonough, head of Science and Engineering in the late 1980s, complained that "you try to matrix people and there aren't enough people to go around, you are always bouncing from here to there. There are fire drills and panics." Sometimes penetration of projects suffered. "With the decline of people and a diversification of projects," Sneed lamented, "automatic responsibility for project integrity diminished somewhat and we tended to get more in a reactive, as opposed to a proactive, mode of operation." Engineers tended to get most involved "when problems occurred or at critical points in the development process such as the key technical design reviews. This mode of operation was not conducive to the most effective management of our projects."[115]

Despite being stretched thin, Marshall recorded important accomplishments. Center personnel diversified a government installation during an era of austerity. This remarkable feat helped preserve an experienced and versatile technical team as a national resource. Marshall's diversification also had social side effects in North Alabama, encouraging Huntsville's economy to become more varied as well.[116]

In addition, the Center made changes in its culture, discovering ways for engineers and scientists to work together. The Center's diversification also contributed to scientific and technological progress. Its hardware and services made possible new discoveries in solar physics, astrophysics, space physics, theoretical physics, chemistry, metallurgy, and biology. Such successes helped the Center gain future projects and operational responsibilities.

Moreover, the dynamism and creativity of Marshall led NASA in new directions. Its entrepreneurship spawned competition and cooperation among field Centers and connected the Agency to other government institutions. The Center undertook commercial ventures, developing marketable technology for mining and performing solar energy research. It also sought to lay foundations for a new industrial sector of materials processing in space. Thus the Center's diversification forced NASA officials and national leaders to define the Agency's mission and refine the role of government in the economy. Within a decade after the first launch of a Saturn V, Marshall had helped conduct many different explorations of outer space.

---

1 Ernst Stuhlinger, OHI by SPW, Huntsville, 8 January 1991, p. 4.

2 William Lucas, OHI by AJD and SPW, UAH, 19 June 1989, p. 26.

3 Barry J. Casebolt, "MSFC to Continue Hunt for Scientific Work," *Huntsville Times* (30 September 1973).

4 Spacelab and Hubble will be discussed in Chapter 9.

5 Andrew J. Stofan, OHI by Adam L. Gruen, 13 November 1987, Space Station History Project, p. 35.

6 Stuhlinger OHI 1991, pp. 18–19; E. Stuhlinger to M. Wright, 29 September 1992, MSFC History Office.

7 David H. DeVorkin, *Science with a Vengeance: How the Military Created the US Space Sciences after World War II* (New York: Springer-Verlag, 1992).

8 John E. Naugle, *First Among Equals: The Selection of NASA Space Science Experiments* (Washington, DC, National Aeronautics and Space Administration Special Publication–4215, 1991), pp. 14–17; Frederick I. Ordway and Mitchell Sharpe, *The Rocket Team: From the V–2 to the Saturn Moon Rocket* (New York: Thomas Y. Crowell, 1979), pp. 354, 383; Stuhlinger OHI 1991, pp. 2–4; Charles A. Lundquist, OHI by SPW, Huntsville, 21 September 1990, pp. 10–11.

9 Roger Bilstein, *Stages to Saturn: A Technological History of the Apollo/Saturn Launch Vehicles* (Washington, DC, National Aeronautics and Space Administration Special Publication–4206, 1980), Appendix H, p. 452.

10 *Marshall Space Flight Center, 1960–1985: 25th Anniversary Report* (MSFC, 1985), pp. 35–36; "Space Science Laboratory," January 1973, Stuhlinger Collection, Record File, Folder 181, Box 3 & 4, ASRC.

11 Dr. George Fichtl, OHI by SPW, 21 July 1992, MSFC, pp. 14–15; William C. Snoddy, OHI by AJD and SPW, 22 July 1992, MSFC, pp. 15, 38–39; ABMA Organization Charts, 14 January 1958, 1 January 1960, MSFC History Office.

12 Quote from Bilstein, p. 325; quote from NASA Press Release 62–237, 13 November 1962; K. Debus, et. al., "Saturn SA–2 Water Experiment," scientific paper, XIII International Astronautical Congress, Sophia, Bulgaria, September 1962; David Woodbridge, et al., "An Analysis of the Second Project High Water Data," NASA Report 02.01, 25 October 1963; E. Stuhlinger OHI, n.d., UAH Saturn Collection, pp. 1–4; G. Bucher OHI, n.d., UAH Saturn Collection, pp. 3–4, 9–10; Wernher von Braun OHI, 30 November 1971, 38–39.

13 NASA Facts: Pegasus, Vol. II, No. 15; MSFC Press Release 66–264, 2 November 1966, Pegasus General folder, NHDD; "Three Pegasus Satellites Tuned Out and Turned Off," *Marshall Star* (21 August 1968), pp. 1, 10; Bilstein, pp. 329–36; Stuhlinger OHI n.d., pp. 4–8, 11.

14 William J. Hines, "An Exercise in Space Futility," *Washington Evening Star* (22 July 1965); Robert J. Naumann, "Pegasus Satellite Measurements of Meteoroid Penetration" scientific paper for Smithsonian Astrophysical Observatory, Cambridge, Mass., 9–13 August 1965, Pegasus General, NHDD; von Braun, p. 39; Bill Snoddy to Mike Wright, 17 December 1992, MSFC History Office; Bilstein, pp. 329–36.

15 E. Stuhlinger, Weekly Note, 18 August 1969, Box 12 & 13, Notes File; R. Shelton, "Science at MSFC," 5 December 1966, Box 1, R&D Administrative File, Stuhlinger Collection, ASRC.

16 See Chapter VI herein.

17 See Chapter V herein.

18 Von Braun, "Adjustment to Marshall Organization, Announcement No. 4," 16 January 1969, Apollo Program Chronological File, Box 70–62, JSC History Office; "Major Reorganization Underway at NASA-Marshall," *Space Daily* (10 December 1968), p. 171.

19 Stuhlinger OHI 1991, p. 1; Stuhlinger to Wright, 29 September 1992, MSFC History Office; E. Rees to H. Gorman, 13 August 1971, Stuhlinger Collection, Record File, Box 2, ASRC.

20 Statistic compiled from information furnished by Jerre Wright, Equal Opportunity Office, MSFC.

21 Unclear authorship, but whomever Lucas as head of Program Development reported to, probably von Braun or Rees, "Discussion regarding Dr. Newell's visit to Marshall on October 24, 1968," 14 November 1968, Box 5–35, Shuttle Chronological Series, JSC History Office.

22 Dr. Charles R. Chappell, OHI by SPW, MSFC, 23 August 1990, pp. 7, 9, 10, 15; Barry Rutizer, "The Lunar and Planetary Missions Board" (NASA–HHN–138, 30 August 1978), cited in Arnold S. Levine, *Managing NASA in the Apollo Era* (Washington, DC, National Aeronautics and Space Administration Special Publication–4102, 1982), p. 251; Dr. Thomas Parnell, OHI by SPW, MSFC, 11 September 1990, pp. 21–22; Fichtl OHI, pp. 15–16.

23 Stuhlinger to Wright, 29 September 1992, MSFC History Office; Stuhlinger, "Relations between NASA and outside scientists," 3 June 1966, R&D Administrative File, Box 1, Stuhlinger Collection, ASRC.

24 F. J. Magliato, "Management of Experiments at MSFC—Draft," 1 November 1965, AAP November-December 1965 folder, MSFC Director's Files; MSFC, "MSFC *Skylab* Corollary Experiment Systems Mission Evaluation," NASA TM X–64820, September 1974, folder 517, MSFC History Office; L. Belew and E. Stuhlinger, *Skylab: A Guidebook* (NASA EP–107, 1973), pp. 114–226; R. J. Naumann, "*Skylab* Induced Environment," in M. I. Kent, et al., eds., *Scientific Investigations of the Skylab Satellite* (New York, AIAA, 1976), pp. 383–96.

25 MSFC, "*Skylab* Student Project: Project Definition Report," 7 May 1972, MSFC History Office, folder 517; H. B. Floyd, "Student Experiments on *Skylab*," in Kent, pp. 471–94;

H. B. Floyd, OHI, by W. D. Compton, MSFC, 5 November 1974, Rice University, pp. 14–15.

26 T. Bannister, "Postflight Analysis of Science Demonstrations," in Kent, pp. 531–50; NASA TM X–64820, Section VII; Dr. Charles Lundquist, OHI by SPW, UAH, 7 July 1991, p. 13.

27 E. Stuhlinger to Mike Wright, MSFC Historian, Comments on MSFC History Draft, Chapter Five, September 1992, p. 4.

28 L. Belew to W. Lucas, "Key Assignments for *Skylab* Experiments," 24 September 1971, AAP 1971 folder, MSFC Director's Files; Ise interview, pp. 10–12; Snoddy OHI, pp. 15–16; Stuhlinger to Wright, p. 4.

29 E. Stuhlinger, Memo for Record, 28 February 1968, E. Stuhlinger to F. Williams, 15 February 1968, Historical File, Boxes 12 & 13, Stuhlinger Collection, ASRC.

30 W. D. Compton and C. D. Benson, *Living and Working in Space: A History of Skylab* (Washington, DC, National Aeronautics and Space Administration Special Publication–4208, 1983), pp. 174–79.

31 NASA OMSF and OSSA, "*Skylab* and the Sun: ATM," July 1973, MSFC Technical Library, 30–34, 38–39; Dr. Anthony DeLoach, OHI by SPW, MSFC, 11 May 1992, pp. 1–8; Lundquist 1992 OHI, pp. 9–10; Snoddy OHI, pp. 17–23.

32 MSFC *Skylab* Program Office, "MSFC *Skylab* Kohoutek Experiments Mission Evaluation," NASA TM X–64879 (September 1974), pp. 1, 3; R. D. Chapman, "Comet Kohoutek," *Space Science and Technology Today I* (September 1973), pp. 30–32; W. C. Snoddy and R. J. Barry, "Comet Kohoutek Observations from *Skylab*," in W. C. Schneider and T. E. Hanes, eds., *The Skylab Results: Advances in the Astronautical Sciences,* Volume 31, Part 2 (Tarzana, CA: American Astronautical Society), pp. 871–94; C. A. Lundquist, ed., *Skylab's Astronomy and Space Sciences* (Washington, DC, National Aeronautics and Space Administration Special Publication–404, 1979), pp. 47–48, quoted, pp. 59–60; Compton and Benson, pp. 390–94.

33 Lundquist 1992 OHI, pp. 2–8, 15; J. A. Eddy, *A New Sun: The Solar Results from Skylab* (Washington, DC, National Aeronautics and Space Administration Special Publication–402, 1979), pp. 67–180, quoted p. vii; J. B. Zirker, *Coronal Holes and High Speed Wind Streams: A Monograph from Skylab Solar Workshop I* (Boulder, CO: Colorado Associated University Press, 1977); P. A. Sturrock, *Solar Flares: A Monograph from Skylab Solar Workshop II* (Boulder, CO: Colorado Associated University Press, 1980); F. Q. Orral, ed., *Solar Active Regions: A Monograph from Skylab Solar Workshop III* (Boulder, CO: Colorado Associated University Press, 1981); Snoddy OHI, pp. 32–34.

34 Snoddy OHI, pp. 15, 34–35.

35 Craig Covault, "Help Abroad Key to Laser Mission Goals," *Aviation Week and Space Technology* (12 May 1975), pp. 36–43; E. Rees to D. D. Meyers, "LAGEOS," 25 January 1973, LAGEOS folder, MSFC Director's Files; "Project LAGEOS Press Kit," NASA Release 76–67, 14 April 1976, pp. 5–7, MSFC History Office Fiche No. 1345.

36 "LAGEOS Task Team formed at MSFC," MSFC Release 73–171, 20 November 1973; Program Development, "Preliminary Project Plan for LAGEOS," November 1973, RHT File No. 495; "LAGEOS Press Kit," pp. 1–5, 9, 12–14; Covault, pp. 36–43.

37 Eugene H. Cagle, Deputy Director, Test Lab to R. G. Smith, "LAGEOS, n.d."; "Summary of Add-on In-House LAGEOS Work for FY–1975," 2 October 1974; R. G. Smith to W. Lucas, 4 October 1974, LAGEOS folder, MSFC Director's Files.

38 Zoller OHI, pp. 5–9; James Kingsbury OHI, by SPW, 22 August 1990, Madison, AL, p. 30.

39 "MSFC/GSFC Memorandum of Agreement relative to the LAGEOS Project Spacecraft Development Support," 28 January 1976, LAGEOS folder, MSFC Director's files; MSFC Release 76–17, 76–82; Covault, pp. 36–43.

40 "LAGEOS: Scientific Results," *Journal of Geophysical Research* 90, B11 (30 August 1985), entire issue.

41 Stuhlinger to Wright, 29 September 1992, MSFC History Office; MSFC, "Project Plan for Gravitational Redshift Space Probe," March 1972; E. Rees to D. D. Myers, "Gravitational Redshift Space Probe," 3 October 1972, MSFC Director's Files.

42 "Relativity Gravity Probe Press Kit," NASA Release 76–106, Gravitational Probe-A folder, NHDD;

43 Quotation from "Discussion on GP–A Management," 11 August 1974; F. Speer, MSFC to A. Schardt, HQ, "Assessment of GP–A Project Requirements," 8 November 1974; quotation from F. Speer to W. Lucas, "Request for Additional GP–A Funding," 14 November 1974, MSFC Director's Files.

44 J. A. Lovingood to R. G. Smith, "Redshift Vibration and Loads Problems," 12 December 1974; quotation from L. Belew to W. Lucas, "Redshift (GP–A) Vibration and Loads Problem," 27 December 1974; "GP–A Dynamics Problem," December 1974; J. Stone to F. Speer, "GP–A Review at OSS," 28 February 1975; F. E. Vreuls to W. Lucas, "Monthly GP–A Review of November 21, 1975"; "Recapitulation of GP–A Funding and Manpower History," n.d.; J. A. Bethay to W. Lucas, routing slip, 2 December 1975; F. Speer to R. Halpern, HQ, "GP–A Reprogramming Recommendations," 12 March 1976, GP–A folder, MSFC Director's Files.

45 "Relativity-Gravity Space Probe Launched," MSFC Release 76–113, 18 June 1976, Gravity Probe-A folder, NHDD; MSFC, "GP–A: The Redshift Experiment," n.d., MSFC Public Affairs Office; R. F. C. Vessot and M. W. Levine to W. Lucas, 22 June 1976, GP–A folder, MSFC Director's Files; R. F. C. Vessot and M. W. Levine, "GP–A Redshift Space-Probe Experiment Project Report," April 1977, MSFC History Office Fiche No. 1768; R. F. C. Vessot and M. W. Levine, "A Test of the Equivalence Principle Using a Space-Borne Clock," *General Relativity and Gravitation* 10 (1979), pp. 181–204.

46 Wallace H. Tucker, *The Star Splitters: The High Energy Astronomy Observatories* (Washington, DC, National Aeronautics and Space Administration Special Publication–466, 1984), chap. 2, 3.

47 Jim Downey OHI, by SPW, Madison, AL, 14 December 1990, pp. 1–6, 9, 14; Herbert Friedman, "High-Energy Astronomy with HEAO," *Astronautics and Aeronautics* (April 1978), pp. 50–51.

48 R. D. Stewart, "HEAO Project Plan," March 1971, MSFC History Office, Fiche 707, quoted pp. 3–1; "HEAO Charter and Staffing Pattern," MSFC Director's Files, HEAO 1971; Downey OHI, pp. 7–8; Tucker, p. 20.

49 "Opportunity for Participation in Space Flight Investigations: HEAO," 3 March 1970, HEAO General Folder, NHDD; "NASA picks experiments for 2 HEAOs," *Huntsville Times* (8 November 1970); Downey, OHI, pp. 5, 10; Tucker, p. 21.

50 "NASA Program Reductions," NASA Press Release 73–3, 5 January 1973, HEAO General, NHDD; "HEAO Program Reoriented," *Space Business Daily* (8 February 1973), p. 1.

51 Dr. Fred Speer, OHI by SPW, 9 October 1990, UAH, pp. 8–11; John Naugle to Petrone, "Revised HEAO Program," 5 March 1973, HEAO 1973 folder, MSFC Director's Files; "Restructured HEAO Contracts Signed," MSFC Press Release 29–74, 29 August 1974; Tucker, pp. 22–24.

52 Petrone memo, 30 March 1973; Speer to Petrone, 12 April 1973; Lucas to Petrone, 13 April 1973; J. P. Noland, "Rationale for retaining current HEAO contractors," 14 March 1973, HEAO 1972–73 folder, MSFC Director's Files.

53 Speer to Rees, "HEAO PI Meeting in Washington (16 January 1973)," 18 January 1973, HEAO 1972–73 folder, MSFC Director's Files; Speer OHI, pp. 8–11; C. Covault, "HEAO Design Program Restarted," *Aviation Week* (7 October 1974); Barry Casebolt, "HEAO to be NASA's 'Major Effort' after 1975," *Huntsville Times* (28 October 1974); Tucker, chap. 4.

54 MSFC, "HEAO Fact Sheet," March 1979, MSFC History Office Fiche No. 819; Tucker, chap. 5.

55 F. A. Speer, "HEAO: Signature of a Successful Space Science Program," 21 September 1980, MSFC History Office Fiche No. 751, p. 8.

56 Speer, "HEAO," pp. 8–9, 11; F. A. Speer, "Prepared Statement on the HEAO Program" for the Subcommittee on Space Science and Applications, Committee of Science and Technology, House of Representatives, 7 February 1977, MSFC History Office Fiche No. 745, pp. 2–3.

57 Speer, "HEAO," pp. 6–7, 10; Tucker, p. 44; Speer OHI, quoted p. 8.

58 Tucker, pp. 38, 44–45.

59 Tucker, p. 44; Speer to R. Halpern, "HEAO–B Experiment Cost Problem," 14 October 1975; R. Giacconi to A. W. Schardt, NASA HQ, 31 October 1975, HEAO 1975 folder, MSFC Director's Files.

60 Speer, "Management Review at AS&E," 30 July 1976; A. Martin to Speer, 17 August 1976; Speer, "AS&E/MSFC Communication Contacts for the HEAO–B Experiment, draft," n.d., HEAO 1976 folder, MSFC Director's files.

61 Speer OHI, pp. 14–16.

62 Tucker, p. 44; Parnell OHI, pp. 8–9, 15–16.

63 Speer OHI, pp. 13–15; Speer to Lucas, "Central Parts Control for HEAO," 2 February 1973, HEAO 1972–73 folder; HEAO Office, "Interim Assessment of the HEAO Central Parts Control," 24 March 1975, HEAO 1975 folder, MSFC Director's Files; Tucker, pp. 48–51; MSFC Press Releases 74–201, 75–100, 76–172, 77–93, 77–186.

64 MSFC, "A Guide to Space Simulators at MSFC," n.d., p. 11, MSFC History Office; Speer, "Prepared Statement"; Tucker, pp. 68–70.

65 Tucker, pp. 75–76; W. Lucas, "Address to HEAO Science Symposium," 8 May 1979, p. 7, HEAO 1979 folder, MSFC Director's Files.

66 Speer to Petrone, "MSFC/GSFC Memorandum of Agreement on Mission Operations for the HEAO Project," 28 September 1977, HEAO 1972–73 folder, MSFC Director's Files; Speer OHI, pp. 11–13; HEAO–A Press Kit, 5 May 1977, p. 15, MSFC History Office, Folder No. 433.

67 "NASA Increases Lifetime Estimates of HEAO Missions," *Aerospace Daily* (27 November 1978), p. 108; Tucker, p. 81, chap. 5.

68 Tucker, pp. 9, 72, chap. 8–23.

69 Speer OHI, pp. 17–18; Dave Dooling, "NASA Planning Large Telescope for X-Ray Astronomy," *Space World* (February 1980), pp. 22–24.

70 Eugene C. McKannan, "A Brief History of the Materials and Processes Laboratory—As I Saw It," n.d., MSFC History Office, Folder No. 14; Dr. Robert Naumann, OHI by SPW, 3 June 1991, UAH, pp. 3–6.

71 R. J. Naumann and H. W. Herring, *Materials Processing in Space: Early Experiments*, (Washington, DC, National Aeronautics and Space Administration Special Publication–443, 1980), pp. 33–34; "A Guide to Space Simulators at the Marshall Space Flight Center," 1987, pp. 8–10; MSFC History Office, Fiche No. 1039; Naumann OHI, pp. 26–28.

72 McKannan, pp. 8–9; R. J. Naumann, "Materials Processing in Space: Review of the Early Experiments," in *Applications and Science: Progress and Potential* (New York: IEEE, 1985), pp. 2–8.

73 Belew and Stuhlinger, pp. 187–196; Naumann OHI, pp. 9–11.

74 E. Stuhlinger, "Technical Processes under Weightlessness: *Skylab* Press Briefing," 8 November 1973, MSFC History Office; E. Stuhlinger, "*Skylab* Results—Review and Outlook," n.d., NASA Cospar S.P. I.4.4.8, MSFC Technical Library, pp. 7–8; C. Lundquist, "*Skylab* Experiments on Metals," paper for ESRO Space Processing Symposium, Frascati, Italy, 25 March 1974; Naumann and Herring, pp. 64, 67, 71–71, 82–83, 86, 92–93, 97–98, 105–106; David Salisbury, "Factories in Orbit may Manufacture Alloys now called Impossible," *Christian Science Monitor*, 14 February 1974; Materials Processing Panel, "*Skylab* Revisit: A Conference in Recognition of the Fifteenth Anniversary," 11 May 1988, Huntsville, videotape, MSFC History Office.

75 L. Zoller to W. Lucas, "GAO Study of Materials Processing in Space," 12 February 1979, MPS 1978–79 folder, MSFC Director's Files.

76 "MSFC to Manage 8 ASTP Experiments," MSFC Press Release 73–147, 15 October 1973; "ASTP Fact Sheet," NASA Press Release 75–9, 1975; C. Covault, "ASTP Research Spurs Medical Benefits," *Aviation Week and Space Technology* (17 November 1975), p. 53; MSFC Space Processing Applications Task Team, "ASTP—Composite of MSFC Final Science Report," NASA TM X–73360, January 1977; Walter Froehlich, "Apollo-Soyuz," NASA EP–109, 1976, pp. 79–80; Naumann OHI, p. 12.

77 Naumann OHI, pp. 28–31; MSFC Commercial Applications Office, "NASA Drop Tube: A Tool for Low-Gravity Research," CMPS 302, n.d., MSFC History Office No. 79; L. Lacy, et. al., "Containerless Undercooling and Solidification in Drop Tubes," *Journal of Crystal Growth* 51 (1981), pp. 47–60; J. Kingsbury, Weekly Note, 20 September 1982, MSFC History Office.

78 "Sounding Rocket Project for Space Processing Experiments," November 1974, Space Processing, 1970–75 folder, MSFC Director's Files; "S&E Labs Play Vital Role in SPAR Program at MSFC," *Marshall Star* (15 December 1976); D. Dooling, "Rockets Test Space Processing: MSFC Engineers Direct Program," n.p., 6 June 1977; Naumann OHI, pp. 12, 32; Roger Chassay, OHI by SPW, 24 July 1992, MSFC, pp. 14–17.

79 Chassay OHI, pp. 18, 21.

80 Chassay to Lucas, "SPAR IV Results," 23 August 1977; J. R. Carruthers, Director MPS

HQ to MSFC Special Projects Office, "Programmatic and Management Recommendations for the SPAR Project," 26 August 1977; J. T. Murphy to Lucas, 15 September 1977 and 29 September 1977; Lucas to Leonard B. Jaffe, HQ, 20 October 1977, Space Processing 1976 folder, MSFC Director's Files.

81 SPAR clippings, Fiche No. 1103, MSFC History Office; Chassay OHI, pp. 15–16.

82 Statement of Dr. Robert A. Frosch, Administrator, NASA on the Space Industrialization Corporation Act of 1979 before the Subcommittee on Space Science and Applications, Committee on Science and Technology, U.S. House of Representatives, 11 May 1979, MPS, 1978–79 folder, MSFC Director's Files; R. Naumann, "Microgravity Science and Applications Program Description Document," February 1984, MSFC Technical Library, I.6–I.9; R. Naumann, "Historical Development," in B. Feuerbacher, et al., eds., *Materials Sciences in Space: A Contribution to the Scientific Basis of Space Processing* (Berlin: Springer-Verlag, 1986), pp. 27–28.

83 "Marshall Studies Materials Processing in Space," MSFC Press Release 74–47, 4 April 1974; J. T. Murphy, "RTOP Commercialization of MPS," 23 November 1974, MSFC History Office, Folder 14E.

84 See Walter A. McDougall, . . . *The Heavens and the Earth: A Political History of the Space Age* (New York: Basic Books, 1985).

85 Bill Sneed, OHI by SPW, 15 August 1990, Huntsville, Alabama, p. 21.

86 MSFC, "Program Plan for Underground Coal Extraction Technology," November 1974. Morton officially asked NASA's assistance in a letter to Fletcher on 9 January 1975, but Marshall's Peter Broussard remembered a story that the challenge first came at a Christmas party, when Morton said in effect, "If you fellows are so smart in automation, . . . why don't you help us in the coal mining business?" Peter Broussard, OHI by SPW, 20 September 1990, Huntsville, Alabama, p. 32. Morton to Fletcher, 9 January 1974, Assistance to Department of Interior (Coal Mining Problems) folder, MSFC Center Director's Files.

87 W. R. Lucas to Joseph P. Allen, "Regional Applications Transfer Activities at Marshall Space Flight Center," 3 November 1977, Fiche No. 1834, MSFC History Office.

88 John Goodrum, "Coal Energy Extraction Program and Related Coal Mining R&D Efforts," 27 June 1974, Coal Energy Extraction Program folder, MSFC Center Director's Files; NASA Release 75–17, 20 January 1975; MSFC Release 75–215, 7 October 1975.

89 MSFC, "Project Plan for Development of Longwall Guidance & Control Systems: Bureau of Mines Advanced Longwall Program," June 1976, pp. III–2 to III–8; Broussard OHI, pp. 15–16; David Cagle, "Space-Program Technology May Boost Coal Production," *Huntsville Times*, 21 October 1976.

90 Broussard, OHI, pp. 19–20.

91 MSFC 1978 Chronology, 9 January 1978; Broussard, OHI, pp. 2, 15–16.

92 Charles W. Mathews to Associate Administrator for Manned Space Flight, 28 June 1973, Solar Energy Heating and Cooling Demo folder, MSFC Center Director's Files. For Lewis involvement with solar energy programs, see Virginia Dawson, *Engines and Innovation: Lewis Laboratory and American Propulsion*, (Washington, DC, National Aeronautics and Space Administration Special Publication–4306, 1991), pp. 206–08.

93 James T. Murphy to Rocco Petrone, 17 October 1973, and Murphy to Petrone and William R. Lucas, 31 October 1973, Solar Energy Heating and Cooling Demo folder, MSFC Center Director's Files.

94 MSFC, "Harnessing the Sun's Energy for Heating and Cooling;" MSFC, "The Development of a Solar-Powered Residential Heating and Cooling System," 10 May 1974; Richard G. Smith, Statement before Subcommittee on Science and Applications, Committee on Science and Astronautics, House of Representatives, May 1974; MSFC Public Affairs Office, "MSFC Solar Energy Demonstration," 15 May 1974; 1974 MSFC Chronology, 22 April 1974, MSFC History Office.

95 Harrison H. Schmitt to Lucas, 18 September 1974, Solar Energy Heating and Cooling Demo folder, MSFC Center Director's Files; MSFC, "Solar Energy for Heating and Cooling: Applied Technology."

96 Lucas memo to Jim Murphy and Dick Smith, 7 November 1974, Solar Energy Heating and Cooling Demo folder, MSFC Center Director's Files.

97 James T. Murphy to Lucas, 27 September 1974, Solar Energy Heating and Cooling Demo folder, MSFC Center Director's Files.

98 Schmitt to Lucas, 18 September 1974; Schmitt to Chris Kraft, 11 November 1974; Lucas to Schmitt, 17 December 1974, Solar Energy Heating and Cooling Demo folder, MSFC Center Director's Files.

99 Donald R. Bowden to Lucas, 19 February 1975, Solar Energy Heating and Cooling Demo folder, MSFC Center Director's Files.

100 Lucas to Harrison H. Schmitt, 4 December 1974; Schmitt to Lucas, 17 December 1974, Solar Energy Heating and Cooling Demo folder, MSFC Center Director's Files. Lucas was careful not to push headquarters too hard, however. He considered informing Washington that because of imminent reductions in force, Marshall would have to suspend its solar energy work unless it received a manpower augmentation, but reconsidered when he decided such a position seemed too much like an ultimatum. Lucas to Schmitt (unsent), 17 January 1975, Solar Energy Heating and Cooling Demo folder, MSFC Center Director's Files.

101 Cited in Philip W. Smith, "'Let Huntsville Give You the Sun,' Urges Dr. Von Braun," *Huntsville Times*, 12 March 1976. Von Braun was speaking to a group of southeastern senators.

102 MSFC, "Solar Heating and Cooling Development Program Plan," SHC–1003, 9 April 1975, Solar Heating and Cooling, January to June 1975 folder, MSFC Center Director's Files.

103 "Marshall is Assigned Solar Role," *Huntsville Times*, 17 April 1979.

104 MSFC Release 76–42, 17 February 1976; MSFC Public Affairs Office Fact Sheet: "Solar Heating and Cooling Program," 10 January 1980.

105 MSFC Release 75–168, 24 July 1976; MSFC Release 76–56, 2 April 1976.

106 Ray Kline, NASA Associate Administrator for Management Operations, to Lucas, 26 April 1979, Energy Research and Development 1979–1980 folder, MSFC Center Director's Files.

107 Donald A. Beattie, Director, Energy Systems Division to Lucas, 5 November 1979, Energy Research and Development 1979–1980 folder, MSFC Center Director's Files.

108 Mr. Sheley, Jr. (first name illegible), GAO Procurement and Systems Acquisitions Division, to Robert A. Frosch, 16, September 1980, Energy Research and Development 1979–1980 folder, MSFC Center Director's Files.

109 Mitzi E. Peterson, Memorandum for the Record Re: OAST Energy Management Council, 1 May 1981, Energy Program, Long Range Mining, 1981 folder, MSFC Center Director's Files; Broussard OHI, p. 23.

110 Burt Solomon, "Hatchet Job: Reagan's OMB Meets the Energy Budget," *The Energy Daily* 9 (Friday, 6 February 1981), p. 6.

111 The Appropriation dropped from $3.9 million to $1.9 million. Walter B. Olstad to Center Directors, 4 May 1981, Energy Program, Long Range Mining, 1981 folder, MSFC Center Director's Files; Jack L. Kerrebrock to NASA Administrator James Beggs, 23 September 1981, Energy Program, Long Range Mining, 1981 folder, MSFC Center Director's Files; A. M. Lovelace to Raymond Romatowski, 26 May 1981, 1981 Energy Program—Longwall Mining folder, MSFC Center Director's Files.

112 Jack L. Kerrebrock to Lucas, 7 October 1981, Energy Program, Long Range Mining, 1981 folder, MSFC Center Director's Files.

113 Lowell Zoller, OHI by SPW, 10 September 1990, Huntsville, Alabama, p. 34.

114 Marshall OHI, pp. 24–25; Parnell, OHI, p. 10.

115 George McDonough, OHI by SPW, 20 August 1990, MSFC, p. 24; Sneed OHI, pp. 16–17; Bill Sneed to Mike Wright, 5 January 1993, MSFC History Office.

116 Jack Hartsfield, "Diversifying is Aerospace Group Aim," *Huntsville Times* (5 April 1970).

Chapter VIII

The Space Shuttle: Development of a New Transportation System

In the aftermath of Apollo, Marshall Space Flight Center increased its research activities, conducted space operations, and engaged in entrepreneurial ventures. But Marshall was still primarily a propulsion Center, and its reputation would rise and fall depending on the success of its rocketry. If the Space Shuttle propulsion system did not dominate Marshall's second two decades in the way that Saturn had in the first, it was nonetheless the Center's preeminent concern, source of its greatest post-Apollo triumphs, and its most sobering tragedy.

Of the four major Shuttle components—solid rocket boosters, external tank, main engines, and orbiter—Marshall bore responsibility for all but the orbiter. Each offered new technological challenges that pushed engineers and administrators beyond Saturn. For the first time the Center developed a rocket that relied on solid fuel. For the first time the Center worked on a reusable vehicle system.

Choosing a Configuration

NASA adopted the Space Shuttle as a formal program in 1969, but the origins of its concepts predate the formation of the Agency. Marshall participated in the earliest Shuttle studies, and the Center's struggle to define its role in the Shuttle program was an important part of its post-Apollo transition.

The Shuttle broke with Apollo technology most significantly as a reusable spacecraft, an idea that had appealed to philosophers, scientists, and rocket engineers for decades. Indeed most 19th century speculation about space travel envisioned reusable vehicles, not because of a systematic approach to technological obstacles, but because of assumptions drawn from familiar systems. German and American theorists suggested the possibility of rocket airplanes in the 1930s and 1940s, and American experimental craft like the X–15 bear more kinship to the Shuttle than to early spacecraft.[1] The Air Force and the Army both pursued studies in the late 1950s that could be considered precursors to the Shuttle. The Army Ballistic Missile Agency in Huntsville, before relinquishing its Development Operations Division to NASA in 1960, contrived various means of recovery for its expendable Redstone and Saturn I rockets including paragliders and parachutes, but none of them were flight tested.[2]

A Shuttle launch.

From the earliest months of its establishment, Marshall began to investigate reusable systems. The first study began in 1961 when the Center's Future Projects Office issued a statement of work calling for winged, reusable launch vehicles including orbital passenger and cargo carriers with easily accessible payload bays in which all stages would be capable of multiple reuse. In December 1963 Boeing, Lockheed, and North American Aviation all conducted studies for Marshall. By December 1963, they concluded that such vehicles were indeed possible.[3] Lockheed and Boeing conducted a follow-on study for the Marshall Future Projects Office in 1964 and 1965 that suggested possible systems criteria for "the design of space launch vehicles similar in operation to today's airplanes."[4]

Hermann Koelle, who headed the Future Projects Office, also pursued studies of high-performance rocket engines. Jerry Thomson remembered Koelle approaching him about engine designs that might surpass the performance of Saturn engines. "Up through the Apollo Program we were only operating about a thousand PSI of chamber pressure, which is what the F–1 ran. But we wanted to go much higher than that," Thomson recalled. "Some of us, sort of on a side track, went off to get some components built and tested for these engines that were later to become the Space Shuttle main engine."[5]

John McCarty, one of Thomson's colleagues, remembered that "When we put the requirements of the aerospace plane together with propulsion rocket engine technology and requirements, it was clear we needed to start a new approach to an engine. We started two or three projects. We started high-pressure turbopumps—one for hydrogen for the fuel and one for oxygen for the oxidizer. We started some engine system design studies to arrive at what was the right configuration. . . . How would you control it? What are some of the fundamental limits in the engine? . . . That was really the beginning, I think, of the SSME [Space Shuttle main engine]."[6]

At the time of these early studies, NASA was far from settling on a major post-Apollo program. When NASA's planners did discuss future goals, they assumed that an orbiting workshop would be the next major manned program. Houston and Marshall already had Space Station Projects Offices. Officials assumed that "the large manned Space Station seems to be the most probable initial mission" for a reusable launch vehicle. In this context, a "Shuttle" would function as a logistics vehicle in support of a Station rather than an independent system. Furthermore, planners would try to minimize development costs for the logistics vehicle in order to avoid compromising station funding. While NASA expected eventual development of a reusable vehicle, planners acknowledged that concrete designs would have to be deferred.[7] The shadow of a presumed Space Station thus constrained investigations, since NASA was already beginning to realize that the post-Apollo era would offer political and economic limits.[8]

Studies at Marshall, Houston, and the Air Force between 1963 and 1967 helped keep plans for a Shuttle-type vehicle alive. People involved in the mid-1960s Shuttle studies acknowledged that they were working in a highly speculative environment. They had no foolproof way of judging the cost of advanced reusable systems, and few precedents for evaluating technical risk, refurbishment costs, abort capabilities, system size, or performance.[9] Since these factors were interrelated, changes in one area could greatly affect others; for example, as size increased, engine performance and thermal protection would both be affected in very complex ways.[10] Frank Williams of the Marshall Future Projects Office suggested that one set of assumptions could lead to hundreds of millions of dollars in savings, while slight changes in these assumptions could lead to hundreds of millions of dollars in losses.[11]

Wernher von Braun helped to keep the idea of a Shuttle-type vehicle before the public. His 1952 Collier's articles envisioned a logistics vehicle to supply an orbital space station. In 1965 he called for a reusable earth-to-orbit vehicle that could service space stations in 10 to 15 years, one in which both launch vehicle and spacecraft would be "capable of returning to Earth in a lifting-flight mode." In one of the optimistic projections of Shuttle use characteristic of early planning, he suggested that a system to deliver a 10,000-pound payload and 10 men to orbit could be developed for $1 billion, and that if it could perform 1 mission per week for 50 to 100 missions, it could lower the cost to lift a payload to orbit to only $50 per pound.[12]

The origins of the Shuttle are disparate, but 27 October 1966 might qualify as the point at which NASA began to define a real configuration for development. On this date representatives of the Office of Manned Space Flight (OMSF), Marshall, and the Manned Spacecraft Center (MSC) met in Houston to discuss logistics systems for the post-Apollo era. Max Akridge, one of the Marshall representatives, called the meetings "the beginning of the Space Shuttle as such." Planning for the Shuttle began at each Center, and engineers began to contemplate possible designs.[13]

Competition between NASA Centers would intensify as Agency resources became scarcer, and competition between Houston and Marshall would be an important factor in Shuttle development. Houston's early configuration study was but an indication of the competition that would characterize post-Apollo relations between the Centers. Houston's Shuttle was a fully reusable two-stage vehicle with straight fixed wings that became the basis for early configuration discussions.[14]

As part of the post-Apollo planning process during 1968, NASA began to pull together concepts developed by Agency and defense contractors. George E. Mueller, NASA's Associate Administrator for Manned Space Flight and the Agency's leading Shuttle advocate, began to argue the merits of a Shuttle independent of a space station.[15] In February Mueller called for a fully reusable low-cost transportation system that might eventually be competitive with other forms of transportation. Marshall helped Mueller's office conduct further econometric and engineering studies examining manned spaceflight options, and among those released midyear was one that offered a cautionary note. It questioned the viability of a fully reusable aircraft-type transportation system before the mid-1980s because of high risks and the necessity for very high annual launch rates over a sustained period of years to amortize high development costs.[16] The very issue of viability showed another difference from Apollo; whereas Apollo's goals were political, Shuttle would always be held to economic criteria. In the fall, NASA directed Marshall and Houston to review their studies on low-cost transportation systems with a view toward reducing costs.[17]

The space program enjoyed a peak of popularity in 1969 as the anticipated Moon landing allowed the nation to divert its attention from the protracted war in Vietnam. Out of the public spotlight, the year saw crucial decisions that would shape the space program for years. In January NASA committed $500,000 to each of four Shuttle feasibility studies and assigned management to field Centers, thus initiating Phase A of Shuttle development.[18] Marshall managed the General Dynamics and Lockheed contracts, Houston monitored McDonnell Douglas, and Langley supervised North American Rockwell. NASA directed each contractor to examine a different design approach and to report their findings at a September appraisal.[19]

On 13 February, President Richard M. Nixon appointed a Space Task Group to give him advice regarding the direction of the space program in the post-Apollo years. Chaired by Vice President Spiro T. Agnew, the task group included NASA Acting Administrator Thomas O. Paine, Secretary of the Air Force Robert C. Seamans, and Lee Dubridge, science adviser to the president, as well as observers from other agencies.[20]

The announcement of the formation of the Space Task Group stimulated planning activity in NASA, for the Agency now had only a few months to influence decisions expected to affect NASA's direction for years. Mueller directed Manned Space planning activities, and in doing so shaped both NASA's commitment to the Shuttle and the role Marshall would play in its development. "The Shuttle business grew out of what I call the Mueller Plan," Huntsville's Bob Marshall recalled. Mueller hired BellComm to aid in planning. "He directed them to plan a program which had in it the Shuttle."[21] Mueller also guarded the Center's interests. Concerned about the traumatic post-Apollo transition in Huntsville, he ensured that the Center received its share of Shuttle development business. The Agency began discussions with the Air Force about possible joint efforts to develop the new vehicle.

Meantime the Centers began jockeying for position. Marshall, in the throes of post-Apollo cutbacks, sensed an opportunity to gain new responsibilities. One of the Houston participants in intercenter meetings noted that "MSFC is really building up to handle the advanced program."[22] Marshall formed an Integrated Launch and Reentry Vehicle (ILRV) task team early in April, two weeks before Mueller did the same at Headquarters, and some speculated that Marshall might win the assignment to manage the Shuttle.[23]

Max Akridge of the Marshall group maintained that the term "Space Shuttle" originated after Mueller's address to the group on 5 May. Akridge recalled Mueller saying that NASA needed "a vehicle that's like a shuttle bus." "I kind of liked the name 'Space Shuttle,'" Akridge recalled, and he directed the Marshall contractors to begin using the term, which soon became common.[24]

Mueller, in one of several actions he initiated to assist Marshall through its post-Apollo reductions, assigned the Center to take the lead in evaluating Shuttle configurations. (Privately, one Houston manager wrote his reaction to the assignment: "MSC losing out."[25]) The baseline characteristics requiring a vehicle that could transport 50,000 pounds to orbit and back and have a payload volume of 10,000 cubic feet eliminated ballistic configurations from consideration, but at least eight options remained open for evaluation in Phase B. Mueller directed that the evaluation be predicated on performance, development risk, cost, and schedule.[26]

In the weeks following the 20 July 1969 Apollo 11 lunar landing, NASA attempted to capitalize on the afterglow of its greatest achievement to gain support for Shuttle and other new starts. Mueller advocated continued development of both Space Station and Shuttle, which would be necessary for Station logistics support; he anticipated that both might be launched by 1975. He also supported development of a space tug that might operate between the Station and other spacecraft, and a nuclear shuttle that could operate between Earth orbit and lunar orbit. The Shuttle, he suggested, could be developed and put into operation for $6 billion, and while NASA's percentage of the Gross National Product might rise slightly during development, it would never reach Apollo-era figures and would decline in the 1980s.[27] NASA was perhaps entitled to a rush of optimism after the Apollo landing.

In September, Vice President Agnew's Space Task Group presented its report, which in effect ratified Mueller's goals for manned space. The report offered guidelines for space operations, and stressed the importance of "three critical factors" of commonality, reusability, and economy. The panel offered President Nixon three alternative courses. The first two were ambitious and expensive, incorporating a manned mission to Mars. The third was more modest, but still supported both a Space Station and a Shuttle. Nixon selected the third option six months later.[28]

In the months that followed the release of the Space Task Group report, NASA made key decisions regarding Shuttle configuration, means of development, and the division of labor between the Centers. During the early months of the year, the Agency saw its future on the line, and battled effectively to influence the Space Task Group report. Now, in the months following the release of the report, the Centers battled to preserve their stake in post-Apollo work. Marshall was fighting this battle on several fronts, and its success in diversifying into space science and maintaining its traditional role as the NASA Propulsion Center ensured the Center's survival.

Marshall and Houston worked out a joint agreement regarding Shuttle contracting and management in a series of meetings in September and October, and referred their plan to Headquarters. Von Braun and Robert R. Gilruth, Center Director at the Manned Spacecraft Center, agreed that the Shuttle was of such complexity that development of the orbiter and booster should be handled by separate contractors. If separate contractors were to be used for the orbiter and booster, different Centers could manage each contract, and their historic roles made it logical that Houston would manage the orbiter, Marshall the booster. The relationship between the Centers would thus be similar to that under Apollo, although the interfaces between the orbiter and booster would be much more complex than those between the Apollo capsule and the Saturn stages.

By the time Mueller resigned as Associate Administrator for Manned Space Flight in December 1969, a general management approach was in place. Task teams had defined general characteristics of the Shuttle; it would be a two-stage fully reusable craft capable of performing for 100 missions. High-performance hydrogen/oxygen engines with throttle capability would provide the vehicle's power. The Shuttle would take off vertically and land horizontally. The orbiter's cargo bay was to be 60 feet long and 15 feet in diameter.[29] Many questions about Shuttle would remain for definition during Phase B of system design.

NASA knew that to win administration approval the Agency would have to build a coalition in support of the Shuttle. Political considerations thus influenced Shuttle planning throughout to a greater degree than they had in earlier NASA programs. NASA needed support from the Department of Defense both for its congressional clout and as a customer that would provide payloads, so DOD had been involved in Shuttle planning from the beginning. Its demands for cross-range (the ability to maneuver in a horizontal plane during reentry) and minimum cargo bay dimensions became inflexible Shuttle requirements that determined Shuttle size and wing configuration.[30]

The aerospace industry would also play a larger role in developing the Shuttle than it had during Apollo. The decline of the arsenal system owed in part to NASA's need for industry support. Contracting created political constituents for the Agency, but as a consequence NASA relinquished its in-house manufacturing capacity, and lost some ability to measure contractor performance. NASA expected competitive development to promote better use of manpower, earlier completion, and lower prices.[31]

Few aspects of the Shuttle program had as much impact on Marshall as NASA's decision to minimize in-house manufacturing. The Center had used in-house manufacturing of prototypes and subsystems to hone its engineering skills. Mueller sought to reassure von Braun that use of contractors offered economic advantages and earlier completion.[32] Von Braun tried to maintain pockets of in-house strength. He warned Headquarters that Marshall would be "more constrained in influencing the contractor's designs and practices," and find it more difficult to "retain its penetration" of contractors. He warned that costs could rise, schedules would be less exact, and contractors would be compelled to take risky shortcuts to maintain a competitive advantage.[33]

Another departure from Apollo was that concern for costs was paramount. George Low put it succinctly: "I think there is only one objective for the Space Shuttle program, and that is 'to provide a low-cost, economical space transportation system.'"[34] Costs became a prime driver of Shuttle development, influencing schedule, prompting design changes, determining development strategies.[35] Unrelenting emphasis on costs led NASA and its contractors to develop over-optimistic projections of anticipated Shuttle performance and low estimates of development costs that precipitated overruns.[36]

With Mueller's departure, some expected that power would shift back to the Centers.[37] In fact the intercenter Shuttle management agreement gave the Centers leverage against Headquarters. Marshall and the Manned Spacecraft Center would continue to quarrel with one another about control of pieces of the Shuttle program, as they did over control of auxiliary propulsion late in 1969.[38] In disputes with Headquarters over Shuttle management, however, the two Centers were in general agreement, defending the autonomy of the field Centers.[39] But Headquarters was reluctant to grant such latitude on Shuttle.

As NASA prepared to initiate Phase B Shuttle studies, it became clear that Mueller's successor, Dale D. Myers, would be aggressive in asserting Headquarters' prerogatives over the Centers. He insisted on the need to "maintain discipline," and stipulated that all changes must be approved at Headquarters.[40] Myers went even further than Mueller in his insistence that contractors be given free rein. He warned Eberhard Rees, who had become Center Director at Marshall when Von Braun accepted a position at Headquarters in January 1970, that "in order to establish the right tenor" the Centers would have to exercise "considerable restraint" in relations with contractors. "We must guard against over-managing and tight control of the contractor's activities," he warned.[41] Three weeks later, he was even more explicit. He told Rees to limit previously approved in-house studies, and informed him that "I hold you responsible to limit the in-house studies to that effort which does not dissipate the contractor or the Center resources and to activities which truly supplement and support the industrial effort."[42]

The concept of a fully reusable Shuttle ran into both technical and fiscal obstacles that forced evaluation of alternatives. A "fly-back" booster would require two piloted stages, one for the orbiter and one for the booster, and would have posed technical difficulties at the point of stage separation and in case of the need for abort. Another critical technical problem involved the challenge of inspecting for reuse large cryogenic tanks that were integral to the Shuttle structure, a problem that led some engineers to champion an expendable external tank.[43]

The problem of controlling costs also forced reconsideration of a fully reusable system. The cost issue became more serious on 7 March when President Nixon retreated from the goals of the Space Task Group. He offered six goals for the space program, of which only the Shuttle survived as a major new start for

NASA. Congressional criticism of the manned space program in general and the Shuttle in particular also forced NASA to reconsider its plans.[44]

Pressed from one side by Air Force requirements to develop a larger and more expensive vehicle than would have been necessary for NASA alone, and from the other by unrelenting pressure to cut costs, NASA had to find a middle way. A fully reusable Shuttle would realize savings over the life of the program, but would be more expensive to develop. By accepting a partially reusable vehicle, NASA might salvage its program by saving development costs, even if it meant that the cost per flight would be higher because of the need to buy expendable parts for each Shuttle flight. Since expendable components were less expensive to develop, their use could save money on the front end of the program by postponing expenses.

NASA thus moved into Phase B Shuttle studies in a very different environment than that immediately following the Apollo 11 moon landing. Headquarters asked Marshall to study the feasibility of a "low cost manned support module which could be transported by the Shuttle."[45] No longer could the Agency rely on the concept of a total manned system linking Shuttle to Station; instead, NASA argued that Shuttle was justified based on reduced payload costs, ironically subordinating the manned space program to unmanned space science.[46]

The plan for Shuttle development became clearer in the spring of 1970 as NASA evaluated Phase B proposals for both the Shuttle and its main engines. The plan for Phase B management represented something of a victory for the Centers, and especially for Marshall Director Rees, who had argued persistently for the "Apollo concept" in which the Centers "were not encumbered with offices and groups to oversee, review, integrate, and coordinate their activities."[47] Headquarters sought to balance management authority between Houston and Marshall, with Houston managing Phase B systems studies, Marshall the main engine studies, and the Centers dividing the Phase A Alternate Space Shuttle Concepts Studies intended to explore alternatives to a fully reusable system.

On 30 April the Agency awarded Phase B Shuttle main engine contracts under Marshall's management to Aerojet, Rocketdyne, and Pratt & Whitney. On 9 May Headquarters announced awards of parallel 11-month Phase B Shuttle contracts to McDonnell Douglas and North American Rockwell to investigate fully reusable concepts employing a two-stage Shuttle with a piloted flyback booster and an orbiter that would carry its payload and fuel internally.[48]

Phase B studies proceeded more slowly than planned, in part because of the constantly shifting fiscal terrain, but largely because of the range of configuration under consideration.

Another important change in emphasis occurred in March. The fully reusable concept began to look untenable. "The OMB [Office of Management and Budget] and the President gave us a budget. And the fully reusable vehicle would not have met that budget," remembered one of Marshall's engineers working on Phase B studies.[49] Discussion of expendable options had become more common with increasing cost pressure. The idea of using an external tank, which apparently originated in the Grumman Phase A study, gained support since it would simplify development of the Orbiter, make the orbiter lighter, and reduce development costs. In a fully reusable system, the orbiter would have carried liquid hydrogen internally. "Because hydrogen is such low density," Marshall's Mike Pessin explained, the orbiter would have required "large hydrogen tanks. It had to protect those hydrogen tanks during reentry, because it was coming back at more of a velocity. It needed the heavyweight, high temperature TPS [thermal protection system]. . . . By going to a drop tank Orbiter, where you had an External Tank, then you ended up bringing the mass fraction of the Orbiter system down, because the Tank no longer had to be protected from the high heating." In March NASA requested all contractors doing definition studies to evaluate use of an external hydrogen tank.[50]

James C. Fletcher became NASA Administrator on 27 April 1971, and soon committed the Agency to the Shuttle. "I don't want to hear any more about a Space Station, not while I am here," he proclaimed.[51] He soon faced budget pressure that made the constraints of previous months seem modest. The Office of Management and Budget announced in May that NASA could not expect any budget increases for the next five years, casting all Shuttle plans in doubt since it would limit funding for the new system to between $5 billion and $6 billion, far below what Paine or Low had anticipated as minimal.[52]

Management of the Shuttle program was another pressing issue when Fletcher took the helm. Houston wanted a Lead Center approach, with the Manned Spacecraft Center responsible for "complete systems engineering, program management and control including financial management," with a Headquarters director "who would review the MSC decisions and concur in these decisions."[53] The Houston plan sought to decrease the authority the Headquarters program office had under Apollo by shifting program and financial management to the Lead

Center.[54] Talk of single-Center management worried people in Huntsville, who feared that Marshall might lose even the propulsion system.[55]

When word leaked out that Myers, the head of the Office of Manned Space Flight, supported the idea of naming Houston Lead Center, the Alabama congressional delegation, led by Huntsville's Senator John Sparkman, requested a meeting with Fletcher. Sparkman dropped his request after receiving assurance that Marshall would get a "sizable portion" of Shuttle work.[56] More than Shuttle work was at stake, however. When Myers sent his organizational plan to Fletcher, he proposed assigning Houston as Lead Center on Shuttle, and assigning Marshall the Research and Applications Module (RAM, the predecessor of Spacelab) and Space Station studies in addition to its Shuttle propulsion.[57]

The Shuttle management plan that Myers announced on 10 June made compromises to minimize Center rivalry. Marshall received responsibility for the booster and the main engines, Kennedy for launch and orbiter implementation. It gave Houston everything it wanted except financial management, which remained in Washington. Christopher Kraft, Houston's deputy director at the time, claimed that leaving financial control in Washington gave Houston technical management but not control. Marshall "got the money for their programs through Headquarters. That was a ploy to satisfy their distrust in the system," Kraft said.[58]

But Headquarters had no intention of relinquishing financial control, particularly when management was seeking to demonstrate its cost-consciousness. As George Low insisted, "We can't let the people at Marshall and Houston solve all their problems by calling up the budget office and saying they were going to let out another contract for $10 or $15 million."[59]

Nor was Marshall satisfied. "That was a very controversial decision, and a decision that I think some people would argue today might not have been a good decision," explained Bill Sneed, who was involved in Shuttle planning as a part of Program Development. "It has been our experience here that it's very difficult for one Center with equal posture to lead and manage another Center. There's a certain amount of competitiveness and parochialism between the Centers that makes it difficult for one Center to be able to objectively lead the other. And perhaps more difficult would be to have one follow the other. That was the real flaw in that arrangement."[60]

Houston's aggressive assumption of its Lead Center responsibilities gave Marshall concern as well. Roy Godfrey, manager of Marshall's Space Shuttle Task Team, attended a meeting of contractors in Houston shortly after the Manned Spacecraft Center became Lead Center, and reported that the contractors received "liberal doses of MSC philosophy from Max Faget and Chris Kraft." When one of the contractors responded to criticism that they were only doing what had been requested in Washington, Kraft told him, "You are in Houston now, not Washington!" Godfrey concluded that "MSC has taken firm hold of Shuttle—they left no doubt in the contractors' minds that they intend to have their way."[61] Two months later, Marshall complained to Headquarters that the Houston Shuttle Program Office was approving its own facility requirements and disapproving Marshall's. Dick Cook, Marshall's Deputy Director for Management, suggested that the facilities issue demonstrated that "no matter how one Center that has been given program management responsibility over other Centers tries, it cannot look at the requirements of another Center in an unbiased manner."[62]

In the summer, as budget pressure increased to the point that the survival of the Shuttle was in question, a configuration breakthrough gave the program new life. The development was so significant that by the end of the year Fletcher could claim that "the cost and complexity of today's Shuttle is one-half of what it was six months ago."[63] The Shuttle orbiter's main engines required both liquid hydrogen and liquid oxygen for fuel. For several months, all four Phase A and B configuration contractors had been looking at designs using an external tank for liquid hydrogen and an internal tank within the orbiter for liquid oxygen. The breakthrough of May 1971 involved putting all of the Shuttle's ascent fuel in external tanks, utilizing one large shell for both liquid hydrogen and liquid oxygen tanks. In addition to lightening the orbiter and allowing for a larger payload bay, the concept allowed the tank to perform the structural function of absorbing the thrust of strap-on boosters.[64] Furthermore, it lowered costs since its development required no new technology. "We went with essentially Apollo technology. We were deliberately not wanting to invest into a high risk technology in the Tank," remembered James Odom, who would later head Marshall's External Tank Program. "That was the way we got the cost down from ten billion down to the five billion. In doing that, we had more expendable hardware. The per launch cost went up, but we got the development cost down to within a range that Congress would support."[65]

Piece by piece NASA had been forced to accept reductions below what it considered necessary to build the Shuttle. From Paine's $10 to $15 billion estimate, Low had accepted a cut to $8.3 billion in the fall of 1970. Fletcher had been able to stave off OMB's goal of $4.7 billion in the protracted battle from May to December 1971. On 5 January 1972, President Nixon approved the Shuttle with a budget of $5.5 billion. Treasury Secretary George Schulz insisted on another cut, and NASA finally had to settle for $5.15 billion.[66]

Nixon approved a Shuttle whose configuration was not yet set. Refinements of the configuration continued until the final decision in March 1972. The expendable external tank concept not only allowed for a more efficient orbiter, but offered new possibilities for booster design. A smaller, lighter orbiter could shoulder more of the burden of attaining orbit; booster separation thus could take place at lower altitude and lower velocity. Budget cutbacks and the external tank thus eliminated the piloted flyback booster from consideration, and forced NASA to examine booster concepts that were simpler and less expensive.

By the fall of 1971, three types of boosters were under consideration: pressure-fed and pump-fed liquid propellant boosters and solid propellant boosters. Marshall had used pump-fed liquid boosters in its Saturn engines. The Center had no peers in their development, testing, and operation. Pressure-fed boosters would have required more technical risk but would have had thicker walls more able to withstand ocean impact, making recovery and refurbishment easier. NASA preferred the lower cost and lower technical risk associated with the pump-fed engine despite recovery disadvantages.[67] So the booster question narrowed to a choice between pump-fed liquids and solids.

No technological issue was as sensitive at Marshall as the debate between liquid and solid rocket engines. With its tradition of conservative engineering and extensive testing, Marshall had always relied on liquid-fueled engines and resisted the use of solids. A liquid system could be tested over and over, "literally thousands of times," according to Marshall's Bill Brown, who had long experience with solids at contractors and Marshall. "The cost of testing large [solid] rocket motors repeatedly is very, very high. . . . They have, I don't know how many, maybe tens of tests rather than hundreds or thousands of tests such as you would have in a liquid system. So, there has to be much more extrapolation of the data" than with a liquid system.[68]

Unlike the Air Force, which had used solid rocket motors, NASA—and Marshall—had experience almost exclusively with liquids. "Solids had never been used in manned space flight before, except the escape rockets on the Apollo and Mercury programs," explained LeRoy Day. "There were people who were not enthusiastic about them. Von Braun was one who didn't think we should go solids."[69]

"The Germans did indeed oppose the solid rocket motors—and not just the Germans. Many of us did," recalled Brown. "The basic problem is that you have your oxidizer and your fuel already mixed. And if you get that started, it is extremely difficult, if not impossible, to stop it from going, unlike the liquid system which mixes the oxidizer and the fuel only at the time you wish to combust them."[70] Ron McIntosh, who spent most of his career at Marshall working on solid rocket motors, explained that "Solid rocket motors are a lot like fireworks or roman candles. Once you light that thing you better be prepared to put up with whatever is going to happen, because you're not going to be able to turn it off."[71]

Recovery of reusable solids posed another problem. According to Day, "There were a lot of skeptics, because the size of the solids is about like a freight train car. . . . It's going to impact the ocean at about 100 miles per hour and . . . the damage would be so severe that it wouldn't be cost effective."[72]

The debate placed Marshall in a precarious position, particularly when Headquarters began to prefer solids. Marshall was opposed to solids, but could not afford to be too persistent for fear of losing the responsibility to manage the booster development. Fletcher had made clear his concern that Marshall would not give solids a fair shake. After a discussion with Headquarters, Rees reflected that "Mr. Myers emphasized again that Marshall Space Flight Center is obviously known as being against solids." Dan Driscoll, preparing to present Marshall's point of view to Headquarters, said that he planned to show that Marshall "understands the advantages of the solids as well as their disadvantages." Rees urged him to convey to Fletcher the Center's "enthusiastic involvement in the configuration of the Shuttle booster with solids."[73]

Aerospace publications perpetuated the widely held conception that Marshall was irrevocably opposed to solids. The *Aerospace Daily* quoted "industry sources" as citing the Center's long history of work with liquids as evidence that Marshall "is not about to put itself out of business."[74] When a report circulated that "directors of certain NASA Centers" were trying to close off debate by selecting pump-fed liquids before competitors had a chance to make their presentations, one of Marshall's executives wrote cynically in the margin: "We will, of course, get full credit for this."[75]

NASA did not decide to go with solids until March 1972, nine weeks after President Nixon approved the Shuttle. In fact, when Fletcher met with the President in January, he took with him a model of the Shuttle graced by pencil-thin liquid boosters.[76] The decision boiled down to two issues: thrust and cost. The Agency anticipated that liquid engines would be used in a series burn configuration, meaning that a liquid booster stage would separate before the orbiter's main engines would ignite. Solids, on the other hand, could be designed in a parallel burn configuration in which the boosters and main engines could fire at the same time, taking maximum advantage of the high performance main engines during early ascent. Solids also would be $700 million less expensive to develop and have a lower unit cost. Since they could withstand impact better, they offered recovery advantages. And since they were less expensive, loss during recovery could be more easily absorbed.[77] For Fletcher the decision was "a trade-off between future benefits and earlier savings."[78]

Selection of a solid propellant booster completed the configuration of the Shuttle. The nation's next generation space vehicle was to be a delta-winged craft with a 60- by 15-foot payload bay. Its main engines were to be powered by liquid hydrogen and liquid oxygen supplied from an expendable external tank. Two reusable solid rocket boosters mounted on the external tank would help power the Shuttle into orbit.

Selecting Contractors

Marshall would manage three Shuttle projects: the main engines, three of which would be arrayed in each orbiter; the solid rocket boosters, two of which would be attached to the external tank below the orbiter; and the external tank itself. Planning for Shuttle contracts clearly showed NASA's new focus on keeping costs to a minimum.

Shuttle was to be a very different program from Apollo. NASA management had to adjust from a program in which there was ample money to one with very tight funding constraints. "The Shuttle presented some new challenges for the Agency that we really had not experienced," remembered Sneed of Marshall's Program Development directorate. With Apollo, the technical requirement was fixed, the schedule was fixed, and cost was a variable. "Any time we got into difficulties with the Apollo program, we had the money to 'buy our way out of it,'" Sneed continued.

"Shuttle program management was more difficult than Apollo in that we had a fixed budget, which significantly influenced every major program decision. Since technical requirements were essentially fixed, it meant that schedules had to be delayed to make dollars available on a near-term basis to solve technical problems. This was an acceptable near-term solution but not a good long-term solution since extended schedules required considerably more total dollars for the program—dollars that were not available to NASA. So there was a conflict built into the program from the outset. It required the Shuttle project managers to complete the development program within a set of fixed technical requirements, fixed budget and a fixed schedule—a most formidable and challenging task. This condition forced our project managers to be more frugal in executing the development program, conducting a minimally acceptable test program, minimizing back-up developments for problem areas, and in general introducing greater risks in the decision making process."[79]

With some 60 percent of the operating costs of each Shuttle mission dependent on components under Marshall's responsibility, Rees realized that the Center would have to place new emphasis on monitoring costs. He decided to establish a Centerwide cost estimating group. "I know that MSFC was never too good in this particular area," he acknowledged.

"Our engineers just are not used to design for low cost. When we awarded the contracts for the Saturn stages, we based them on Work Statements which never spelled out unit costs. These contracts were rather spelling out a development program for those stages and incidentally included in the price was the delivery of so and so many stages within a certain time."[80]

The constant threat of recurring reductions-in-force reinforced programmatic demands that Marshall monitor costs carefully. Fletcher made the connection between Shuttle costs and personnel reductions explicit in August 1973 when he insisted that if the Program Office made a decision that increased the cost of the Shuttle, Marshall would have to lose another 150 people.[81]

The impact of costs on Shuttle development affected the negotiation of development contracts, a process already underway during the evolution of the Shuttle configuration. Development of the Shuttle main engine preceded Marshall's other Shuttle programs. An integral part of the orbiter, the main engine was the pacing component of the Shuttle; its development had to proceed in tandem with Houston's work on the orbiter.[82] Thus the main engine moved through Phase B program definition and preliminary design while shuttle configuration studies were still underway. Three aerospace contractors—Aerojet, Pratt & Whitney, and North American Rockwell's Rocketdyne Division—participated in the preliminary design studies. Marshall planned to follow Department of Defense procurement strategy and have a "shoot-out" in which, as Frank Stewart explained, "we'd go up to a few engine-level test firings with two contractors, and then we'd make a final selection." Stewart remembered having set aside $25 million to execute the plan.[83]

Then tightening budgets intervened. Headquarters decided that rather than continue two main engine contracts into Phase C/D development and then have a "shoot-out" to select the better design, NASA would select one contractor at the conclusion of Phase B definition studies. Marshall's program management office worried that "once we choose a company and a configuration, we are locked in," and that "the 'benefits of competition' must be realized at the negotiation table."[84] Nor was the approach necessarily less costly in the long run. Richard L. Brown, who helped evaluate the main engine proposal, claimed "there were economic studies that indicated it would actually be cheaper to run the competition because of its influence on price" and to arrive at "a better definition of cost, and therefore less overrun."[85]

The Center issued its Request for Proposals for Phase C/D in March 1971, and the three companies that had participated in definition studies all responded. On 13 July, Marshall announced selection of Rocketdyne for negotiations leading to a contract worth perhaps $500 million for design, development, and delivery by 1978 of 36 engines, each capable of 100 missions.[86] Pratt & Whitney protested, initiating what one report termed "a savage fight between two giants in the economically depressed aerospace industry."[87] Pratt & Whitney filed charges with the General Accounting Office (GAO), claiming experience superior to that of Rocketdyne, and complaining of the selection as "manifestly illegal, arbitrary and capricious, and based upon unsound, imprudent procurement decisions."[88] Both Alabama senators joined seven colleagues from the Southeast protesting selection of a California company over one from Florida:

"It seems inconceivable that Pratt & Whitney's low risk design based on flightweight hardware testing can be matched by limited boilerplate testing and paper studies of the bidding competition."[89] Rocketdyne, which built Saturn engines for Marshall, claimed better experience in building large liquid-rocket engines.[90]

The protest had several ramifications. In the short term, it delayed work on the main engines, which NASA considered the pacing item for the Shuttle. The GAO allowed Marshall to continue to negotiate with Rocketdyne with the understanding that no definitized contract could be signed until resolution of the protest, which took seven-and-one-half months.[91] The Center issued a series of interim level-of-effort contracts to Rocketdyne pending resolution. On 31 March 1972 GAO ruled in favor of NASA. Rocketdyne worked under a letter contract until completion of the formal contract in August—more than a year after NASA first selected the company for negotiations.[92]

The long-term ramifications of the protest were more serious. With NASA still worried about winning approval of the Shuttle late in 1971, the Agency could ill afford another protest. NASA needed the support of aerospace contractors. Top manned spaceflight and Shuttle administrators met late in November and discussed ways to bolster the depressed aerospace industry. Marshall's Shuttle Program Manager Roy Godfrey reported to Rees:

"George [Low] and his people were very concerned about handling the selection and subcontract awards so we minimized the possibility of a protest. This led to a discussion of dividing up the orbiter and Booster into subcontracts, such as avionics, structures, etc. . . . This way, all the major primes would get enough Shuttle business to support the Shuttle and not protest."[93]

NASA thus adopted a strategy of spreading out Shuttle business among as many aerospace contractors as possible, a pragmatic approach that raised no dissent. Sound politics does not necessarily lead to sound engineering, however. The test of the plan would come as NASA negotiated contracts for other Shuttle components; it would affect in particular the way in which the solid rocket motor (SRM) would be contracted, developed, and assembled.

Negotiations for the solid rocket motor contract were as laden with controversy as the main engine deliberations. The first disagreement was internal, as NASA prepared to request proposals from industry. NASA envisioned the solid rocket booster (SRB) as a system comprised of a steel case (the SRM), and several other elements such as forward and aft skirts, nose cone, attachment structures, thrust vector control, separation, and recovery devices.[94] Rather than contract the solid rocket booster and require industry to be responsible for the entire system, Fletcher decided to contract only the solid rocket motor and give Marshall integration responsibility.

Fletcher's decision did not have unanimous support at Headquarters. NASA comptroller Bill Lilly proposed making one contractor responsible for the entire system, including design, recovery, and refurbishment. To break bidding into contracts for separate components would double the price of the booster, he argued. Fletcher chose to ignore Lilly's warning, hoping to spread business around and fend off OMB's threat of closing Marshall.[95] When NASA developed a list of 19 internal ground rules before initiating booster procurement, the first guideline gave Marshall the sort of protection it had been seeking since the peak of Apollo: "SRB to be designed in-house with the exception of the SRM."[96]

NASA also made a key decision affecting the configuration before letting the SRM contract. The Program Office in Houston, supported by prime Shuttle contractor Rockwell, decided in April 1973 to eliminate a baseline (minimum) requirement for an abort procedure called thrust termination. Thrust termination would have required a means of shutting down the solid rocket boosters within a specified period of time (which had not yet been determined). It would have been designed to protect against failure of the SRB to ignite before launch, loss of two or three main engines, or burnthrough of the casewall of the sort which caused the *Challenger* disaster.

But thrust termination would have been costly. No abort procedure could be a hundred percent risk-free. Three years earlier, when NASA first considered abort procedures for the Shuttle, Max Faget had commented on one proposal that suggested a 0.999 guaranteed probability of success, "This is going to greatly increase cost if carried to nauseating extreme." Faget argued that system redundancy requirements might be waived "where common sense indicates the risks are low and the cost high." Thrust termination might have added as much as 8,000 pounds to the external tank and increased the orbiter load from two-and-one-half times the force of gravity to three times. Rockwell argued that the concept had too high a system penalty for too little return, and the Program Office believed that the system had sufficient design redundancy.

At meetings in Houston and Washington, Marshall agreed to eliminate thrust termination, but argued for retaining an option to implement it later. Houston considered allowing the option, but Headquarters determined to disallow "scar penalties" (weight allowances held in reserve) that might have made later addition of thrust termination possible, but did allow for SRB separation studies that were never executed. Marshall made one last attempt to revive the thrust termination option in August, but in reality the Headquarters decision ended any possibility of reconsideration.[97]

So when NASA requested proposals for its major booster contract on 13 July 1973, the request involved only the solid rocket motor and lacked provision for thrust termination. Four aerospace companies responded: Aerojet Solid Propulsion Company, Lockheed Propulsion Company, Morton-Thiokol Chemical Corporation, and United Technology Center. The SRM was to include the case, flexible nozzle, ignition system, case liner and insulation, and propellant. Aerojet seemed to have an advantage, since it planned to use a large tract in Florida for assembly and could have constructed one-piece motors for water shipment to Michoud in Louisiana and to Kennedy, whereas the other companies would build segmented boosters for shipment by rail.[98]

After evaluation of proposals by teams involving 289 people representing five NASA Centers, Headquarters, and the three military services, NASA selected Thiokol Chemical Corporation to develop the solid rocket motor. The top three competitors ranked closely on mission suitability criteria; Thiokol won the competition principally on the basis of cost. Thiokol's proposal anticipated the lowest costs for the early years of the program and for development and production, an advantage gained by virtue of lower expenses for facilities and labor.[99] Cost weighed heavily, and indeed Congress had lauded Fletcher's pledge that solid rocket motor procurement "would be accomplished in the manner considered most cost effective."[100]

The selection of Thiokol prompted controversy for two reasons. Critics alleged that Fletcher had pushed business to his home state of Utah, where Thiokol had its headquarters. Fletcher vehemently denied the charge, and others on the Source Evaluation Board defended him. The rationale announced for the selection and the close competition also raised questions, and Lockheed filed a formal protest. Once again NASA feared that its schedule would slip while the Agency sought to defend its decision. Marshall's analysts estimated that the delay would cost $60,000 per day if the dispute was not resolved by 1 February 1974, and

$400,000 per day if it was not settled by 15 March.[101] On 24 June, the General Accounting Office ruled against Lockheed, and two days later NASA awarded the contract to Thiokol.[102]

As a result of the decision to separate the SRM from the rest of the booster, Marshall managed the SRB differently from either the other Shuttle components or other large programs. In addition to the Thiokol contract, Marshall's SRB Program Office managed a contract with United Space Boosters, Incorporated (USBI) for booster assembly in a conventional contractual arrangement. What was unusual was that the Science and Engineering Directorate (S&E) performed as a third prime contractor, and subcontracted other elements of the SRB including the recovery system, booster separation motors, and integrated electronic assembly. The arrangement not only gave Marshall more business than it would have had if all SRB work had been given to a single contractor, but required less money in the early years of development.[103]

The expendable external tank was the third Shuttle component under Marshall's supervision. Rees considered the tank "something very challenging to work on, but also very complex and difficult. I want to go even so far as to state that an optimum drop tank design is one of the key factors for the whole Shuttle Program not only from the viewpoint of performance but also as to economics."[104]

As with all Shuttle components, cost was of primary importance in tank design. James Kingsbury, who headed Marshall's Science and Engineering Directorate during tank development, explained that "the challenge with the Tank was to get it built at minimum cost. There was nothing really challenging technologically in the Tank. . . . The challenge was to drive down the cost."[105] The tank was nonetheless as complex as Rees anticipated. The contractor selected for external tank development would be responsible not only for the liquid hydrogen and liquid oxygen tanks themselves, but for an intertank section, avionics equipment, a thermal protection system, and the assemblies connecting the tank to other Shuttle systems. And the tank would be more than just a container for fuel: it would be the critical structural component of the Shuttle system, the base to which the boosters and orbiters would be attached during ascent.[106] Kingsbury explained that "whereas in the original concept it was a big dumb tank that just kind of carried fuel, it became the structural backbone of the stack."[107]

After the selection of the Shuttle configuration in March 1972, the Center began to devise a strategy for tank development. Since the power systems for the Shuttle were interdependent, and since the tank required less new technology than other Shuttle systems, one school of thought in NASA held that the tank should be the variable element and its development should be deferred until other systems were defined and sized.[108] Rees disagreed, and wanted Marshall's laboratories to start work immediately. "We can initiate immediately all kinds of necessary parametric and trade-off studies, help in clarifying requirements, look into possible tank designs, select best materials, establish tank pressure ranges," he directed.[109] Industry studies confirmed Rees's approach, suggesting that once system weight estimates were set, basic tank design could be frozen and solid rocket motor diameter established.[110]

Selection of the contractor for the external tank went smoothly. In August 1973, NASA named the Denver Division of the Martin Marietta Corporation (MMC) for negotiation of a contract for the design, development, test and evaluation of three ground test tanks and six developmental flight tanks. NASA stipulated that assembly would take place at Marshall's Michoud Assembly Facility in New Orleans.[111]

Developing the Elements

By the time Marshall completed negotiation of contracts for its Shuttle projects, NASA's system for Shuttle program management was in place. NASA established three levels of management. Level I resided in the Office of Manned Space Flight at Headquarters, where the Space Shuttle Program director administered overall planning and allocated resources. Level II resided at Houston's Johnson Space Center, where Robert F. Thompson exercised Lead Center responsibilities as the Space Shuttle Program Manager.[112] Project offices comprised Level III management, and each of Marshall's three Shuttle projects had its own project manager. Marshall also had a Shuttle Projects Office to oversee the three Huntsville projects. Roy Godfrey headed the Marshall projects office during most of the contract negotiation period; in March 1973 Robert Lindstrom took his place. Marshall's Shuttle Projects Office thus had two lines of responsibility: to the Program Office in Houston, and to the Marshall Center director.[113]

Marshall's experience in *Skylab* led the Center to initiate a means to exercise independent engineering judgment on its Shuttle projects through which the Science and Engineering Directorate could make technical decisions unencumbered by managerial responsibilities. Larry Mulloy, who worked on both the external tank and the solid rocket boosters, explained that, "in a project office you're balancing budgets and schedules against technical requirements. And growth in technical requirements leads to growth in budget, leads to growth in schedule. The Project Manager is often under pressure to not grow budget and schedule. His decision process relative to technical matters might be clouded a little bit by those other factors. So they decided to set up a separate Associate Director for Engineering in the Science and Engineering Directorate and have chief engineers who have an autonomy from the project office in terms of technical courses of action."[114]

Each Marshall project had both a project manager and a chief engineer. Project managers were responsible for schedules, budgets, contractor oversight, and contract changes. But the chief engineer had technical authority. Project offices "didn't want the lab making engineering decisions for them," Kingsbury explained, but they "were not staffed with the engineering talent to make those decisions. So they had to depend on the labs."

Thus in addition to their direct lines of authority to the program manager in Houston and the Center Director at Marshall, project managers had to weigh input from Science and Engineering. William Lucas wanted to ensure that "S&E talent will be used as an influential part of the team, not in a second-guessing or trouble-shooting role."[115] As head of Science and Engineering, Kingsbury had the same concern. If the project manager "didn't pay any attention to my engineers, then he was accountable to me," Kingsbury insisted. "If he didn't pay any attention to me there was another guy he would pay attention to, that was his boss and mine. We never had any confrontations."[116]

Their mutual boss was of course the Center Director. The Center Director was technically not part of program management, but NASA recognized his responsibilities by differentiating between "programmatic relationships" and "institutional relationships."[117] Since the Shuttle was the largest program involving Marshall personnel, it would have been inconceivable for the Center Director not to be involved in Shuttle management. This was particularly true of Lucas, who became Center Director in June 1974 when Rocco Petrone returned to Washington as Associate Administrator. Lucas had been involved in propulsion throughout his career. He had founded the Program Development directorate, and participated in Shuttle planning as its Director and as Deputy Center Director after his appointment in 1971.

Lucas insisted that the project manager and chief engineer on each Marshall project keep him informed. He used the Weekly Notes initiated during von Braun's directorship as a management tool. "It was a technique that encouraged communication," Lucas explained.

"People in the laboratories could introduce these notes. They were read and annotated and sent back. . . . They did not supplant any other thing in terms of communication, any of the more formal things. It was an information exchange, to help the top management understand other views."

"In top management, it is pretty easy to get isolated. You are totally dependent upon what other people tell, you can't be everywhere. It gave you a little better feel for what the disagreements were. . . . I always read the notes; even if I had to leave off something else, I would do that."[118]

Lucas used the Weekly Notes as both a means of gathering information and as a means of communication. In marginal comments he responded to the remarks of his managers, often promulgating policy in the process. His comments thus often set the tone for Marshall's response to problems, often highlighting, for example, deficiencies with contractor management.

Lucas's long experience in engineering and administration prepared him to direct both technical and managerial aspects of Marshall's Shuttle projects. "His technical participation in Shuttle development was as significant as any engineer at the Center," according to Bob Marshall. "His participation in Shuttle was more from a chief engineer role than the senior manager."[119] His role in guiding Marshall's participation in Shuttle development also grew as a result of changes at Headquarters. Over time Level I management became more active; a 1979 internal NASA report concluded that the Associate Administrator for the Shuttle program had become the de facto program director, and demanded more direct participation by Center Directors.[120]

With its management structure in place, Marshall began to move its Shuttle projects into development. The Space Shuttle main engine (SSME), the first of Marshall's projects to begin development, was "the real challenge in Shuttle," according to Kingsbury. "It was an unproven technology. Nobody had ever had a rocket engine that operated at the pressures and temperatures of that engine."[121] The engine had to develop 470,000 pounds of thrust for eight and one-half critical minutes of each flight, and although this was less thrust than Saturn engines, those had not been reusable. It was to be lighter and more efficient than previous spaceflight engines, requiring the use of new materials and welding techniques. Operation would generate very high temperatures, so an efficient cooling system utilizing the engine's own hydrogen fuel had to be employed or the engines could melt down. The engines had to withstand reentry and still be reliable enough to make 55 flights without overhaul.[122] "The SSME was by far the most challenging and difficult of all the Shuttle elements," according to Bob Marshall. "Nearly every engine test run contributed a 'first' time test for a fix of a failure in the previous test."[123]

Since the main engine was the pacing development project in the Shuttle program, there was great concern throughout the Agency when the project began to encounter problems. By mid-1974, the main engine project was in trouble, experiencing delays in construction of facilities and in development of critical components, management problems at the contractor, schedule slippage, and substantial cost overruns. Fletcher warned in May that Rocketdyne's projected cost increases were "unacceptable and pose serious threats to the Space Shuttle Program."[124] An internal company report a month later acknowledged that several things were going wrong, including "technical, schedule and cost problems in the Honeywell controller, delays and overruns in the construction of the facilities at Santa Susana, serious material shortage and vendor delivery problems."[125]

Some of Rocketdyne's problems derived from its management of subcontractors for the main engine controller and facilities at Santa Susana. The controller was an electronic computer meant to monitor the functions of the engine such as pressure, temperature, and flow, and then to translate these readings to direct a predetermined sequence of events. Honeywell's controller experienced design and fabrication problems related to the power supply and line noise in the interconnect circuits. For a time these problems were so troubling that Fletcher expressed "serious doubts about the capability of Minneapolis-Honeywell to develop the engine controller for reasonable cost under Rocketdyne management."[126] Rocketdyne even considered development of an alternate backup system, but by the summer of 1975, Marshall was confident that remaining difficulties could be solved.[127]

The Santa Susana facility issue was perhaps more troubling, since it raised questions about Rocketdyne's management of its main engine responsibilities. Rocketdyne operated a test area at Santa Susana in the mountains north of the San Fernando Valley near Los Angeles. Bovee and Crail, another subcontractor, had responsibility for constructing test positions at the cluster of Santa Susana test sites designated COCA–1 through COCA–4. Rocketdyne's schedule had already slipped by the beginning of 1974 when the company requested an additional $2.7 million to complete construction. For the next several months, things only got worse. Marshall, hoping to keep main engine development on track, requested an accelerated construction schedule. Instead, the schedule slipped again and again, and NASA cited Rocketdyne for "failure to perform." Rocketdyne and Bovee and Crail agreed to work 10-hour days and 6-day weeks in order to finish the facilities by an "absolutely necessary" deadline of 15 December.[128]

Cost overruns plagued facilities construction, controller development, and labor expenses. Fletcher called the increases in wages and fringe benefits resulting from a new labor agreement "staggering," and warned that "the funding level for the Space Shuttle Budget is essentially fixed and will not accommodate inflationary growth of this projected magnitude."[129] A Rockwell internal review of Rocketdyne acknowledged poor morale and criticized a $70.3 million cost overrun, a six-month slip in schedule, and excessive overtime. The report observed that "working relationships between Rocketdyne and NASA at all working levels have deteriorated," and judged that both Rocketdyne and the government had underestimated the complexity of the project.[130]

Marshall responded aggressively to Rocketdyne's problems, and increasingly focused on the company's management as their source. As soon as the Santa Susana cost and schedule problems surfaced, the Center formed a "Facilities Tiger Team."[131] In May, two Marshall reviews cited management shortcomings. One said that while there had been improvements in scheduling, "good control is not yet evident."[132] The other, from Program Development, made recommendations, the first two of which were to "get the company integrated" and "make the VPs accountable and measure their performance against hard criteria."[133] When Rocketdyne mislabeled equipment, Lucas considered it symptomatic, an indication that "discipline is still lacking in the Rocketdyne organization."[134] Rockwell complained that Marshall was "so concerned over the Honeywell situation that it appears to have 'taken over' technical management of the controller program."[135]

Rocketdyne made changes, naming a new program manager for its main engine program, bringing in other new people, and conducting program reviews.[136] By the end of summer, improvement was apparent. The new program manager seemed "keenly aware of the need for good morale and a team spirit."[137] Facilities problems continued, but engine development was now moving along, and Marshall's main engine Project Manager J. R. Thompson told Lucas that "we probably understand and have better control over the engine powerhead in terms of cautious, safe operation than we have over the facilities."[138] By late October, Marshall's assessment of the Rocketdyne operation was even more positive. "Tests now occur when planned" noted one comment, and morale among test personnel, where there had been so many problems with facilities, was "now one of the highest at Rocketdyne." Problems remained, for the cost overrun continued to grow and Marshall still expected improvement in management, but the engine program had passed through a difficult early shakedown.[139]

Space Shuttle main engine test in Mississippi.

In March 1975, Rocketdyne completed the first main engine a month ahead of schedule. The engine was intended for testing, not flight. Rocketdyne shipped it to the National Space Technology Laboratories (NSTL) in Bay St. Louis, Mississippi, a facility operated by Marshall and used to supplement tests conducted at the COCA site at Santa Susana.[140]

Cost considerations forced Marshall to apply a different approach to testing Shuttle than had been used in Apollo. First, during Apollo more money was available during the design phase. "The heritage of the Germans was conservatism always," Marshall engineer Robert Schwinghamer explained, "and if there was any question or any doubt on the Saturn, you just overdesigned it." Shuttle had less money for robust designs.[141] Second, component testing on Shuttle was more limited than in the Apollo Program, where Marshall applied extensive independent component testing before assembling and testing the whole engine. "We didn't have that luxury on the Shuttle," according to Schwinghamer. "We just never really had enough money to go into a components test program on the Shuttle. And so, I think some of the problems that we had with the Engines in the early days had to do with wringing out the bugs. . . . That did give us some problems."[142]

Test activity at both the California and Mississippi sites was intense. "We worked harder on that program than on any program that I have ever been associated with," according to Jerry Thomson, the chief engineer on the main engines. "It was a 60-hour a week job. . . . We were running tests late into the night, and worrying about getting everything fixed that we failed, and we were trying to make schedule. . . . None of the Apollo activities ever had the challenge and the difficulties that we had with the SSME."[143]

The first major technological challenge involved a rotor instability problem that caused vibration, limited the speed of the turbopump, and caused bearing failures. In March 1976 turbine end bearings failed as a result of high temperatures and violent rotor instability known as subsynchronous whirl. "The rotor was orbiting within its bearing supports," according to J. R. Thompson, who later remembered this as "one of the more elusive problems we had." A joint NASA-Rocketdyne team used mathematical models, consultation with universities and industry as well as laboratory tests to derive design changes. These adjustments eliminated the whirl problem.[144]

Four explosions associated with testing high-pressure oxidizer turbopumps occurred before the first Shuttle flight. Rocketdyne's project engineer described liquid oxygen explosions as "nightmarish events in rocket development programs." Not only did they take equipment out of commission and thereby disrupt schedules, but the explosions often destroyed equipment, leaving no evidence of the cause of the failure. At least two of the fires resulted from failure to keep liquid oxygen separate from the hydrogen-enriched steam that drives the turbine, the "overriding design concern" with the turbine pump. Design changes included modifications to shaft seals and turbine end bearings.[145]

Ron Tepool remembered the first time an engine blew up at the Mississippi test area. The accident took place at the time of a main engine quarterly review in Huntsville, so J. R. Thompson and a "planeload" of Marshall executives went to inspect the damage. "About two in the morning, J. R. wanted to see the engine. So, we went up to the test stand, just he and I. And he stalked this thing. He just walked around it, looking. It was just ashes basically. And he said, 'We ain't never going to do *this* again.' I told him then that in the F–1 program, we blew up about 15 engines or something like that. I told him this was just the first of many. He didn't believe that, but he believes it now."[146]

Tepool was right. As time went on Thompson became more sanguine about engine tests, and Ron Bledsoe remembered that "J. R. always indicated that whenever we had a failure, it was an opportunity."[147] John McCarty explained that "we always used to say that an engineer didn't learn anything until we had a failure. There's a lot of truth to that, because if you're just operating and everything's performing as predicted, all you know is that it's performing as predicted. It could mean your prediction is perfect or could mean that your prediction is off."[148]

Failures were to be expected in a high-risk developmental project, but they were nonetheless costly. On 4 February 1976, an oxygen flowmeter failed at the COCA–1A Test Site at Santa Susana. Parts broke loose and hit a liquid oxygen discharge valve, causing an explosion and igniting a fire that lasted 20 minutes. The machinery under test and the test stand suffered significant damage, and Marshall had to divert $1.2 million from the Mississippi facility to make repairs.[149] Four months later a fuel subsystem test at the neighboring COCA–1B site resulted in another major fire.[150] Fires, lack of resources, and the expense of operating two main engine testing facilities finally forced NASA to phase out component testing at the COCA site by September 1977, although other areas at Santa Susana would be used for main engine testing. "They just couldn't afford to keep both Mississippi and COCA open, so they closed COCA down," according to McCarty. "We couldn't get a reliable enough test frequency out of it."[151] New NASA Administrator Robert Frosch rationalized that "the best and truest test bed for all major components . . . is the engine itself."[152]

The Mississippi facility was just as susceptible to test accidents. Tests involving the liquid oxygen pump system resulted in three fires at the National Space Technology Laboratory in 1977 and 1978.[153] Each incident delayed development.

With each failure, said Herman Thomason, "there's an investigation. Put a freeze on and go in and do a complete investigation and find out what happened. You've got to report all the way up to the Administrator. And everybody takes a rap on the knuckles and go fix that. You go test for another week and something else goes wrong, and you've got to go through it all again."[154]

Fortunately Headquarters gave strong backing to Marshall's main engine team. Chief engineer Jerry Thomson recalled that "When J. R. Thompson and I were blowing up the engines every few months and wondering how soon would we be dismissed, John Yardley was giving us encouragement, 'You guys will get it fixed. Just keep trying.'"[155]

Main engine development proceeded more slowly than planned, but NASA still hoped to launch a first manned Shuttle flight before the end of the decade. The engine performed well for several months of successful tests, including one at 100-percent power, before the July 1978 Bay St. Louis fire.

With the main engines operating at higher temperatures and pressures than any previous engine, turbine blade problems became a recurring challenge. The first instances of blade failure occurred in two separate tests late in 1977. In the second and more serious accident, debris from a shattered blade caused the pump to seize up causing loss of the engine. Engineers attributed both accidents to blade fatigue and insufficient damping of the blades. In 1978, as J. R. Thompson remembered, "We really started getting cranked up and running the engine." More fatigue-related problems developed in the main injector and main oxidizer valve. Early in 1979 cracks in the blade platforms and the blades themselves threatened to delay again the oft-postponed first Shuttle launch. But Thompson insisted that in the late phase of development, "the failures predominantly are those associated with fatigue which one would expect in this development program of extended life."[156]

Unlike the main engine, the external tank did not require major technological breakthroughs. Mulloy explained that "The ET [External Tank] was state of the art. There was no technological challenge in the building of the External Tank. The only challenge was building it to sustain the very large loads that it has to carry, and the thermal environment that it is exposed to during ascent within a weight bogie that was assigned as some 75,000 pounds."[157]

The relative simplicity of the tank ironically prompted the tank to go through more design changes than any other Shuttle element. Kingsbury explained that when there was a structural problem with one Shuttle element, engineers studied possible design trade-offs: "Which does it cost the least to modify, that element or the Tank? And more often than not, like 95 percent of the time, the answer came back that the Tank was easiest to modify. So the Tank went through design change and design change—hundreds of them."[158]

Marshall's Shuttle elements entered development during a period of national economic instability that affected all contractors. Like the main engine, the external tank project ran into cost problems immediately. After its selection for negotiation of the external tank contract, Martin Marietta presented cost projections to Marshall that exceeded the company's original proposal by $8 million over the life of the contract. Martin Marietta blamed inflation for the increases, but also explained that the aerospace industry had declined less in New Orleans than expected, making local hiring difficult. Marshall speculated about underbidding, worried about the unreliability of using Martin Marietta figures for planning purposes, and suggested issuing only a short-term contract to guard against future overruns.[159] Fletcher sent a stern letter to Martin Marietta, as he had done two months earlier to Rockwell about main engine cost growth, regarding "alarming increases in the external tank work," warning that the Shuttle budget "will not accommodate a cost growth of this magnitude."[160] In spite of disagreements over costs, by January 1975 Marshall and Martin Marietta agreed on terms for a $152,565,000 cost-plus-award-fee contract for design, development, and test of the external tank.[161]

That the external tank was the only expendable Shuttle element made its development different from other Shuttle projects. As Project Manager James Odom explained, "One of the unique things about the Tank project was that it was a production program, which was new to NASA." Other NASA programs might involve production of perhaps twenty or thirty units at most, but "we had tooled up to build 400 tanks over the next twenty years."[162] Porter Bridwell, who headed Odom's Project Engineering Group, remembered that "we had a Production Readiness Review. We went back to the Army, went to industry, and patterned [the production plan] after what they had done with respect to assuring that when you do start into production, you have the tooling, automation systems, and software on line and ready to go."[163]

Marshall also used an unusual approach in designing for production. "I did something that's a bit unique in a production program," said Odom. "Typically, you will design an article and you will build what you call a prototype. . . . In my case, I wanted to make sure the Tanks I qualified were built on the same tooling that I was going to build the flight Tanks on. I took the risk and put a $200 or $300 million investment into tooling up front that normally gets invested later in a program."[164] Mike Pessin, who assisted Odom, said that "we took the risk of going ahead with production tooling from scratch. The tools that we're building the Tanks on today, in most cases, are the same tools that we built the very first test items, with modifications that you walking by would never notice."[165]

Workers in the liquid hydrogen tank, part of external tank, in May 1977.

In a production program, Odom insisted "you have to go in and really look at the plant layout." Michoud's proximity to the Gulf of Mexico gave access for barge transportation of the 154-foot-long, 28-foot-diameter Shuttle tanks to the Kennedy Space Center. The assembly facility spread over 833 acres, and Odom remembered that "we had one building that was literally forty-two acres under just one roof." Expecting to produce 24 tanks a year initially, Martin Marietta assembled a work force of 4,300.[166]

While assembly would take place at Michoud, approximately 70 percent of the funds committed to the external tank went to subcontractors scattered around the country, most of whom supplied materials to Martin Marietta. Odom believed that one of Martin Marietta's strengths was its ability to manage subcontractors. "We would go and visit each subcontractor before we would sign a contract with him: get to know the management, get to know their capabilities, . . . what their financial posture was. We knew every one of those contractors literally on a first name basis almost before we signed a contract."[167] Trucks carrying oversize loads streamed into New Orleans from Dallas, San Diego, Baltimore and other cities around the country, and by the spring of 1976, Michoud was operating at near capacity.[168]

Although the external tank may not have required the cutting-edge technology necessary in the development of other Shuttle elements, the project nonetheless presented formidable engineering challenges. Two requirements in particular, weight and insulation, demanded constant attention throughout development, and further modification after the first Shuttle flight.

Weight was the most significant design issue affecting the external tank. The Houston program office lowered the control weight requirement from 78,000 pounds to 75,000 pounds in 1974.[169] Marshall and Martin Marietta experimented with lighter materials, but found that they were not suited for use with cyrogenic fuels. Marshall reduced weight by using an aluminum alloy with exterior foam insulation and reducing NASA's mandatory manned flight safety factor for the tank.[170] Nevertheless design changes mandated as a result of alterations in other elements forced the weight of the tank to creep up again. By mid-1980, less than a year before the first Shuttle flight, the tank had edged back up to 76,365 pounds.[171]

Another trying design challenge on the external tank was insulation. "In the case of the tank," Odom explained, "you are looking at a tank at the top that's got about a million and a quarter pounds of liquid oxygen at about minus 297 degrees. The whole bottom two-thirds of the Tank is liquid hydrogen. It's much less dense—it only has about a quarter of a million pounds—but it's three times the volume at minus 423 [degrees]."[172] Without proper insulation, ice could form on the tank that might shear off and damage the orbiter tiles during flight. The tank surface and every line and bracket on the outside of the tank had to be insulated to keep the exterior temperature above 32 degrees. Furthermore, insulation had to be as light as possible; but in the initial tank design, insulation contributed to the weight problem. "At the time that we built the first six flight Tanks," remembered Bridwell, "we had a superlight ablator which we put on the sub-strata Tank. Then we sprayed an inch of foam all over the Tank."[173] Paint then covered the foam insulation. "Just imagine how much paint it takes to fill a third of an acre," said Odom. "That insulation really soaks up a lot of paint."[174] The paint proved unnecessary, and its later elimination reduced weight significantly.

Complicating external tank engineering concerns was the fact that Marshall harbored doubts about Martin Marietta's management of the project. Early in the project, Marshall worried about the ability of the company's Denver division to supervise operations in New Orleans, and urged Martin Marietta to establish a separate Michoud division.[175] The company delayed, and management issues soon became a point of contention. In a performance review early in 1977 Marshall criticized the company's failure to give effective direction.[176]

A tooling incident at Michoud in June brought matters to a head. The dome spray system used to apply insulation to the tank malfunctioned, causing the carriage drive assembly to fall 80 feet to the floor. The company blamed the accident on a software error and mechanical problems, but Marshall claimed Martin Marietta "completely overlooked the lack of management discipline required to preclude this type of incident from occurring." Top Marshall project and engineering managers gave Martin "a pretty rough going over.'"[177] Lucas concluded that "we need to be firm with Martin in our requirement for better management discipline in the daily operation of the activity at Michoud."[178]

Odom and Lindstrom worked with Martin Marietta to improve what Marshall considered weaknesses in Michoud's workforce and supervisory management, using Rocketdyne as an example of strong project management. Martin restructured, running its Michoud operations as if they were a separate division as Marshall had long wanted. Lindstrom reported early in 1978 that Martin had agreed to establish a project manager and had developed an organizational plan that was "perhaps better" than the one he had proposed.[179]

But Marshall's concerns about Martin's management did not go away. From time to time incidents revived old worries, most seriously when the Center learned that the contractor had designed forward orbiter struts below the required factor of safety. "What else has MMC failed to do that we haven't caught yet?" Lucas wondered.[180]

Marshall ran an extensive test program on the external tank, with tests conducted at Michoud, the National Space Technology Laboratories at Bay St. Louis, Mississippi, and in Huntsville. Tests at Michoud and Marshall examined the tank's structural integrity, its ability to withstand cyrogenic temperatures, and its thermal protection system; those at the NSTL checked the Shuttle main propulsion system by integrating the tank and the Shuttle main engines. Where possible, the Center modified existing test stands; the pneumatic test facility at Michoud, which checked for leaks, was the only new structure built for testing the tank.[181]

Tests conducted in Huntsville revived memories of the 1960s, when Saturn rockets fired on the giant test stands at Marshall shook the city. The Center modified some of the Saturn test stands for external tank tests, changing platforms, instrumentation, and the control system. The Test Laboratory also planned to use modified Saturn test stands for mated vertical ground vibration tests (MVGVT) in which all elements of the Shuttle would be assembled for the first time. The Center used barges along the Mississippi, Ohio, and Tennessee Rivers to transport the tank from New Orleans to Huntsville, just as it had done during Apollo.[182]

The technology of testing, however, was entirely new. "We instrumented these test articles probably heavier than any other test article I've ever seen," according to test manager Chuck Verschoore. "On the intertank alone, we had close to 2,000 measurements, . . . on the hydrogen tank we had 4,000, and on the LOX tank, we had another 2,000. . . . Old technology would have taken us forever to monitor all that."[183]

Before testing the assembled external tank, Marshall separately tested the liquid hydrogen and oxygen tanks and the intertank structure. The Center ran four major tests: structural and vibration tests on the LOX tank, and structural tests on the intertank and the hydrogen tank. The test lab contrived a unique way to simulate G-forces for liquid oxygen tank tests. "LOX and water are about the same density, but we get three Gs on the Tank which means it's three times heavier," explained Jack Nichols. "So we mixed up driller's mud and hauled it [from] Mississippi. . . and filled that thing with driller's mud. We had trucks running day and night. But that simulated the pressure from the propellant at maximum G level."[184]

Verschoore and Garland Johnston remember one test that had them both "sweating blood." "This big old LOX tank had 100,000 gallons of fluid in it," according to Verschoore. "One of the conditions we had to test was in the pitch condition just before burnout, and it was 13 degrees [of inclination]. So, we had that whole Tank full of water at 13 degrees . . . floating on airbags because we had to decouple it from any solid structure. . . . And the airbags were not positioned exactly right."[185] Garland Johnston, the test engineer, continued the story:

"No one can imagine 1,400,000 pounds sitting on 33 airbags. It's a huge thing. And we have the thing sitting out there, and we try to raise it on the airbags, and she starts walking north like it's going go right out through the north side of [Building] 4619. And there wouldn't have been anything we could have done to stop it if it did. So, you do an emergency dump, and you slam it down, and you start sweating blood. So, that's what we did for seven days. We measured; we calculated; we raised; we did everything we could think of. And finally, just finally, I found on the airbag set on the southeast corner, I don't recall now how it was overlooked by quality, but somebody had mismeasured. [It] was 7/10 of an inch off."[186]

External tank loaded aboard NASA barge Orion at MSFC in August 1981.

Marshall and Martin Marietta conducted tests on tank components throughout 1977, culminating with a test of the entire tank on 21 December. Successful completion of the sequence meant that the external tank was ready for Shuttle systems tests at Marshall in the spring of 1978.

The final Shuttle element under Marshall's umbrella was the solid rocket booster. Unlike the main engines, Marshall remained within technological frontiers in the development of the boosters; instead, the goal was to apply state-of-the-art solid booster knowledge to ensure reliability. Unlike the expendable external tank, the booster was a reusable element, and as such posed different development issues. The booster had to be designed not only for performance, but for what project manager George Hardy called the "four R's": recovery, retrieval, refurbishment, and reuse.[187]

Reusability influenced the in-house design approach used on the boosters. Engineers considered cost analyses for individual components to determine design characteristics and replacement frequency. "We would put that into our models and decide how strongly we need to make this part in order to keep the attrition rate at the right level," explained Clyde Nevins. "It was a very unique design approach. Usually, you design something not to fail at all. And here we were designing it to fail a certain percentage of the time, because that was the cheapest way to design the hardware."[188]

Preparations for the SRB recovery system began long before Thiokol won the solid rocket motor contract. Marshall conducted impact studies dropping a 77-percent scale model from heights of up to 40 feet in California's Long Beach harbor in February 1973.[189] Later in the year, the Center used another scale model to test a parachute recovery system in drops on the Tennessee River south of the Center.[190] From these tests evolved a recovery system comprised of pilot and drogue parachutes to ensure descent stability, and three main ribbon chutes, the largest of their type ever used in flight operations. The pilot and drogue chutes nestled in the booster nosecone, the three main chutes in the frustum immediately behind.[191]

Although the Thiokol solid rocket motor was its heart, the booster was much more complex than indicated in labels like "giant firecracker" or "Roman candle." Subassemblies had to be integrated with the solid rocket motor to build a booster. The thrust vector control system, commanded by a sophisticated guidance system external to the booster, steered the booster by directing its nozzle. The booster incorporated subsystems for instrumentation, separation from the external tank, range safety, and recovery. Its aft skirt, which housed the thrust vector control system, also served as a platform for four points at which the booster was attached to the rest of the Shuttle. Similarly, the forward skirt provided hardware for connection to the external tank, as well as housing most booster avionics. A large flexible bearing swiveled the nozzle, which penetrated into the motor case.

Contracts for these subsystems spread Shuttle business around the country. McDonnell Douglas, the most active subcontractor, held responsibility for the forward and aft skirts, the frustum and nosecap, and the systems tunnel that housed cables for electrical connections.[192] Marshall began systems integration in-house, and contracted it to United Space Boosters, Incorporated, late in 1976.

Like other Shuttle elements, the SRB recorded historic "firsts." Not only was it the first solid rocket booster designed for human space flight, but it was the biggest gimballed solid ever built. Bigger than any other solid in use, it carried 1.1 million pounds of fuel, or three times the fuel of the Titan III. Thiokol ignited the solid rocket motor for the first time on 18 July 1977 on its Utah proving grounds, 2 miles from the closest building and 24 miles from Brigham City, the nearest town.[193]

Mixing SRM propellant at Thiokol near Brigham City, Utah, in 1980.

The successful first test of the solid rocket motor was particularly welcome. Marshall's Shuttle projects, and indeed the entire program, were entering a crucial phase. Marshall's projects were all maturing, and were about to enter a period of intense testing. Unfortunately, at a time when ample resources were essential to execute a rigorous testing program and complete development of all three elements, pressure again began to mount from several quarters. The

Carter administration was even more frugal in its approach to space than its predecessors. President Jimmy Carter was a supporter of space science, but had questions about the value of an expensive manned space program, and asked Frosch, his new NASA Administrator, to evaluate the Shuttle program to determine whether it ought to continue. Vice President Walter Mondale had been a vociferous critic of NASA as a senator, and put people who shared his views in the Office of Management and Budget where they could challenge NASA's budget.[194]

Static firing of the solid rocket motor in northern Utah in February 1979.

The new environment had an immediate impact at Marshall. The impending test series meant that the Center's support requirements were expanding as budget pressures became more confining. At a Center performance review in June 1977, Headquarters informed Marshall that its next budget submissions would have to "contain very explicit descriptions of the program requirements" in order to meet new Carter zero-based budget requirements. Headquarters acknowledged related pressures on the Huntsville Center: increasing schedule pressure, lack of sufficient travel funds, reductions in support contractors, and an increasing skill mix imbalance in civil service personnel as a result of reductions-in-force.[195]

To make matters worse, Marshall had begun to experience problems in administration of its SRB contracts, and the constraints enumerated at the Center review compounded them. Cost, schedule, and processing problems hindered the McDonnell Douglas structures fabrication contract. Marshall worried that it had insufficient penetration to monitor the contractor's corrective action. Marshall implemented daily reviews, assigned more personnel, and insisted that "MDAC [McDonald Douglas Astronautics Company] must resolve hardware processing problems [and] MDAC must provide MSFC some visibility into these resolutions."[196]

Even more troubling were problems with the Thiokol solid rocket motor contract. During the summer and early fall, seven material handling incidents took place; none of them caused serious damage, but as Hardy reported, "the trend is disturbing."[197] Incidents continued. By the next summer, Marshall conducted its own investigation and demanded a Thiokol review of 26 incidents over an 18-month period. Thiokol blamed insufficient training, schedule pressure, and human error; but Hardy suggested that lack of adequate management attention was behind all incidents. Lucas agreed, and questioned whether Thiokol had "strong management determination" to improve.[198] Thiokol and Marshall both took corrective action. Marshall initiated a three-shift quality assurance program at the contract site.[199] Nonetheless Lindstrom, head of Marshall's Projects Office, told Thiokol of his concern that "the conditions and circumstances contributing to these incidents may exist with SRM manufacturing and quality control operations."[200]

An incident in December 1978 caused an estimated $750,000 damage to a segment in one of the development motors, and triggered an investigation.[201] Although Marshall and Thiokol agreed on the findings and recommendations of the investigating team, they disagreed on an essential point. John Potate, the Center's acting deputy director, explained that Thiokol blamed "equipment design as primary cause of problem with procedural inadequacy as a contributor. Our report just reverses these two conclusions."[202] Marshall gave precedence to managerial shortcomings, Thiokol to material deficiencies.

Thiokol began a training program and instituted stricter controls. Still, improvement was slow, and the Center worried eight months later that "negligent events ...continue to plague the program." Marshall considered using "severe penalties" in award fee evaluation to pressure Thiokol management.[203]

Marshall's management of all three major Shuttle element contractors bore similarities. Since Marshall often blamed problems on weak management, the contractors' project managers sometimes became reluctant to report problems. Despite formal lines of communication, information often did not flow as intended, and problems took too long to surface. Marshall's William P. Raney summarized the problem:

"In principle, there was a hierarchical responsibility to MSFC, which was supposed to make sure it fit and worked together. In practice, there were lateral responsibilities for exchanging information, specifications, and jointly working out technical solutions. There was a heavy dependence on documentation to make that work, rather than hands-on contact. However, none of the contractors had any authority to force adequate communication or experience, and MSFC didn't force it."[204]

Houston's Kraft described Marshall's approach as "a hands-off management, an arms-length management of their contractors." In Kraft's view, Marshall "wanted to let the contractor do his thing and then hit them in the head to do it right if they screwed up. And they expected them to screw up."[205]

Once a problem surfaced, Marshall took aggressive action with its three major Shuttle contractors—on-site visits in which high-level managers gave the contractor "a pretty rough 'going over,'" with demands for changes in personnel or organization, or threats to impose award fee penalties. Several factors contributed to the approach. Constant budget reductions and reductions-in-force had eroded Marshall's ability to monitor contractors. Unlike Apollo, where Marshall had skills that often exceeded those of the contractor and ample personnel for effective oversight, in the Shuttle Program the Center had to rely on post-facto action, which was often forceful but less involved.

"MSFC worked to the limit of their manpower to see that the various elements were coming along satisfactorily," Raney said, but manpower was indeed limited. Budget constraints also reduced testing, decreased travel funds and manpower for on-site inspections, and forced revisions in schedules. Rather than working side-by-side with contractors, Marshall had little choice but to rely on quality assurance teams, which worked as inspectors rather than co-workers or on-site evaluators. And the number of people involved in quality and reliability work fell by 71 percent from the mid-seventies to the mid-eighties, more than twice the rate of decline of the rest of NASA's workforce.[206] Contractors resisted penetration, so Marshall had to be firm to keep abreast of problems.

Marshall's relations with its contractors underscored a communications problem that plagued the program throughout the Agency. As Raney observed, "For a combination of semi-political reasons, the bad news was kept from coming forward. Contractors didn't want to admit trouble; Centers didn't want Headquarters to know they hadn't lived up to their promises; and Headquarters staffs didn't want to risk the program funding with bad news."[207]

Marshall's management of contractors also reflected broader trends characteristic of NASA management in general. Kraft argued that similarly high-pressure methods under James Beggs, Hans Mark, and James Abrahamson drove NASA Centers to create "an underground decision-making process" that ran counter to the Agency's traditions and prevented open discussion.[208]

High-pressure management was not always characteristic of Marshall contract management. Marshall regularly worked cooperatively with contractors to derive creative solutions. Plasma arc welding (an improvement introduced for use on the external tank and discussed below) was one such case. As Schwinghamer explained, "We brought the contractor in with us and we developed that thing together. And when it was finished, there was no NIH [not invented here] factor—it wasn't invented here. We had done that together. And [Martin Marietta] felt very comfortable with that."[209]

Ultimately technical problems required technical solutions. Chief Engineer Bob Marshall argued that the Center emphasized technical solutions over managerial ones. "It is true that if you have a technical problem, management is to blame because they are responsible programmatically and technically," he explained. But "these problems were strictly technical and could not be resolved without correct technical analysis and action."[210]

One advantage that Marshall did have in monitoring the work of its contractors was its vast test complex on the southern sector of the Center. And early in 1978, attention of all of NASA—indeed of the nation—shifted to Huntsville and Marshall's test stands. For the first time all Shuttle elements would be assembled and Americans would get a first look at the new Space Transportation System. NASA's purpose was to run the mated vertical ground vibration tests (MVGVT) in which the vehicle would be subjected to different types of stress to determine its structural integrity.

March 1978 was a festive month in Huntsville as residents turned out to celebrate the arrival of Shuttle components. The orbiter *Enterprise* garnered the most attention. It arrived at the Redstone Arsenal atop a Boeing 747 on 13 March. After "demating" the orbiter from the aircraft, technicians towed it at a walking pace along the road that bisects the Center and past the Headquarters building as Marshall employees watched. Over the weekend Huntsville residents turned out in "throngs" to view the *Enterprise*. One small boy asked his father, "Is this the same one that's on Star Trek?"[211]

Technicians modified the Dynamic Test Stand used a dozen years earlier for Saturn V tests in preparation for the vibration tests. For the first phase, which began in May, they used air bags and cables to suspend the *Enterprise* and the external tank from a truss structure high in the 360-foot-high test stand, simulating the configuration of the Shuttle after separation of the boosters and before separation of the tank. The vibration tests did not involve physically shaking the Shuttle; rather, the test laboratory used amplifiers similar to those used on home stereo sets to generate vibrations through shaker rods attached to the vehicle. The first phase went well, slowed only when the dome on the LOX tank buckled while it was being filled with fluid early in the test sequence. The test team repressurized the tank and it returned to its original shape.[212]

Shuttle Enterprise *rolls past MSFC office complex, March 1978.*

On 11 October, Marshall completed the first assembly of the entire Space Shuttle, with the orbiter and tank now attached to two solid rocket boosters in launch configuration. The Center modified the test stand, and now the Shuttle stood with its boosters resting on a cylinder-piston platform with bearings on top that gave the vehicle freedom of motion. In the first tests on these hydrodynamic stands, the boosters were filled with inert propellant, bringing the weight of the Shuttle to over four million pounds. Later, in the final phase of vibration tests, the Center measured the system with boosters empty as they would be just before separation, reducing system weight to 1.5 million pounds.

The Center completed the MVGVT tests on 23 February 1979. Results from the tests prompted some modifications, including strengthening of brackets at the forward section of the boosters. Eugene Cagle, director of the Test Laboratory, reported that "from a structural dynamics standpoint, we are confident that the Space Shuttle will perform as expected."[213]

The tests at Marshall verified only the structural integrity of the Shuttle, and tests continued concurrently on other Shuttle elements. NASA Associate Administrator John Yardley told Congress in September 1978 that "the only significant Shuttle problems [are] with the main engine and the vehicle's weight." Yardley thought that the main engine could be ready within a year, and that the weight problems would not impact the program until after the early flights.[214]

Shuttle Enterprise *suspended at Marshall's Dynamic Test Stand, July 1978.*

The biggest threat to the Shuttle in the weeks following the Marshall tests was budgetary rather than technical. In May 1979, NASA predicted that the Shuttle might have a cost overrun of $600 million over the course of four years.[215] The announcement touched off a barrage of criticism, precipitated further schedule delays, and put the already fiscally constrained Shuttle program in jeopardy. NASA "is in deep trouble," said one commentator. Congress worried that "serious mismanagement" of the Shuttle program was threatening defense plans dependent on the Shuttle.[216] NASA Administrator Frosch defended NASA program management, arguing that the Agency had done well operating under stringent limitations.[217] But three months later, a NASA panel blamed the cost overruns and schedule slips on insufficient funds, unrealistic schedules, and inadequate long-range planning.[218]

The final preparations for the first Shuttle launch also encountered technical problems. As Houston worked to repair the ceramic tiles comprising the Shuttle thermal protection system, Marshall worried about cracks in the SRB propellant, external tank shrinkage (1.5 inches when loading cyrogenic hydrogen), and uprating the main engines to achieve greater thrust and payload capacity.[219] Damage to the O-rings used to join segments of the solid rocket motor appeared for the first time, but tests on intentionally damaged O-rings seemed to demonstrate their effectiveness.[220] None of the difficulties threatened the mission, and in the last months people focused more on orbiter tiles than on any problems associated with Marshall's elements.

So despite budgetary threats, schedule slippage, and nagging technological difficulties, NASA moved toward the first manned orbital flight of the Shuttle. In December 1979, the main propulsion system successfully fired for 9 minutes and 10 seconds, longer than would be required to lift the Shuttle into orbit.[221] Two months later the solid rocket motor completed a series of seven test firings begun in July 1977, and Marshall deemed the tests "highly successful."[222] By early 1980, J. R. Thompson could compare main engine testing favorably with that of the J–2 in the 1960s: the main engine had undergone nearly three times as much operating time, had a comparable success rate, and would soon surpass the J–2 in numbers of tests.[223]

In November 1980, personnel at Kennedy Space Center began stacking and integrating the first Shuttle in the Center's Vehicle Assembly Building (VAB), preparing for a launch the following spring. On 3 November, they attached the external tank to the solid rocket boosters. Three weeks later, they conducted the "rollover" of the orbiter *Columbia*, moving the vehicle into the VAB for mating with the tank and boosters.[224] On 29 December, workers moved the entire Shuttle assembly along a three and one-half mile route from the VAB to launch pad 39A. In February 1981 the

Complete Space Shuttle mated for first time in the Marshall Center's Dynamic Test Stand, 6 October 1978.

Center ran flight readiness firing tests on the flight hardware, briefly running the main engines and gimballing their nozzles, concluding that "all MSFC hardware performed as designed."[225]

First Flight and Post-Flight Development

On 12 April 1981, the Shuttle embarked from Kennedy Space Center on its maiden flight, a trip of two days. Marshall engineers monitored the anxious early minutes of flight, during which the Shuttle propulsion system would face its test. "Any time you build a big vehicle like this," Odom said recalling his feelings at the time of launch, "and you put it together for the first time, especially with a man on board, you really worry, 'Have I really tested everything that that vehicle is going to see in that first flight?'"[226] Relief spread through the Cape and the communications center in Huntsville as the boosters shut down after 2 minutes, and jettisoned 12 seconds later. At 520 seconds the main engines shut down on schedule, and 30 seconds later the external tank jettisoned. Less than 10 minutes after liftoff, Huntsville's elements had accomplished their part of the mission. Former Center Director Rees leaned against a console at Marshall and reflected how much this day would have meant to Wernher von Braun. Deputy Center Director Jack Lee told reporters with a smile, "We were on the high side of performance."[227]

The Shuttle returned to Earth two days later, landing on a long runway at Edwards Air Force Base in California. After a week of analyzing data, Lindstrom, Marshall's Shuttle Projects Manager, declared the performance of the Center's elements "flawless."[228]

Marshall had ample reason for pride in the performance of its Shuttle elements, but a satisfying first mission did not mean its development task was complete. Even before the first flight Marshall had begun to plan design changes, and each successive flight exposed new targets for fine tuning. "After we started to fly, there were development efforts to improve performance and increase life," according to J. Wayne Littles. "A lot of our effort after we started flying was keeping the vehicle flying: getting each set of hardware to fly a mission; reviewing it and making sure it was ready to fly; reviewing the data of each flight [and] making sure there were no anomalies . . . and get[ting] rid of latent defects that caused us to change parts out more frequently than we would like to."[229] And after a measured analysis of the first flight, it was clear that some components needed immediate attention.

Space Mission Operations Control Center.

Recovery efforts after the first Shuttle flights demonstrated, Mulloy admitted, that NASA was "far from reaching the operational goal of a recoverable, reusable booster that could rapidly be refurbished and put back into line." Indeed the recovery system qualification program included the first flights since the elements were so large that there was no other way to test them, and the damage sustained far exceeded expectations.[230]

The boosters sustained too much damage upon ocean impact to achieve the quick turnaround necessary to support the planned 24 flight-per-year schedule, let alone the long-term goal of 48 flights per year. Clyde Nevins, who headed an investigation of the recovery system, said that "After the first launch we had excessive damage on the aft skirts. It just tore the heck out of the aft skirt inside—the stiffener rings on the outside, inside the cone, on the back end on the aft skirt. Very severe damage in there. It just wasn't like we predicted at all."[231]

The damage occurred when "it hit the water tail first, nozzle end first at about 88 feet per second, which is 60 miles an hour," Herman Thomason related. "It drove itself into the water, . . . the water was like a hydraulic ram. It comes up inside and you had compression taking place inside where the fuel had burned out." Impact damaged the aft skirt and the thrust vector control system. Compression forced salt water into parts of the rocket not designed to withstand its effects. Then, according to Thomason, "that thing comes back out of the water . . . and it slaps down on its side. And you get all these slap down loads, and even if the thing was five inches thick across, that's not going to be able to take those kinds of loads."[232]

Nevins's investigation showed that the reinforcement rings used in test models had differed slightly from those used on the booster. "So we ended up having to go in and put up a lot of reinforcement. And we also put in some light density foam, which smoothed out the internal contours of the ring. . . . The foam would get damaged, but it was sacrificial."[233]

On the fourth flight, explosive bolts attaching parachutes to the boosters fired prematurely, and the boosters could not be recovered. After locating the boosters on the ocean floor, searchers had to abandon plans to survey them.[234] NASA made improvements to strengthen the boosters. By the 11th flight, Marshall was putting "big gobs of spray foam" on the skirt, and using deflectors called "cow catchers" to keep water away from sensitive components.[235] Along with changes to the parachute recovery system, these changes improved the condition of recovered boosters.[236]

All of Marshall's Shuttle elements continued development after the Shuttle's maiden flight. Jerry Thomson said of work on the main engines, "We had to make design changes to improve the life of the Engine and improve the reliability. So we made some design changes even after we had made the first flight."[237] The types of changes included "basic changes in internal components, like improvement in blades to improve blade life in turbines, and making improvements in bearings."[238]

Limits on the life of components proved to be one of the most persistent challenges in main engine development. Bearings, turbopumps, and turbine blades were the sources of greatest concern. Bearing failure was a problem in the main engine from the early days of the program. The engines ran at about 30,000 rpm, generating heat that always threatened the integrity of the bearings. "The bearings are cooled with liquid hydrogen," explained Jud Lovingood, main engine project manager in the 1980s, and because temperatures are so high, "when you're trying to cool them the [liquid] hydrogen changes to a gas [and] it doesn't cool as much. You end up with bearings overheating and that weakens them. It also changes clearance because of the expansion you get. . . . NASA has just gradually improved them over the years . . . but they still have life limits."[239]

Greater than expected damage to pumps and turbine blades came from dynamic stress, cavitation erosion (caused when cavities of gas developed and collapsed in liquid fuels), and high temperatures. Using technology unavailable when main engine development began, Marshall worked with NASA's Lewis Research Center, Rocketdyne, and contractors at Aerojet and Pratt & Whitney. A combination of approaches including powerhead redesign, thermal coatings on the blades, computer modeling to study fluid mechanics, and new metal alloys enabled the Center to gradually extend the life of the main engine.[240]

Anticipated increases in payload demands dictated a need for increased main engine performance, and even before first flight NASA determined to improve engine performance to 109-percent of rated power.[241] The 109-percent rating was "what we originally started calling emergency power level," Bob Marshall explained, and "ultimately it grew to be full power level, FPL."[242] The Center's goal was "to get more performance out of the fuels and higher performance out of the engine," Herman Thomason recalled. Better performance was "a function of temperature and pressure."[243]

The challenges involved in increasing power rating were enormous; power increase of merely four percent would nearly double cavitation erosion, for example.[244] Thus Marshall's efforts to bolster the main engine rating had to overcome persistent obstacles. The most serious accident occurred on 7 April 1982 at NSTL. As the test team pushed the engine to the 109-percent level, vibration forces inside the main oxidizer pump increased to 38 times the force of gravity, causing an explosion that ripped the pump apart. "There were pieces scattered all over," according to Lovingood.[245] Development continued in the months that followed, and by the time of the *Challenger* accident, "we were within one test of qualifying the Engine and within about two weeks of starting the Main Propulsion Test of three engines running at 109%," according to Bob Marshall. The *Challenger* tragedy forced the Center to reconsider the 109-percent rating and look for other ways to improve performance.[246]

No Shuttle element underwent more changes after the Shuttle's first flight than the external tank. Design changes in other elements had increased the weight of the tank, limiting potential Shuttle payloads. In the summer of 1980, 10 months before the Shuttle's maiden flight, Marshall initiated a plan to lighten the tank. Martin Marietta had already produced six flight tanks, and the redesigned lightweight tank would not be used until those earlier models were expended. The two-year redesign program trimmed 7,000 pounds from the 71,000-pound tank used on the first Shuttle flight.[247]

Not painting the tank produced the most visible change. "That saved a couple thousand pounds of weight," explained Odom. "The first time we rolled out one that was that brown color, . . . a lot of people just said that just doesn't look like NASA hardware, it's not pretty. It was pretty to me, because it was economical."[248] After the first six flights, the Center learned that it could eliminate the superlight ablator that coated the tank before application of the foam insulator. "The significant change from a processing standpoint is that it reduced the cost of the Tank significantly," according to Bridwell.[249]

Marshall also introduced structural changes. The Center modified support structures, altered production techniques, and changed materials. Dome caps, previously milled on one side, now had metal shaved from both sides.[250] Nichols explained two other methods used to trim weight: "You start off with an inch-and-a-half thick plate and you machine it down and you leave little stiffeners in it. We took those out. There's a huge ring frame that takes the kick load from the Solid Rocket Motor and also from the orbiter. By going back and sculpturing it, rather than making it uniform all the way around, we took some weight out."[251]

Production of the tank became easier and less costly in 1984 after engineers at Marshall adopted new welding technology.[252] As Kingsbury explained, "The welding of aluminum, historically, has been a reasonably difficult process because aluminum oxidizes very quickly. You really can't have unoxidized aluminum in our atmosphere. It oxidizes too fast. And that oxide becomes a problem when you weld it."[253] Marshall engineers developed a process called plasma arc welding that minimized weld defects. "There are about five miles of weld on the Tank, and any defect that you might find has to be repaired," according to Bridwell. "Once again, that is labor intensity, and if you eliminate that, the cost of the tank goes down."[254] That savings accrued is clear from a comparison with Apollo; according to Schwinghamer, when engineers welded the same alloy on the Saturn V, "about every six feet or so we would get defects in the welding. And the X-ray would show a flaw. We would have to go in and grind the flaw out and reweld."[255]

Redesign of the external tank alone was insufficient to meet Shuttle payload requirements. The Department of Defense planned to launch a satellite in polar orbit from Vandenburg Air Force Base in California. Such a mission required payload lifting capability beyond that of the first flight configuration. Beginning in 1976, NASA conducted systems engineering studies with a view toward improving Shuttle performance. These studies identified three candidates for modifications to achieve increased capacity, all of which would have affected Marshall's Shuttle elements: increase the thrust of the main engines from a rating of 109 percent to 115 percent; attach a liquid boost module comprised of four propellant tanks to the external tank; or develop a lighter solid rocket booster by replacing the steel casing with a filament wound case.[256]

In the early 1980s NASA rejected the augmented thrust rating for the main engines and the boost module after conducting cost, technical, and schedule analyses on all three options. The Agency decided that the filament wound case for the SRB had the shortest development schedule, least cost, and least technical risk.[257] Most other solid-fueled missile systems except the Titan already employed such cases, so the technology was not new.[258] Marshall planned to use plastic reinforced with graphite fiber, winding it into a cylinder that would reduce the weight of an empty booster by one-third, from 98,000 pounds to 65,000 pounds, and increase the Shuttle's payload capacity by 4,600 pounds. The plan would also simplify booster assembly, replacing 8 of the 11 steel cases with 4 filament cases.[259]

Manufacture of the SRM filament wound case.

By 1985, development of the filament wound case was proceeding well, and NASA was using the program to improve the joints between booster segments. Following the *Challenger* accident, however, the Agency decided to eliminate the filament wound case project.[260]

The booster project faced another challenge from erosion of the nozzle. "We were seeing some very bad, greater than predicted erosion of the motor nozzle insulation, which is a carbon phonolic about three inches thick that is on the inside of the metal part of the nozzle," explained Mulloy, Marshall's project manager for the booster. The insulation was designed to ablate while the motor burned, but the amount of erosion was unpredictable. Marshall instituted design changes. As McIntosh described it, the Center "made a few changes in the configuration of propellant and grain a little bit and changed the size of the nozzle throat and increased the exit cone length."[261]

NASA's Shuttle problems in the early 1980s were not all technological. The cost decisions of the early 1970s now began to catch up to the Agency. The decision to abandon a fully reusable Shuttle a decade earlier had traded savings in development for larger costs for expendable elements later. In 1982, NASA reduced by nearly 200 the number of Shuttle flights projected over the next decade. Inflation drove costs higher, refurbishment time was longer than planned, all elements needed more additional development work than expected, and each mission required extensive post-flight analysis.

Soon it became apparent that even NASA's reduced projections were too optimistic. In 1983 a National Research Council panel told Congress that NASA's goal of increasing to 24 launches in 1988, 30 in 1990, and 40 in 1992 was unattainable because "major pieces" of the Shuttle would not be available. To have enough solid rocket boosters to achieve even 18 launches in 1990 seemed "marginal." "Because of very strict budgetary constraints in the program," the report continued, "NASA has had to concentrate on near term needs, and its capacity to deal with the longer term requirements was inevitably curtailed."[262]

NASA's rosy expectations for the Shuttle found critics even within the Agency. Noel W. Hinners, director of Goddard Space Flight Center, wondered about projections for reuse, commercialization, and costs. He argued that NASA was too optimistic in its expectations for reusing Shuttle components before encountering "structural integrity problems," and cautioned against expecting "routine" operations in a high-risk venture. "The Orbiter is a subsidized operation," he warned. "I see no way anyone can make a profit at this point without the government being accused (validly in my mind) of a giveaway of its R&D investment."[263] Instead of the early visions of orbiting payloads for as little as $100 a pound, by 1983 the cost was over $5,000 a pound.[264] Criticism of NASA for failing to make the Shuttle commercially viable continued, however, and the Agency even considered relinquishing management of the Shuttle to a private concern before abandoning the idea in 1985.[265]

Shuttle development had faced formidable limitations from the beginning. Cost pressures influenced every step from configuration planning through development and flight. In comparison to Apollo, NASA's only previous program of comparable scope, Marshall worked with less money, fewer people, and reduced skills.

The Center had to learn a new way of doing business. Marshall, without the arsenal system, had to rely on contractors and reduce its testing. And unlike Apollo, the Shuttle was a program of ongoing development in which major improvements continued even after operations began; even as evaluators certified components for flight, engineers were working on improvements.

These new approaches raised new problems, for Marshall and NASA were stretched to the limit of their manpower, skills, and resources. Within the environment of financial and political pressures, the Center and the Agency could no longer afford the conservative engineering approaches of the Apollo years, and had to accept risks that never confronted an earlier generation of rocket engineers.

If the Shuttle fell short of expectations, it may have been because expectations were unrealistic. NASA made extravagant claims for the Shuttle while seeking congressional approval, promising frequent flights, low cost to orbit, rapid refurbishment, and decreasing costs as expendable components entered mass production. The Shuttle was to be a space truck; it would soon pay for itself by providing routine operations. But as Schwinghamer insisted, "it's never going to be like driving a truck. And I guess some people kind of forgot that somewhere in the middle of this thing. But it is a fine-tuned machine. It's a wonderful machine. It's an engineering triumph in terms of efficiency, performance, and in every respect."

Schwinghamer expressed a common sentiment at the Center when he said that "in the context of the limitations imposed, that's an elegant design. That's the finest machine in existence today."[266]

1 Max Akridge, "Space Shuttle History," 8 January 1970, p. 2, Box 0375, Shuttle Series—MSFC Documents, JSC History Office.

2 Akridge, pp. 3–4.

3 H.H. Koelle to Ewald Diemer, 8 February 1963, and Minutes, Meeting of Government Agency Representatives at NASA Marshall Space Flight Center, 24 January 1963, Box 005–13, Shuttle Chronological Series, JSC History Office; "Marshall Studies Advanced Large Vehicles," *Aviation Week and Space Technology* (2 December 1963), 33–34; Akridge, p. 5.

4 Aero-Space Division of the Boeing Company, "System Criteria for Launch Vehicle Systems," May 1965, Box 005–21, Shuttle Chronological Series, JSC History Office.

5 Jerry Thomson, OHI by Jessie Whalen/MSI, 16 March 1989, Huntsville, Alabama, p. 2.

6 John P. McCarty, OHI by Sarah McKinley/MSI, 29 March 1990, Huntsville, Alabama, pp. 2–3.

7 Minutes, Meeting of Government Agency Representatives at NASA Marshall Space Flight Center, 24 January 1963; John F. Guilmartin, Jr., and John Walker Mauer, Management Analysis Office, JSC, "A Shuttle Chronology, 1964–1973," Volume I, December 1988, p. I–4.

8 Guilmartin and Mauer, Volume I, pp. I–2 to I–5; J.H. Brown and D.S. Edgecombe, "The Relevance of Launch Vehicle Recovery and Reuse to NASA OSSA Launch Vehicle and Propulsion Programs," Report Number BMI–NLVP–TM–67–3, Battelle Memorial Institute, 24 February 1967, Box 005–31, Shuttle Chronological Series, JSC History Office; Akridge, pp. 16–21.

9 M.V. Ames, Chairman, Ad Hoc Subpanel on Reusable Launch Vehicle Technology, "Comments on Economic Aspects of Reusability Requested by SSRT Panel," 30 January 1967, Box 005–31, Shuttle Chronological Series, JSC History Office.

10 Loftus, et. al., "The Evolution of the Space Shuttle Design," p. 3.

11 Cited in Brown and Edgecombe, "The Relevance of Launch Vehicle Recovery and Reuse to NASA OSSA Launch Vehicle and Propulsion Programs."

12 Cited in Charles D. LaFond, "Von Braun Urges Reusable Transport," *Missiles and Rockets* (8 November 1965), 16.

13 Akridge, pp. 25–26.

14 Guilmartin and Mauer, Volume I, pp. I–12 to I–14; Jesse C. Phillips to Thomas Grubbs, undated but catalogued under 31 December 1967, Box 005–34, Shuttle Chronological Series, JSC History Office.

15 Akridge, p. 36.

16 NASA, Office of Manned Space Flight, "MSF Vehicle Studies, 1 July 1968, cited in Guilmartin and Mauer, Volume I, pp. I–85 to I–87.

17 Mueller to Charles Donlan, 23 September 1968; and Douglas Lord to Frank Williams, MSFC, and John Hodge and Maxime Faget, MSC, 14 October 1968, Box 005–35, Shuttle Chronological Series, JSC History Office.

18 The contracts developed from a statement of work written largely at MSFC and issued in May 1968, and an RFP (Request for Proposals) developed jointly by MSFC and MSC in October 1968. Guilmartin and Mauer, Volume I, pp. II–1 to II–2.

19 The ILRV (Integrated Launch and Re-entry Vehicle) studies assigned each contractor specific approaches for examination. General Dynamics (MSFC) was to compare flyback and expendable lower stage configurations. Lockheed (MSFC) was to study the stage-and-one-half configuration, in which a strap-on stage would drop off before the stage itself would be fully expended. McDonnell Douglas (Langley) was to analyze several specific design configurations. North American Rockwell (MSC) was to examine configurations with expendable lower stages and a reusable spacecraft. Akridge, pp. 48–53.

20 Observers included U. Alexis Johnson, under secretary of state for political affairs; Glenn T. Seaborg, chairman of the Atomic Energy Commission; and Robert P. Mayo, director of the Bureau of the Budget. Guilmartin and Mauer, Volume I, p. II–11.

21 Bob Marshall, OHI by Jessie Whalen and Sarah McKinley/MSI, 22 April 1988, Huntsville, Alabama, p. 4.

22 Andre J. Meyer, 7 March 1969, cited in Guilmartin and Mauer, Volume I, p. II–12.

23 Meyer, 21 April 1969, cited in Guilmartin and Mauer, Volume I, p. II–30.

24 Max Akridge, OHI by Jessie Whalen/MSI, 1 August 1988, Huntsville, Alabama, p. 10; Akridge, "Space Shuttle History," p. 65. Akridge acknowledged that Mueller had used the term in his address to the British Interplanetary Society nine months earlier. But at the time of the task group meeting in May, the vehicle was still known as the IRLV.

25 Andre J. Meyer, handwritten notes, cited in Guilmartin and Mauer, Volume II, p. III–18.

26 Mueller to von Braun, 22 July 1969, Box 005–45, Shuttle Chronological Series, JSC History Office. Guilmartin and Mauer, Volume I, p. II–154.

27 U.S., Congress, Senate, Committee on Aeronautical and Space Sciences, *Future NASA Space Programs*, hearing, 91st Cong, 1st sess, 5 August 1969, pp. 41–43.

28 Space Task Group, *The Post-Apollo Space Program: Directions for the Future*, Report to the President, September 1969; McDougall, p. 421.

29 L.E. Day and B.G. Noblitt, "Logistics Transportation for Space Station Support," Presented at the IEEE EASCON Session of Earth Orbiting Manned Space Station, Washington, DC, 29 October 1969, 10s Apollo: Space Shuttle Task Team, NASA folder, MSFC History Archives; Space Shuttle Phase B Request for Proposal (RFP), 20 February 1970, cited in Guilmartin and Mauer, Volume II, pp. III–103 to III–108.

30 Richard P. Hallion, "The Path to the Space Shuttle: The Evolution of Lifting Reentry Technology," (Air Force Flight Test Center, 1983), 56–57; George M. Low to Dale Myers, 27 January 1970, 10s Apollo: Space Shuttle 1970 folder, MSFC History Archives; Joseph J. Trento, *Prescription for Disaster* (New York: Crown Publishers, Inc., 1987), pp. 100–01.

31 To make his case for competition, Mueller relied on Richard L. Brown, "A Study of the Use of Competition in the Space Shuttle Program," August 1969, 10s Apollo: Space Shuttle Task Team, NASA folder, MSFC History Archives.

32 Mueller to von Braun, 25 September 1969, 10s Apollo: Space Shuttle Task Team, NASA folder, MSFC History Archives.

33 Von Braun to Charles W. Mathews, Acting Associate Administrator, NASA, 30 December 1969, 10s Apollo: Space Shuttle Task Team, NASA folder, MSFC History Archives.

34 Low to Dale Myers, 27 January 1970, 10s Apollo: Space Shuttle 1970 folder, MSFC History Archives. The requirement had been spelled out in the Space Task Group report.

35 See for example John M. Logsdon, "The Space Shuttle Program: A Policy Failure?" *Science* 232 (30 May 1986), 1099.

36 Humboldt C. Mandell, Jr., "Management and Budget Lessons: The Space Shuttle Program," National Aeronautics and Space Administration Special Publication–6101, Autumn 1989, p. 44.

37 Andre Meyer of MSC, in a personal handwritten note, wrote: "With Mueller leaving direction will return to the Center." Andre J. Meyer, handwritten notes, 3 November 1969, cited in Guilmartin and Mauer, Volume II, p. III–63.

38 The centers finally agreed that each would administer two of four parallel contracts, with MSFC chairing the Source Evaluation Board. Gilruth to Von Braun, 16 January 1970, 10s Apollo: Space Shuttle 1970 folder, MSFC History Archives.

39 Rees to Myers, 17 February 1970; Rees to Myers, 2 April 1970, 10s Apollo: Space Shuttle 1970 folder, MSFC History Archives.

40 Myers to Rees, 17 June 1970, 10s Apollo: Space Shuttle 1970 folder, MSFC History Archives.

41 Myers to Rees, 28 May 1970, 10s Apollo: Space Shuttle 1970 folder, MSFC History Archives.

42 Myers to Rees, 17 June 1970.

43 Guilmartin and Mauer, Volume III, p. IV–2.

44 Guilmartin and Mauer, Volume III, pp. IV–7 to IV–8.

45 Myers to Rees, 1 April and 22 April 1970, 10s Space Shuttle 1970 folder, MSFC History Archives.

46 Andre J. Meyer, 16 March 1970, cited in Guilmartin and Mauer, Volume III, p. IV–26.

47 Rees to Myers, 17 February 1970, 10s Space Shuttle 1970 folder, MSFC History Archives. Myers acknowledged that "the organizational arrangement presented by MSFC reflects the concept of the Management Plan." Myers to Gilruth, 13 March 1970, cited in Guilmartin and Mauer, Volume III, p. IV–17.

48 Rees to Myers, 3 June 1970, and Lucas to Cook, 9 June 1970, Shuttle Phase A Studies—Lucas Notes folder; David Baker, "Evolution of the Space Shuttle," pp. 204–212; Hallion, pp. 57–58; Guilmartin and Mauer, Volume III, pp. IV–2 to IV–5, IV–70 to IV–71. Marshall had responsibility for the McDonnell Douglas contract, Houston the Rockwell contract. The backup partially reusable Phase A contracts went out in June. Marshall had responsibility for a Lockheed study of an expendable tank orbiter and a Chrysler study of a single-stage reusable orbiter. Houston would manage a Grumman/Boeing study of a stage-and-a-half vehicle with expendable tanks, strap-on boosters and a reusable orbiter.

49 Mike Pessin, OHI by Jessie Whalen and Sarah McKinley/MSI, 18 December 1987, Huntsville, Alabama, p. 13.

50 Guilmartin and Mauer, Volume IV, pp. V–7 to V–8, V–41; Mike Pessin, OHI by Jessie Whalen and Sarah McKinley/MSI, 18 December 1987, Huntsville, Alabama, pp. 12, 13.

51 Cited in Howard E. McCurdy, *The Space Station Decision: Incremental Politics and Technological Change* (Baltimore: The Johns Hopkins University Press, 1990), p. 29.

52 Logsdon, "The Space Shuttle Program: A Policy Failure?", p. 1101.
53 Christopher C. Kraft, Jr., Deputy Director, MSC, memorandum for the Record, 8 April 1971, Shuttle Management/Lead Center folder, Box 1, Shuttle Series—Mauer Source Files, JSC History Office.
54 "NASA Space Shuttle Program Management: Manned Spacecraft Center's Recommended Plan," 21 April 1971 Shuttle Management/Lead Center folder, Box 1, Shuttle Series—Mauer Source Files, JSC History Office.
55 Jack Hartsfield, "MSFC Fighting to Keep Shuttle," *Huntsville Times*, 21 January 1971.
56 Dale Grubb, Assistant Administrator for Legislative Affairs, to Fletcher, 14 May 1971, Shuttle Management/Lead Center folder, Box 1, Shuttle Series—Mauer Source Files, JSC History Office. *Huntsville Times*, 9 June 1971.
57 Myers to Fletcher, 20 May 1971, Shuttle Management/Lead Center folder, Box 1, Shuttle Series—Mauer Source Files, JSC History Office. Myers to OMSF Center Directors, 10 June 1971 cited in Guilmartin and Mauer, Volume III, pp. V–56 to V–57. By the time Myers announced the lead center decision on 10 June, Marshall's grip on the space station had slipped, apparently without knowledge of Center management. Regarding station, Myers had originally recommended that "MSFC would be designated 'lead Center' at the end of the present Phase B studies." When he made the lead Center announcement, Myers promised Marshall only responsibility for "any further space station studies at the end of the current Phase B studies."
58 Christopher C. Kraft, Oral History Interview by Andrew J. Dunar and Stephen P. Waring (hereinafter OHI by AJD and SPW), 28 June 1991, Webster, Texas, pp. 15–16.
59 Low, quoted by Willis Shapley, cited in Trento, p. 112.
60 Bill Sneed, OHI by Jessie Whalen/MSI, 8 April 1988, Huntsville, Alabama, p. 3.
61 Godfrey to Rees, Lucas, and Murphy, 24 June 1971, 10s Space Shuttle 1971 folder, MSFC History Archives.
62 R.W. Cook, draft of Rees to Myers, 7 September 1971, DM01 files, Shuttle 1973 folder; Cook to Harry Gorman, 20 August 1971, DM01 files, Shuttle 1973 folder, MSFC History Archives.
63 Fletcher, memo to Jonathan Rose, Special Assistant to the President, 22 November 1971, NASA Headquarters History Office.
64 Guilmartin and Mauer, Volume IV, p. V–14.
65 James B. Odom, OHI by Jessie Whalen/MSI, 9 February 1988, pp. 6–7.
66 McCurdy, p. 31; Logsdon, "The Space Shuttle Program: A Policy Failure?", p. 1104; Trento, pp. 112–13.
67 "Space Shuttle System Definition Evolution," pp. 4–5.
68 William Brown, OHI, p. 24.
69 LeRoy Day, OHI by Sarah McKinley/MSI, 7 June 1990, Huntsville, Alabama, p. 3.
70 William Brown, OHI by Jessie Emerson/MSI, 15 August 1989, Huntsville, Alabama, p. 23.
71 Ron McIntosh, OHI by Jessie Emerson/MSI, 8 August 1990, Huntsville, Alabama, p. 2.
72 LeRoy Day, OHI, pp. 1–2.

73 Dan Driscoll to Rees, 27 January 1972; Rees to Driscoll, 31 January 1972, Space Shuttle 1972 folder, MSFC History Archives.

74 "Shuttle Booster," *Aerospace Daily* (24 January 1972), 53.

75 MSFC Public Affairs Memo, 24 February 1972, reporting on "Problems Loom in NASA on Shuttle Booster Selection," 23 February 1972, p. 298, Space Shuttle 1972 folder, MSFC History Archives.

76 McCurdy, pp. 30–31.

77 "Space Shuttle System Definition Evolution," pp. 5–7; NASA Special Release, "Space Shuttle Evolution," 1 August 1972, Shuttle History folder, NASA Headquarters History Office.

78 Fletcher to Caspar W. Weinberger, Deputy Director, OMB, 6 March 1972, Shuttle History folder, NASA Headquarters History Office; "Space Shuttle Evolution."

79 Sneed, OHI, pp. 5–6; Sneed to Mike Wright, 5 January 1993, MSFC History Office.

80 Rees, memo to Dick Cook, 7 November 1972, Space Shuttle 1972 folder, MSFC History Archives. Rees estimated that the external tank and the solid rocket boosters would consume 60 percent of the operation costs of each flight.

81 Roy E. Godfrey to Petrone, 8 August 1973, Space Shuttle—Solid Rocket Booster 1973 folder, MSFC History Archives.

82 "NR's Rocketdyne Picked for Shuttle Main Engine," *Space Business Daily*, 57 (14 July 1971), 52.

83 Frank Stewart, OHI by Jessie Whalen/MSI, 4 February 1988, p. 10–11.

84 Myers to Low, 22 October 1970, Shuttle Booster Stage folder, NASA Headquarters History Office; W.D. Brown to Colonel Mohlere, 16 October 1970, 10s Space Shuttle folder, MSFC History Archives.

85 Richard L. Brown, OHI by Jessie Whalen/MSI, 27 June 1988, Huntsville, Alabama, p. 12.

86 "Space Shuttle Engine Negotiations," NASA Release No. 71–131, 13 July 1971, Shuttle Propulsion System 1971 folder, NASA Headquarters History Office.

87 Arthur Hill, "Contract to Develop Shuttle Engine Sparks Savage Fight," *Houston Chronicle*, 29 August 1971.

88 "NASA Charged with Improper Conduct in Shuttle Engine Award," *Space Business Daily*, 57 (6 August 1971), 169; "P&W Files Formal Protest on Shuttle Engine Award," *Space Business Daily*, 57 (19 August 1971), 230.

89 Edward J. Gurney, et. al. to Fletcher, 14 July 1971, Space Shuttle Main Engine Working File, 1971, Mr. Cook, Tab 6, MSFC History Archives.

90 "NR Challenges P&W Engine Experience Claim," *Space Daily* (3 November 1971), pp. 16–17.

91 R.W. Cook, Deputy Director , Management, MSFC to J.T. Shepherd, 19 August 1971, Space Shuttle Main Engine Working File, 1971, Mr. Cook, Tab 6, MSFC History Archives.

92 MSFC, "Chronology: MSFC Space Shuttle Program: Development, Assembly and Testing Major Events (1969–April 1981)," MHR–15, December 1988, pp. 15–21; "GAO Rejects P&W Protest on Shuttle Main Engine," *Space Business Daily* (1 April 1972), 175.

93 Godfrey, memo to Rees, 1 December 1971, Space Shuttle Main Engine 1970–1971 folder, MSFC History Archives. Godfrey describes a meeting held on 29 November. Among those in attendance were Low, Myers, Headquarters Shuttle manager Charles Donlan, and Houston's Shuttle Program manager Robert Thompson.

94 "Proposals Asked for Shuttle Solid Rocket Motors," MSFC Release No. 73–133, 18 July 1973, Shuttle Booster 1972–1973 folder, NASA Headquarters History Office.

95 Trento, pp. 114–15. See also Lucas to Petrone, 2 July 1973, Solid Rocket Motor (SRM) folder, MSFC History Archives.

96 "Groundrules and Assumptions," 25 May 1973, Solid Rocket Booster (SRB) 1974 folder, MSFC History Archives.

97 Robert F. Thompson, "Elimination of SRB Thrust Termination Requirement," Control Board Directive—Level II, 27 April 1973; H. Thomason, "SRB Thrust Termination" Briefing Charts, 27 April 1973; H. Thomason, "Thrust Termination" Briefing Charts, 3 May 1973; Godfrey to Petrone, 8 August 1973; Godfrey to Robert F. Thompson, 8 August 1973; Godfrey to Petrone, 15 August 1973; Thompson to Lindstrom, 18 September 1973; Space Shuttle—SRB 1973 folder, MSFC History Archives. Maxime Faget to M. Silveira, 2 February 1970, Box 005–63, Shuttle Program Chronological Files, JSC History Office. Estimates for the weight penalty of adding thrust termination varied widely; figures cited ranged from 2,400 pounds to 8,000 pounds. Godfrey maintained that Rockwell's principle motive was to prevent "the serious weight penalty with early separation," and that Marshall had fought a "clean, hard decision," only to be blocked by the Headquarters decision. Godfrey to Petrone, 15 August 1973.

98 "Proposals Asked for Shuttle Solid Rocket Motors," MSFC Release No. 73–133, 18 July 1973. "NASA Needs Firecracker," *Washington Post,* 14 October 1973.

99 James C. Fletcher, "Selection of Contractor for Space Shuttle Program Solid Rocket Motors," 12 December 1973, Space Shuttle-SRB 1973 folder, MSFC History Archives. Lockheed, UTC and Thiokol were the top three firms, with Lockheed leading in technical areas and trailing in management, Thiokol leading in management and trailing in technical areas, and UTC falling between the others in both areas. NASA considered these results "essentially a stand-off," leading to Thiokol's selection on the basis of cost.

100 NASA Headquarters TWX to MSFC, 6 November 1973, Shuttle (General) 1973 folder, MSFC History Archives. Headquarters quoted a report from the Subcommittee on HUD, Space Science and Veterans.

101 Richard C. McCurdy, letter to the editor, *New York Times*, 3 January 1987; Godfrey to Petrone, 23 January 1974, Solid Rocket Booster (SRB) 1974 folder, MSFC History Archives.

102 The GAO found NASA's evaluation process fair and reasonable, although it did cite a major error in evaluating costs that would have reduced the difference between the Lockheed and Thiokol proposals from $122 million to $54 million. "GAO Report Clearly Leaves SRM Decision Up to NASA, *Defense/Space Daily,* 26 June 1974, p. 316. NASA, Press Release, 26 June 1974, Solid Rocket Booster (SRB) 1974 folder, MSFC History Archives.

103 George Hardy, OHI by Jessie Emerson/MSI, 1 March 1990, Huntsville, Alabama, pp. 2–3. Subcontracts under S&E management included McDonnell Douglas "for structural, MMC for the recovery system, UTC Chemical Systems Division for booster separation motors, Bendix for integrated electronic assembly, and Moog and Sunstrand for the thrust vector control system." Emerson notes.

104 Rees to Jim Murphy, 30 March 1972, Space Shuttle 1972 folder, MSFC History Archives.

105 James Kingsbury, OHI by Jessie Whalen/MSI, 17 November 1988, Huntsville, Alabama, p. 9.

106 External tank basic requirements were first set forth in External Tank Development Branch, "Preliminary External Tank Requirements and Description," 12 December 1972, Box 0375, MSFC Documents, Shuttle Series, JSC History Office.

107 Kingsbury, OHI, p. 9.

108 Max Faget was one who thought the external tank should be the variable element. Low to Fletcher, 24 August 1972, Shuttle External Fuel Tank folder, NASA Headquarters History Office.

109 Rees to Jim Murphy, 30 March 1972.

110 Chrysler Corporation, Space Division, "Space Shuttle External Tank Sizing Analysis: Basic Questions and Recommendations, 11 August 1972; Godfrey to Rees, 21 August 1972, Space Shuttle—External Tank No. 1, 1972–73 folder, MSFC History Archives.

111 NASA Release No. 73–163, "Martin-Marietta to Develop Space Shuttle Tank," 16 August 1973, Space Shuttle—External Tank No. 1, 1972–73 folder, MSFC History Archives. The unsuccessful bidders included the Chrysler Corporation Space Division, McDonnell-Douglas Astronautics Company, and Boeing.

112 The Manned Spacecraft Center was renamed Lyndon B. Johnson Space Center on 27 August 1973.

113 NASA Management Instruction 8020.18B, "Space Shuttle Management Program," 15 March 1973, Shuttle Program Management folder, NASA Headquarters History Office.

114 Larry Mulloy, OHI by Jessie Whalen/MSI, 25 April 1988, Huntsville, Alabama, pp. 4–5.

115 Lucas to J. R. Thompson, 19 December 1974, Space Shuttle Main Engine (SSME) 1974 folder, MSFC History Archives.

116 James Kingsbury, OHI by SPW, 22 August 1990, Madison, Alabama, pp. 31–32.

117 NASA Management Instruction 8020.18B, "Space Shuttle Management Program," 15 March 1973.

118 William Lucas, OHI by AJD and SPW, 19 June 1989, Huntsville, Alabama, pp. 28–29.

119 Bob Marshall to Mike Wright, comments on chapter draft, 5 January 1993, MSFC History Office.

120 Cited in Craig Covault, "Changes Expected in Shuttle Management," *Aviation Week & Space Technology* (24 September 1979).

121 James Kingsbury, OHI by Jessie Whalen/MSI, 17 November 1988, Huntsville, Alabama, p. 6.

122 "Engines for the Space Shuttle," *American Machinist* (1 February 1975), 36–39; Trento, pp. 139–40.

123 Marshall to Wright, comments on chapter draft, 5 January 1993.

124 Fletcher to Robert Anderson, President and Chief Executive Officer, Rockwell International, 16 May 1974, Space Shuttle Main Engine 1974 folder, MSFC History Archives.

125 J.P. McNamara, internal Rockwell letter to R. Anderson and W. B. Bergen, 25 June 1974, Space Shuttle Main Engine 1974 folder, MSFC History Archives.

126 Fletcher to Robert Anderson, 16 May 1974.

127 Robert E. Lindstrom to Robert F. Thompson, 9 July 1975; Robert F. Thompson to Director, Space Shuttle Program, NASA Headquarters, 25 July 1975, Space Shuttle Main Engine (SSME) 1975 folder, MSFC History Archives; John P. McCarty OHI, pp. 8–10.

128 W.J. Brennan, President, Rocketdyne Division, Rockwell International, internal letter to Robert C. Bodine, 6 May 1974; Robert E. Lindstrom, memo to M.S. Milkin, NASA HQ, 20 June 1974; E.B. Crain, twx to W. Brennan, Rocketdyne, 23 September 1974; "Sequence of Events: SSME Test Positions," 25 September 1974; "SSME Facilities Contract NAS8–5609(F)," 7 October 1974; Space Shuttle Main Engine (SSME) 1974 folder, MSFC History Archives.

129 Fletcher to Robert Anderson, 16 May 1974.

130 J.P. McNamara, internal Rockwell letter to R. Anderson and W. B. Bergen, 25 June 1974.

131 "Sequence of Events: SSME Test Positions," 25 September 1974. MSFC formed the tiger team in January 1974.

132 R.E. Pease to James R. Thompson, Jr., 15 May 1974, Space Shuttle Main Engine (SSME) 1974 folder, MSFC History Archives.

133 J.T. Murphy, "Review of SSME Program at Rocketdyne, 23 May 1974, Space Shuttle Main Engine 1974 folder, MSFC History Archives.

134 Lucas to Lindstrom, 20 May 1974, Space Shuttle Main Engine 1974 folder, MSFC History Archives.

135 J.P. McNamara, internal Rockwell letter to R. Anderson and W. B. Bergen, 25 June 1974.

136 Presentation viewgraphs attached to J.P. McNamara, internal Rockwell letter to R. Anderson and W. B. Bergen, 25 June 1974.

137 John A. Chambers to Lucas, 12 August 1974, Space Shuttle Main Engine (SSME) 1974 folder, MSFC History Archives.

138 Thompson to Lucas, Potate, Smith, and Lindstrom, 3 September 1974, Space Shuttle Main Engine (SSME) 1974 folder, MSFC History Archives.

139 "SSME Project Summary: April 1974–Present," 21 October 1974, Space Shuttle Main Engine (SSME) 1974 folder, MSFC History Archives.

140 Formerly the Mississippi Test Facility (MTF).

141 Robert J. Schwinghamer, OHI by Jessie Emerson/MSI, 30 August 1989, Huntsville, Alabama, p. 12.

142 *Ibid.*, p. 15.

143 Thomson, OHI, p. 6.

144 Robert Biggs, "Space Shuttle Main Engine: The First Ten Years," 2 November 1989, SHHDC–1552; "Shuttle Engine Four Months Behind Schedule," *Defense/Space Daily* (17 September 1976), 93–94; James R. Thompson, Jr., "The Space Shuttle Main Engine and the Solid Rocket Booster," briefing with George B. Hardy, 14 October 1980, MSFC, Shuttle Booster 1980 folder, NASA Headquarters History Office.

145 Robert Biggs, "Space Shuttle Main Engine: The First Ten Years."
146 Ron Tepool, OHI by Jessie Whalen/MSI, 31 October 1988, Huntsville, Alabama, p. 1
147 Bledsoe, OHI, p. 9.
148 McCarty, OHI, p. 20.
149 Joseph P. Allen to Dr. John V. Dugan, 24 March 1976, and R.H. Curtin to NASA Comptroller, 9 February1976, Shuttle Propulsion System 1976 folder, NASA Headquarters History Office; "Annual Report to the NASA Administrator by the Aerospace Safety Advisory Panel on the Space Shuttle Program," June 1976, pp. 34–35, Space Shuttle 1976 folder, MSFC History Archives.
150 Robert Biggs, "Space Shuttle Main Engine: The First Ten Years," 2 November 1989, SHHDC–1552.
151 McCarty, OHI, p. 19.
152 Cited in Biggs, "Space Shuttle Main Engine: The First Ten Years."
153 "Investigation Board Report: SSME 0003 Oxygen Fire on Test Stand A–1, National Space Technology Laboratory, 24 March 1977," Space Shuttle Main Engine, 1977 folder, MSFC History Archives; "Main Engine Fire May Delay Shuttle," *Huntsville Times*, 29 July 1978.
154 Herman Thomason, OHI by Sarah McKinley/MSI and Tom Gates/MSI, 29 July 1988, Huntsville, Alabama, pp. 21–22.
155 Thomson, OHI, p. 18.
156 Craig Covault, "Further Shuttle Launch Slip Forecast," *Aviation Week and Space Technology* (26 February 1979), 17; Thompson, "The Space Shuttle Main Engine and the Solid Rocket Booster."
157 Mulloy, Larry, OHI by Jessie Whalen/MSI, 25 April 1988, Huntsville, Alabama, pp. 7–8.
158 Kingsbury, OHI by Jessie Whalen, p. 9.
159 Garland G. Buckner and J.P. Noland, Memo for the Record, 9 July 1974; Lucas to Yardley, 9 July 1974; Robert S. Williams, Martin Marietta, to James B. Odom, External Tank—Shuttle, January-June 1974 folder; Robert E. Lindstrom to Robert F. Thompson, 16 August 1974, Space Shuttle 1974 folder; James B. Odom, "External Tank Cost Assessment for MSF, 6 September 1974, External Tank, August–December 1974 folder, MSFC History Archives.
160 Fletcher to Thomas G. Pownall, President Martin Marietta Aerospace, 26 July 1974, Shuttle—External Tank, August–December 1974 folder, MSFC History Archives.
161 "External Tank Definitized Contract Issued," MSFC Release No. 75–28, 31 January 1975; "Martin Awarded Seven-Year $152.6 Million External Tank contract," *Defense/Space Daily* (31 January 1975), 166. The contract was to run through 30 June 1980 and was to cover fabrication of six flight model tanks and development of capability for manufacturing production tanks.
162 James B. Odom, OHI by Jessie Whalen/MSI, 9 February 1988, Huntsville, Alabama, pp. 2–3.
163 Porter Bridwell, OHI by Jessie Whalen and Sarah McKinley/MSI, 18 December 1987, Huntsville, Alabama, pp. 2–3.
164 Odom, OHI, pp. 14–15.

165 Pessin, OHI, pp. 9–10.

166 Mary Elaine Lora, "Fueling the Force Behind the Space Shuttle," (1983); Odom, OHI, p. 3.

167 Odom, OHI, p. 4.

168 MSFC Release No. 76–60, 6 April 1976.

169 Lindstrom to Robert F. Thompson, 28 June 1974, External Tank—Shuttle, January–June 1974 folder, MSFC History Archives.

170 Lindstrom to Robert F. Thompson, 14 March 1977, External Tank 1977 folder, MSFC History Archives.

171 "NASA Signs Contract to Reduce Shuttle External Tank Weight," MSFC Release No. 80–102, 1 July 1980.

172 Odom, OHI, p. 7.

173 Bridwell, OHI, p. 3.

174 Odom, OHI, p. 8.

175 Thomas G. Pownall to Fletcher (unsent), 14 August 1974, Shuttle-External Tank, August–December 1974 folder, MSFC History Archives.

176 Award Fee Performance Evaluation of Martin Marietta Corporation, 10 March 1977, External Tank 1977 folder, MSFC History Archives. The report stated that "The management and integration of the project have yet to reflect effective direction of diverse activities and requirements among and within the functional disciplines of the project."

177 R.G. Smith to Lucas, 10 August 1977, External Tank 1977 folder, MSFC History Archives. Details of the incident are in Industrial Mishap Report, 17 June 1977, External Tank 1977 folder, MSFC History Archives.

178 Lucas to Smith, 17 August 1977, External Tank 1977 folder, MSFC History Archives.

179 Lindstrom to Dr. Malkin, 19 January 1978, Shuttle—External Tank 1978 folder, MSFC History Archives.

180 Lucas, handwritten note on Odom to Lucas, 30 July 1979, Shuttle—External Tank 1979–80 folder, MSFC History Archives.

181 MSFC Release Nos. 76–137, 76–142, and 76–143.

182 MSFC Release Nos. 76–195, 29 October 1976, and 77–30, 25 February 1977.

183 Chuck Verschoore, OHI by Sarah McKinley/MSI, 27 June 1988, p. 5.

184 Jack Nichols, OHI by Jessie Whalen and Sarah McKinley/MSI, 12 December 1987, Huntsville, Alabama, pp. 1–2.

185 Verschoore, OHI, p. 9–10.

186 Garland Johnston, OHI by Sarah McKinley/MSI, 7 November 1988, Huntsville, Alabama, pp. 15–16.

187 George B. Hardy, "The Space Shuttle Main Engine and the Solid Rocket Booster," briefing with James R. Thompson, Jr., 14 October 1980, MSFC, Shuttle Booster 1980 folder, NASA Headquarters History Office.

188 Clyde Nevins, OHI by Jessie Emerson/MSI, 7 March 1990, Huntsville, Alabama, p. 16.

189 *Astronautics and Aeronautics*, 1973, 10 February 1973.

190 *Marshall Star*, 21 November 1973.

191 Craig Covault, "Solid Rocket Booster Nears Milestones," *Aviation Week and Space Technology* (8 November 1976), pp. 84–85.

192 Covault, pp. 84–89; "SRB," Chronology, September 1984, MSFC Documents, Shuttle Series, SRB Information 1973–1984 folder, Box 0379, JSC History Office.

193 MSFC Release 77–133, 18 July 1977.

194 Trento, pp. 146–55.

195 Richard H. Weinstein, minutes of MSFC Center Performance Review of 22 June 1977, 5 July 1977, Visit to MSFC by Mr. Groo, 6/22/77 folder, MSFC History Archives.

196 R.E. Lindstrom, briefing paper for Lucas, 22 March 1977, Solid Rocket Booster 1977 folder, MSFC History Archives.

197 Hardy to Lucas, 28 October 1977, Solid Rocket Booster 1977 folder, MSFC History Archives.

198 Ben D. Bagley, memorandum for the record, 15 June 1978; Hardy to Lucas, 7 July 1978; Hardy, memorandum for the record, 13 July 1978; Solid Rocket Booster (SRB) 1978 folder, MSFC History Archives.

199 Kingsbury Weekly Notes (Brooks), 13 August 1978, MSFC History Archives.

200 Lindstrom to Antonio L. Savoca, VP and General Manager, Thiokol Wasatch Division, 1 June 1978, Solid Rocket Booster (SRB) 1978 folder, MSFC History Archives.

201 "Summary of Board of Investigation Report: SRM DM–4 Aft Center Segment Accident, 2 December 1978," 24 January 1979, Solid Rocket Booster (SRB) 1979 folder, MSFC History Archives.

202 Thiokol/Wasatch Division, "Investigation Report SRM DM–4 Aft Center Segment Incident LMCP Pad No. 8, 2 December 1978," 14 December 1978; Potate to Lucas, 15 January 1979; Solid Rocket Booster (SRB) 1979 folder, MSFC History Archives.

203 Kingsbury Weekly Notes (Brooks), 13 August 1978, MSFC History Archives.

204 William P. Raney, note to Ted Speaker, 17 December 1984, Shuttle—Lessons Learned folder, NASA Headquarters History Office.

205 Chris Kraft, OHI by AJD and SPW, 11 July 1990, Johnson Space Center, Houston, Texas, pp. 7–9.

206 "Senator Says NASA Cut 70% of Staff Checking Quality," *New York Times*, 8 May 1986, SHHDC–1895; and "NASA Cut Quality Monitors Since '70," *Washington Post*, 8 May 1986, SHHDC–1896, MSI Shuttle Collection, MSFC.

207 Raney, note to Ted Speaker, 17 December 1984.

208 Christopher C. Kraft, OHI by AJD and SPW, 28 June 1991, Webster, Texas, pp. 14–15.

209 Schwinghamer, OHI, pp. 9–10.

210 Marshall to Wright, 5 January 1993.

211 *Huntsville Times*, 13, 14, 15, and 19 March 1978.

212 "Shuttle Testing at the Marshall Center," MSFC Release No. 78–33, March 1978; Dave Dooling, "Space Shuttle Enterprise Tests to Be Unexciting, Crucial," *Huntsville Times*, 12 March 1978; A.A. McCool, Chairman, MVGVT Incident Investigation Team, "Final Report on MVGVT LOX Tank Incident," External Tank 1978 folder, MSFC History Archives.

213 "Shuttle Ground Vibration Tests End," MSFC Release No. 79–21, 28 February 1979.

214 NASA News Release No. 78–145, 25 September 1978.

215 "$600-Million Shuttle Cost Overrun Startles Congress," *Aviation Week & Space Technology* (7 May 1979), 18.

216 Dick Baumbach, "Shuttle Problems May Spark Shakeup," *Today* (20 May 1979); Bob Mungall, "Shuttle Program Hangs by Thread," *Today* (24 May 1979); "Shuttle Problems," *Aviation Week & Space Technology* (11 June 1979), 25.
217 Robert A. Frosch, Statement before Subcommittee on Space Science and Applications of the Committee on Science and Technology, House of Representatives, 28 June 1979, Shuttle Program Management folder, NASA Headquarters History Office.
218 Craig Covault, "Changes Expected in Shuttle Management," *Aviation Week & Space Technology* (24 September 1979); "Shuttle Problems Assessed," *Washington Post*, 19 September 1979.
219 Craig Covault, "NASA Assesses Shuttle Engineering," *Aviation Week & Space Technology* (23 June 1980), 16; Covault, "Shuttle Engine, Tile Work Proceeding on Schedule," *Aviation Week & Space Technology* (15 September 1980), 26; "Final Shuttle Engine Tests Set, *Aviation Week & Space Technology* (27 October 1980), 53.
220 "NASA Daily Activities Report," 23 July 1980.
221 MSFC Release No. 79–133, 18 December 1979.
222 MSFC Release No. 81–18, 18 February 1980.
223 Thompson, "The Space Shuttle Main Engine and the Solid Rocket Booster." The SSME had operated for 20,000 seconds compared to 7,000 for the J–2. The SSME had a success rate of over 0.9, the J–2 slightly less.
224 MSFC, "Chronology: MSFC Space Shuttle Program: Development, Assembly and Testing Major Events (1969–April 1981)," MHR–15, December 1988, pp. 76–78.
225 MSFC Shuttle Projects Flight Evaluation Working Group, "Space Shuttle Flight Readiness Firing Evaluation Report, 30 March 1981, Box 0376A, Shuttle Series, MSFC Documents, JSC History Office.
226 Odum, OHI, p. 16.
227 Jack Hartsfield, "The Thread of History Still Runs at Marshall," *Huntsville Times*, 13 April 1981; Dave Dooling, "Columbia Shakes Off Lingering Questions, *Huntsville Times*, 13 April 1981.
228 MSFC Release No. 81–49, 21 April 1981.
229 Dr. J. Wayne Littles, OHI by Jessie Emerson/MSI, 15 August 1989, Huntsville, Alabama, p. 5.
230 James Kingsbury to Mike Wright, 16 December 1992, comments on chapter draft, MSFC History Office.
231 Nevins, OHI, p. 15.
232 Mulloy, OHI, pp. 10–11; Thomason, OHI, pp. 18–20.
233 Nevins, OHI, p. 19.
234 "NASA Abandons SRB Retrieval," Aerospace Daily, (13 September 1982).
235 Mulloy, OHI, pp. 13–14.
236 Edward H. Kolcum, "NASA Pinpoints Space Shuttle's Launch Damage," *Aviation Week & Space Technology* (13 July 1981), pp. 57–63; "Shuttle Boosters in Good Condition," *Aviation Week & Space Technology* (5 April 1982), p. 47; "NASA to Test Shuttle Parachute System," *Aviation Week & Space Technology* (28 February 1983), p. 61.
237 Jerry Thomson, OHI, p. 8.

238 Littles, OHI, p. 7
239 Dr. Judson A. Lovingood, OHI by Jessie Emerson/MSI, 16 August 1989, Huntsville, Alabama, pp. 21–22.
240 "Cracks and Pitting Hit Shuttle's Turbopumps," *New Scientist* (5 April 1984), p. 21; "NASA Wants More SSME Life," *Flight International* (5 May 1984), p. 1205; "Industry Observer," *Aviation Week & Space Technology* (15 October 1984), p. 13; "Main Engines Get Longer Lives," Ralph Vartabedian, *Los Angeles Times*, 30 December 1984; Research and Technology to Improve Space Shuttle Main Engine," NASA Release Nos. 85–122, 4 September 1985.
241 Robert A. Frosch, statement to Subcommittee on Science, Technology and Space of the Senate Committee on Commerce, Science and Transportation, 4 June 1979.
242 Marshall, OHI by Whalen and McKinley, p. 24.
243 Herman Thomason, OHI by Sarah McKinley/MSI and Tom Gates/MSI, 29 July 1988, Huntsville, Alabama, p. 23.
244 "Cracks and Pitting Hit Shuttle's Turbopumps," *New Scientist* (5 April 1984), p. 21.
245 New York News, 23 April 1982.
246 Marshall, OHI by Whalen and McKinley, pp. 24–25.
247 MSFC Release No. 80–93, 30 June 1980; MSFC Release No. 82–81, 8 September 1982; Kingsbury, OHI by Jessie Whalen/MSI, p. 9.
248 Odom, OHI, pp. 8–9.
249 Bridwell, OHI, pp. 2–3.
250 Edward H. Kolcum, "Space Shuttle Lightweight Tank Production Begins," *Aviation Week and Space Technology* (16 November 1981), p. 79.
251 Nichols, OHI, pp. 2–3.
252 *Marshall Star*, 16 May 1984.
253 Kingsbury, OHI by Jessie Whalen/MSI, p. 9.
254 Bridwell, OHI, p. 3.
255 Schwinghamer, OHI by Jessie Emerson/MSI, p. 10.
256 James M. Beggs to Edwin (Jake) Garn, Chairman, Subcommittee on HUD-Independent Agencies, 3 February 1982, Shuttle Boosters 1982 folder, NASA Headquarters History Office; Mulloy, OHI by Jessie Whalen/MSI, pp. 17–19; Littles, OHI by Jessie Emerson/MSI, pp. 5–7; Craig Covault, "Decision Near on Shuttle Payload Boost," *Aviation Week & Space Technology* (10 August 1981), pp. 50–52.
257 Beggs to Garn, 3 February 1982.
258 Ron McIntosh, OHI by Jessie Emerson/MSI, 8 August 1990, Huntsville, Alabama, p. 4.
259 "Filament Booster Case," MSFC Fact Sheet, 28 February 1984, Fiche No. 1464, MSFC History Office.
260 See Chapter 8.
261 Mulloy, OHI, pp. 15–16; McIntosh, OHI, p. 4.
262 Gerald S. Schatz, "Panel Questions Logistics of Space Shuttle Schedule," *News Report* (May–June 1983). The article reported on a report of the National Research Council's Committee on NASA Scientific and Technological Program Reviews entitled "Assessment of Constraints on Space Shuttle Launch Rates."

263 Noel W. Hinners to Benjamin Huberman, 22 September 1982, Box 4, Center Series—STS Management Studies, JSC History Office.

264 Trento, p. 237.

265 Craig Covault, "NASA, Defense Department Drop Idea of Private Shuttle Management," *Aviation Week & Space Technology* (29 April 1985), p. 42.

266 Schwinghamer, OHI, pp. 24–25.

Chapter IX

The *Challenger* Accident

On the morning of 28 January 1986, the Space Shuttle *Challenger*, mission 51–L, rose into the cold blue sky over the Cape.[1] To exuberant spectators and breathless flight controllers, the launch appeared normal. Within 73 seconds after liftoff, however, the external tank ruptured, its liquid fuel exploded, and *Challenger* broke apart. Stunned spectators saw the explosion and the trails from the spiral flights of the solid rocket boosters, but the vapor cloud obscured how the orbiter shattered into large pieces. The crew cabin remained intact, trailing wires and plummeting to the Atlantic; the six astronauts and one school teacher aboard perished.[2]

Over the next three months, a presidential commission led by former Secretary of State William P. Rogers and a NASA team investigated the accident. Television images of the flight revealed an anomalous flame from a joint between segments of the right-hand solid rocket motor. Photographs showed puffs of black smoke escaping from the joint during the first moments of ignition. Wreckage of the motor recovered from the Atlantic floor demonstrated the failure of the joint and proved that propulsion gases had melted surrounding metals and caused the explosion of the external tank. Propulsion engineers from Morton-Thiokol Incorporated, the Utah company responsible for the solid rocket motors, testified that for years they had been discussing problems with the joints and their O-ring seals, especially in cold weather. The night before the launch they had warned Marshall officials that the anticipated cold weather could freeze the rubber O-rings and trigger disaster, but company executives and Marshall project managers had rejected calls for a launch delay.

The Rogers Commission concluded that managers at Marshall and Thiokol had known (or should have known) that the case joints were hazardous. They had failed to inform senior officials in the Shuttle program or to act promptly to reduce risks, and thus had failed to prevent a predictable accident. The commission decided that since Marshall officials had prior knowledge of the hazard, the accident primarily resulted from ineffective communications and management at the Center.[3]

The commission's interpretation has dominated discourse about *Challenger*. Journalists and academics have relied on the commission's evidence, and have mainly added analysis to confirm its "bad communication" thesis. "Instant histories" often treated the scenarios in the Rogers Report as quasi-crimes, with journalist-authors reporting dirty deeds in the Shuttle program and telling scabrous stories about NASA officials with "the wrong stuff."[4] Academic studies tried to show why the Rogers scenario occurred, explaining how communications problems could have emerged from the interdependence of Marshall and Thiokol, the lapses in statistical analysis by propulsion engineers, the groupthink of the preflight reviews and last minute teleconference, and authoritarian management patterns at Marshall.[5] Two scholars have also discussed why the interpretation of the presidential commission seemed persuasive to the media and the public while the point of view of Marshall officials did not.[6]

Unfortunately, the commission's interpretation oversimplified complex events. The oversimplifications emerged mainly because the commission and later pundits dismissed the testimony of Marshall engineers and managers and distorted information about hazards in written sources from the Shuttle program prior to the accident. Allowing Marshall engineers and managers to tell their story, based on pre-accident documents and on post-accident testimony and interviews, leads to a more realistic account of the events leading up to the accident than that found in the previous studies. The story of the Marshall engineers and managers was that they had carefully studied the problems of the motor case joints and had concluded that the joints were not hazardous, that they had taken steps to improve the joints, and that they had communicated their conclusions and actions to superior Shuttle officials. Because they believed the joints were not hazardous, they did not predict the accident and could not have prevented it.

Design and Development

From the beginning of the design and development phase of the Solid Rocket Motor (SRM) project in 1973, Marshall had trouble with Morton-Thiokol and the joints.[7] Several Center engineers worried that the joint and seal designs were deficient and Center managers fretted about the contractor's quality systems. But after improvements, reviews, and many successful tests, senior project managers and engineers decided that the design was successful and the joints were safe to fly.

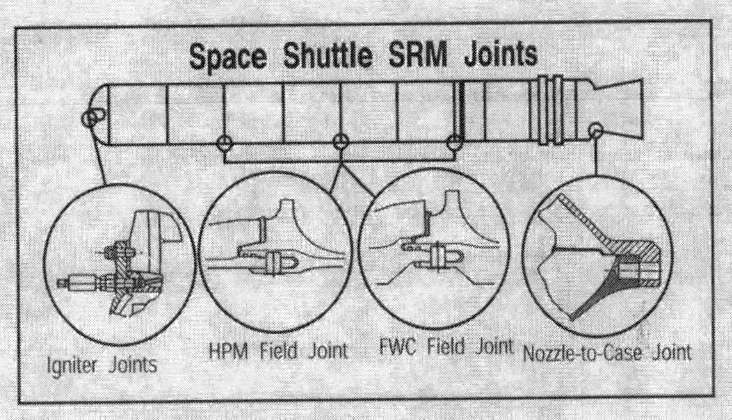

Space Shuttle solid rocket motor joints.

Because Thiokol would ship the motor by rail between its Utah production site and the Florida launch site the solid rocket motor case was divided into several segments (as shown in illustration of Shuttle SRM joints). This meant that the design required joints and seals to prevent leaks of the high-temperature, high-pressure propulsion gases. Thiokol's engineers used the Titan III–C rocket, considered state of the art for solid motors and very reliable, as their model. The Shuttle motor, however, differed from the Titan because the SRM was larger and intended for refurbishment and reuse. The differences in design showed in the field joints connecting the motor case segments.[8] The Titan had insulation along the interior wall of the steel case to prevent hot gases from penetrating the joint and damaging its rubber seals (see the SRM field cross section and the comparison illustration if Titan III and SRM joints); the SRM used an asbestos-filled putty. The segments had upper and lower parts; Thiokol engineers expected that motor pressure would push together the "tang" (the tongue on the rim of the upper segment) and the inner flange of "clevis" (the groove on the rim of the lower segment) and facilitate sealing. Motor pressure would also push the primary O-ring, a quarter-inch diameter rubber gasket, against the steel case and seal the joint. Thiokol sought redundancy by placing a second O-ring behind the primary O-ring.

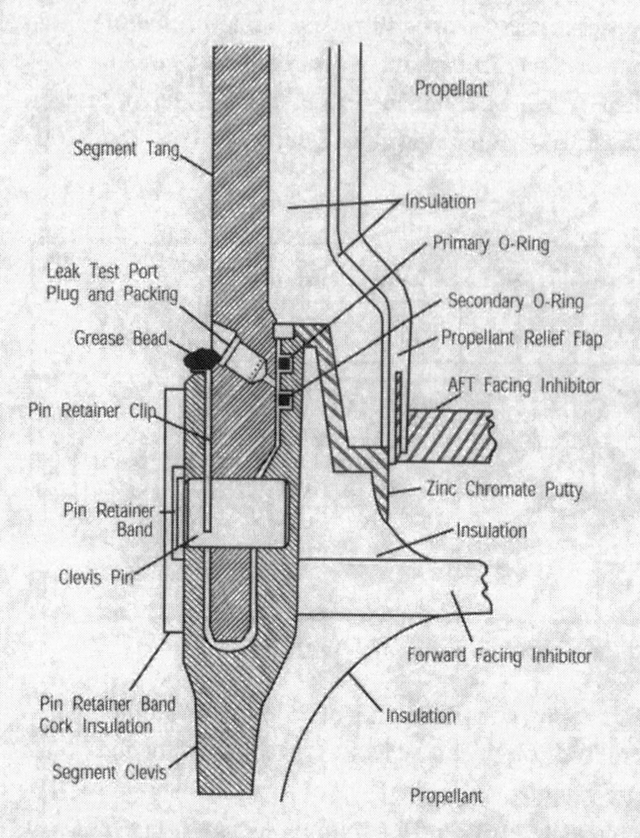

Cross section of SRM field joint.

The second O-ring forced another departure from the Titan, requiring that the SRM have a longer tang and a deeper clevis. The longer joint of the SRM was more flexible than the Titan, since the combustion pressure in the SRM was one-third higher than the Titan and its case had a greater diameter. Moreover, the SRM clevis pointed up, rather than down like that of the Titan. Finally, to test the seals after connecting the segments,Thiokol engineers added a leak-check port so that compressed air could be forced into the gap between the O-rings and verify whether the primary O-ring would seal. The leak-check, however, pushed the primary O-ring to the wrong side of its groove.[9]

Thiokol and Marshall evaluated the SRM and its case joints in structural, pressure, and static firing tests beginning in 1976. Because tests of the large solid rocket were more expensive than liquid engines, engineers ran fewer tests.[10] From the beginning of the test program, they showed confidence in their design, perhaps stemming from the success of Titan. They scheduled static firings of the entire motor before completion of subsystem tests such as pressure tests of the joint and case. The first static firing of DM–1 (Development Motor 1) confirmed that the hardware met design requirements, including the integrity of the steel case and the thrust of its motor. Marshall's Weekly Notes reported on DM–1 that "all case joints were intact and showed no evidence of pressure leaking" and that "all test objectives were met."[11]

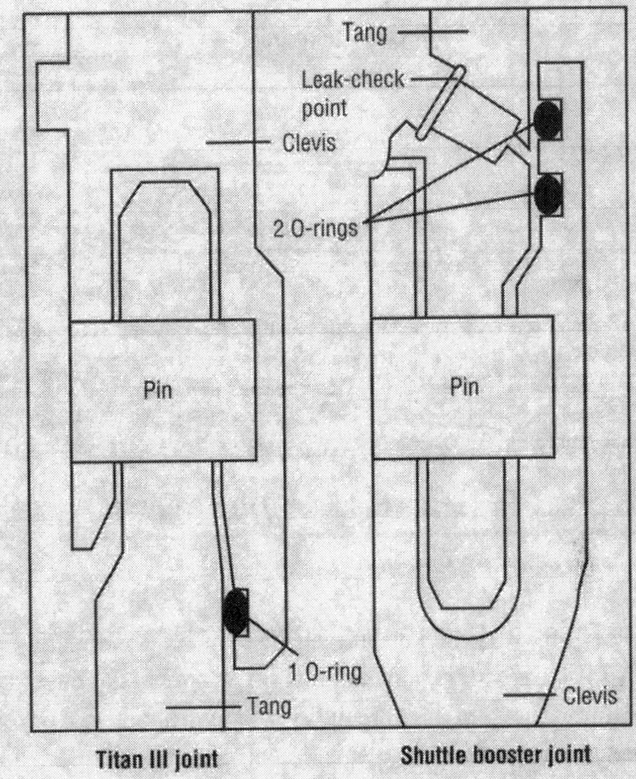

Comparison of Titan III and SRM joints.

In a September 1977 hydroburst test, however, the field joints and O-rings performed contrary to expectations. Engineers simulated a launch by filling a motor with fluid under 50 percent more pressure than during ignition. Thiokol had expected pressure to force the tang and the inner flange of the clevis to bend toward each other and squeeze the O-rings tighter. The company's final report of the test concluded that "the burst test was a complete success and met all the design requirements. Failure occurred in the joint seals. The leakage was caused by the clevis spreading and not providing the required O-ring squeeze." The engineers were perplexed and reported that the joint opened more than they predicted.[12]

In Weekly Notes, Marshall's SRM project engineers said the burst test revealed "excessive O-ring leakage." Both the primary and secondary O-rings leaked, and disassembly revealed each had pinches and cuts. "The most logical explanation," the MSFC motor engineer observed, "is joint rotation which allowed both O-rings to lose compression."[13]

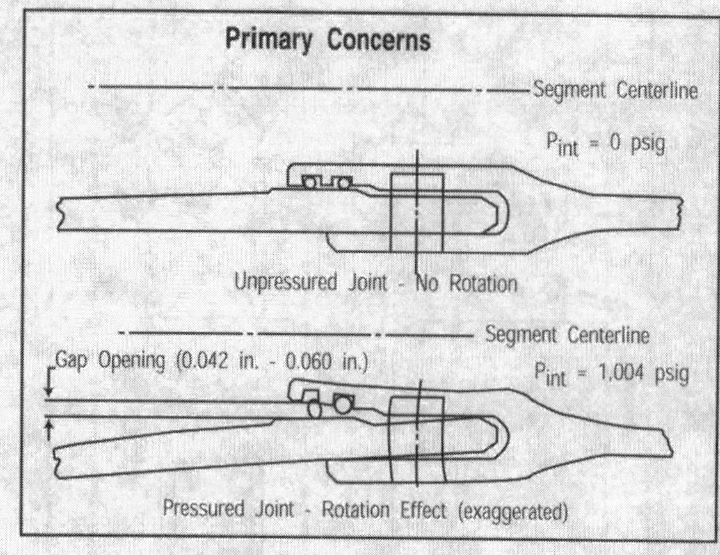

Joint rotation of SRM field joint.

Joint rotation meant that under pressure the tang and the inner flange of the clevis bent apart and the joint opened (as shown in the joint rotation of the SRM field joint). Rotation occurred because the motor joint was thicker and stiffer than the case walls on either side; as the flexible case wall expanded outward, it spread the tang and clevis and opened the joint. The joint opening during the hydroburst test unseated the O-rings and created a gap too large to seal.[14]

Thiokol denied that the tests revealed design flaws. The test subjected the same hardware with the same O-rings to 20 cycles of pressure and release; only in the final cycles did the rings leak. Consequently, Thiokol engineers believed that with each cycle, the O-ring was pushed into the gap, then released, then pushed in farther, and so on until the rubber condensed, cut, and failed. Rather then interpreting the tests as indications of bad design, Thiokol engineers argued that the joint had withstood many cycles without failure and so test results showed the soundness of the joint. They believed that potential leaks on flight motors could be avoided through careful assembly of the joints and by inserting dozens of shims, which were U-shaped clips, between the outer clevis and the tang. The shims would maintain the centricity of the case and the compression of the O-rings; this would prevent any "gathering" or bunching of the O-ring that could cause a leak.[15]

Some engineers in Marshall's laboratories disagreed with the contractor and believed the joint design was flawed. In September 1977, Glenn Eudy, the Center's chief engineer for the SRM, expressed his doubts to Alex McCool, director of the Structures and Propulsion Lab, and argued that refined assembly methods alone could not fix the problem. "I personally believe," Eudy wrote, "that our first choice should be to correct the design in a way that eliminates the possibility of O-ring clearance." He requested that the director of Science and Engineering review the problem. In October, Center engineer Leon Ray argued that shims allowed for error during assembly and hence were "unacceptable." He advised that the "best option for a long-term fix" was a "redesign of the tang" to prevent joint opening.[16]

By January 1978 Ray and his boss, John Q. Miller, chief of the Structure and Propulsion Lab's solid rocket motor branch, believed that the joint issue required the "most urgent attention" in order to "prevent hot gas leaks and resulting catastrophic failure." Alarmed that Thiokol was trying to lower requirements for the joint, they saw "no valid reason for not designing to accepted standards." Miller and Ray recommended "redesign of clevis joints on all oncoming hardware at the earliest possible effectivity to preclude unacceptable, high risk, O-ring compression values."[17]

Not only did Thiokol reject the analysis of the Marshall rocket engineers, but so did Center managers. Marshall management accepted the existing design, complemented by shims, mainly because of the continued successes of static motor firings. In the Weekly Notes following the firing of DM–2 in January 1978, McCool wrote that "all major test objectives were met" and "no leaks were observed in the case during the firing and post-test examination revealed no discolorations nor other evidence of leakage." Robert Lindstrom, Shuttle Projects Office manager at Marshall, wrote that preliminary analysis of DM–2 indicated "no problems which require immediate attention of NASA."[18]

A Thiokol report on the October firing of DM–3 said "all case joints were intact and showed no evidence of pressure leaking." The report acknowledged that "the relative movement of the sealing surfaces is much more than indicated," but this evidence of joint rotation was not presented as anything ominous.[19] In November, Thiokol's SRB (Solid Rocket Booster) project manager wrote George B. Hardy, Marshall's project manager, that the static firings "confirmed the capability of the O-rings to prevent leakage under the worst hardware conditions."[20]

Results from Structural Test Article–1 (STA–1), however, were less optimistic. Hydroburst tests through the summer of 1978 on STA–1 again revealed the dangers of joint rotation. Thiokol's report concluded that "the relative movement between the clevis and the tang at the interior of the case joints was greater than expected. This resulted in some oil (pressurizing fluid) bypassing the O-ring seals at the case joints." The engineers decided that the O-rings unseated as the joint opened. Nevertheless, company engineers dismissed the leaks, arguing that test pressure was higher than flight pressure and "the amount of oil loss on any one occasion or totally was very small and motor case pressurization was never lost or affected by this phenomenon." As on the tests from the previous year, they concluded that the repressurization cycles had caused the failures rather than a faulty design. They acknowledged that imprecise calibration devices prevented accurate measures of the joint opening, but denied that the joint opened so wide as to be unsafe.[21]

STA–1 data led Miller and Ray to call Thiokol's design "completely unacceptable." In January 1979 they wrote another memo to Eudy and Hardy, explaining that joint rotation prevented the design from meeting contractual requirements. The contract specified that seals operate through compression, but the opening of the joint caused the primary O-ring to seal through extrusion. As a sealing mechanism, extrusion used ignition pressure to push the O-ring across the groove of the inner flange of the clevis until it distorted and filled the gap between clevis and tang. This, they said, "violates industry and Government O-ring application practices." In addition, Miller and Ray for the first time questioned whether the secondary O-ring provided redundancy. Although the contract required verification of all seals, tests had proven the secondary O-ring design to have been "unsatisfactory" because the opening of the joint "completely disengaged" the O-ring from its sealing surface.[22]

In February 1979 Ray sought advice from two seal manufacturers. One manufacturer said that the design required the O-ring to seal a gap larger than that covered by their experience. The Parker Seal Company, the contractor for the SRB O-rings, reacted the same way and "expressed surprise that the seal had performed so well." Ray reported that Parker engineers believed that "the O-ring was being asked to perform beyond its intended design and that a different type of seal should be considered."[23] However, Ray and Miller failed to convince Thiokol and Marshall to change their commitment to the existing design. The contractor reported that the static test of DM–4 on 19 February had "no indication of joint leakage" and the case showed "no evidence of structural problems." Thiokol's summary of the development motor firings concluded that after each test "all case joints were intact and showed no evidence of pressure leaking" and measurements revealed "no stresses that indicate design problems or that compromise the structural integrity of the case."[24] In 1979 and 1980 three qualification motors fired successfully and had no leaks.[25]

When in 1980 the Shuttle program underwent its final qualification for flight, the Center and contractor presented their data and conclusions to NASA's Space Shuttle Verification/Certification Propulsion Committee on the motor. During briefings from May to September and in its report, the committee fretted over O-ring leaks, assembly problems, and joint rotation. Members were concerned that the leaks "could grow in magnitude and could impinge on the ET [External Tank] during flight." Moreover the Propulsion Committee pointed out that testing on new assembly configurations "does not appear to exist and sensitivity data on O-ring damage is lacking." For the design to function, assembly procedures had to be perfect; although the O-ring leak check put the secondary O-ring in position to seal, it pushed the primary O-ring in the wrong direction (as evidenced in the illustration comparison of Titan III and SRM joints). Accordingly the panel recommended "an up-to-date rigorous and complete verification package covering safety factor on sealing at ignition," including purposely testing to failure and static firings at temperatures from 40 to 90 degrees F.

NASA did not conduct such a test program before the first Shuttle flight. The booster office at Marshall, the Level II Shuttle Program Office at Johnson Space Center (JSC), and the Level I associate administrator for Space Flight at NASA Headquarters all believed that previous tests showed the primary O-ring was an effective seal and that the secondary O-ring provided redundancy. Marshall and Thiokol offered reassurances that readings about joint rotation were misleading because of faulty measuring devices and that corrections were underway using shims and bigger O-rings. NASA believed that careful assembly procedures would ensure safety and that ongoing tests on a new lightweight motor case fulfilled the Propulsion Committee's intent.[26]

The committee accepted this response, and the flight certification phase of Shuttle development closed when the Agency assigned a "criticality" designation to the field joints and O-ring seals. A criticality rating in the Shuttle critical items list categorized the reliability of important hardware; the designation affected the attention the item received in flight preparations and reviews. Thiokol's November 1980 report for the critical items list, which NASA approved, designated the joint as criticality 1R, meaning that the component had "redundant hardware, total element failure of which would cause loss of life or vehicle." The report justified the redundancy rating by the design's similarity to the successful Titan III–C joints and the solid rocket motor's successes in structural and burst tests and seven static motor firings.

Nonetheless, the criticality report contained a contradiction. It admitted that the "redundancy of the secondary field joint seal cannot be verified after motor case pressure reaches approximately 40 percent of maximum expected operating pressure." At that point, joint rotation created a gap too large for the secondary to seal. The report added that it was "not known if the secondary O-ring would successfully re-seal if the primary O-ring should fail after motor case pressure reaches or exceeds 40 percent of maximum expected operating pressure." In other words, the report classified the seals as redundant despite incomplete data.[27]

Throughout the design and development period of the solid rocket motors, Marshall had sufficient oversight of its contractor to discern technical and managerial problems. For this reason, the presidential commission concluded not only that the joint design was flawed, but that "neither Thiokol nor NASA responded adequately to internal warnings about the faulty seal design," and that neither made "a timely attempt to develop and verify a new seal after the initial design was shown to be deficient." In addition, NASA's internal investigation teams for Development and Production and for Data and Design Analysis, which included many Marshall personnel, faulted the test program. Testing was not realistic; dynamic loads of launch and flight conditions were not adequately simulated; the putty configuration during static firings differed from the launch configuration because after assembly of the test motors, engineers crawled through the bore of the propellant and packed in extra putty to fill voids; tests did not evaluate performance under temperature extremes; subsystem tests did not yield realistic information about putty performance, joint rotation, and O-ring compression and resiliency.[28]

Even so, the 1986 accident investigations tended to read history backwards and to ignore the positive information about the joint. Looking back after *Challenger*, Marshall managers believed that they had studied, tested, reviewed, and verified the joint design. Lindstrom, Marshall's Shuttle Projects manager, Hardy, the SRB project manager, Bill Rice, the SRM project manager, Eudy, the SRM chief engineer, and McCool, the director of the Structures and Propulsion Lab, explained that they had no data showing serious problems. The positive data from all the static firings far outweighed negative data from parts of the pressure tests. Because of positive test data, McCool said, "no one really took it with seriousness, and I say all of us collectively, as serious as we should have." James Kingsbury, the head of Marshall's Science and Engineering Directorate, believed that before 51–L NASA had not fully understood the design. Lack of information "posed a real problem for us safety-wise—obviously one we did not fully resolve. There were some things about the Motor that had never been done before. It was a very big motor. It was being reused. And so there were some complications."[29]

Similarly, Hardy thought that the tests of the solid motors compared unfavorably to the Saturn system of testing. For the Saturn rockets, the Center had conducted many tests and had tested components and subsystems before hardware reached a final design. The solid rocket motor tests, in contrast, had been too few and too mild to return realistic and complete information. To save money in the short-term, Marshall had moved away from testing of subsystems and toward testing of whole systems. Restricting tests to late stages of development, Hardy said, had locked NASA into one joint design and boosted costs in the long-term.[30]

Marshall's engineers in the SRM Branch of the Structures and Propulsion Lab had a different recollection of the design and test phase. Looking back after the accident, they vouched for the openness of communication channels. Believing he had opportunity to present his criticisms, Ray told investigators for the presidential commission that Marshall differed from the military and allowed dissenters to bypass the chain of command and disagree with superiors. "Communication is very good," he said. "I feel at ease in picking up the phone and talking to anybody. It doesn't make any difference who it is. And I have—many times."[31] They had kept arguing that the joints had failed to meet contract requirements and that Thiokol had underestimated the width of the joint opening.

Although the engineers stopped short of recommending that the solid rocket motors not be flown, they recommended during the design and evaluation phase that new hardware built for later Shuttle missions incorporate a redesigned joint. They were unable to convince senior managers and engineers however. Miller, chief of the SRM Branch, recalled that "when you present something, a concern to someone, and nothing is perhaps immediately done, you don't—in the position I was in, you don't push the point and try to back them in a corner." Consequently the engineers pushed for tight requirements in the assembly of the joints. Ben Powers, a propulsion engineer, recollected that "after we had been turned down over and over, I think we just accepted the fact that as the hardware kept being manufactured, you know, that there was not going to be a change in the joint design. So we, I think, accepted that fact that we were not successful in getting that change that we recommended and had to do the best we could with the joint that we had."

When evaluating the cases for and against the design, the engineers concluded, Center officials showed more trust in the contractor than in laboratory personnel. Center management recognized that Thiokol's engineers had greater expertise in solid rockets than the Center's liquid propulsion specialists and so depended on the contractor's interpretation of the data.[32] Center Director William Lucas recalled that "We did not consider ourselves expert in what Thiokol was doing. In fact we were not, so we relied heavily on Thiokol to bring the expertise of solid rocket propulsion to the program. We were not able to assess the details of what they were doing."[33]

Flight Review and Response

Beginning with early flights, the solid rocket motor experienced recurrent problems with its joint and O-ring seals. Thiokol and Marshall regarded the problems as aggravating but acceptable anomalies; successful flights and ground tests gave the engineers confidence that the joints were not hazardous. They recommended that flights continue, improved motor assembly configurations, and initiated redesign studies in the summer of 1985. After *Challenger*, Marshall's response seemed too little, too late, and the presidential commission faulted NASA's management structure and flight readiness review process, and criticized the Center's judgment and communications.

The primary system of communication and decision-making during the flight phase of the Shuttle program was the flight readiness review. In formal inquiries, contractor and government officials discussed the preparedness of hardware, paperwork, and personnel for the upcoming flight. They also discussed problems and anomalies encountered in the previous flight, solutions that had been implemented or planned, and rationales that confirmed safety and reliability. Marshall reviews proceeded up from the Level IV at the prime contractor, to Level III at the project office, to the Shuttle Projects Office (which Center personnel called "Level 2-and-A-Half"), to the Center director, to the Level II pre-flight readiness review at JSC, and finally to the Level I flight readiness review at Headquarters held two weeks before launch. Officials from all space flight Centers attended the Level I review.

Decisions coming out of the reviews were only as sound as the information going in. Flight would proceed only after the reviews had certified each element safe and reliable. The success of the process depended on an upward flow of information and a downward flow of probing questions. Presentations were most detailed at the low levels, but with each step up the ladder, time constraints due to reports from additional project organizations normally led to increasingly general discussions.

At the Level I review, NASA rules required that project managers discuss problems with criticality 1 hardware that could cause loss of mission or life. In practice, however, the amount of detail varied. Project managers gave meticulous presentations of new problems or problems considered as major. For problems considered minor or routine, project managers often gave brief comments; they frequently proceeded like a pilot reading through items in a preflight checkoff sheet and listed such problems as "closed out," meaning verified safe.[34]

Rocket engineers first noted field joint O-ring problems in November 1981 on STS–2, the second Shuttle flight. When they took the recovered motor apart and examined the O-rings, they found one scorched primary O-ring. They interpreted this as a failure of the zinc chromate putty to protect the ring from combustion gases. This impingement erosion of a sealed O-ring, the deepest found on a primary ring in a field joint before 51–L, had resulted from a hole the diameter of a pencil in the putty. Marshall's project office reasoned that the void came during "lay-up" of the putty; high humidity and temperature during joint assembly had made the putty "tacky" and caused gaps. They expected that refrigerating the putty before assembly would eliminate the problem. Marshall notified NASA Headquarters of the flight anomaly but did not report the condition in the Level I flight readiness review before the next mission or in the Center's problem assessment system.[35]

In early 1982 Marshall and Thiokol concerns about the seals led to new studies of putty lay-up and the joints of the new lightweight steel motor case. Tests, especially high-pressure tests of the O-rings, convinced Marshall management that Ray and Miller had been right; joint rotation could prevent the secondary O-ring from sealing. Consequently Marshall reclassified the joint from criticality 1R to criticality 1, meaning no redundancy, and received approval for the change from the Shuttle Level II Office at Houston and Level I at Headquarters. The new critical items report explained that leakage beyond the primary O-ring was "a single point failure due to possibility of loss of sealing at the secondary O-ring because of joint rotation." Failure could result in "loss of mission, vehicle and crew due to metal erosion, burn-through, and probable case burst resulting in fire and deflagration."

Despite reclassification to criticality 1, the criticality report argued for the reliability of the design. Virtually all Marshall and Thiokol engineers believed that the joint was safe and redundant most of the time. The report explained that the joints had no leaks in eight static firings and five flights, the primary O-ring alone provided an effective seal, and the joint was similar to the safe Titan III which had one O-ring. In addition, some tests showed that the secondary O-ring would seal and so the joint often had redundancy. Accordingly some documents continued to designate the seals as Criticality 1R five weeks after the *Challenger* accident. This mislabeling, the presidential commission charged, confused decision-makers and made it "impossible" for them to make informed judgments.[36]

In fact, Marshall and Thiokol engineers were convinced that the joint still had redundant seals. Given the criticality 1 designation, such claims confused the presidential commission. In commission interviews and testimony, the Center's institutional managers, project managers, and chief engineers explained their understanding of joint dynamics during ignition. They expected that combustion gases would almost always seat the primary O-ring. In the rare event that the primary O-ring would not seal, gas would almost instantly flow to the secondary O-ring and seal it; later the joint rotation would widen the gap but the secondary O-ring would flatten enough to seal.[37]

According to Marshall, the joint lacked redundancy only under exceptional circumstances and these necessitated the documentary change. Dr. Judson Lovingood, deputy chief of the Shuttle Projects Office, said assembly errors could be "such that you get a bad stackup, you don't have proper squeeze, etc. on the O-ring so that when you get joint rotation, you will lift the metal surfaces off the [secondary] O-ring." Lawrence Mulloy, project manager for the solid rocket booster after November 1982, described another "worst case" scenario. If the primary O-ring sealed and then failed after joint rotation, he said, the joint could have "a worst case condition wherein the secondary seal would not be in a position to energize." Such a circumstance, Mulloy believed, was very unlikely. Hardy, Mulloy's predecessor as SRB project manager and later deputy directory of Science and Engineering at Marshall, agreed, saying "the occasion for blow-by on the secondary O-ring, in my opinion, would be extremely nil or maybe not even possible."[38] Similarly, propulsion engineers and managers at Thiokol considered the joint to have redundant seals, and the company's paperwork continued to categorize the joint as criticality 1R even after the *Challenger* accident in January 1986.[39]

Other NASA managers accepted the judgment of Marshall and Thiokol even after the criticality change. Glynn Lunney, former manager of the Level II Shuttle Program at JSC, believed "there was redundancy." L. Michael Weeks, the Level I associate administrator for Space Flight (Technical) at Headquarters said that "we felt at the time—all of the people in the program I think—felt that this Solid Rocket Booster was probably one of the least worrisome things we had in the program." Only a few engineers in the Center's Solid Rocket Motors Branch believed that joint rotation could jeopardize the secondary seal.[40]

Even so Marshall and Thiokol began working on a long-term fix on a new lightweight plastic SRB case. To increase the Shuttle's lifting capacity for military payloads, NASA decided to develop a filament-wound case with graphite fiber-epoxy matrix composite casewalls and steel joints. The joints would incorporate a "capture feature," a steel lip on the tang that would fit over the inner flange of the clevis and eliminate joint rotation (fig. 1). Hercules Incorporated proposed the design for the capture feature, which became one of the primary reasons why NASA in May 1982 chose the company to develop the filament wound case as a subcontractor to Morton-Thiokol. Marshall's Ray remembered "there was a lot of opposition" to the capture feature from engineers who "didn't understand" joint rotation, especially from those at Morton-Thiokol. NASA's choice of Hercules not only meant less business for Thiokol, but also indicted the firm's design of the steel case joints. According to Ray, Marshall's engineers debated whether to add the capture feature to the steel motors, but decided to wait for test results from the filament wound case. The first full-scale static firing of the new design occurred in October 1984.[41]

Meanwhile NASA confidently proceeded with Shuttle launches, and successes seemingly justified belief in the technology. After the fourth flight in June 1982, the Agency declared the Shuttle system "operational," meaning that the spacecraft and propulsion system had passed their flight tests. In seven static firings and nine launches, O-ring problems occurred on only four joints.[42]

In early 1983, however, NASA made changes in the solid rocket motor that would exacerbate O-ring problems. The Fuller O'Brien Company, which had manufactured the original asbestos-filled putty, ceased making the product and NASA substituted a putty made by the Randolph Products Company. The Randolph putty, used first on STS–8 in the summer of 1983, proved to be more difficult to pack in the joint during assembly and less able to provide thermal protection during launch.[43]

In addition the Center and its contractor believed that success depended on careful assembly and they sought to improve procedures, including the O-ring leak check. To ensure that the O-rings could hold a seal, the leak check compressed nitrogen in the cavity between the rings much like inflating a tire. The rocket engineers had initially used pressure of 50 pounds per square inch (psi). Since the Randolph putty alone could withstand this low pressure and hide a faulty O-ring, they raised the pressure to 100 psi on STS–9 in November 1983. Still the Randolph putty hampered the tests and produced leak check failures. After a failed check, assembly crews had to destack the solid rocket motors, and reassemble the joint. To minimize this expensive procedure and to verify the O-rings, the engineers decided to raise the leak check pressure to 200 psi for case-to-case joints on STS 41–B in January 1984 and to 200 psi for all joints on STS 51–D (the 16th flight) in April 1985.[44]

Unfortunately the high pressure necessary for a leak check also blew gaps in the putty. These voids, normally about one inch wide, would direct jets of combustion gas to sections of the primary O-ring and produce erosion. Thiokol and Marshall engineers found the gaps after disassembling recovered motors. Nonetheless the joint design created a conundrum; the engineers wanted high-pressure tests to verify O-ring assembly, but verification of the O-rings could create dangerous gaps in the putty, which could jeopardize the O-rings.[45]

Greater leak check pressure led to increased incidence of O-ring anomalies. Before the January 1984 increase in the pressure for case-to-case joints, post-flight inspection had found only one anomaly for nine flights. After the increase, over half the missions had blow-by or erosion on field case joints. The changes were even more dramatic for the nozzle joint. At 50 psi, 12 percent of the flights had anomalies; at 100 psi the rate rose to 56; and at 200 psi anomalies occurred on 88 percent of the flights. Unfortunately the engineers did not fully analyze this pattern, and no one performed an elementary statistical analysis correlating leak check pressure and joint anomalies.[46]

Worries over the O-rings mounted in February 1984 after 41–B, the 10th mission, when primary O-rings eroded in two case-to-case joints and one case-to-nozzle joint. The erosion on one case-to-case joint O-ring was 0.050 inch of the 0.250-inch diameter. Thiokol and Marshall recorded the incidents and conducted studies. On 29 February 1984, Keith Coates, an engineer in the Center's SRB Engineering Office, fretted that Thiokol was overconfident and so had very weak plans to resolve the problem. On 28 February, John Miller of the Center's SRM branch observed that environmental conditions during assembly and leak check were creating voids in the putty. Finding a solution was an "urgent matter," he said, because the putty was a thermal barrier which prevented "burning both O-rings and subsequent catastrophic failure."[47]

Although acknowledging problems on 41–B, Center engineers recommended that flights proceed. In Weekly Notes on 12 March, McCool wrote that in spite of the large number of occurrences, no hot gas had leaked past the damaged O-ring seal.[48] The Center Flight Readiness Review Board for the next mission, STS–13 (41–C), decided to fly based on the following rationale:

"Conservative analysis indicates that the maximum erosion possible on STS–13 is 0.090 inch. Laboratory testing shows the O-ring to maintain joint sealing capability at 3000 psi with a simulated erosion of 0.095 inch. The Board accepted a recommendation to fly STS–13 as is, accepting the possibility of some O-ring erosion."[49]

In other words, Marshall created a new performance criteria, diluting its original standard of no erosion to a new one of "acceptable erosion" with a numerical margin. In a presidential commission interview, SRB manager Mulloy explained "there was a very clear recognition that this was something that we couldn't be proud of. It was working but it wasn't performing to the standards that we set for ourselves." Ironically erosion created false confidence, since erosion meant that the joints were sealing even with weaknesses. It was "an O-ring erosion problem not a joint leak problem," Mulloy explained. "It was not a perceived problem of this design won't work," he said. "It was this design won't work unless we do these things" to improve putty lay-up and O-ring installation. Mulloy remembered that "nobody ever" recommended that flights cease until O-ring erosion could be eliminated.[50]

In March 1984 Marshall and Thiokol presented their rationale for accepting erosion to a Level I flight readiness review at NASA Headquarters. Hans Mark, NASA deputy administrator, and General James Abrahamson, associate administrator for Space Flight, attended the review and agreed with the rationale. In April, however, Mark issued an "Action Item" for May that required Marshall to perform "a formal review of the Solid Rocket Motor case-to-case and case-to-nozzle joint sealing procedures to ensure satisfactory consistent close-outs." This followed Abrahamson's January 1984 request for a Marshall plan to improve the design and manufacture of the solid rocket motors. The Center's project office passed these directives to its contractor.[51]

In May 1984 Thiokol issued a preliminary proposal for improvements and the next month Marshall assured NASA Headquarters that the Center would carefully monitor the situation. But for more than a year, until August 1985, NASA allowed the contractor to proceed without a plan to eliminate erosion. Headquarters dropped pressure after Mark and Abrahamson left the Agency later that spring. Other NASA Headquarters administrators followed the guidance of Marshall and Thiokol and accepted the anomalies. As Mulloy later explained, "we never perceived that we had to make a radical design change."[52] In other words, the engineering consensus that the joints were safe slowed the responses of Marshall and Thiokol.

The consensus came not only from the success of flights with O-ring anomalies, but also from successful ground tests conducted at Thiokol beginning in the spring of 1984. The company's engineers created a subscale model of an SRM joint, and fired a five-inch solid rocket in three-second burns into a chamber housing a section of O-ring. The model tested various putty and O-ring materials and configurations, and demonstrated O-ring erosion. Although the engineers debated how realistic the subscale tests were, they concluded that erosion primarily occurred because of voids in the putty. In fact, they found that in tests without putty, there was no O-ring damage. Consequently the tests led to continued efforts to improve assembly procedures and to study various putty materials and new puttyless configurations. More importantly, the engineers believed that subscale tests confirmed the safety of the existing design, showing that combustion gas would not melt an O-ring enough to produce a leak.[53]

By fall 1984, Marshall's reports in flight readiness reviews had become routine. A "quick look" bulletin on Mission 41–D dismissed heat distress as "typical of O-ring erosion seen on previous flights." When Lucas asked Mulloy about the problem during the Center flight readiness review for Mission 41–G, the project manager reviewed the problem, concluded that it was "an acceptable situation," and explained that a search for an alternative putty was underway. At the Shuttle Projects Office review for 41–G Mulloy said the "maximum erosion possible" was "less than erosion allowable."[54]

Concerns arose again in late January 1985 after Mission 51–C. The launch was the Shuttle's coldest and O-ring temperature was 53 degrees F. Two primary O-rings in case-to-case field joints had erosion, and primary rings in two field and both nozzle joints had soot blow-by. A form of erosion appeared that differed from previous impingement erosion of a sealed O-ring; blow-by erosion resulted from combustion gases burning an unsealed O-ring and flowing beyond it. Not surprisingly a secondary O-ring in a field joint experienced heat damage for the first time.[55]

The incidents startled Marshall, and Mulloy sent Thiokol a "certified urgent" request for an erosion briefing at the next flight readiness review.[56] At the 8 February 1985 SRB Board, Thiokol engineers discussed in detail the new types of O-ring damage and for the first time described the effects of temperature on the resiliency of the O-rings. For the joint to seal, the rubber rings had to be resilient because the primary O-ring had to travel rapidly across its groove and both rings had to flatten quickly to fill the opening gap. Low temperature, the contractor observed, made the putty "stiffer and less tacky" and made the O-ring smaller and harder. Thus cold could slow the sealing process and produce an "enhanced probability" of erosion.

Thiokol's engineers admitted that similar events could happen again, but concluded that the joints were "acceptable for flight." Cold was not a concern because Mission 51–C had experienced very rare weather, the "worst case temperature change in Florida history." Even so erosion remained "within the experience data base" and the margin of safety. Consequently the contractor did not request a new launch commit criterion based on temperature for the O-rings. Finally, the engineers decided that damage to the secondary O-ring, rather than revealing flaws in the primary, proved the joint had a redundant seal.[57]

Because Marshall's project office and Thiokol had faith in the seals, they downplayed bad news about 51–C as they went up the levels of flight readiness review. At the Center Shuttle Projects Office Board, the presentation barely mentioned temperature effects and listed as "closed" all motor problems requiring action before the next launch. The meeting's Final Report identified "no SRB failures and anomalies." At the Center board, one sentence covered the 51–C joint incidents. Mulloy's presentation to the Headquarters Level I board ignored temperature issues and briefly explained that thermal distress beyond the primary O-ring was an "acceptable risk because of limited exposure and redundancy." With this positive information, Level I administrators approved Marshall's recommendation to keep flying.[58]

Since the *Challenger* disaster occurred in cold weather, the 51–C reviews in retrospect seemed a lost opportunity to examine temperature effects on O-ring dynamics. But in the SRB review, Marshall had disputed the theory of Thiokol engineers that cold increased the possibility of O-ring failure. Mulloy, recalled Roger Boisjoly, an O-ring expert at Thiokol, had "objected to some of our original statements in our charts that temperature had an effect on the joints." Robert Crippen, an astronaut attending the Level I meeting, said Marshall presented the 51–C "as an anomaly" but failed to explain that the joint was a single point failure; "it wasn't considered that much of a big deal, and it wasn't like we had a major catastrophe awaiting in front of us." Mulloy later told the presidential commission, "I can't get a correlation between O-ring erosion, blow-by [around] an O-ring, and temperature" because anomalies had occurred at warm as well as cold temperatures.[59]

Neither the contractor nor the Center conducted a statistical analysis of existing data. In 1987 Bob Marshall, the manager of MSFC's Shuttle Project Office, regretted that 51–C had not moved motor experts to reinterpret the available data. After 51–C, he said, "we should have been thinking more. . . . The analyses of the tests we were doing just wasn't enough. We weren't finite enough in what we were doing." No one performed a statistical analysis correlating past O-ring performance with either temperature or leak check pressure.[60] Lacking such analysis, both Thiokol and Marshall had been correct about 51–C; the contractor engineers had rightly surmised that low temperature increased the probability of erosion, and Center managers had rightly questioned Thiokol's demonstration of the correlation.

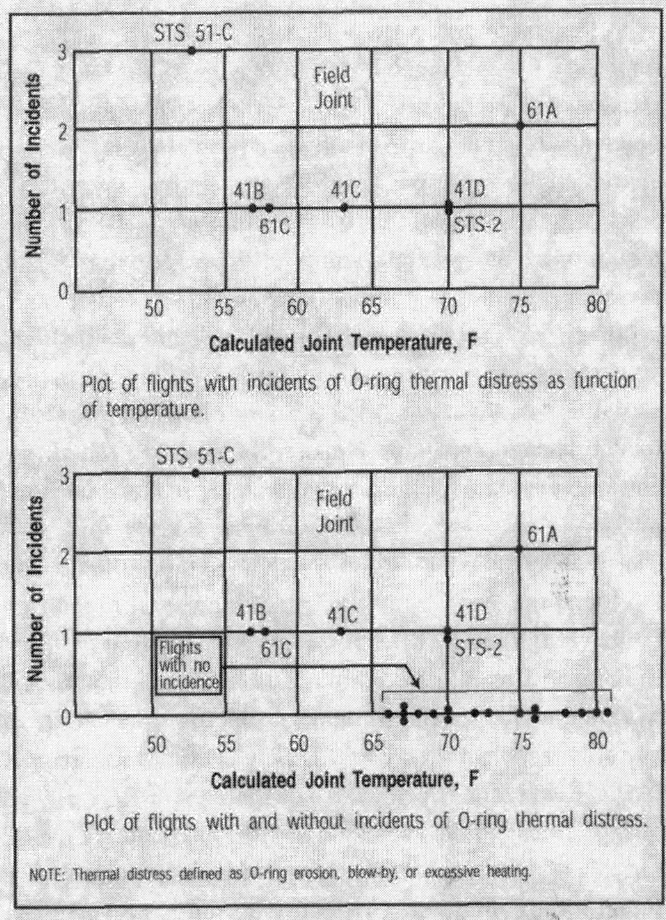

Plots of incidence of O-ring distress as function of temperature.

At the 51–D review Marshall Director Lucas confidently observed that "we are maturing. There are fewer action items than last time, and we are getting better hardware." Even so the four flights after 51–C had O-ring problems with the most extreme occurring on the April 51–B mission. When Thiokol disassembled the segments in late June, the engineers found that the left nozzle joint's primary O-ring had not sealed and had eroded severely and its secondary O-ring had eroded as well. The 51–B findings were doubly troubling because motor engineers had always expected the primary to seal and had never experienced erosion of a secondary.[61]

With their predictions disproven by 51–B, Marshall engineers in July 1985 *imposed a formal launch constraint on the motor nozzle joint for all subsequent* missions including 51–L. According to NASA requirements, a formal constraint prevented flight until a technological problem was fixed or verified safe. Flights continued, however, because SRB project manager Mulloy filed formal waivers lifting the constraint for each of the six flights through 51–L. NASA required review and approval of each waiver by organizations responsible for project management, engineering, and quality. After the *Challenger* accident, however, several NASA and Thiokol officials claimed ignorance of the formal constraints and waivers. The claims by Thiokol managers are difficult to explain given that the company's records listed Marshall's document number for the launch constraint. Apparently Marshall failed to report the constraint and waivers to Level II Shuttle managers in Houston.[62]

The presidential commission condemned this failure to communicate bad news but found that NASA lacked clear guidelines for reporting problems. In 1983, the Level II Shuttle Office had changed reporting requirements from Level III in order to streamline communications for the Shuttle's "operational" phase. Marshall no longer had to report problems on hardware elements for which it had sole responsibility. Level II only required reports which dealt with interface hardware for which Marshall shared responsibility with Houston. Consequently Marshall only sent one copy of its monthly Open Problems Report to a Level II flight control engineer and a statistical summary to Rockwell, the Shuttle integration contractor. Criticality 1 items, however, were supposed to be reported to Level II.[63]

Moreover, a NASA 51–L investigation team determined that after several Shuttle flights, the Level I flight readiness reviews adopted a built-in bias that limited the flow of information. Since the Shuttle had proven flight worthy and was designated "operational," and the experts in lower levels had already certified flight readiness, the Level I review became increasingly ritualistic. Reviews were often short and key officials failed to attend.[64]

Nonetheless, the commission severely criticized Marshall's reports and response to the clear evidence of technological flaws from 51–B. At the Level I review on 2 July, Marshall did not mention the launch constraint, accepted erosion to the secondary, and presented 51–B O-ring problems as "closed," meaning acceptable for flight. The Center, the commission charged, had lowered standards, neglected to report problems, and failed to implement actions necessary to ensure safety.[65]

The Center's motor officials denied these charges and defended their judgment. Mulloy admitted his failure to inform Level II managers directly of the formal launch constraint, but pointed out that Marshall and Thiokol continued to discuss the joint problems in flight readiness reviews. Motor officials also made a thorough presentation to Headquarters in August.[66] Mulloy explained that he had lifted the launch constraint because motor engineers had again reviewed the problems and found the situation acceptable.[67]

Particularly encouraging were results from a computer model that Thiokol created to evaluate the risks of O-ring erosion. The model, called ORING, used data from flights, static firings, and subscale tests. It predicted that chances were "improbable" that hot gases would burn through a sealed primary O-ring or that hot gases blowing past a primary would melt through the secondary O-ring. The model had limitations as an analysis of the potential danger; it defined the hazard based on evidence from previous missions and tests, none of which had resulted in catastrophic failure, and hence drew the obvious conclusion that there was no proof of a hazard. Nevertheless Thiokol's ORING, first presented to Marshall in April 1985 and updated to include the nozzle joints in July, helped bolster confidence among NASA and contractor officials.[68]

Moreover, engineers working on the solid rocket motor concluded that the 51–B problems had resulted from a faulty leak check procedure. They believed that leak check pressure on 51–B had been too low; putty in the joint had withstood 100 psi and thereby had masked a faulty primary O-ring. The engineers decided that increasing the pressure to 200 psi would prevent recurrence of the problem. The Thiokol report also took solace from how the primary O-ring erosion had been "within historical levels" and the damage on the secondary had been "within the demonstrated sealing capability of eroded O-rings." The company concluded that "this anomaly is not considered a launch constraint." As Mulloy told the commission in 1986, the motor experts had reviewed and responded to the situation and "it was not just a matter of nothing was done."[69]

Two years later, however, during a retrospective interview, Mulloy questioned the engineering evaluation of 51–B. "I truly believe that if there was a fatal error made . . . among a lot of people in engineering judgment, it was accepting that kind of condition where you've completely destroyed a primary O-ring and accepted damage to the second and concluded that that was an acceptable thing to fly with." He added that "there's something drastically wrong when something that you think isn't supposed to get any damage at all sustains that kind of damage, and you conclude it's okay."[70]

After the *Challenger* accident, space flight veterans also doubted NASA's decision to keep flying. In 1991, Chris Kraft, former director of JSC, said that "The creed of manned spaceflight is you never fly with a known problem. Never. Get that word never. So . . . when the main ring is burned and the back-up ring is scorched in a joint and you don't stop the goddamn thing right there and fix it, regardless of whether it be a band-aid fix or any other kind of fix, you have made a cardinal sin. You many times fly with unknown unknowns, but you do not fly with known unknowns."[71]

Concerns about erosion on 51–B led the Center to seek a permanent solution. In July Marshall asked Thiokol to go beyond improving assembly procedures and begin studying new hardware designs. The contractor established an O-ring Task Force whose goal, according to Mulloy's Weekly Notes, was finding "a longer term, a design solution to the O-ring erosion" and to joint rotation. By the end of August the task force had proposed 63 possible joint modifications, including 43 for the field joints. The proposals included a capture feature lip similar to the filament wound case. Indeed in July Marshall ordered from the Ladish Company 72 steel case segments with the capture feature.[72]

Meanwhile Thiokol continued to verify the safety of the existing design. In early June 1985 the contractor performed bench tests to evaluate the effects of temperature and joint rotation on the performance of the secondary O-rings. Thiokol reported to Marshall on 9 August that "at 100 degrees F the O-ring maintained contact [with the metal sealing surface]. At 75 degrees F the O-ring lost contact for 2.4 seconds. At 50 degrees F the O-ring did not re-establish contact in ten minutes at which time the test was terminated." The tests also indicated that joint rotation made the secondary O-ring more likely to fail late in the ignition phase. The company report reassured Marshall, however, that it had "no reason to suspect that the primary seal would ever fail."[73]

NASA Headquarters knew about Marshall's O-ring worries. The Propulsion Division at Headquarters had held monthly reviews on the motor joints since

51–C in March 1984. In July 1985 a Headquarters' engineer visited Huntsville for a briefing on 51–B and reported that Marshall was concerned about the putty and joint rotation. The Center was working on solutions, but he recommended a briefing to Level I. Moreover Headquarters budget officials, using information from the Propulsion Division, had discussed how motor problems and solutions could impact the Agency's FY 1987 budget.[74]

In a briefing at NASA Headquarters on 19 August 1985 Marshall and Thiokol finally responded to the April 1984 action item and presented their engineering evaluation and redesign plan. The experts observed that only 5 of 111 primary O-rings in field joints and 12 of 47 primary O-rings in nozzle joints eroded. O-ring erosion resulted from blow-holes in the putty, increased frequency of voids, and heat damage resulted from defective putty, higher leak check pressure, and greater engine pressure. Nonetheless, Thiokol argued that data from static firings, Shuttle flights, subscale tests, and the ORING computer model verified the safety of the design. Erosion could be no worse than 51–B; even "worst-on-worst case predicted erosion" was "within [the] demonstrated sealing capacity of [an] eroded O-ring."

The review rated the field joint as the "highest concern" and described the criticality change from 1R to 1. Erosion could damage the primary seal and joint rotation could cause the secondary O-ring to fail. The experts believed that "the primary O-ring in the field joint should not erode through but if it leaks due to erosion or lack of sealing the secondary seal may not seal the motor." They warned that "the lack of a good secondary seal in the field joint is most critical and ways to reduce joint rotation should be incorporated as soon as possible to reduce criticality." Nozzle joints were of less concern because of the greater rigidity of the case and because 51–B proved that its secondary O-rings would seal even if eroded.

The motor engineers and managers also presented plans for improving the joints. Marshall and Thiokol planned to introduce short-term changes for the field joint; they would qualify an alternate putty source, use thicker shims to ensure O-ring compression, and replace the 0.280-inch-thick O-rings with thicker 0.292-inch rings that would provide an extra safety margin, add insulation strips in the joint to prevent hot gas circulation, and insert a third O-ring. NASA would introduce long-term changes in 27 months, including the capture feature already proven on the filament wound case; this would reduce joint rotation and ensure redundancy. The review concluded that leak checks and careful assembly made it "safe to continue flying [the] existing design." Nevertheless NASA's reconfiguration and redesign efforts needed "to continue at an accelerated pace to eliminate SRM seal erosion."[75]

Marshall's presentations to Agency officials during July and August would become controversial after the *Challenger* accident. Several top Agency officials said Marshall had not brought problems to the surface and Center managers said they had. Both were right. Marshall had failed to discuss O-ring resiliency at the August briefing, and evidently told Thiokol to delete from the conclusion a sentence that said "data obtained on resiliency of the O-rings indicate that lower temperatures aggravate this [sealing] problem." Center managers continued to deny that temperature was a factor because erosion had occurred at cool and warm temperatures. Moreover Mulloy pointed out that in the reviews "the effect of temperature never came across [from Thiokol to Marshall] as the overwhelming and most important concern on that joint." Temperature excepted, the August presentation was thorough and the presidential commission concluded that "the O-ring erosion history presented to Level I at NASA Headquarters in August 1985 was sufficiently detailed to require corrective action prior to the next flight."[76]

As the work of Thiokol's O-ring task force proceeded in the summer and fall, members became frustrated by a lack of support from corporate management. Thiokol O-ring expert Boisjoly explained to engineering management in July that joint rotation could yield a "catastrophe of the highest order—loss of human life." He protested that the problem required "immediate action" but that support was "essentially nonexistent at this time." The task force had only 5 full-time engineers out of the 2,500 employed at Thiokol. On 1 October Robert Ebeling, manager of the group, signaled "HELP! The seal task force is constantly being delayed by every possible means" and "this is a red flag." He thought "MSFC is correct in stating that we do not know how to run a development program." On the same day another project engineer complained that the group was "hog-tied by paperwork every time we try to accomplish anything" and requested "authority to bypass some of the paperwork jungle." A few days later Boisjoly wrote that Morton-Thiokol's "business as usual attitude" prevented progress and that "even NASA perceives that the team is being blocked in its engineering efforts." He believed that "the basic problem boils down to the fact that ALL MTI [Morton-Thiokol Inc.] problems have # 1 priority and that upper management apparently feels that the SRM program is ours for sure and the customer be damned."[77] These bureaucratic obstacles slowed purchase of equipment and manufacture of test hardware, and thus delayed tests.[78] As an example of inertia at Thiokol, Boisjoly noted in his log on 13 January 1986 that O-ring resiliency tests requested in September 1985 were now scheduled for January 1986.[79]

Throughout the fall, Marshall motor engineers maintained close contact with their Thiokol counterparts. They had teleconferences every week and face-to-face reviews every few weeks, and the Center regularly sent experts to Utah to monitor the contractor's work. Although these contacts mainly discussed technical problems, Marshall technical personnel were aware of the organizational and financial obstacles faced by the O-ring task force and of the delays in procurement and testing. Officials from Marshall's Solid Rocket Motor Branch and SRB Chief Engineer's Office offered to help the task force get more authority and resources.[80]

In late August, after years of argument between Marshall and Thiokol about whether joint performance was within design specifications, the Center convinced the company to accept a "referee test"; Marshall hoped that an independent expert would settle the controversy and pave the way for a redesign. In early September, Kingsbury wrote to Mulloy that the task force efforts "do not appear to carry the priority that I attach to this situation. I consider the O-ring problem on the SRM to require priority attention of both Morton-Thiokol/Wasatch and MSFC." The Center's project office tried to speed problem-solving by allowing Thiokol to make the first public description of the joint problems to the Society of Automotive Engineers on 7 October.[81] Even so, Marshall's efforts did little to accelerate the progress of Thiokol's O-ring task force.

A primary reason for the slow progress was Thiokol's incentive-award fee contract. After 51–L, congressional investigators found that the contract offered the corporation no incentives to spend money to fix problems believed unlikely to cause mission failure.[82] Based on this information, a sociologist concluded that, "The incentive fee, rewarding cost savings and timely delivery, could total as much as 14 percent of the value of the contract; the award fee, rewarding the contractor's safety record, could total a maximum of 1 percent. No provisions existed for performance penalties or flight anomaly penalties. Absent a major mission failure, which entailed a large penalty after the fact, the fee system reinforced speed and economy rather than caution."

Not only did Thiokol have disincentives to fix problems that would cause flight delays, but Marshall had little means to sanction the firm's pace. In fact NASA imposed no penalties on Thiokol for the anomalies and at the time of 51–L the Agency was contemplating awarding the company a near maximum incentive fee of 75 million dollars.[83]

After the accident, Thiokol task force members explained how the contract, corporate policy, and government regulations created obstacles. Because preparations for upcoming missions had higher priority than redesign activity, work on flight hardware came before work on test hardware. The company paid the costs of the redesign activities without additional money from Marshall. To get the extra money necessary to speed progress for the O-ring task force, Thiokol would have had to submit an engineering change request and thus acknowledge the failure of its design. Consequently, the task force had responsibility without authority or resources.[84]

Looking back on the fall of 1985, Marshall motor officials maintained that they had no information that indicated urgency. Jim Smith believed Thiokol was "working the problem in a timely manner." He and other Marshall officials claimed that no Thiokol engineer had communicated serious concerns about safety or bureaucratic obstacles. No Marshall official saw the memos drafted by O-ring task force members that expressed alarms about the delays to Thiokol management. Smith said that if the task force had informed him of the need for flight delays or for extra resources, he would have presented and defended their position to Marshall management.

Lawrence Wear, manager of the SRM, said the consensus was that the problem was "troublesome" and "contrary to design." But at the time "there was no discussion and no revelation on anybody's part that what we're doing here is flying something that is in an absolutely unsafe condition and you ought to stand down until you get it fixed." Leslie F. "Frank" Adams, deputy SRB manager, said the communications from Thiokol were "not in the context of a safety of flight kind of concern." Stanley Reinartz, manager of the Shuttle Projects Office, believed the contractor's position after August 1985 was that the motor was "completely safe and reliable for launching while these concerns about O-rings were being worked on in a parallel fashion."[85]

Marshall's confidence in the joints was evident in many ways. In comparison to the Saturn POGO problem or Space Shuttle main engine development, the Center devoted minimal attention and resources to the SRM joints. Jerry Peoples observed in a presidential commission interview that the Marshall Center task force was organized at a "low level." When briefed on the O-rings, he said, Marshall project and institutional managers would "politely listen to our presentation, but seemed to give no response or heed no warning as to what we were saying and seemed to . . . be in certain times bored with what we were saying."[86]

Moreover Marshall and Thiokol continued Shuttle flights while delaying by several months the static firing of Qualification Motor–5 which would test the filament wound case and the capture feature. In the Weekly Notes, Mulloy said delay was needed to prepare for modifications that could "alleviate the joint O-ring erosion experienced." Eventually Marshall scheduled the firing for 13 February 1986. The Center informed Level I officials at Headquarters of the progress in a November briefing.[87]

Marshall and Thiokol's confidence in the joint also showed in flight readiness reviews in the fall and winter. Thiokol continued to verify that the case joints were not hazardous. In the Level I review in late September on mission 51–I Marshall dismissed two cases of O-ring nozzle erosion as being "within experience base." Mission 51–J had no damage and the Shuttle Project review on 15 October said its O-ring performance was "nominal." Mission 61–A had nozzle erosion and blow-by past the primary O-ring on two field joints which Mulloy described to Level I on 18 November as "within previously accepted experience." Flight 61–B had primary O-ring erosion of both nozzle joints and blow-by past one, but he informed Headquarters on 11 December there had been "No 61–B Flight Anomalies." Similarly mission 61–C had nozzle joint erosion and blow-by and field joint erosion; nevertheless at the Level I review on 15 January 1986, the meeting which certified *Challenger* 51–L for flight, Mulloy's presentation listed "No 61–C Flight Anomalies" and "No Major Problems or Issues."[88]

The presidential commission concluded that by late 1985 Marshall's flight readiness reviews only discussed problems that were "outside the database" and dismissed O-ring problems as routine and hence insignificant.[89] Mulloy later admitted that "since the risk of O-ring erosion was accepted and indeed expected, it was no longer considered an anomaly to be resolved before the next flight."[90]

Acceptance of the anomalies helped lead to formal "closure" of the O-ring problems in Marshall's Problem Assessment System. In this system, engineers with an open problem would write monthly reports and conduct flight-by-flight reviews until they implemented a correction. Then they would report their solution to a review board and the board would "close out" the problem and no longer require reports or reviews in the system.

Although Morton-Thiokol was working on the problems and Marshall still had a launch constraint on the nozzle joint, Kingsbury, Marshall director of Science and Engineering, requested that the firm reduce its open items, including O-ring items. Hence on 12 December 1985 Thiokol's project manager requested that monthly problem reports on the O-rings be discontinued because a task force was working on a correction and regular reports were proceeding through group's reports and flight readiness reviews. Consequently on 23 January 1986 a Marshall problem report stated that the problem was "closed" because Thiokol had filed a plan to improve the seals.[91] A close-out of an open problem perplexed the presidential commission. To commissioner Robert W. Rummel the closure signified that "somebody doesn't want to be bothered with flight-by-flight reviews, but you're going to continue to work on it after it's closed out." MSFC's SRB project managers said the closure was "in error" and that they had not approved it.[92]

At the same time as the closure, Morton-Thiokol's contract was coming up for renewal, and NASA asked aerospace contractors for preliminary proposals for a second source for the solid rocket motors. This was not done out of specific dissatisfaction with Morton-Thiokol's performance, and indeed Marshall believed the firm was improving. Mulloy noted in October 1985 that the average number of problems per flight set was decreasing. Instead the initiative resulted from lobbying by Thiokol's competitors for a piece of NASA's solid rocket market and from desires by Congress to ensure a steady supply of motors for the Shuttle's military payloads.[93]

NASA's bidding rules for the second source threatened Morton-Thiokol. The rules, announced on 26 December 1985, assumed the motor joints were operational and so the government would not give "qualification funds" for rocket redesign to the competitors. Consequently each firm would have to invest as much as $100 million in production facilities, test equipment, and prototypes without any guarantee of a contract. However, since NASA required no redesign, the Agency could encourage competition by publishing Thiokol's blueprints and asking competitors for lower bids. NASA was also stimulating competition by proposing a "split buy" rather than a "shoot out." Thus even if Morton-Thiokol would retain considerable motor business, the firm would face competition. NASA's initiative, which the presidential commission overlooked, threatened Morton-Thiokol's monopoly and so corporate managers had incentive to please their customer during negotiations in January 1986.[94]

Meanwhile, Thiokol's task force continued work. After the accident, Robert Ebeling, manager of the SRM task force, told the commission that he had discussed with team members the possibility that "we shouldn't ship any more motors until we got it fixed." Regardless of these discussions, formal presentations by the task force to Thiokol management and Marshall officials in mid-January described its activities and long-term schedules without any expression that the existing joint was too hazardous to fly.[95]

The central theme in the history of O-ring erosion before 51–L was that officials at Marshall and Morton-Thiokol had confidence that the joints were not hazardous. Based on static firings, flight data, and laboratory tests, they concluded that the primary O-rings provided effective seals, that thermal damage was limited and acceptable, and that the secondary O-ring normally offered redundancy. "Neither Thiokol nor the Marshall Level III project managers," concluded the presidential commission, "believed that the O-ring blow-by and erosion risk was critical" and both thought that "there was ample margin to fly with O-ring erosion."[96]

Confidence in the joint affected communications. Because their overall evaluation of the joints was positive, officials sometimes failed to communicate contradictory information. Marshall, the presidential commission observed, minimized problems in flight readiness reviews and failed to report the launch constraint and waivers, the controversy about temperature and O-ring resiliency, and the O-ring anomalies of later flights.[97] This silence, however, evolved from confidence that the joint was not hazardous rather than from some conspiracy to cover up problems.

Unfortunately the certitude rested on weak engineering analysis. Presidential commission member Richard P. Feynman, a physicist and Nobel prize winner, drove this point home after the fact. He observed that although the Center and its contractor used tests, analyses, and computer models, the standards of project officials showed "gradually decreasing strictness." They assumed, Feynmen said, that risk was decreasing after several successful missions and so they lowered their standards. The standard became the success of the previous flight rather than the danger of erosion and blow-by. Thus a successful flight with erosion was proof of the reliability of the O-rings and justification for another launch, rather than a warning of a potential catastrophe and a sign to stop and fix the problem.[98]

Once decision-makers at Marshall and Thiokol accepted the problems, they failed to facilitate deeper analysis. Project engineers failed their managers, neglecting to perform even elementary statistical analysis of the relationships between O-ring anomalies and such factors as temperature and leak check pressure.[99] Had they done so, they may have understood the risk better than they did, and that flying the Shuttle was, in Feynman's words, like playing Russian roulette.[100]

The Teleconference and Launch

On the evening of 27 January 1986 before the scheduled launch of 51–L the next morning, Center and contractor project managers and engineers held an impromptu flight readiness review over the telephone. Thiokol engineers argued that cold temperatures, projected to be the coldest recorded in Florida, would aggravate the O-ring problem. Neither Thiokol nor Marshall managers accepted their arguments that the cold was hazardous, and the managers decided to launch 51–L.

Earlier in the day, high crosswinds at the Kennedy Space Center (KSC) forced NASA to postpone Flight 51–L for the fourth time. Launch managers, tired from lots of work and little sleep, rescheduled launch for the next morning. Even so the weather forecast predicted an overnight low of 18 degrees F, and early in the afternoon Marshall asked Morton-Thiokol to consider the possible effects of cold. At the plant in Wasatch, Utah, Thiokol's SRM engineers decided that the temperatures were far below previous experience and could make the O-rings too stiff and hard to seal the joints. While the engineers prepared a presentation, their managers arranged a teleconference with Marshall personnel. The teleconference, which connected Wasatch, Huntsville, and Cape Kennedy, began at 5:45 P.M. Eastern time. Because of hissing phone lines and missing officials, however, participants decided to postpone. In the interim, Marshall's Stanley Reinartz, the Shuttle Projects manager, informed Center Director Lucas of the impending discussions.[101]

After Thiokol had faxed hand-written charts, the teleconference began at 8:45 P.M. Eastern time. Thiokol participants included motor engineers and project managers and the vice presidents for engineering and space motor programs. Also attending in Utah were the senior vice president for Wasatch operations and the vice president and general manager for space programs; no Marshall official in Alabama or Florida knew of their presence or their participation in the engineering discussions. The senior participant in Huntsville was George Hardy, the deputy director for Science and Engineering, who had support from several project officials and laboratory engineers. Reinartz and SRB project manager Mulloy participated from KSC. As usual for a Level III review, no Houston or Headquarters officials were present.[102]

The Thiokol engineers wanted to show that cold temperature could stop the O-rings from sealing. They observed that cold temperature would thicken the grease surrounding rings, and stiffen and harden the O-rings; these factors would slow the movement of the primary O-ring across its groove and reduce the probability of a reliable seal. Sealing with a cold O-ring, the contractor reasoned, "would be likened to trying to shove a brick into a crack versus a sponge." If hot gases blew past the primary O-ring after the joint had opened, the probability of the secondary O-ring sealing would decrease.[103]

The engineers also presented a history of erosion in field case joints. They pointed out that the previous coldest launch, 51–C in January 1985, had occurred at 53 degrees, and that the predicted launch-time temperature of 29 degrees was far outside Shuttle experience. Moreover 51–C had eroded O-rings and its blow-by deposits of charred grease and O-ring rubber had been jet black, which was an ominous sign that the primary O-ring had nearly failed. Some of Thiokol's evidence, however, appeared contradictory. It showed that four static motors fired between 47 degrees and 52 degrees had no blow-by and that the October 1985 flight of 61–A had blow-by at 75 degrees. The Thiokol engineers dismissed this contrary evidence, rationalizing that the vertical static-fired motors had had a putty packing method unavailable for the horizontal 51–L motors, and that the blow-by deposits of the 61–A flight were less dark and more innocuous than 51–C. In conclusion, Thiokol argued that NASA should stay within the experience of 51–C. Air temperature should be at least 53 degrees at launch time (see page 359, plots of incidence of O-ring distress as a function of temperature).[104]

Marshall officials immediately questioned Thiokol's ideas. Hardy said that he was "appalled" by the contractor's reasoning. Reinartz observed that the recommendation violated the Shuttle requirement that the motor operate between 40 and 90 degrees. Mulloy noted that NASA had no launch commit criteria for the joint's temperature and that the eve of a launch was a bad time to invent a new one. He asked, "My God, Thiokol, when do you want me to launch, next April?"[105]

Marshall's institutional and project managers doubted that cold increased risk over previous flights. Test data showed, they believed, that the O-rings would have to be colder than the expected temperature before resiliency and reliability declined significantly. Moreover, motor pressure was so great and increased so rapidly that combustion would almost instantly force even a cold primary O-ring into place. Even if the primary was too cold to seal, gas would blow past quickly, before the joint opened, and seal the secondary. "We were counting," Mulloy said later, "on the secondary O-ring to be the sealing O-ring under the worst case conditions."[106]

Most importantly, however, the Center's managers saw no causal connection between temperature and O-ring damage and believed that 61–A proved their case. During the teleconference Mulloy, the project manager, and Hardy, the senior Marshall engineer, criticized Thiokol's proofs. Hardy told the presidential commission that the temperature data were not conclusive because blow-by had occurred at 75 degrees. He added that "I do not believe that temperature in and of itself induces the blow-by, and I think that is kind of obvious because we have occasions for blow-by at all temperatures."[107] Thiokol admitted that that they lacked a statistical analysis to verify the relationship. O-ring expert Boisjoly remembered, "I was asked to quantify my concerns, and I said I couldn't,

I couldn't quantify it, I had no data to quantify it, but I did say I knew that it was away from goodness in the current data base."[108]

Despite Thiokol's failure to demonstrate the causal connection, it existed and was easily quantifiable. Thiokol's charts did not juxtapose temperature and O-ring damage in elementary two variable plots. If done, this would have shown that in the 24 flights before 51–L, 20 missions had temperatures of 66 degrees or above and of these only 3 had problems in field joint O-rings. In contrast, all four flights with temperatures below 63 degrees had problems in field joint O-rings. Moreover the predicted temperature at launch time was 29 degrees, 3.6 standard deviations below the average launch temperature of 68.4 degrees. With this information, the engineers would have known that the launch would be far outside Shuttle experience and very risky.[109]

Given the history of success and the confidence in the joint, Thiokol's engineers needed hard, quantitative information and not to believe what they had been believing so long in order to persuade top corporate and NASA officials to postpone. Boisjoly later observed that Marshall engineers, following the lead of Center Director Lucas, would only make decisions based on a "complete, fully documented, verifiable set of data."[110] Unfortunately Thiokol's data were inconclusive. After the accident, NASA investigators concluded that "the developed engineering knowledge base, and the interpretation of available engineering data, were inadequate to support the STS 51–L launch decision process." The presidential commission believed that "a careful analysis of the flight history of O-ring performance would have revealed the correlation of O-ring damage and low temperature. Neither NASA nor Thiokol carried out such an analysis; consequently they were unprepared to properly evaluate the risks of launching the 51–L mission in conditions more extreme than they had encountered before."[111]

During the teleconference Thiokol and Marshall were distracted by comparison of dissimilar data. They equally weighted static tests and Shuttle flights although each had different forms of putty packing. They pooled erosion data for the two case-to-case and case-to-nozzle joints, thereby confusing different causal systems, since case joints were sensitive to temperature, but not to leak check pressure, and nozzle joints were sensitive to leak check pressure but not to temperature. Without distinguishing between fundamental sources of O-ring damage, Thiokol's rationale seemed insubstantial. Ultimately the teleconference focused on only two data points, 51–C and 61–A, and the contradictory evidence caused debate to dwindle after more than an hour.[112]

As participants sought a conclusion, Allan McDonald, Thiokol's SRM project director, observed from the Cape that the leak check shoved the primary O-ring on the wrong side of its groove and put the secondary in a position to seal. Although he later said he had intended to show dangers for the primary, most participants, including Thiokol management, understood that he believed the secondary would provide redundancy. Hardy then remarked that the data did not prove the O-rings were hazardous, but said he would not overrule his contractor's recommendation to hold the launch. After Reinartz requested a response, Joe Kilminster, Thiokol's vice president for booster programs, took Utah off the line for a five-minute caucus to reassess.[113]

The Utah caucus lasted for 30 minutes and initially two engineers from the O-ring task force repeated their warnings. When they realized that Thiokol's upper management was not listening, the engineers stopped talking and the others stayed silent. Kilminster and Robert K. Lund, vice president for engineering, hesitated to overrule the engineers. Jerald E. Mason, vice president for Wasatch operations then told Lund, "Take off your engineering hat and put on your management hat." Mason later explained that "we didn't have enough data to quantify the effect of the cold" and so "it became a matter of judgment rather than a matter of data." Lund agreed that no correlation existed between temperature and risk and the four Thiokol vice presidents in Utah recognized that they could not prove that 51–L was more dangerous than previous launches.[114]

When the teleconference resumed at 11:00 P.M., Kilminster said that the data were inconclusive and therefore the company recommended that the launch proceed. The rationale was the same as previous launches: despite the problems of joint rotation and cold temperature, the primary O-ring could withstand three times the erosion of 51–C and the secondary O-ring provided redundancy. Level III manager Reinartz asked for dissenting comments, and, hearing none, ended the teleconference.[115]

At the time, two Marshall participants believed the teleconference was unusual. In Huntsville, William Riehl, a materials engineer, wrote in his notes that "Mulloy is now NASA-wide deadman for SRB/SRM" and "did you ever expect to see MSFC want to fly when MTI-Wasatch didn't?" At the Cape, Cecil Houston,

Marshall's resident manager, told Jack Buchanan, his Thiokol counterpart, that "he was surprised because MSFC was usually more conservative than the contractor and in this instance, the roles were reversed."[116]

In testimony to the presidential commission, Thiokol officials complained that Marshall had pressured them to launch and had reversed the normal roles of contractor and government during a flight readiness review. McDonald said that normally "the contractor always had to get up and prove that his hardware was ready to fly. In this case, we had to prove it wasn't, and that is a big difference. I felt that was pressure." Boisjoly affirmed that "this was a meeting where the determination was to launch, and it was up to us to prove beyond a shadow of a doubt that it was not safe to do so. This is in total reverse to what the position is in a preflight conversation." Lund said he and the other Thiokol managers changed their recommendation because "we had to prove to them that we weren't ready, and so we got ourselves in the thought process that we were trying to find some way to prove to them it wouldn't work, and we were unable to do that. We couldn't prove absolutely that that motor wouldn't work."[117]

The presidential investigators largely accepted Thiokol's explanation. Commissioner David C. Acheson, an attorney, argued the company should have backed its engineers and ordered NASA to launch only under specific conditions. But the commission's final report stated that "Thiokol management reversed its position and recommended the launch of 51–L, at the urging of Marshall and contrary to the views of its engineers in order to accommodate a major customer."[118]

Throughout the hearings, the Marshall managers tried to refute these charges. Reinartz thought Marshall had conducted the teleconference "in a thorough and professional manner and in the NASA tradition of full and open participation." The discussions, he said, were "deliberate and intense" but "not highly heated or emotional." Marshall managers denied that their questions and challenges constituted "pressure." They needed hard data to overturn a rationale that had been in place since the second Shuttle launch and to request a delay from Level I and Level II. After discussion, both the contractor and the Center concluded, Mulloy said, "there was no significant difference in risk from previous launches. We'd be taking essentially the same risk on January 28 that we have been ever since we first saw O-ring erosion." Marshall's top managers and engineers challenged Thiokol's arguments, but never asked the firm to retract its original recommendation, and Hardy had stated that he would not launch without the contractor's concurrence.[119]

Mulloy believed that Marshall had maintained the traditional government-contractor roles in flight reviews and had never asked the firm to prove the O-ring would fail. But even if the Center had done so, the goal remained flight safety. Both NASA and Thiokol had wanted safety; the firm had an incentive fee contract that rewarded them for success and penalized them for a launch failure. If anyone had abandoned NASA traditions, the Marshall officials argued, it had been Thiokol. The firm had not informed the Center that Thiokol's top managers had been present in Utah or that these managers had recommended the cold launch over the objections of the motor engineers. Moreover Thiokol's dissenters remained silent when Reinartz asked for comments.[120] Center Director Lucas told the commission "the responsibility rests with Thiokol, but I'm not trying to shake the responsibility of the Marshall Space Flight Center. Thiokol reports to us. But I do rely upon the contractor, the prime contractor, to recommend launch" and "I don't recall that we have ever . . . knowingly overridden a go/no-go decision by a contractor."[121]

At least two Marshall engineers also opposed a cold weather launch. Before the teleconference, Keith Coates, a former chief engineer for the solid rocket motor, had expressed concerns about the cold to project officials. Ben Powers, a motor engineer, informed his boss, John McCarty, deputy director of the propulsion lab, and Jim Smith, the SRB chief engineer, that "I support the contractor 100 percent on this thing. I don't think we should launch. It's too cold." But no objections went over the wire.[122]

The presidential commission decided the flight review had "a serious flaw" because it stifled the expression of "most of the Thiokol engineers and at least some of the Marshall engineers."[123] Center engineers who participated offered mixed evidence. Frank Adams believed that the same sort of "questioning that went on" during the teleconference was "the same as any I have sat in thousands of times over the years that I've been here." Lawrence Wear said "it is an open world at Marshall" and "in our system, you are free to say whatever you wish, to recommend whatever you wish. But you've got to be able to stand the heat, so to speak, based on what you have said."[124]

Some engineers said they had been reluctant to bypass the chain of command and inform Hardy of their concerns. Although Hardy had consulted with his senior advisors and said at one point, "for God's sake, don't let me make a dumb mistake," he did not poll all his engineers and was unaware of divergent views. Coates did not "lay down on the tracks," he later explained, because he lacked formal responsibility. McCarty did not forward Power's objections to Hardy, and later said he did not "really believe I had a decision as to whether . . . the temperature concerns were valid or not" and that Powers could have spoken for himself. Powers said "you don't override your chain of command. My boss was there; I made my position known to him; he did not choose to pursue it. At that point it's up to him; he doesn't have to give me any reasons; he doesn't work for me; it's his prerogative." Wear admitted that at Marshall "everyone does not feel free to go around and babble opinions all the time to higher management." The definite statements from Center officials could have intimidated dissenters; he acknowledged that "when the boss had spoken, they might quiet down."[125]

Mulloy, in testimony to a Senate committee, best summarized the circumstances. "We at NASA," he said, "got into a group-think about this problem. We saw it, we recognized it, we tested it, and we concluded it was an acceptable risk. . . . When we started down that road we were on the road to an accident."[126] Indeed the teleconference was a classic case of "groupthink," a form of decision-making in which group cohesion overrides serious examination of alternatives. Top level Marshall and Thiokol officials, believing the joint was safe, rationalized bad news from experts, and refused to consider contingency plans. Recognizing consensus among superiors, some subordinate engineers exercised self-censorship. Consequently participants in the teleconference failed to communicate and find useful ways to analyze the risks of cold temperature.[127] Two personnel experts, who conducted management seminars at NASA from 1978 to 1982, argued that groupthink was not unique to Marshall and was inherent in NASA culture. They believed that internal career ladders, homogeneous professional backgrounds, masculine management styles, political pressures to downplay problems, and over-confidence resulting from a history of success had produced a quest for harmony that was often dysfunctional.[128]

At 11:30, SRB project manager Mulloy and Shuttle projects manager Reinartz of Marshall telephoned Level II Manager Arnold Aldrich of JSC. They discussed the effects of cold weather, especially ice on the launch pad and the status of the booster recovery ships, and agreed that the launch should proceed. The Marshall officials did not mention the teleconference or discuss O-rings. At 5:00 A.M. on January 28, Reinartz met with Lucas, and Kingsbury, chief of the Center's Science and Engineering, informing them of Thiokol's concerns about the O-rings, the firm's initial recommendation to delay, and the final decision to launch.[129]

The presidential commission criticized these exchanges in some of its strongest language, finding that "Marshall Space Flight Center project managers, because of a tendency at Marshall to management isolation, failed to provide full and timely information bearing on the safety of flight 51–L to other vital elements of Shuttle program management" and they "felt more accountable to their Center management than to the Shuttle program organization."[130] Commissioner Donald J. Kutyna, a major general in the Air Force, said going outside the "reporting chain" to describe the O-ring concerns to Lucas rather than Aldrich was like reporting a fire to the mayor rather than the fire chief.[131]

Marshall's project managers, of course, never thought the O-rings were hazardous. Reinartz told the commission that they did not report the teleconference or Thiokol's concerns because the question had been "successfully resolved," the experts had decided that the launch was safe, and the final decision "did not violate any launch commit criteria." Agreeing that Marshall had not violated any "formal documentation," Aldrich wished he had been informed anyway. In hindsight Reinartz acceded the wisdom of notifying Level II, but he doubted that this would have stopped the launch of 51–L since both Thiokol and Marshall had agreed to proceed. Mulloy said "it was clearly a Level III issue that had been resolved," and "it did not occur to me to inform anyone else then nor do I consider that it was required to do so today."[132]

The project managers' responses, however, did not explain why they notified Lucas rather than Aldrich. Cecil Houston, Marshall's resident manager at the Cape, believed that Center rivalry affected their decision. Reinartz and Mulloy, he told commission investigators, "didn't want to mention" the matter to a JSC official. "There is between Centers a certain amount of 'them' and 'us,' you know. It's not overt and we don't make a big deal out of it, but they [MSFC's project managers] do feel like some things are not necessarily their [JSC's] business." The discussion should have been reported to Aldrich, Houston thought, and "we had always done it before."[133]

Between 7:00 and 9:00 the next morning, the ice crew at the Cape inspected the icicle-draped Launch Pad 39B and measured temperature. They recorded a temperature of eight degrees near the aft joint of the right solid rocket motor. They did not report this finding because it fell outside their directives. At 9:00, the NASA mission management team, which included the Level I, II, and III managers, discussed the ice and decided conditions were safe. No one discussed the O-rings. In Huntsville that morning, Powers told a fellow motor engineer of his fear for *Challenger's* astronauts, worrying that "these guys don't have more than a fifty-fifty chance." At 11:38, the boosters fired, helping to lift mission 51–L off the pad. In little more than a minute, the aft field joint on the right motor failed and destroyed *Challenger*.[134]

---

1 NASA initially designated each Shuttle mission by sequential numbers. After 1983, each flight had two numbers and a letter; the first number referred to the fiscal year, the second the launch site (1 for Kennedy), and the letter designated the sequence of the flight in the fiscal year.

2 *Presidential Commission on the Space Shuttle Challenger Accident: Report to the President*, Washington, DC: US GPO, 6 June 1986, Vol. I, pp. 19–21, 31, 40, hereinafter, PC Vol., page; Dennis E. Powell, "The Challenger Deaths: What Really Happened?" *Huntsville Times*, 13 November 1986.

3 See PC I, espec. p. 148.

4 Malcolm McConnell, *Challenger: A Major Malfunction, A True Story of Politics, Greed, and the Wrong Stuff* (Garden City, New York: Doubleday, 1987); Joseph Trento, *Prescription for Disaster* (New York: Crown, 1987).

5 Diane Vaughn, "Autonomy, Interdependence, and Social Control: NASA and the Space Shuttle Challenger," *Administrative Science Quarterly* 35 (June 1990); Frederick F. Lighthall, "Launching the Space Shuttle Challenger: Disciplinary Deficiencies in the Analysis of Engineering Data," *IEEE Transactions on Engineering Management* (February 1991); Gregory Moorhead, Richard J. Ference, and Chris P. Neck, "Groupthink Decision Fiascoes Continue: Space Shuttle Challenger and a Revised Groupthink Framework," *Human Relations* 44 (June 1991); Phillip K. Tompkins, *Organizational Communication Imperatives: Lessons of the Space Program* (Los Angeles: Roxbury Publishing, 1993).

6 Thomas F. Gieryn and Anne E. Figert, "Ingredients for a Theory of Science in Society: O-Rings, Ice Water, C-Clamp, Richard Feynman, and the Press" in Susan E. Cozzens and Thomas F. Gieryn, eds., *Theories of Science in Society* (Bloomington: University of Indiana Press, 1990).

7 The Morton Salt Company purchased Thiokol Incorporated in 1982, thus creating Morton-Thiokol.

8 SRB segments had joints of two types, factory and field. The Thiokol plant near Brigham City, Utah connected factory joints, and contractors at the Kennedy Space Center connected four field joints. The field joints in turn were of two types, one which connected case segments and one which connected case and nozzle segments.

9 PC I, pp. 120–22; Trudy E. Bell and Karl Esch, "The Fatal Flaw in Flight 51–L," *IEEE Spectrum* (February 1987), pp. 40–42.
10 William Brown, MSFC, OHI by Jessie Whalen, 15 August 1989, pp. 23–24.
11 Lindstrom, "SRB DM–1 Status," and Marshall, "SRM DM–1 Static Firing Test," 25 July 1977, Weekly Notes, MSFC Center Archives.
12 Thiokol, "Case Burst Test Report," 21 December 1977, TWR–11664, Sections 2.0, 4.2, 4.3.4, MSFC Documentation Repository.
13 Lindstrom, "SRM Case," Weekly Notes, 3 October 1977; Marshall, "SRM Case Hydroburst," Weekly Notes, 11 October 1977, quote; McCool, "SRM Case—SRM Hydrostatic Burst Test Vehicle," Weekly Notes, 17 October 1977, MSFC Archives; PC I, pp. 122–23.
14 Thiokol, "Analytical Evaluation of the SRM Tang/Clevis Joint Behavior," 6 October 1978, TWR–12019, MSFC Documentation Repository; PC I, pp. 122–23; Lawrence Mulloy Testimony, 10–11 February 1986, PC IV, pp. 286–87, 354–55
15 PC I, pp. 122–23; Bell and Esch, p. 40; Howard McIntosh, Manager Steel Case and Refurbishment for SRM, MTI, PC I by Patrick Maley, 4 April 1986, Wasatch, Utah, pp. 7–13: Arnold Thompson, MTI, PCI by Robert C. Thompson, 4 April 1986, MSFC, pp. 11–21.
16 Eudy cited in PC I, pp. 123–24; Leon Ray, "SRM Clevis Joint Leakage Study," 21 October 1977, PC I, p. 233.
17 J. Q. Miller and L. Ray to G. Eudy, "Statement of Position on SRM Clevis Joint O-ring Acceptance Criteria and Clevis Joint Shim Requirements," 9 January 1978, PC I, pp. 234–34.
18 McCool, "DM–2 SS SRM Static Firing" and Lindstrom, "SRB DM–2 Static Test," Weekly Notes, 23 January 1978, MSFC Archives; Thiokol, "DM–2 Flash Report," TWR–11704; Thiokol, "Test Report SS SRM DM Static Firing Test," Vol. III, DM–2, TWR–12348, MSFC Documentation Repository.
19 Thiokol, "Test Report, DM–3," TWR–12348, Vol. IV, pp. 108, 103.
20 Cited in Bell and Esch, p. 40.
21 Thiokol, "SRB Structural Test, STA–1 Motor Case, Final Test Report," 2 November 1978, TWR–12051, Sections 2.0, 4.6, 5.0; McIntosh PCI, 4 April 1986, pp. 22–39; Thompson PCI, 4 April 1986, pp. 2–11; Maurice Parker, MTI, PC I by E. Thomas Almon, 14 May 1986, MSFC, pp. 6–27.
22 J. Q. Miller and L. Ray to G. Eudy, "Evaluation of SRM Clevis Joint Behavior," 19 January 1979, PC I, p. 236.
23 L. Ray, "Visit to Precision Rubber Products Corporation and Parker Seal Company," 6 February 1979, PC I, p. 238.
24 Thiokol, "DM–4 Flash Report," p. 3, TWR–12168; Thiokol, "Test Report, DM–4," p. 124, TWR–12348, Vol. V.; Thiokol, "DM Test Report: Introduction and Summary," pp. 36–37, TWR-12348, Vol. I, MSFC Documentation Repository.
25 Thiokol, "QM–1 Flash Report," TWR–12359, MSFC Documentation Repository; "First SRM Qualification Test 'Quite Satisfactory,'" *Marshall Star*, 20 June 1979; "Second SRM Qualification Test Passed," *Marshall Star*, 3 October 1979.

26 Space Shuttle Verification/Certification Propulsion Committee, 5th Meeting, 10 July 1980, pp. C.7.21–22; Robert F. Thompson and John F. Yardley, "Space Shuttle Program Office Response to Propulsion Committee," 10 August 1980, pp. 58–60, Rogers' Commission Records, National Archives, Washington, DC (hereinafter RCR); PC I, pp. 124–25; Bell and Esch, p. 42.

27 Maurice Parker, Thiokol, "SRB Critical Items List," 24 November 1980, PC I, pp. 239–40; PC I, p. 125.

28 PC I, p. 148; J. Sutter and T. J. Lee, "NASA Development and Production Team Report," Appendix K, PC II, pp. K–30–31; STS 51–L Data and Design Analysis Task Force, "Preliminary Lessons Learned Report," April 1986, pp. D.1, D.4–9, 51–L Investigation Documents, Box 5, JSC History Office.

29 Glenn Eudy, Presidential Commission Interview (hereinafter PCI) by Ray Molesworth, 26 March 86, pp. 1–19, RCR; Glenn Eudy, OHI by Jesse Emerson, 30 January 1990, pp. 2–4; MSI SSHP; James Kingsbury, OHI by Jessie Whalen, 17 November 1988, pp. 5–6, MSI SSHP; Alex McCool, PCI by Emily Trapnell, 27 March 1986, pp. 1–25, quoted p. 28, RCR; Bob Lindstrom, OHI by Jessie Whalen, 26 June 1986, pp. 8–9; George Hardy, PCI by Robert C. Thompson and E. Thomas Almon, 3 April 1986, pp. 15–16; Bill Rice, PCI by Ray Molesworth, 26 March 1986, MSFC, pp. 10–12; Fred Uptagraff, PCI by Robert C. Thompson and E. Thomas Almon, 1 April 1986, pp. 1–3, 6–20, 33, RCR.

30 George Hardy, OHI by Jessie Emerson, 1 March 1990, MSI SSHP, pp. 3–4.

31 William Leon Ray, PCI by Emily Trapnell and Bob Thompson, 25 March 1986, pp. 50–51.

32 John Q. Miller, PCI by Ray Molesworth, 27 March 1986, pp. 24–26, 29–31, 34–36, quoted p. 35; William Leon Ray, PCI by Emily Trapnell and Bob Thompson, 25 March 1986, pp. 10, 48–42; Ben Powers, PCI by Ray Molesworth, 25 March 1986, pp. 8–9, 19, 38–39, quoted p. 35, RCR.

33 William Lucas, Oral History Interview by Andrew J. Dunar and Stephen P. Waring (hereinafter OHI by AJD and SPW), UAH, 27 October 1992, p. 41.

34 PC I, pp. 82–83. For examples of the process, see W. Lucas, "MSFC FRR Board for MSFC Elements for Mission 61–C," 3 December 1985, 61–C folder, MSFC Archives; "Flight Readiness Review Treatment of O-ring Problems," PC II, Appendix H.

35 MSFC SRB Office to Paul Wetzel, NASA HQ, "STS–2 SRM O-ring-Effected Condition," 9 December 1981, STS–2 folder; James A. Abrahamson, HQ Asst. Admin. for Space Flight, "STS–3 FRR Minutes," 10 March 1983, STS–3 folder, MSFC Archives; PC I, p. 125.

36 Maurice Parker, Thiokol, "SRB Critical Items List," 17 December 1982; MSFC, "SRB and RSCD System CIL Change Notice No. 23," 21 January 1983; Glynn Lunney, Shuttle Program Manager, "SRB CIL Requirements for SRM Case Joint Assemblies," 2 March 1983; L. Michael Weeks, NASA HQ, "SRB CIL Requirements," 28 March 1983 all in PC I, pp. 241–44; PC I, pp. 126–27, 159.

37 Larry Mulloy and George Hardy testimony, 26 February 1986, PC V, pp. 834–38; William Lucas testimony, 27 February 1986, PC V, pp. 1036–37; William P. Horton, PCI by Ray

Molesworth, 27 March 1986, pp. 20, 22–23; Robert E. Lindstrom, PCI by P. J. Maley, 5 May 1986, p. 5, RCR.

38 Judson A. Lovingood testimony, 26 February 1986, PC V, pp. 925–26; Mulloy and Hardy testimony, 26 February 1986, PC V, pp. 834, 838, 880.

39 MTI, "SRM Redundant Seals: Summary of FMEA/CIA and Certification Rationale," TWR–15538, March 1986, PCR; Robert K. Lund, MTI, PCI by Pat Maley, 1 April 1986, p. 14; Wesley L. Hankins, PCI by Pat Maley, 2 April 1986, pp. 9–11; Roger Boisjoly, PCI by John Molesworth, 2 April 1986, pp. 33–34; Howard McIntosh, MTI, PCI by Pat Maley, 4 April 1986, pp. 60–62, 71; Joe Kilminster, PCI by E. Thomas Almon, 4 April 1986, pp. 26–29.

40 PC I, pp. 126–28, 159; Ben Powers, PCI by Ray Molesworth and Emily Trapnell, 12 March 1986, pp. 36–40, RCR.

41 "Filament Cases Proposed for Solid Rocket Motor," *Marshall Star*, 10 March 1982; "SRB Modification Contract Awarded by Marshall," *Marshall Star*, 19 May 1982; MSFC Structures Engineering Branch, "Structural Test and Instrumentation Requirements for the Space Shuttle FWC SRB," December 1982, MSI Space Shuttle History Project, hereinafter, SSHP; Ray quoted in William J. Broad, "NASA had Solution to Key Flaw in Rocket when Shuttle Exploded," *New York Times*, 22 August 1986.

42 PC I, pp. 5, 128.

43 "SRB Stacking Begins on STS–6 Vehicle," *Marshall Star*, 6 October 1982; "STS–6 Vehicle Moves to Pad," *Marshall Star*, 1 December 1982; Budd McLaughlin, "Prober: Faulty Rocket Joint, Other Factors, Exploded Challenger," *Huntsville News*, 9 April 1986; PC I, p. 125; William H. Starbuck and Frances J. Milliken, "Challenger: Fine-tuning the Odds until Something Breaks," *Journal of Management Studies* 25 (July 1988), pp. 325–26, 337. In April 1983 on STS–6 NASA began using a lightweight steel booster case with thinner case walls and began filling the booster with a more powerful fuel.

44 PC Staff, "Leak Check," 1986, RCR, PC 102597, Reel 46, National Archives; PC I, pp. 133–34.

45 PC Staff, "Leak Check," RCR; PC I, pp. 133–34; Bell and Esch, pp. 42–43.

46 PC I, pp. 133–34; Frederick F. Lighthall, "Launching the Space Shuttle Challenger: Disciplinary Deficiencies in the Analysis of Engineering Data," *IEEE Transactions on Engineering Management* (February 1991), pp. 68–74.

47 Coates cited in PC I, p. 128; J. Q. Miller to G. Hardy and K. Coates, "Burned O-rings on STS–11 (41–B)," 28 February 1984, in PC I, p. 245.

48 McCool, "Burned O-Rings on the SRM," Weekly Notes, 12 March 1984, MSFC Archives.

49 James A. York, "Minutes of the MSFC FRR Boards for Mission 41–C (STS–13)," 12 April 1984, p. 7, 41–C folder, MSFC Archives.

50 PC I, p. 145; Larry Mulloy, PCI by Robert C. Thompson, 2 April 1986, pp. 63, 56, 16, 43, RCR.

51 David E. Sanger, "Top NASA Aides Knew of Shuttle Flaw in '84," *New York Times*, 21 December 1986; James A. Abrahamson, PCI by Emily Trapnell, 17 April 1986, p. 28; PC I, pp. 128–32, quote p. 132.

52 Thiokol, "Program Plan: Protection of Space Shuttle SRM Primary Motor Seals," TWR–14359, 4 May 1984, RCR; Mulloy, MSFC to W. F. Dankhoff, HQ, "Response to Action Item–3–30–4–2 from 41–C FRR," 84–06–6, RCR; PC I, p. 132; Mulloy PCI, 2 April 1986, p. 40.

53 MTI, "SRM Sub-scale O-ring Test Program Status Report," 7 May 1984; Brian Russell, MTI, "Minutes of Vacuum Putty/O-ring Meeting, 31 May 1984," 1 June 1984; Brian Russell to R.V. Ebeling, MTI, "Minutes of Telecon with MSFC on 7 June 1984," 13 June 1986; MTI, "Vacuum Putty Telecon," 29 June 1984; MTI, "Vacuum Putty/O-Ring Erosion Study and Program Plan," TWR–14563, 16 October 1984; Joe Kilminster, PC I by E. Thomas Almon, 4 April 1986, pp. 37–44, RCR.

54 MSFC, "STS–41D Quick Look Bulletin," 4 September 1984, 41–D folder; P. Lovingood, "Minutes of MSFC FRR for 41–G," 19 October 1984, 41–G folder, MSFC Archives; "FRR Shuttle Projects Office Board," 19 September 1984, PC II, p. H–18.

55 MSFC, "STS–51C: Flight Evaluation Report, Shuttle Projects,"14 February 1985, 51–C folder, MSFC Archives.

56 L. Mulloy to L. Wear, "51C O-ring Erosion Re: 51E FRR (TWX)," 31 January 1985, PC I, p. 247.

57 "FRR SRB Board," 8 February 1985, pp. H–20–32.

58 "FRR Shuttle Projects Office Board," 12 February 1985, PC II, pp. H–32–40; "FRR Marshall Center Board," 14 February 1985, p. H–41; "FRR Level I," 21 February 1985, PC II, pp. H–41–42.

59 Roger Boisjoly, PCI by John Ray Molesworth, 2 April 1986, p. 38; Robert Crippen Testimony, 3 April 1986, PC V, p. 1418; Mulloy Testimony, 10 February 1986, PC IV, p. 290; Mulloy Testimony, 26 February 1986, PV, pp. 845, 847.

60 Bob Marshall, OHI by Jessie Whalen and Sarah McKinley, 22 April 1988, p. 32; PC I, pp. 145–146; Lighthall, pp. 63–74, espec. pp. 68–72.

61 Paulette D. Lovingood, "Minutes of MSFC FRR for Mission 51–D," 19 April 1985, p. 9; 51–D folder, MSFC Archives; PC I, pp. 136–37; Morton Thiokol, "STS–51F FFR, SRM Special Topic: Nozzle O-ring Anomaly Observed on STS–51B," Combined MSFC Review Boards, 1 July 1985, PC II, pp. H–49–66; James W. Thomas, Chief SRM Technical Operations Office, MSFC, PCI by Patrick J. Maley, 10 April 1986, pp. 45–46, RCR.

62 PC I, pp. 137–39, 84, 102–103; Brian Russell, Allen McDonald, J. Kilminster, Thiokol, Testimony, 2 May 1986, PC V, pp. 1578–79, 1581, 1590; Kathy Sawyer and Boyce Rensberger, "6 Waivers Let Shuttle Keep Flying," *Washington Post*, 3 May 1986.

63 PC I, pp. 153–156, 159; "The Silent Safety Program," *Professional Safety* 31 (November 1986), pp. 10a–10g.

64 STS 51–L Data and Design Analysis Task Force, "Lessons Learned Report," June 1986, Section P–2, P–4.

65 PC I, pp. 137–39, 84, 102–103; PC V, pp. 1513, 1516, 1519–20, 1531–32; "FRR STS–51B Level I," 2 July 1985, PC II, pp. H–66–68.

66 Mulloy Testimony, 10 February 1986, PC IV, p. 303; PC I, pp. 137–39.

67 Mulloy and Wear Testimony, 2 May 1986, PC V, 1513, 1514, 1519.

68 Mark Salita, MTI, "Prediction of Pressurization and Erosion of SRB Primary O-rings during Motor Ignition," TWR–14952, 23 April 1985, RCR PC 102505; Mark Salita, MTI, PCI by John Molesworth, 2 April 1986, Wasatch, pp. 1–14; Arnold Thompson, MTI, PCI by Robert C. Thompson, 4 April 1986, MSFC, pp. 27–29.

69 MTI for MSFC, "SRM Significant Problem Report No. DR–4–5/49; Secondary O-ring Erosion in the Nozzle to Case Joint on Mission 51–B," TWR–15091–1, 16 July 1985, RCR, Reel 29; "Level I SRM FRR for STS– 51F," 2 July 1985, PC II, pp. H–66–68; Mulloy and Wear Testimony, 2 May 1986, PC V, 1513, 1519, quote p. 1514.

70 Larry Mulloy, OHI by Jesse Whalen, 25 April 1988, pp. 16, 17.

71 Kraft believed that "flying with that goddamn thing known as a problem and then compounding it with the weather conditions [of 51–L] . . . is the worst engineering decision that has ever been made in the space program." Kraft, OHI by AJD and SPW, 28 June 1991, p. 21.

72 Mulloy PCI, 2 April 1986, pp. 33–34; Mulloy, "MTI O-ring Task Force at MSFC," 9 September 1985, Weekly Notes, RCR; Mulloy Testimony, 26 February 1986, PC V, pp. 885–86; T. M. Gregory, MTI, "Preliminary SRM Nozzle/Field Joint Seal Concepts," TWR–15117, August 1985, RCR, PC 067761; Donald Christiansen, "One Flight Too Many," *IEEE Spectrum* (February, 1987), p. 27; William J. Broad, "NASA Had Solution to Key Flaw in Rocket when Shuttle Exploded," *New York Times*, 22 August 1986.

73 B. G. Russell, MTI to J. W. Thomas, MSFC, "Actions Pertaining to SRM Field Joint Secondary Seal," 9 August 1985, PC V, pp. 1568–69.

74 HQ Propulsion Division, "SRM O-ring Charring," review notes from March 1984-February 1985, RCR, PC 071188–071196; D. Winterhalter, HQ to L. M. Weeks, HQ, "Secondary O-ring Seal Erosion" 28 June 1985, RCR; Irving Davids, HQ, to Associate Administrator for Space Flight, "Case to Case and Nozzle to Case 'O' Ring Seal Erosion Problems," 17 July 1985, PC I, p. 248; Richard C. Cook to Michael B. Mann, Chief STS Resources Analysis Branch, HQ, "Problem with SRB Seals," 23 July 1985, PC IV, p. 273; Richard C. Cook, "Why I Blew the Whistle on NASA's O-ring Woes," *Washington Post*, 16 March 1986; Richard C. Cook Testimony, 11 February 1986, PC IV, pp. 377–94; "SRM FY 1987 Budget Recommendation to the Administrator," 16 August 1985, PC IV, p. 276.

75 Quotations from MTI, "Erosion of SRM Pressure Seals: Presentation to NASA HQ," 19 August 1985, RCR; L. Mulloy to W. Lucas, "SRM Briefings to Code M," 16 August 1985, Space Shuttle Projects Office, SRB Investigation (51–L), folder 2, MSFC History Office.

76 Mulloy Testimony, 26 February 1986, PC V, quote p. 847, chart p. 908; PC I, p. 148.

77 R. M. Boisjoly to R. K. Lund, MTI VP for Engineering, "SRM O-ring Erosion/Potential Failure Criticality," 31 July 1985; R. Ebeling, to A. J. McDonald, MTI Director SRM Project, "Weekly Activity Report," 1 October 1985; S. R. Stein to MTI distribution, "Potency of O-ring Investigation Task Force," 1 October 1985; Roger Boisjoly, "Activity Report," 4 October 1985, PC I, pp. 249–55.

78 A. R. Thompson to S. R. Stein, MTI Project Engineer, "SRM Flight Seal Recommendations," 22 August 1985, PC I, p. 251; A. R. Thompson Testimony, 14 February 1986, PC IV, pp. 696–97.

79 Russell P. Boisjoly, Eileen Foster Curtis, Eugene Mellican, "Roger Boisjoly and the Challenger Disaster: The Ethical Dimensions," *Journal of Business Ethics* 8 (August 1989), p. 221.

80 James Smith, MSFC, PCI by Robert C. Thompson, 13 March 1986, MSFC, pp. 57–62; Powers PCI, 12 March 1986, pp. 39–40; MSFC Solid Rocket Motor Branch, Weekly Notes, 3 October 1985, RCR, PC 102548; B. Russell, MTI, PCI by Emily Trapnell, 3 April 1986, pp. 13–14; B. Russell, MTI, PCI by Emily Trapnell, 19 April 1986, pp. 61, 70; Biosjoly, Curtis, and Mellican, p. 220.

81 PC I, p. 140; J. E. Kingsbury to L. Mulloy, "O-ring Joint Seals," 5 September 1985, PC I, p. 256.

82 U.S. House of Representatives, Committee on Science and Technology, *Investigation of the Challenger Accident* (Washington, DC: US GPO, 1986), pp. 179–80.

83 Vaughn, pp. 247–48, quote p. 247.

84 Brian Russell, PCI by Emily Trapnell, 19 March 1986, Wasatch, pp. 61–70; Arnold Thompson, PCI by Robert C. Thompson, 4 April 1986, MSFC, pp. 52–54; Bill Ebeling, PCI by John Molesworth, 19 March 1986, Wasatch, pp. 6–8; Boisjoly PCI, 2 April 1986, p. 21; Bell and Esch, p. 45.

85 J. Smith, PCI by Robert C. Thompson, 13 March 1986, pp. 45–46, 63, 58; Wear PCI, pp. 70–72, 75–76; L. Adams, PCI by John R. Molesworth, 12 March 1986, pp. 24, 27, 29, 32, RCR. S. Reinartz, Testimony, 26 February 1986, PC V, p. 921. See also John P. McCarty, PCI by John R. Molesworth, 19 March 1986, pp. 30–34, RCR.

86 Jerry Peoples, PCI by John R. Molesworth, 12 March 1986, pp. 22–24.

87 Mulloy Testimony, 26 February 1986, PC V, pp. 885–86; Mulloy, "SRM O-Ring Seal Erosion," 4 November 1985, Weekly Notes, MSFC Archives; "Summary of 11/18/85 [FWC] Briefing to Jesse Moore," 19 November 1985, FWC 1985–86 folder, MSFC Archives.

88 "Flight Readiness Review Treatment of O-ring Problems," Appendix H, PC V, p. H–3.

89 PC I, pp. 148, 103.

90 Mulloy quoted in Bell and Esch, p. 43.

91 PC I, pp. 142–44; A. J. McDonald, MTI to L. O. Wear, MSFC, "Closure of Critical Problems [on] Nozzle/Segment Joint," 10 December 1985; D. Thompson, MTI to J. Fletcher Rockwell International, Huntsville, 3 January 1986; L. O. Wear, MSFC to J. Kilminster, MTI, 24 December 1985; MSFC Problem Assessment System, 23 January 1986, PC V, pp. 1550, 1553, 1554, 1547–49.

92 PC I, p. 143; Mulloy and Wear Testimony, 2 May 1986, PC V, pp. 1522–23.

93 Mulloy, "SRM Quality Improvements," 21 October 1985, Weekly Notes, MSFC Archives; Bell and Esch, pp. 46–47; NASA Press Release, "NASA Proposes Solid Rocket Motor Second Source," PR 85–178, 26 December 1985.

94 NASA PR 85–178; Adams PCI, pp. 9–14; Mulloy, "SRM Negotiation," 27 January 1986, Weekly Notes, MSFC Archives; Bell and Esch, pp. 46–47; Michael Isikoff, "Thiokol Was Seeking New Contract When Officials Approved Launch," *Washington Post*, 27 February 1986.

95 Robert V. Ebeling Testimony, 2 May 1986, PC V, p. 1594; MTI, "O-Ring Task Force Report to MSFC," 16 January 1986, RCR, NA, 067838.

96 PC I, p. 85.

97 PC I, pp. 145–148, 82, 102–104.

98 R. P. Feynman, "Personal Observations on Reliability of Shuttle," Appendix F, 1–2, PC II; R. P. Feynman with R. Leighton, *"What Do You Care What Other People Think?" Further Adventures of a Curious Character* (New York: Norton, 1988), pp. 179–85.

99 Lighthall, pp. 63–74.

100 PC V, p. 1446.

101 "Human Factors Analysis," Appendix G, PC II, p. G5; PC I, pp. 85–88. For a chronology of events on January 27–28, see PC I, pp. 104–110.

102 PC I, pp. 88, 111; Cecil Houston, PCI by E. Thomas Almon, 10 April 1984, p. 67, RCR; Wear PCI, pp. 56–57.

103 Roger Boisjoly Testimony, 25 February 1986, PC IV, pp. 785–93, quoted 791; MTI, "Temperature Concern on SRM Joints," 27 January 1986, PC IV, pp. 645–51.

104 "Temperature Concern," PC IV, pp. 645–51; Roger Boisjoly and Allan McDonald Testimony, 14 February 1986, PC IV, pp. 617–27.

105 George Hardy Testimony, 26 February 1986, PC V, pp. 865, 878; Stanley Reinartz Testimony, 26 February 1986, PC V, pp. 912–22; Lawrence Mulloy Testimony, 26 February 1986, PC V, p. 843; Allan McDonald Testimony, 25 February 1986, PC IV, p. 741.

106 Mulloy and Hardy Testimony, 26 February 1986, PC V, pp. 850–53, 855–56, 862, 872; Mulloy Testimony, 14 February 1986, PC IV, pp. 604–607, quoted, 607.

107 Hardy Testimony, 26 February 1986, PC V, pp. 861–62.

108 Boisjoly Testimony, 25 February 1986, PC IV, pp. 785–86.

109 PC I, pp. 145–46; Lighthall, pp. 63–74, espec. pp. 68–73.

110 Boisjoly PCI, 2 April 1986, pp. 39–40.

111 STS 51–L Data Design and Analysis Task Force, "Lessons Learned Report," P–2, 51–L Investigation Documents, Box 5, JSC History Office; PC I, p. 148.

112 Lighthall, pp. 63–74; Mulloy Notes for Testimony, 14 February 1986, PC IV, pp. 610–15.

113 PC I, pp. 96–100; McDonald Testimony, 25 February 1986, PC IV, pp. 719–22; Hardy Testimony, 26 February 1986, PC V, pp. 864–65, 872.

114 PC I, pp. 90–98, 108; Boisjoly Testimony, 14 February 1986, PC IV, p. 680; Jerald E. Mason Testimony, 14 February 1986, PC IV, quoted p. 632, 25 February 1986, PC IV, pp. 763–73, quoted p. 764.

115 "MTI Assessment of Temperature Concern on SRM–25 (51L) Launch," 27 January 1986, PC I, p. 97; PC I, pp. 96–100, 108; Reinartz, 26 February 1986, PC V, p. 915.

116 William A. Riehl, Personal Documentation, 27 January 1986, PC 000465; Jack Buchanan, MTI, "MSFC/MTI Telecon 27 January 1986, p. 6, PC 000404, RCR.

117 McDonald Testimony, 25 February 1986, PC IV, p. 729; Boisjoly Testimony, 25 February 1986, PC IV, p. 793; Robert V. Lund Testimony, 25 February 1986, PC IV, p. 811.

118 Hearings, 14 February 1986, PC IV, pp. 697–98; PC I, p. 104.

119 Reinartz Testimony, 26 February 1986, PC V, pp. 912–13, 917; Hardy Testimony, 26 February 1986, PC V, pp. 866–67; Mulloy Testimony, 26 February 1986, PC V, pp. 839–41, 848; Mulloy quoted in Bell and Esch, p. 47.

120 Mulloy Testimony, 26 February 1986, PC V, pp. 840, 856–57; Hardy Testimony, 26 February 1986, PC V, pp. 866–67; Reinartz Testimony, 26 February 1986, PC V, pp. 915–16.

121 William Lucas Testimony, 27 February 1986, p. 1046.

122 Powers PCI, 12 March 1986, p. 10; Coates PCI, 25 March 1986, pp. 1–7.

123 PC I, pp. 104, 82.

124 Adams PCI, 12 March 1986, pp. 34–35; Wear PCI, pp. 83–85, 24.

125 Coates PCI, 25 March 1986, pp. 16, 24–25, Hardy quoted, pp. 13–14; John P. McCarty, PCI by John R. Molesworth, 19 March 1986, pp. 9–11; Powers PCI, pp. 10, 25; Wear PCI, 12 March 1986, pp. 85, 49, RCR.

126 Mulloy cited in Howard Benedict, "Senators Blast Commission Report," *Huntsville News*, 18 June 1986, p. A1.

127 Gregory Moorhead, Richard J. Ference, and Chris P. Neck, "Group Decision Fiascoes Continue: Space Shuttle Challenger and a Revised Groupthink Framework," Human Relations 44 (June 1991), pp. 539–50. Although plagued with factual errors about the teleconference, this article has a useful conceptual framework. For more on the concept, see Irving Janis, *Groupthink* (Boston: Houghton Mifflin, 1983).

128 See Kenneth A. Kovach and Barry Render, See "NASA Managers and Challenger: A Profile and Possible Evaluation," *Personnel* 64 (April 1987), pp. 40–44.

129 PC I, p. 98; Arnold Aldrich Testimony, 14 February 1986, PC IV, pp. 636–37, 27 February 1986, 3 April 1986, PC V, pp. 1019–20, 1490–91; Reinartz Testimony, 26 February 1986, PC V, pp. 917–18; Lucas Testimony, 27 February 1986, PC V, pp. 1035–40, 1042–43.

130 PC I, pp. 200, 198.

131 Hearings, 26 February 1986, PC V, pp. 917–18.

132 Reinartz Testimony, 14 February 1986, PC V, pp. 633–34, 640, 26 February 1986, PC V, pp. 917–18, 920; Lucas Testimony, 27 February 1986, PC V, pp. 1042–43; Mulloy Testimony, 14 February 1986, PC IV, 615, 26 February 1986, PC V, p. 849; Aldrich Testimony, 27 February 1986, PC V, pp. 1055–56.

133 Cecil Houston, PCI by E. Thomas Almon, 10 April 1986, pp. 74, 78, RCR.

134 Peoples PCI, 12 March 1986, p. 39; PC I, p. 110.

Chapter X

The Recovery: Investigation and Return to Flight

For every Shuttle launch, technicians in Marshall's operations support Center watched consoles showing continuous updates of data. For the ill-fated 51–L launch, they were stunned when the screens froze shortly after liftoff. Initially suspecting a telemetry problem rather than a catastrophe, the technicians turned to television screens and saw the vapor cloud caused by the destruction of the external tank. They sat in complete silence hoping to see the orbiter come out of the cloud, but instead they saw contrails of burning, falling debris. Working silently, they began collecting the data necessary for the post-accident investigation.

The weeks after the *Challenger* accident were the most traumatic in the first three decades of the Marshall Space Flight Center. Marshall people felt shock and a deep sense of loss. They had dedicated themselves to the Shuttle program, identified with its accomplishments, embraced the astronauts as colleagues and friends, and so experienced the accident as personal failure. Many wondered if their anguish would ever go away.[1]

Marshall personnel began investigating within moments after the disaster. Serving on task force panels and on laboratory teams, many worked 12-hour days for months. Their dedication paid off as Center employees played the major role in finding the technical cause of the accident and in fixing the problem. This effort, which Marshall people called "the recovery," enabled the Center and the Agency to return the Shuttle to flight within three years.

While Marshall worked on technical matters, however, independent investigations made Marshall the Center of controversy. In the first half of 1986 official groups and congressional committees studied the events and decisions before the accident, and journalists provided running commentary. Although investigations often made useful examinations of technical causes and organizational circumstances and suggested improvements in NASA and the Shuttle program, the process sometimes degenerated into an inquisition. The inquiries, and especially the scapegoating, were agonizing. The months of investigation and preparation for flight showed the ability of Marshall and NASA not only to fix technical flaws, but also to address sensitive questions, accept criticism, overcome organizational weaknesses, and reorient cultural patterns.

Center of Controversy

After the death of three astronauts in the Apollo 204 fire, NASA had used an internal investigation board which largely confined itself to technological issues and ignored organizational and political factors that contributed to the accident. The narrow technical approach reflected the congressional and presidential commitment to the Apollo end-of-decade deadline and NASA Administrator James Webb's ability to protect the space program from outside criticism.

Challenger not only had an internal investigation by NASA technical panels, but also an independent inquiry by a presidential commission. In part this happened because NASA leaders did not protect the Agency. Administrator James Beggs, subject of an investigation by the Justice Department (which was unrelated to his NASA services and which eventually cleared him of all charges), had surrendered authority over NASA. Deputy Administrator William Graham was new to the Agency and deferred the question of the nature of the investigation to the White House. President Ronald Reagan's Chief of Staff, Donald Regan, worried about allegations that the White House had pressured NASA to launch on 28 January to ensure that the first teacher-in-space would fly on the day of President Reagan's State-of-the-Union message. The charges were groundless, but the Reagan administration was in the midst of numerous scandals and Regan wanted a thorough inquiry to avoid any hint of a cover-up. Consequently President Reagan decided to appoint a special investigatory commission.[2]

The commission, established on 3 February and headed by former Secretary of State William P. Rogers, began directing NASA investigation teams by mid-February. Rogers was a lawyer and he later told reporters that he wanted a thorough and accurate investigation in order to avoid the sort of controversy that had followed the Warren Commission. One way of achieving this was to keep the inquiry open. Rogers said that "full disclosure has advantages over indictments. You don't want to punish. You just want to make sure it doesn't happen again."[3]

NASA implemented its contingency plan and established several technical panels to study various scenarios that could have caused the accident. James R. Thompson (called "J.R." by his colleagues), formerly Marshall's Shuttle main engine project manager and later a university research administrator, headed the NASA investigation. Since the disaster occurred during launch, a phase during which Marshall had primary responsibility, Center personnel played key roles on the technical panels. Propulsion engineers gathered in the Huntsville Operations Support Center to check prelaunch and flight records. With this data, teams led by Center Deputy Director Thomas "Jack" Lee, began to identify possible failure modes and isolate causes. Preliminary analysis pointed to anomalies in the right solid rocket booster (SRB). John W. Thomas, manager of the Spacelab Program Office, headed a team that performed tests on the case joint, and James Kingsbury, head of the Center's Science and Engineering labs, led another team that planned design improvements. Other Marshall employees worked on the parts recovery team to help salvage pieces of 51–L from the ocean floor. Several hundred Marshall employees participated in these teams and worked more than 12 hours a day from February until mid-May.[4]

An unclear division of labor between NASA and the presidential commission contributed to problems that Marshall had with the media. NASA Headquarters directed that no one serving on the NASA task force give media interviews and referred questions about the accident and the investigation to the commission. Marshall personnel with expertise on the subject areas, moreover, were working long hours and had little time for talking with the press. The Center's Public Affairs Office handled technical inquiries from 25 news organizations, including most of the major national outlets, which had set up shop at Marshall when attention focused on the solid rocket boosters. The office relayed answers from Marshall experts, but the reporters were not satisfied by the limited access and information. The Center's public information officers believed that the Headquarters' policy left Marshall defenseless and, by depriving the media of news, encouraged an adversarial posture toward Marshall and the entire Agency. Reporters searched for stories by hanging out in the Marshall cafeteria and camping outside the homes of Center officials.[5]

Meanwhile on 10 February in a closed session of the presidential commission, Morton-Thiokol officials described the history of the joints and their original recommendation to delay the launch because of the dangers of cold weather. During a lull in testimony on 11 February, Commissioner Richard Feynman performed a dramatic demonstration with a section of O-ring, a clamp, and a glass of ice water; this showed that a cold, compressed O-ring material only slowly returned to normal shape when the pressure was released. The demonstration showed how temperature could inhibit the sealing of O-rings and helped reporters explain the cold weather thesis and move easily from technical causation to managerial responsibility.

Afterwards, the commission increasingly challenged Marshall officials. Rogers described NASA's decision process as "flawed" because the eleventh-hour teleconference had allowed a launch with a known hazard; he asked the Agency to exclude SRB project officials, Shuttle managers, and Center directors from internal investigation teams.[6] Rogers became very critical, saying Marshall personnel had lacked "common sense" and had "almost covered up" the joint problems. Feynman called the joint design "hopeless" and said that poor communication between engineers and managers at Marshall was symptomatic of "some kind of disease."[7]

After 15 February the national media also began finding fault with NASA and regarded the ban on interviews as an attempt to cover up a scandal. Marshall officials wanted to talk to the media to correct what they believed was an inaccurate interpretation of the launch decision. They decided to keep silent, however, fearing that the commission would regard press interviews as crude attempts to influence proceedings.[8]

On 26 and 27 February the commission took testimony from Marshall officials involved in the teleconference. Center Director William Lucas said the tone of questioning was "very sharp." Center officials complained of difficulty explaining how they had experienced events and believed the commission did not listen sympathetically. Judson Lovingood, deputy manager of the Shuttle Projects office, said, "we're engineers . . . and that makes me tend to think one way and try to communicate one way. I found it difficult to communicate with some members of the commission. And that's not critical of them. But . . . an engineer does not think like a lawyer might think."[9]

After they had testified, Marshall officials held a series of press interviews. Defending his people, Lucas said "in my judgment, the process was not flawed," and "given what they say they knew, what they testified they knew, I think it was a sound decision to launch."[10] Managers defended the launch process which allowed decisions to be made by low-level experts. They exonerated the joint's design, argued that they had lacked hard evidence that the cold was a hazard, disputed the claim that cold weather was the technical cause of the O-ring failure, and suggested that assembly errors could have damaged the O-ring and caused the accident.[11]

The Marshall strategy of openness backfired. Media reports interpreted their statements as attempts to discredit the commission and as signs of an arrogant refusal to admit mistakes. Marshall public information officers later complained that the media had twisted information and lamented that Marshall had been "gang-banged by the media."[12] The commission's response was just as critical. One commission member believed that the Marshall managers' defense of the flight readiness review process and their decisions was "totally insensitive." Commissioner Joseph F. Sutter believed Center managers were "pretty defensive." After reading the stories and after the commission requested tapes of the interviews, Marshall officials concluded that talking to the media did more harm than good.[13]

In retrospect, Marshall leaders challenged the wisdom of a public investigation. Bill Sneed said NASA should have tried "to understand what went wrong and tried to make it right, rather than almost put the people on trial." Lucas argued that a public investigation was "clearly a gross error." The commission, he believed, was "totally politically motivated" and "its genesis almost determined its outcome." Its purpose "was never to find out technically what went wrong, but to find out where we could put some blame that would deflect it as far from the [Reagan] administration as possible." Lucas worried that the public inquiry had been "counter-productive entirely" and "could close NASA up." An internal investigation would have discovered as much without the side effect of making people "more inclined to protect their own tail, so to speak, rather than have a purely open situation."[14]

The presidential commission and its NASA investigation teams published a common report on 6 June 1986. The report contained four major conclusions: the SRM (solid rocket motor) joint had a flawed design; NASA's safety and quality systems had been inadequate; the Shuttle flight schedule had been too demanding; and Marshall had poor communications, especially with the Level II Shuttle Program office.[15]

Plume of flame from aft field joint of right SRM of STS 51–L, approximately 60 seconds after ignition.

The accident analysis team, led by Thomas and supported by Marshall personnel, studied flight data and wreckage, performed 300 tests on 20 different joint configurations, and concluded that the O-rings had failed and caused the disaster. In addition, the team concluded that the joint design was flawed and that the weaknesses had not been fully understood before the accident. Only after the accident had ground tests thoroughly checked joint behavior and shown that the design was very sensitive to many factors, including joint rotation, cold temperature, hard O-rings, ice in the O-ring grooves, leak check displacement of the primary O-ring, delay of O-ring pressure actuation by the putty, blow-holes in the putty, misfit of the

Photo of recovered fragment of aft center segment of right SRM of STS 51–L, showing hole burned through the case wall.

tang and clevis caused by out-of-round and reused segments, excessive compression on the O-ring by the tang and clevis, and structural stress on the joint caused by an external tank strut and launch dynamics. Thomas's team concluded that NASA must "modify the SRM joint to preclude or eliminate the effect of all these factors and/or conditions."[16]

While accepting that post-accident tests had revealed the inadequacies of the design, most Marshall officials observed that they had had confidence in the design before the accident. Keith Coates, former SRM chief engineer, said, "We knew the gap was opening. We knew the O-rings were getting burned. But there'd been some engineering rationale that said, "It won't be a failure of the joint." And I thought justifiably so at the time I was there. And I think that if it hadn't been for the cold weather, which was a whole new environment, then it probably would have continued. We didn't like it, but it wouldn't fail."[17] Lovingood, former deputy director of Shuttle projects, brooded that "we thought we had thoroughly worked that joint problem. And, you know, I just see it as an error in judgment—a terrible error in judgment."[18]

Some Center officials, however, sought to discredit any simplistic cold weather interpretation. They believed that the design was adequate in cold weather if the joint was properly assembled. Kingsbury doubted that temperature alone had caused failure of the O-rings; if conditions had been so severe, he asked, why had the other five field joints sealed?[19]

Instead, Kingsbury and others pointed to misassembly of the fateful joint as a possible technical cause of the accident. The Accident Analysis Team had found that the joint that failed had been one of the most difficult to assemble in the entire Shuttle program because the upper and lower segments were out-of-round. Ovality of the reusable segments was caused by the sagging of the case walls as the segments lay on their side during rail shipment from Thiokol's plant in Utah to Kennedy Space Center. The Thiokol assembly team at KSC had failed to mate the segments for the 51–L aft right joint several times and succeeded only after using a rounding tool to force the upper segment into shape. While the assembly process followed the correct procedures and the mate was within NASA's numerical specifications, the fit was extremely tight with possible metal-to-metal contact of the tang and clevis. The accident analysis team's report observed that the fit could have compressed the O-rings so tightly that they could not slide across the groove and seal the joint. The report noted that the tightest fit of the segments was in the same location as where gases burned through the joint.[20]

This evidence implied that the tight fit alone could have caused a leak and that the accident could have occurred even in warm weather. Obviously if cold did not cause the accident and if launch managers had not known of the assembly problem, then criticism of the launch decision process and the decision to launch in cold weather was misplaced and more scrutiny should have fallen on the assembly process. Kingsbury believed that the Rogers Commission had made conclusions too early in the investigation, put too much emphasis on cold weather as the technical cause of the accident, paid too little attention to assembly factors, and then made unfair accusations against Marshall managers. Chairman Rogers made up his mind, Kingsbury said, he "quit investigating and became prosecutor" and "we were hanging on the cross and bleeding and hoping it would end quickly."[21]

The official reports of the investigations had different conclusions about the tight fit of the fateful joint. The presidential commission's report devoted an appendix to the issue, and acknowledged the danger of a metal-to-metal fit. The commission concluded, however, that assembly records and flight experience showed no causal connection between tight joints and O-ring problems either on 51–L or on previous launches. The NASA accident analysis team's report described the tight fit only as one of many factors that contributed to the leak. The team's report did not single out any single factor that had caused the joint failure, and instead showed problems in the entire design.[22]

J.R. Thompson, overseer of the day-to-day work of the NASA investigation, faulted the whole design and its sensitivity to many factors. Thompson said that "we were walking right on the edge of a cliff and several of these factors just pushed us over." He lamented that, "We missed it in the design, and some of the prior flight anomalies just really were not taken seriously. Looking back on it, that joint has several shortcomings and it is quite marginal, so if things are not just right it is very susceptible to a leak. It did leak on some prior successful launches… This was just the first time it propagated to a failure. The conditions were marginal enough that it just fell over the edge."[23]

Thompson later denied that the joint had been improperly assembled, but observed that cold was not the only factor that had contributed to the accident.

If temperature had been the single cause, then NASA could have introduced a launch rule that prohibited cold weather launches. The NASA accident team believed the culprit had been an inadequate design and so had recommended redesign of the joints and seals.[24]

The presidential commission also faulted the Agency and the Center for their "silent safety program" and failure to uphold "the exactingly thorough procedures" of the Apollo Program. The Agency and Center had safety, reliability, and quality assurance offices that were responsible to chief engineers in Washington and Huntsville. Marshall's quality office, the commission charged, had failed to maintain a consistent listing of the change of O-ring criticality from 1 to 1R, to perform statistical analysis of trend data, to attend key reviews, and to report critical problems and launch constraints to officials outside the Center. Without knowledge of hazards, managers could not make informed decisions. The commission attributed these problems to an inadequate number of personnel, lack of independence for the quality office, and unclear communications guidelines.[25]

In commission interviews, Marshall's quality officials described how their work had changed greatly from the Saturn era. In 1965 the Center's Quality Laboratory had 629 people; the lab independently analyzed and tested hardware built by the Center. After abandoning Arsenal practices in the seventies, the contractors oversaw quality, and NASA relied on inspectors from the Air Force or the Defense Logistics Agency. In 1985 Marshall's quality office had only 88 inspectors who tracked problems reported in formal documents, and checked that the Center and contractors were addressing anomalies. Center officials acknowledged some lapses in documenting criticality and launch constraints. Nonetheless, Center Director Lucas said the safety program "wasn't silent. It might not have been as noisy as it should have been" and "probably was not as strong as it should have been because we didn't have the personnel."[26]

Lucas and Wiley Bunn, director of the quality office, agreed that the commission misunderstood quality practices in Marshall's matrix organization. Rather than merely the responsibility of special inspectors, quality and safety were the primary charges of the Center's Science and Engineering Directorate. Lab specialists were studying the joint problem, project officials were reporting it in flight readiness reviews, and both had determined that no hazard existed. However, the quality office lacked resources to duplicate research and therefore it depended on the labs for engineering analysis and accepted their judgment that the joint was safe. Quality officials had no reason to "lay down in front of the truck," Bunn explained, because "the truck wasn't even coming." He regretted that 51–L had resulted from incorrect judgments rather than an inspection or reporting error. "Had the problem with 51–L been a clear quality escape," Bunn said, "in other words the area I'm responsible for had overlooked something that had resulted in the tragedy, it would have been better for NASA, it would have better for this Center, and better for the people involved in the decision to fly."[27]

Bunn also regretted that no one in his office or the labs made statistical correlations of O-ring damage with leak check pressure or temperature.[28] Indeed the presidential commission had ignored how this failure was symptomatic of NASA's antipathy to "numerical risk assessment." Here the Agency's technical engineering practices lagged behind the military and the nuclear industry which had routinely used statistical methods since the 1970s. Developed by Bell Labs and the Air Force, the system sought to help decision-makers by providing a probabilistic statement of risk. This computer-aided technique traced the causes of potential malfunctions back through every subsystem to identify parts most likely to fail.

During the lunar program, however, the Agency had bad experiences with probabilistic risk assessment. When General Electric, using primitive techniques, determined that the chance of a successful landing on the Moon was less than five percent, NASA abandoned the practice. Will Willoughby, the head of the Agency's quality office during Apollo, said "Statistics don't account for anything. They have no place in engineering analysis anywhere." NASA engineers were uncomfortable with probabilistic thinking and argued that meaningful risk numbers could not be assigned to something as complicated and subject to changing stresses as the Space Shuttle. Thus the Agency did not normally require statistical assessments for its hardware.

NASA used a more qualitative approach called "failure mode effects analysis," or FMEA, developed by the Agency and Boeing in the 1960s for the Apollo Program. It emphasized engineering analysis during the design stage rather than risk assessment in the operational stage. Rather than assign probability estimates to parts or systems, failure mode analysis identified worst case problems. Engineers could then design critical parts for reliability. Failure mode analysis worked well during the Apollo Era because NASA had the money to develop several different designs and then could choose the best.[29]

When NASA began using numerical techniques, assessments of the solid rocket boosters became political. In 1982 the J.H. Wiggins Company determined that the boosters were the highest risk on the Shuttle and likely to fail on 1 of 1,000 flights. Challenging this, the Space Shuttle Range Safety Ad Hoc Committee said the study had included data from primitive military solid rockets and that improvements made the Shuttle's boosters likely to fail on 1 of 10,000 flights. In 1983 Teledyne Energy Systems estimated the probability of failure was 1 in 100 flights, but a 1985 study by JSC (Johnson Space Center) put the failure rate at 1 in 100,000 launches, a prediction which was 2,000 times greater than the performance of any previous solid rocket.[30] Presidential commission member Feynman compared informal estimates from NASA engineers and managers and found that the engineers expected failure in 1 of every 200 or 300 launches while the managers expected failure in 1 of every 100,000. Feynman concluded that the manager's "fantastic faith in the machinery" precluded realistic judgments.[31]

Some Marshall veterans attributed the poor judgments to a decline in the technical culture of the Agency. The abandonment of the Arsenal system and the adoption of contracting, the retired German rocket engineers observed, had meant a loss of "dirty hands engineering" at Marshall. Karl Heimburg, who had headed the Test Lab, believed that the in-house design and development of prototypes produced more reliable technology than contracting and ensured that civil servants understood the hardware. Walter Haeussermann, former chief of the Guidance Lab, said that "if the engineer has only to supervise, without going and directing experiments, he is not as familiar with it. Finally, you get a paper manager." A 1988 survey of NASA employees found that less than 4 percent of professional workers spent most of their time at hand-on jobs and 76 percent worked most of the time at office desks.[32]

The presidential commission attributed some of the risky decisions to an "optimistic schedule" for Shuttle launches imposed by NASA and the Reagan administration. The commission found no "smoking gun" that showed that the Reagan administration had applied pressure to any NASA official to launch 51–L on 28 January. However the administration and Agency had maximized total flights in order to minimize the cost per flight and please commercial customers. The Shuttle had flown 9 missions in 1985, and officials had been confident that they could fly 15 in 1986 and 24 in 1990. Consequently they had assumed the Shuttle was "operational" and safe rather than experimental and risky, reduced tests to free up money for flying, accepted problems rather than apply costly fixes, and subordinated reviews of past performance to planning future missions.[33] After the accident, some in the news media acknowledged that they had applied pressure to NASA by criticizing the Agency for missing its schedules.[34]

Marshall personnel were very aware of schedule pressures. The RIFs of the 1970s had made Center personnel sensitive to meeting schedule and budget requirements.[35] Personnel evaluations in the Agency were based in part on schedule criteria and several Shuttle officials at Marshall and other Centers received salary bonuses for staying within time constraints.[36] Marshall engineers used the expression "get under that umbrella" to show desire to finish a task on time.[37] Moreover, when the Center had been the source of delays, such as with development work on the Space Shuttle main engines or launch postponements due to propulsion problems, NASA Administrator Beggs had been critical.[38]

Time pressure affected the mentality and decisions of Center officials. Sneed, assistant director for Policy and Review, recalled that Marshall had been "budgeting to fly" rather than to make long-term improvements. "Because we were flying the thing at the rates we were," he recalled, "most of our attention—our management attention, our engineering attention—was on flying the next vehicle. Maybe more so than looking and saying, 'Well, how did that last one fly?' and 'What is wrong with the last one, and what do we do to make it better, to make it more reliable?'" The Center, Sneed said, "didn't have time to stop and fix and end flight; you had to continue to fly and try to get your fixes laid and incorporated downstream."[39]

The pressures had intensified by late 1985. In December 1985, Jesse Moore, Level I Shuttle manager, set a goal of 20 flights per year by FY 1989 and requested that this objective be the principal item for discussion at the February Management Council Meeting. In the meantime Moore suggested that between flights NASA should only make modifications that were "mandatory for reliability, maintainability, and safety." After Marshall had delayed launch of 61–C because of a troublesome auxiliary power unit in the SRB, Arnold Aldrich, the Level II manager, wrote that the Shuttle program was "proud of calling itself 'operational.' In my view one of the key attributes of an operational program is to be able to safely and consistently launch on time."[40] During the 27 January teleconference, Allan McDonald of Thiokol recalled, Lawrence Mulloy observed that the 53-degree criteria would jeopardize NASA's plans to launch 24 shuttle flights per year by 1990, especially those scheduled from Vandenberg Air Force Base in northern California.[41]

Nevertheless, Marshall officials denied that they had sacrificed safety to meet the schedule. They believed that they had carefully reviewed the joint problems throughout the Shuttle's flight history and that schedule pressures had not affected their decisions. No Center employee who participated in the 51–L teleconference believed that schedule pressure had affected decisions. George Hardy, the highest ranking engineer present, said Science and Engineering was responsible for safety, not for schedule or the flight manifest. Ben Powers said that lab engineers referred to the schedule and money concerns of the program office as "bean counting." Center Director Lucas observed that "there is always schedule pressure," but "I don't know of anybody at Marshall who would deliberately, knowingly, take a chance just for the sake of schedule. We had never done that before. We'd been called down from launches, and I didn't feel any pressure and I didn't think that [for 51–L] there was any pressure."[42]

Finally, the presidential commission attributed the accident to Marshall's "management isolation" and a failure to communicate bad news, especially with the Level II office in Houston. The commission found it "disturbing" that "contrary to the testimony of the Solid Rocket Booster Project Manager [Mulloy], the seriousness of concern was not conveyed."

Aldrich, and Jesse Moore, the Level I manager, said they had not been informed of the launch constraint, the O-ring anomalies on flights late in 1985, the temperature concerns, or the teleconference. They admitted that NASA had confusing communications requirements, but thought the NASA custom was to report concerns about criticality 1 hardware. Aldrich also said he had not known that the Center had ordered steel SRB cases with the capture feature lip in July 1985; the budget channel for Marshall's Shuttle work came through Headquarters rather than the Shuttle Program Office at JSC.[43]

Although the commission report did not explain the communications problems, Commissioner Feynman did in his autobiography. Center rivalry and budget pressures, he reasoned, led NASA managers to think like businessmen who wanted only good news.[44] In any event, the commission recommended that NASA improve its communications requirements, strengthen Shuttle management, and "take energetic steps to eliminate this tendency [to isolation] at Marshall Space Flight Center, whether by changes of personnel, organization, indoctrination or all three."[45]

The notion that Marshall had a closed culture and had tried to hide the O-ring problems was believed throughout the Agency. Given the long-standing rivalry between the Centers, the view was prevalent at Houston. Astronaut Story Musgrave said "the trail goes on and on and on, and it turns out that the trouble is endemic to a major part of the organization." One JSC official said, "Nothing was ever allowed to leave Marshall that would suggest that Marshall was not doing its job. Everything coming out of that Center had to have 'performance' written all over it." Moreover, Marshall's culture was not open enough to detect and solve problems; superiors had been unwilling to hear bad news and subordinates had been unable to make themselves heard. Jack James, an astronaut instructor, said "if you have too closed a shop, you get in-grown and convoluted." Chris Kraft, the former director of JSC, wondered if Marshall had decided to keep problems to itself because the authoritarian management of Administrator Beggs, his Associate Administrator Hans Mark, and Associate Administrator for Space Flight, General James Abrahamson had created "underground decision-making" throughout the Agency. Marshall officials, Kraft speculated, "knew that if they made it [the O-ring problem] visible it would be hell to pay."[46]

Aerospace scholars used long-standing stereotypes to explain Marshall's apparent provincialism. Alex Roland, a space historian at Duke University, said "von Braun set up Huntsville as a feudal state with himself as lord of the manor. He insisted on a high degree of autonomy, and as a result Huntsville was and is highly defensive and combative, almost a bunker-style mentality." John Logsdon, an aerospace policy expert at Georgetown University, thought "there is a certain closed character about Marshall, an unusual arrogance, and at the same time a paranoia, perhaps because it has been a place that the Office of Management and Budget wanted to close."[47]

The presidential commission sought evidence of a cover-up and Marshall's closed culture. Investigators found no evidence of an after-the-fact cover-up and little clear evidence of closed communications within the Center. Investigators never found the anonymous middle manager who penned a vituperative attack on the "feudalistic" management of Director Lucas. Signed "Apocalypse," the letter said Lucas was intolerant of dissent, used a "good old boy" promotion system, and tried to "cover up" O-ring problems. Lucas allegedly had a flawed flight readiness philosophy; "for someone to get up and say that they are not ready is an indictment that they are not doing their job." Problems, the letter said, were "glossed over simply because we were able to come up with a theoretical explanation that no one could disprove," and "if no one can prove the hardware will fail, then we launch." The commission, however, never found Apocalypse.[48]

The theme of bad communications was taken up in a management study by Phillip K. Tompkins. In interviews conducted in January 1990, he asked middle and high ranking Marshall managers, almost none of them from inside the Shuttle organization, about communications under Center Director Lucas. Tompkins believed that subordinates felt intimidated by Lucas; they feared his tendency to "kill the messenger" bringing bad news and so they censored bad news or sugarcoated problems. The result was a "paranoid organization" that could not discuss problems or communicate them to outsiders.[49]

In interviews with commission investigators in 1986, however, Marshall personnel defended the openness of the Center. Engineer after engineer said that Marshall management was open, but insisted on facts to corroborate opinions. Bunn told the commission "if there's one thing that Dr. Lucas really doesn't like, it's for somebody to tell him something that they don't know. He can't stand that. Or somebody to know something and not tell him."[50] In later statements, Marshall personnel and contractors defended Lucas. Bob Marshall, a Center propulsion engineer, said that "the institution takes on the character of the lead manager because his style is emulated in those who work with him" and "we are a disciplined organization. We are also a driven organization." Joe Moquin, president of Teledyne Brown Engineering, said "He was demanding. He demanded the facts and substantiation of the facts. He could be tough on the experts." The president of Rockwell International's Rocketdyne division wrote Lucas that "You have set standards that we must maintain. After all our internal reviews, we always asked the final question, 'Will Dr. Lucas accept our logic?'"[51]

Marshall personnel also denied that their Center had failed to communicate the O-ring situation to the rest of the Agency. Center officials believed they had reported what they knew about the booster joint to "everyone" and Mulloy said he had told the truth during reviews and commission hearings. Kingsbury argued "I don't want to take exception to the commission's report," but "I don't know how they came to the conclusion that we are autonomous. . . . I don't believe we're autonomous or isolated." Lucas later said that the charge of isolation was "probably one of the most hurtful things because it's the furthest from the truth. The readiness reviews were held in the presence of Headquarters and everybody else" and "the weakness in that particular design joint had been recognized by Marshall, by Johnson, by Headquarters, including the Administrator." He believed that "the only thing I know of that was not common knowledge was the description of what occurred the night before, the so-called very hard arguments about whether we're ready to fly or not and apparent fact that the management of Thiokol applied pressure to their engineering people."[52]

The disagreement between the presidential commission and Marshall was essentially a matter of chronology: When did responsible Center officials know that the booster joint was unsafe? The commission's answer, stated baldly, was that the joint had always been hazardous and that Marshall had hard evidence of the danger from the beginning. Rather than admit failure, the Center discretely began repairs, and deliberately glossed over bad news through the launch of 51–L. If Marshall had communicated the bad news, the commission implied, wiser heads in Houston or Headquarters would have stopped flight until the joint was fixed. This assumption that more complete communications would have produced solutions or stopped the launch of 51–L was pure speculation. Would officials without expert understanding have stopped flying a joint verified safe by experts from the contractor and NASA's propulsion Center? No.

The response of the Marshall engineers and managers was that the joint was always "safe" in the sense that they lacked convincing contrary evidence. Successful launches had confirmed its reliability, and so the Center had little bad news to report and much good news to believe in. Even so the Center had continued studies, introduced short-term improvements, and begun long-term redesign. Although the Center had no excuse for not always communicating all the information and minority views, Marshall officials had typically described the strengths and weaknesses of the joint and their rationale for believing in its safety. When had they known the joint was unsafe? After 51–L.

When the commission published its report on 6 June, Center workers naturally had mixed feelings. John Q. Miller said "I personally have not seen any indications that there has been any lapse in concerns over safety here" and "we thought the necessary precautions had been taken." Feeling betrayed, one engineer, an 18-year NASA veteran, said "we were working overtime to give Mr. Rogers everything he wanted," but the commission criticized the Center unfairly and "nobody in NASA has stood up to defend us." Dr. Lucas said he viewed the report "as an assessment of a mistake that was made, or mistakes, perhaps, and it's going to enable us to fix problems and move on with the program as it should be" and he promised that "not one single word will be taken lightly."[53]

Scarcely had the Center absorbed the commission report, when Congress held its own hearings. The hearings before the House Committee of Science and Technology and Senate Space subcommittee mainly duplicated the anachronisms of the commission and assumed that decision-makers had known the joint was unreliable before 51–L. The main congressional contribution was in making second-guessing and scapegoating explicit. Congress complained that the commission report should have named names. Representative James H. Scheurer (D–NY), wanted to "find out what NASA officials knew and when they knew it." Senator Donald Riegle (D–MI) said "every single person that didn't behave and function properly has got to be identified and some kind of disciplinary action has to be taken." They wanted irresponsible civil servants held accountable and removed from the chain of command; this would ensure that in the future NASA officials would follow procedures. The most challenging questions were directed at Marshall officials. Senator Ernest Hollings (D–SC) blamed "Lucas policy" for creating "a cancer at Marshall" and said "that fellow [Mulloy] either misled or lied" to the commission.[54]

Faced with such comments, NASA officials said if they knew then what they knew now they would have stopped flight, but they did not doubt the joint then. Mulloy explained that 51–L happened because "I wasn't smart enough, the people who advised me weren't smart enough, the contractor wasn't smart enough . . . the people who review my activities weren't smart enough. . . . No one was smart enough to realize what was necessary." After the accident, he said, "knowing that something has failed, one might be able to recognize better what might have precluded it." Some Headquarters officials, including the Level I Deputy Director L. Michael Weeks, acknowledged that they had known of the O-ring problems from the August 1985 briefing. Dr. James Fletcher, who again became NASA Administrator in June 1986, told Congress that "Headquarters was at least as much to blame as other parts of the organization. I don't think all the responsibility should reside at the Marshall Space Flight Center."[55]

Other NASA veterans questioned putting the blame only on Marshall. Kraft said, "You have to fault the Johnson Space Center just as much as the Marshall Space Flight Center. They knew the goddamn thing was bad. It was written up in their files over and over again. That came out in the Rogers' Commission explanation. I don't know why the whole system allowed that to continue to fly. They are all to blame. Every goddamn one of them are to blame."[56]

The pressures helped several Marshall officials decide to leave the Agency. By the end of 1986, Hardy, Mulloy, Reinartz, Kingsbury, and Lucas had retired. Kingsbury said of his long-time friend and boss that Lucas had received a "bum rap" for 51–L. Instead Lucas should have gotten credit for initiatives that had diversified Marshall. "Before Lucas we had just been a propulsion Center. We built rockets. But under his direction we have branched out into Spacelab, the Space Telescope, a major role in the Space Station—all the things that have made Marshall a more viable, more important part of the American space program." Lucas, Dr. Ernst Stuhlinger recognized, had "directed more space accomplishments than almost any other NASA director." Kraft, a rival and ally from Houston, recognized the constraints on Marshall and NASA, writing Lucas that "those of us in the forefront of NASA, particularly the Center directors in the manned space flight programs, have an insight into the management of NASA over the last 10 years which no one else has even an inkling of. Maybe someday, when all the present trauma passes, we will be able . . . to tell the real history of the situation. At any rate, you and I know what had to be endured and the accomplishments that were brought about in spite of these inadequacies."[57]

In summary, the conclusions of the presidential commission were a mix of fact and fallacy. On the positive side, they revealed real problems about technology, resources, schedule, and communications and helped NASA find solutions. Revelation of the problems, and NASA's promise to fix them, removed suspicions and allowed the Agency to win the congressional support necessary to return the Shuttle program to flight. On the negative side, the commission engaged in scapegoating that put unfair blame on a few individuals. While this may have satiated the psychological needs of the nation and the political needs of powerful people inside and outside the Agency, scapegoating led to widespread misunderstanding of the accident, the Space Shuttle, and the process of development of high technology by complex organizations. Scapegoating also damaged the reputation of Marshall and NASA and left a legacy of bitterness and perceived injustice among many Center veterans. Only time would tell whether such sentiment would actually close the culture that the investigations had sought to open.

Recovery and Redesign

The recovery from the disaster and preparations for a return to flight began almost immediately after 51–L. The Center and the Agency reorganized Shuttle management, improved communications, and revitalized safety and quality programs. Bolstered with extra appropriations, Marshall redesigned and tested the SRM joints and improved other Shuttle hardware. The recovery culminated in the launch of STS–26 in September 1988.

As the 51–L investigation progressed, NASA administrators recognized that the Shuttle flights would be delayed for a considerable time. With the overall goal of a "conservative return to operations," NASA began studies of problems in the Shuttle program and in the Agency as a whole.[58] Organizational studies conducted by committees led by astronaut Robert Crippen and former Apollo manager Sam Phillips complemented the recommendations of the presidential commission and the House Committee. By the fall of 1986, the implementation of the recommendations was well underway.

NASA's organizational changes sought to open communication and centralize direction by copying parts of the Apollo Program. Dale Myers, a former Apollo manager who returned to the Agency as deputy administrator, said the reforms would "reduce the trend toward parochialism that tended to grow at the Centers under the pre-*Challenger* accident management style." The reforms strengthened the Management Council and established an independent quality and safety office. Headquarters devoted more full-time personnel to the Shuttle program; a deputy director for Shuttle operations, a new official, would work from the Cape; he would have a small staff at each Center, manage the flight readiness reviews, and direct the launch decision process.

Many of the reforms helped Headquarters and the Centers exchange information. The reforms increased the authority and access of the JSC Level II office. A Level II deputy director managed the day-to-day Shuttle program and directly supervised the manager of the Shuttle projects office at Marshall; both officials would be responsible to Headquarters rather than to any Center director. In addition, the Level II office was brought into the budget process of Marshall and all other space flight Centers; Marshall's director would still submit requests for Shuttle funding to the Headquarters program director, but the Level II manager would offer an assessment. The Level II office also penetrated deeper into the Shuttle organization by strengthening its engineering integration office and by using astronauts as liaisons with technical teams at the Marshall Center. Bob Marshall, the new manager of Center's Shuttle projects office, said that the new structure would "assure that in our discussions and in the problems that we have to address that we have not left someone out or bypassed them."[59]

In addition to a new Shuttle projects manager, the Marshall Center had personnel changes in several offices including the SRB project manager, director of Science and Engineering, and Center director. Marshall's new director was J.R. Thompson, who had worked at the Center from 1963 to 1983. Thompson had managed development of the Space Shuttle main engines; "I've blown up more engines," he said, "than most of those guys have seen." After leaving NASA in 1983, he went to the Princeton Plasma Physics Laboratory before returning to direct the technical aspects of the 51–L investigation for the presidential commission.

To restore Marshall's reputation and recover the Shuttle program, Thompson recognized that improved technical analysis and communications were necessary. He believed "they've done it better at Marshall than anybody else had been able to—but that's still not near good enough." In reference to the commission's charge that Marshall had been isolated and closed, Thompson said, "When I was there, I was not aware of it. If you go back through the twenty years I was there and that was true, then I was part of the problem. But in the spirit of accepting the commission report, I'm going to assume there's probably some substance there and we're going to fix it. . . . We will open up that communication."[60]

Thompson later recalled that when he became director in 1986, NASA had lost some of the "internal tensions between Centers and within a Center" that he had remembered from the early 1980s. During Shuttle design, development, and testing, experts from within Marshall and across the Agency had quarreled about technical issues. The conflicts, which often seemed like wasteful infighting to outsiders, were actually sources of strength which had deepened thought and improved technology. When the Shuttle became "operational," however, Thompson believed that all of NASA "got too comfortable" with the Shuttle and stopped looking for problems and arguments. Headquarters had imposed the goal of making the Shuttle pay for itself and so ground tests were reduced and criticism muted. One of Thompson's goals as Center director was to cultivate openness and allow free discussion of problems.[61]

His candor showed in a reply to Aaron Cohen, the director of JSC. Cohen had forwarded a memo from John Young, chief of the astronaut office, that had criticized Marshall's solid rocket tests. Thompson reassured Cohen with a technical explanation, and then, in a hand-written note, he said, "I appreciate John's assessment on this and other items. We'll keep him informed of our progress and where we're wrong. JR."[62]

Thompson improved the Center's internal and external communications. He made impromptu visits to Center work sites, ended the executive luncheons on the ninth floor of Marshall's Headquarters building and ate in the cafeteria, initiated more employee socials and old-timers gatherings, improved media access, facilitated exchanges and meetings with other Centers, and encouraged Marshall employees to take temporary assignments at Headquarters. To open decision-making, Thompson created a Marshall Management Council and expanded attendance at meetings. The Center fostered participative management, offered monetary rewards for suggestions, and established quality control circles called NASA Employee Teams.[63]

Alex McCool, who became director of Marshall's quality office, said Thompson wanted to make "a cultural change" at the Center by trying "to keep us talking together, working closer together, communicating." McCool explained that "Prior to *Challenger*, we had a kind of 'kill the messenger' syndrome. In other words, [if] somebody brings bad news, man, shoot him. We had that. The Agency had that, particularly at this Center" and "if you'd bring bad news, first thing you know all the bosses would jump on you. And there you are on the defense."[64] Accordingly the Center's management training program sought to teach openness. In one such program in April 1987, middle managers, after hearing a Thompson speech, offered anonymous comments on what they had learned: "survey results at MSFC indicates worst Center in NASA for communications; separate technical differences from personal relationships; taking a position is not as important as surfacing all sides; be prepared to defend and support positions with both the pros and cons; don't allow ourselves to become 'comfortable' in our technical and managerial jobs to the point that 'feedback' data is either ignored, overlooked, or not evaluated."[65]

The Center director tried to be the model of a participative leader. He began meetings by asking the question, "What are the problems?" Robert Schwinghamer, director of the Materials Lab, said that Thompson ran meetings much like von Braun and both men created a climate in which people said what they thought and "nobody feels like he's inhibited anymore."[66]

The new Center Director reorganized Marshall to improve rocketry engineering and management. On the laboratory side of the Center, he sought to bring propulsion specialists together. He divided the old Office of the Associate Director for Engineering into offices for Space Systems and Propulsion Systems, and gathered rocket engineers from several laboratories into a new Propulsion Laboratory. On the project side, the Shuttle Projects Office reorganized for the recovery and for later return to flight. Two offices merged to form the Space Shuttle Main Engine Office which began developing an alternate turbopump and testing the main engine. The SRB Project Office created a Systems Management and Integration Office to handle project control and contractor management.[67]

Center and Agency programs in flight safety and technical quality also restructured in the post-*Challenger* reforms. People throughout the Agency recognized that safety functions had to be strengthened. McCool said that after 51–L the Agency developed "an obsession" to "do the job right." Everyone recognized that "we can't have another *Challenger*. The nation can't stand it. I'm saying... we probably wouldn't have NASA with another *Challenger*." McCool kept a billiard ball on his desk to remind him that he was "behind the eight ball" and had to do a good job.[68] As part of the reforms, NASA opened a confidential hotline for reporting safety problems, trained engineers in quality control, increased use of statistical risk and trend analysis, and standardized procedures for tracking significant problems. The Shuttle program developed a computerized database to support trend analysis and problem reporting. NASA moved away from cost-plus-incentive-fee and cost-plus-fixed-fee contracts that subordinated quality standards to cost and schedule requirements, seeking to enhance safety and quality by using cost-plus-award-fee contracts with specific quality requirements and incentives, and putting quality experts on Award Fee Boards.[69]

The Center established a new Safety, Reliability, and Quality Assurance Office to consolidate the old Marshall Safety Office, the Reliability and Quality

Assurance Office, and part of the Systems Analysis and Integration Laboratory. The office gained greater capacity to make independent judgments by employing more civil servants with expertise in quality, hiring a support contractor, separating from the Science and Engineering Directorate, and reporting directly to the Center Director and the new associate administrator for Quality at NASA Headquarters. The office employed an astronaut as liaison to facilitate communication with the astronaut office at Johnson Space Center.[70]

Two oversight panels, one of NASA personnel and another from the National Research Council (NRC), studied the Agency's quality control programs and proposed improvements. The independent panels worried that NASA still had not corrected some flaws in the quality organization that had contributed to the accident. They worried that NASA performed hazard analysis after-the-fact rather than as part of the design process, implemented quantitative risk assessment too slowly, and clogged communications between flight managers and organizations responsible for inspection, tests, and repair. They fretted that NASA's matrix organization could jeopardize the independence of quality engineers and that the proliferation of Shuttle boards and committees could lead to "collective irresponsibility." This complicated, multilayered organization, the National Research Council worried, could "lead individuals to defer to the anonymity of the process and not focus closely enough on their individual responsibilities in the decision chain." Nonetheless, both committees decided the quality and safety systems were sound and represented progress over pre-*Challenger* days.[71]

During the reorganization, the Shuttle program reviewed the safety of all Shuttle flight hardware, software, and ground support equipment. The work was painstaking and Marshall people met the challenge with a spirit of self-sacrifice. Many Center employees delayed retirement to help. Many more worked 60- or 70-hour weeks for the 32-month recovery effort. Special teams implemented the recommendations of the presidential commission. System design reviews identified problems for redesign and improvement. As if the Shuttle was flying for the first time, new design certification reviews verified that all hardware met contract requirements, passed qualification tests, and had proper documentation. The Shuttle Projects Office reviewed the external tank, Space Shuttle main engines, and the solid rocket boosters. With the assistance of Level I and Level II, the office also reevaluated all failure mode and effects analyses, critical items lists, and hazard analyses. New rules for the critical items lists substantially increased the number of items designated as criticality 1 or single point failures. Rather than just designating a subsystem like a turbopump, the new rules included several parts of the pump. Bob Marshall, manager of the Shuttle Projects Office, said "I agree with the new ground rules because it has put more potential failure modes under more controlled approach and review."[72]

Although the external tank and Space Shuttle main engines had not caused the 51–L accident, the Center performed reviews and introduced improvements. Marshall and Martin Marietta made few modifications to the external tank, but changed test and checkout procedures. They improved the tank's lightning grounding system and studied proposals for its use with an unmanned Shuttle.[73] For the Space Shuttle main engine, Marshall and Rockwell International's Rocketdyne division enhanced safety and reliability by increasing performance margins and durability. They modified the vibration damper system for the turbine blades in the turbopumps, strengthened the main combustion chamber, redesigned a temperature sensor, ensured redundancy in the hydraulic actuators, improved the electronic engine controller, and added latches to hold open the fuel disconnect valves between the main engines and the external tank. Performance rules became more conservative with power levels of 104 percent during a normal launch; the previous norm of 109 percent power would be used only during emergencies and tests would be run at 113 percent. Ground tests became very rigorous and included tests with built-in flaws and margin tests to destruction to determine weak links. Static firings totaled more than 83 hours, the equivalent of 50 Shuttle missions. Although most of the firings occurred at NASA's Stennis Space Center in Mississippi and at Rocketdyne's Santa Susana Field Laboratory in California, some took place in Huntsville, where Marshall's Saturn SI–C test stand, rechristened as the Technology Test-Bed, became a site for main engine tests.[74]

The solid rocket motors, of course, underwent the greatest modification, and the Marshall members of the SRM redesign team deserve the greatest credit for the successful return to flight. Particularly important were personnel from the Structures and Propulsion Lab. Not only did Marshall personnel determine the technical cause of the accident and analyze the weaknesses in the motor joints, but the Center also conceived the solution.

Marshall, in response to presidential and congressional directives and technical imperatives, adopted an unusual organization for booster redesign. To prevent the redesign teams from becoming isolated, a problem that the presidential commission had believed contributed to the accident, Marshall sought openness through an elaborate system of cross-checks which gave overlapping responsibility to numerous organizations. The Center, according to SRB Project Manager Gerald Smith, "probably violated every management rule that you would ever have the occasion to violate in trying to do the program." Marshall's SRB accident investigation team under John Thomas and its SRB redesign team under Kingsbury merged in April under Thomas. To generate the best ideas, Thomas's team in Huntsville worked separately from a Morton-Thiokol team in Utah; the teams met regularly to compare ideas and select the best designs. The Marshall team included about 100 Center specialists and engineers from other NASA Centers, and another 200 experts from Martin Marietta, Lockheed, Wyle Labs, Teledyne Brown, United Space Boosters Incorporated, Rockwell International, McDonnell Douglas Technical Services, and Morton-Thiokol in Huntsville. The entire redesign process came under scrutiny of experts from Headquarters, JSC Shuttle program and astronaut offices, other NASA Centers (Langley, Lewis, and KSC), the solid rocket industry, the Jet Propulsion Laboratory, the Air Force Rocket Propulsion Lab, the Army Missile Command, and the National Research Council (NRC). Also maintaining surveillance were officials from congressional committees, the General Accounting Office, and the Federal Bureau of Investigation. Public interest in the program was intense, and the redesign team responded to more than 2,300 letters offering criticism and advice.[75]

The most important oversight came from a National Research Council panel for SRM redesign. The NRC panel, which had been formed at the suggestion of the Rogers Commission, monitored the entire redesign effort and participated in nearly 100 meetings, technical interchanges, reviews, conferences, and site visits. The panel drafted reports with criticisms and recommendations about all aspects of the redesign, and pressed NASA to conduct a thorough test program. Oversight by the NRC played a determining role in the success of the redesign.[76]

Managing a program with so many overlapping responsibilities and so much political interest was very difficult. Many people also felt depressed, Smith observed, because they felt responsible for 51–L, and "were absolutely devastated from the accident." Marshall's solution was the "open door policy." Thomas and other managers of the redesign team, Smith said, "made it very clear at meetings that if anyone had a concern or issue, let's raise it. Do not, do not hold back. If you've got a problem, let's say it. If you don't like a decision, let's hear it, let's talk about it." Solving technical problems required that people communicate bad news and that they know they were "not going to get punished for it."[77]

In order to expedite the recovery, NASA renegotiated its contract with Morton-Thiokol. After the accident, the company had been willing to accept a $10 million penalty for failure of its hardware, but had refused to sign a document admitting legal liability. Consequently, NASA and Morton-Thiokol negotiated a deal that would avoid litigation and return the Shuttle to flight as quickly as possible. The company accepted a $10 million reduction of its incentive fee and admitted no legal liability. It would perform at no profit approximately $505 million worth of work to redesign the field joint, reconfigure existing hardware, and replace motor hardware lost with 51–L.[78] Congressmen questioned this agreement, which seemingly rewarded Morton-Thiokol for its deficiencies. But NASA had few choices given the pre-existing contract and pressures to return to flight quickly.[79]

Throughout the redesign period, NASA quality experts remained troubled with Thiokol's organization. A June 1986 review of the firm's management by Air Force and Marshall inspectors rated all functions as "satisfactory," except for Safety and Engineering, which rated "marginal." Even though Marshall established a resident

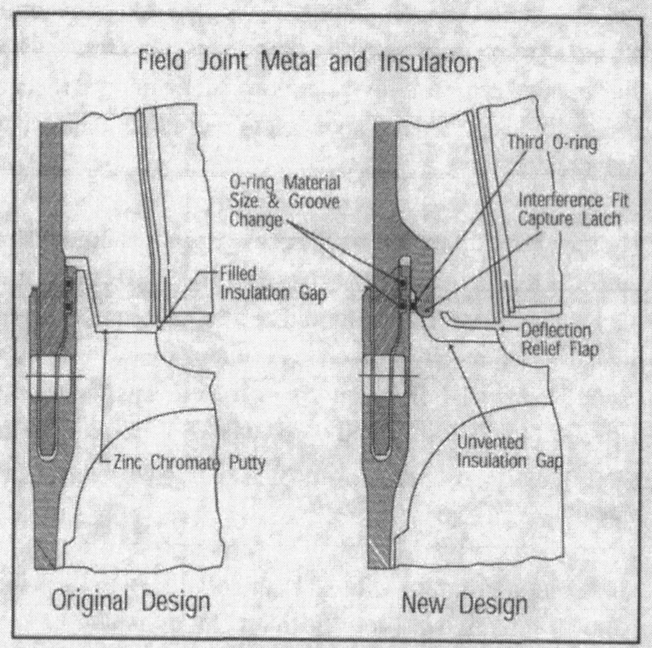

Comparison of original and redesigned SRM case field joints.

quality office in Utah, troubles continued. A March 1987 Marshall review concluded that Thiokol quality and manufacturing personnel paid "an inordinate amount of attention to schedule." In August the Marshall Center resident manager for quality worried that the firm's quality program was "in a mode of complete capitulation to schedule pressure" and told Thiokol management that "quality and safety will not be compromised blindly to meet a 'schedule.'" Marshall ordered the firm to give quality managers more authority, track information more carefully, and surface bad news more readily. A JSC quality inspector complained that the Thiokol attitude was "just tell me what you want me to do and I'll do it" and attributed the company's lack of initiative to "NASA's constant criticism and overmanagement."[80]

Marshall imposed strict requirements for the redesigned motor joints and changed the design from a dynamic seal activated by ignition pressure to a quasi-static seal that was not pressure dependent. The technical requirement specified that the seals be redundant, verifiable, and perfect; the redesign would tolerate no blow-by or erosion.[81]

By August 1986, Thomas, as leader of the redesign team, announced the new concepts. The case-to-case field joints had several improvements that added redundancy and safety margin (see the illustrated comparison of original and redesigned SRM case field joints). The engineers deleted putty from the design and protected the joint from hot gases with insulation formed into a rubber J-seal, a flap inside the case that closed with motor pressure. The steel capture feature lip reduced joint deflection, created an extremely tight fit between tang and clevis, and maintained contact between the O-rings and sealing surfaces. By changing only tang segments, NASA saved money by using its clevis segment inventory. The capture feature also housed a third O-ring and a silicon filler to protect the primary O-ring. The combination of the J-seal, capture feature, and third O-ring prevented combustion gases from reaching the primary O-ring.

In addition, a second leak check port added above the primary O-ring ensured it was in sealing position. Custom shims between the outer surfaces of the tang and clevis maintained proper compression on the O-rings. External heaters maintained joint temperature at 75 degrees; rubber and cork sealed the heater bands to the case and kept rain out of the joints. Longer pins that joined the segments and a reconfigured retainer band increased the margin of safety.[82]

Joint Environment Simulator test at Morton-Thiokol, November 1986.

The redesign team also improved other parts of the solid rocket motors (SRMs). They reworked the case-to-nozzle joints on much the same principle as the case-to-case joints; they deleted putty, used radial bolts to join the metal of the case and nozzle more tightly, modified and bond the insulation, incorporated a third O-ring, and inserted an additional leak check port. Moreover, modifications improved the factory joints, nozzle, propellant contours, and ignition system. The team also redesigned ground support equipment at KSC to minimize case distortion during handling, improved the measurement of segment diameters to facilitate stacking, minimize risk of O-ring damage during assembly, and enhance leak tests.[83]

Although experts from NASA, the solid rocket industry, and the National Research Council questioned the complexities of the design, they gave preliminary approval.[84] Marshall and Morton-Thiokol then began tests to verify their ideas. The test program for redesign was much more thorough and realistic than the original test program and this rigor was the key to the successful return to flight. The tests proceeded in a hierarchy from tests of components to subsystems to full-scale motors. Laboratory and component tests verified the properties of the joint parts. Subscale tests simulated gas dynamics and thermal conditions for components and subsystems. Hydraulic tests of full-scale segments tested the new joint and seal configuration.

Unlike the original test program, both the Center and its contractor built simulators to study joint behavior and test designs. Marshall's Transient Pressure Test Article (TPTA), built in 1987, used a short SRM stack with two field joints, a nozzle joint, 400 pounds of fuel, a motor, and an igniter. During the two-second firing, the simulator added one million pounds of weight to simulate the rest of the solid rocket booster and applied stress from three struts to duplicate the loads from the external tank. This recreated the dynamic loads on the joint during ignition and allowed engineers to gather information from 1,500 data channels. Morton-Thiokol operated a similar apparatus called a Joint Environment Simulator. Motor engineers conducted 16 simulator tests under different temperatures and with intentionally flawed configurations. The introduction of deliberate flaws was also a departure from the original test program.[85]

The recovery also had five full-scale, full-duration static firings, including two development motor tests and two qualification motor tests. Because of problems simulating flight conditions in static tests of solid motors, the National Research Council initially questioned whether the firings verified the design. After the second firing, John Young, JSC special assistant for Engineering, Operations, and Safety complained that "the motors were fired with dubious conditions which MSFC maintained would not have been allowed in the flight motors. This attitude, which accepts uncertain conditions, cannot be tolerated if we wish to be successful in space flight with humans." He argued that allowing phenomena that were "not fully understood and where we are not convinced beyond any doubt that the seal in its application will stop the flow, we could be back in the STS 51–L mode." Gerald Smith, SRB project manager, recalled that Marshall tried to duplicate flight situations by testing with intentional flaws. Introducing deliberate flaws was also controversial, however, because many worried that a failure would delay the program. They developed confidence in their designs by first testing with flaws in simulators. After such tests the Center used a production verification motor to test the flight configuration in August 1988. Royce Mitchell, SRM project manager, said "the hardware and data show that the booster is ready to fly. We demonstrated that the motor is fail-safe."[86]

Indeed the tests made Marshall very confident in the redesign. Gerald Smith said that "the testing we've conducted has been unprecedented and our understanding of the system is thorough. We've established the testing standard for the entire solid rocket industry. NASA's solid rocket booster program, I feel, is the yardstick against which future programs will be measured." As early as January 1987, J.R. Thompson told Congress that the tests showed "that the insulation does not leak hot gas even if not bonded, and that gapping is so small that any candidate O-ring material, even the old fluorocarbon material, can remain sealed with a 200 percent factor even with two of the three O-rings missing."[87] To ensure proper assembly of the first redesigned flight motor, Thompson dispatched a Marshall team headed by John Thomas to Kennedy Space Center to direct the process.[88]

Firing room celebration after launch of STS–26.

During the test program, NASA reformed its launch rules and procedures. Crippen, a former astronaut and the first NASA deputy director for Shuttle operations, wanted to eliminate ambiguities in launch criteria and "make sure we had clean lines of responsibility and authority." The Agency reviewed all Launch Commit Criteria and established a clear one for temperature. J.R. Thompson suggested that ambient temperature should not fall below 40 degrees at any time during the 24 hours prior to launch; "the specific temperature," he said, "is not magic, but near the spirit." The Level I Flight Readiness Review now required discussion of launch constraints and waivers. A Launch-Minus-Two-Day Review formally verified any changes after the Level I review. For the first time project managers from the contractors joined the Mission Management Team and had authority to stop the countdown without permission of a field Center. A Space Shuttle Management Council, composed of the associate administrator for space flight and the directors of Johnson, Kennedy, Marshall, and the National Space Technology Laboratories, became senior launch advisors. In early summer 1988 a launch simulation checked the new system. In addition, safety and budget concerns led NASA to constrict the Shuttle's flight schedule, which would escalate over several years to a maximum of 16 Shuttle flights per year, 8 fewer than pre-*Challenger* goals.[89]

On 29 September 1988, 32 months and $2.4 billion after 51–L, the recovery came to a close with the final countdown for STS–26. During that conclusive interval, "the biggest change," according to Lovingood, "was people were frightened, including me" and he was too afraid to watch the launch. Most members of the redesign team, however, were confident and were eager for the flight. Employees in Huntsville locked their eyes on the television and some dressed in "green for Go!" Dr. Wayne Littles, head of Marshall's Science and Engineering, watched the launch from the Huntsville Operations Support Center and said, "for the first two minutes [of ascent] you could hear a pin drop."

At the Cape when the solid rocket boosters ignited and *Discovery* lifted off the pad, even staid project managers shouted "Go!" and released months of tension. Cary Rutland, manager for booster assembly, said "I hollered when it lifted off, and I hollered when the solids separated" from the Shuttle. Gerald Smith gushed "this was probably the most exciting day of my life. It was unbelievable. When the solids ignited, I was probably holding my breath. When they separated, I think I yelled 'War Eagle.' I'm not sure." The launch was flawless and in the post-launch press conference, J.R. Thompson said "one good launch doesn't make a space program, but it's a damn good start." He then pulled out a foot-long Jamaican cigar and said, "I'm going to get me a cigar, light my pipe, and get a little glass of bourbon."

In the flush of success, some engineers became philosophical. Garry Lyles, chief of liquid propulsion at Marshall, observed that the Center would probably not get much credit for the successful launch even though they received most of the blame when *Challenger* failed. "We do a lot of patting each other on the back," he said, "We have a very professional organization. Whether anyone outside pats us on the back, it really doesn't matter."[90] Thompson

Pallet-Mounted Instrument Pointing System, first used on Spacelab 2.

derived lessons from 51–L and the recovery, believing that space exploration required that everything be "perfect" and without that, "we're gonna end up back on the beach."[91] Because of the improvements in technology and Center culture, Marshall people believed they and the Shuttle were stronger than before the accident. The successes of the post-*Challenger* Shuttle flights gave supporting evidence for their assessment.[92]

---

1 Deborah Hoop and Martin Burkey, "Marshall Workers 'Sickened,'" *Huntsville Times*, 28 January 1986; Francine Schwadel, Matt Moffett, Roy J. Harris, and Roger Lowenstein, "Thousands Who Work on Shuttle Now Feel Guilt, Anxiety, and Fear," *Wall Street Journal*, 6 February 1986.

2 Wayne Biddle, "Two Faces of Catastrophe," *Air and Space Smithsonian* (August/September, 1990), pp. 46–49; Joseph Trento, *Prescription for Disaster* (New York: Crown, 1987), chap. 10–11; Ronald Reagan, Executive Order 12546, 3 February 1986, PC I pp. 212–13; "Commission Activities," PC I, pp. 206, 208; Gaylord Shaw and Rudy Abramson, "Days Following Disaster Damage NASA Image: Did Graham Botch the Aftermath?" *Huntsville Times*, 2 March 1986.

3 Kathy Sawyer and Boyce Rensberger, "Tenacious Challenger Commission Likely to Set New Standard," *Washington Post*, 8 June 1986, pp. A1, A14.

4 "Lee, Thomas Named to Task Force," *Marshall Star*, 12 March 1986, p. 1; Bob Lessels, "Return to Flight Began Minutes After Accident," *Marshall Star*, 21 September 1988, p. 3; Ron Tepool, Oral History Interview (hereinafter OHI) by Jessie Whalen, 31 October 1988, pp. 12–13.

5 William Lucas, "Remarks to Employees," 30 January 1986, MSFC History Office; Robert Dunnavant, "Marshall tells Workers not to Talk," *Birmingham News*, 31 January 1986; Robert Dunnavant, "Space Center has withdrawn into its Shell," *Birmingham News*, 2 February 1986; Charles Redmond, NASA HQ, Public Affairs Office, "Challenger Chronology Input from MSFC," n.d., probably fall 1986, NASA History Office.

6 PC I, p. 206; Thomas Gieryn and Ann E. Figert, "Ingredients for a Theory of Science in Society: O-Rings, Ice Water, Clamp, Richard Feynman, and the Press," in Susan E. Cozzens and Thomas Gieryn, eds., *Theories of Science in Society* (Bloomington: Indiana University Press, 1990), pp. 67–97; Philip Boffey, "NASA Taken Aback by Call to Screen its Inquiry Panel," *New York Times*, 17 February 1986, pp. A1, A14.

7 See Kevin Klose and Boyce Rosenberger, "Rogers Says Process at NASA 'is Flawed,'" *Washington Post*, 28 February 1986, Rogers quoted; Philip Boffey, "Space Agency Hid Shuttle Problems, Panel Chief Says," *New York Times* 11 May 1986; Martin Burkey, "SRB Joint Said 'Hopeless,'" *Huntsville Times*, 14 March 1986; Charles Fishman, "Questions Surround NASA Official," *Washington Post*, 24 February 1986; Philip Boffey, "Zeal and Fear Mingle at Vortex of Shuttle Inquiry," *New York Times*, 16 March 1986.

8 Redmond, "Challenger Chronology," NASA History Office.

9 Martin Burkey, "Lucas: Launch Process Not Flawed," *Huntsville Times*, 28 February 1986; Charles Fishman, "5 NASA Officials Stand by Actions," *Washington Post*, 1 March 1986.

10 *Ibid.*

11 Charles Fishman, "NASA Data May Alter Seal Theory," *Washington Post*, 7 March 1986, pp. A1, A20.

12 Redmond, "Challenger Chronology"; Charles Redmond, [Notes from Interviews with MSFC PAO on 51–L], n.d., probably fall 1986, NASA History Office.

13 Maura Dolan, "MSFC Key Officials Termed 'Insensitive,'" *Huntsville Times*, 4 March 1986; Redmond, "Challenger Chronology," NASA History Office.

14 Bill Sneed, OHI by Jessie Whalen, 8 April 1988, p. 25; William Lucas, Oral History Interview by Andrew J. Dunar and Stephen P. Waring (hereafter OHI by AJD and SPW), Huntsville, 3 November 1992, pp. 12–14, 29–30. Richard G. Smith, director of KSC and a former MSFC official, criticized the commission for sensational comments that were misleading and unfair to people who had been doing their best. 51–L, Smith observed, had been "an accident" rather than a crime or a fraud. See R.G. Smith, "Statement," 15 March 1986, PC 008211–12, RCR.

15 PC I, pp. 198–201.

16 "NASA Accident Analysis Team Report," Appendix L, PC II, quoted p. L–49. See also PC I, pp. 40–72.

17 Keith Coates, OHI by Jessie Emerson, 1 August 1988, p. 8.

18 Judson A. Lovingood, OHI by Jessie Whalen, 16 August 1986, p. 23.

19 Martin Burkey, "MSFC Official Doubts Booster Theory," *Huntsville Times*, 2 March 1986.

20 "NASA Accident Analysis Team Report," PC II, Appendix L, pp. L.4, L.37–L.45.

21 James Kingsbury, OHI by AJD and SPW, 3 March 1993, Huntsville, pp. 1–14, quoted pp. 7, 8.

22 PC I, Appendix C, pp. 219–223; PC II, pp. L.45–L.49.

23 Michael Isikoff, "Fatal Flaws in Shuttle Pinpointed," *Washington Post*, 9 April 1986, pp. A1, A24; "Booster Certification Process Raises Further Shuttle Design Flaw Concerns," *Aviation Week and Space Technology*, 14 April 1986.

24 J.R. Thompson, OHI by AJD and SPW, 26 August 1994, Huntsville, p. 8.

25 PC I, pp. 152–61, quote p. 152.

26 Lucas OHI, 3 November 1992, p. 17; Wiley C. Bunn, Presidential Commission Interview (hereinafter PCI) by E. Thomas Almon, 3 April 1986, pp. 5–18, quote p. 18; Henry P. Smith, PCI by E. Thomas Almon, 2 April 1986; Arthur Carr, PCI by E. Thomas Almon, 2 April 1986; Jackie C. Walker, PCI by E. Thomas Almon, 3 April 1986; James O. Batte, PCI by Robert C. Thompson, 2 April 1986, pp. 1–28; George Butler, PCI by E. Thomas Almon, 2 April 1986, RCR; Kathy Sawyer, "NASA Cut Quality Monitors in Seventies," *Washington Post*, 8 May 1986, pp. 1, 8; Mark Tapscott, "Chief Engineer Says Cuts Hurt NASA's Safety," *Washington Times*, 5 March 1986; Michael Brumas, "NASA Defends Quality Assurance Methods," *Birmingham News*, 22 May 1986.

27 Wiley Bunn, PCI by John Fabian, Randy Kehsli, Emily Trapnell, and E. Thomas Almon, 17 April 1986, pp. 26–28, 56, 64–66; Bunn PCI, 3 April 1986, pp. 27, 35, 42, quote p. 29, 18.

28 Bunn PCI, 17 April 1986, pp. 37, 64–66.

29 Haggai Cohen, PCI by Robert C. Thompson, 14 April 1986, pp. 19–23; "Failure Mode and Effects Analysis and Critical Items List," 17 March 1986, 51–L Criticality 1 file,

NHDD; Stuart Diamond, "NASA's Risk Assessment Isn't Most Rigorous Method," *New York Times*, 5 February 1986, p. A25; Kevin McKean, "They Fly in the Face of Danger," *Discover,* April 1986, pp. 48–58; Trudy E. Bell and Karl Esch, "The Space Shuttle: A Case of Subjective Engineering," *IEEE Spectrum* 26 (June 1989), pp. 42–46.

30 Bell and Esch, "Space Shuttle," pp. 42–46; "NASA Ignored Booster Study?" *Huntsville News*, 2 June 1986.

31 Feynman, PC II, p. F–1; Feynman, *What Do You Care*, pp. 179–83.

32 J. Michael Kennedy, "It Takes Pride in being 'Excellence Oriented,'" *Huntsville Times*, 3 March 1986; Michael Tackett, "NASA Experts Insist Shuttle Blast Wasn't in the Data," 16 March 1986; Howard E. McCurdy, "The Decay of NASA's Technical Culture," *Space Policy,* (November 1989), pp. 301–310, statistics 309–310; Howard E. McCurdy, *Inside NASA: High Technology and Organizational Change in the U.S. Space Program* (Baltimore, Johns Hopkins, 1993), chap. 5.

33 PC I, pp. 164–76; Ray Molesworth, "Report on Alleged White House Pressures to Launch 51–L," 8 May 1986, PC 148946–149154; Comptroller General, "Report to Congress: NASA Must Reconsider Operations Pricing Policy to Compensate for Cost Growth on the STS," 23 February 1982, PC 023753, RCR.

34 David Ignatius, "Did the Media Goad NASA into the Challenger Disaster," *Washington Post,* 30 March 1986, pp. D1, D5.

35 Jerry Cox, MSFC, PCI by Ray Molesworth, 25 March 1986, pp. 13–15.

36 Martin Burkey, "Besieged NASA officials Received Bonuses," *Huntsville Times*, 16 March 1986.

37 Coates PCI, 25 March 1986, pp. 31–32; 43–44.

38 Rudy Abramson (*Los Angeles Times*), "Shuttle Probe Looks at 'Human Frailty,'" *Huntsville Times*, 16 February 1986; Jim Leusner and Dan Tracy (*Orlando Sentinel*), "Was Center 'Beat Up,' Pressured to Launch?" *Huntsville Times*, 3 March 1986.

39 Bill Sneed, OHI by Jessie Whalen, 8 April 1988, pp. 12–13.

40 J. Moore, HQ, "Continued Improvements to Shuttle Turnaround," 23 December 1985, Shuttle, General 1986 folder; Mulloy, "STS 61–C Abort Problem," 6 January 1986, Weekly Notes; A. Aldrich, JSC, "STS 61–C Launch," 14 January 1986, 61–C folder, MSFC Archives; PC I, pp. 174–75.

41 McDonald Testimony, 25 February 1986, PC I, p. 741.

42 Hardy Testimony, 26 February 1986, p. 867; Powers PCI, 13 March 1986, pp. 48, 60–69; Adams PCI, 13 March 1986, p. 55; Wear PCI, 12 March 1986, p. 80; Lucas OHI, 3 November 1992, p. 15.

43 PC I, pp. 148, 200, quote p. 85; Moore and Aldrich Testimony, 27 February 1986, PC V, pp. 1047–58; Aldrich Testimony, 4 March 1986, PC V, pp. 1490–91.

44 Feynman, *What do You Care*, pp. 212–19.

45 PC I, p. 200.

46 Henry F.S. Cooper, "Letter from the Space Center," *The New Yorker* (10 October 1986), pp. 83–114, quote pp. 92, 96; Christopher Kraft, OHI by AJD and SPW, 28 June 1991, pp. 19–21. See also "Summary of March 24–25, 1986 Meeting with the Mission Planning and Operations Panel of the Presidential Commission," 1 April 1986, 51–L Mission Documents, JSC History Office; Trento, pp. 260–61.

47 Laurie McGinley and Bryan Burrough, "After-Burn: Backbiting in NASA Worsens the Damage," *Wall Street Journal*, 2 April 1986; Rudy Abramson (*Los Angeles Times*), "Tragic Price: Competition Between NASA Centers Seen as Factor in Accident," *Huntsville Times*, 8 June 1986.

48 Philip M. Boffey, "Alleged Destruction of Data at NASA Center is Studied," *New York Times*, 17 May 1986; Ben Powers and Robert Gaffin, PCI by Ray Molesworth, 16 May 1986, RCR; Martin Burkey, "Commission Wants to Talk to Letter Writer," *Huntsville Times*, 12 March 1986; "Apocalypse" letter, 6 March 1986, PC 167128–30; Bill Bush, PCI by Ray Molesworth, 26 March 1986, 4–5, RCR; Robert Dunnavant, "Challenger Explosion Clouds Close of Lucas' Stellar Career with NASA," *Birmingham News*, June 1986, Bush quoted.

49 Phillip K. Tompkins, *Organizational Communication Imperatives: Lessons of the Space Program* (Los Angeles: Roxbury, 1993), chap. 10.

50 Bunn PCI, 3 April 1986, p. 65. See also Wear PCI, 12 March 1986, pp. 83–84; Miller PCI, 13 March 1986, pp. 28–32; Smith PCI, 13 March 1986, pp. 52–53; Riehl PCI, 13 March 1986, pp. 21–23; Coates PCI, 25 March 1986, pp. 46–47; Ray PCI, 25 March 1986, pp. 48–52; Cox PCI, 25 March 1986, pp. 5; Lovingood PCI, 23 April 1986, p. 33ff; Lindstrom PCI, 5 May 1985, pp. 32–34.

51 Robert L. Hotz, "Space Shuttle Disaster Casts Lasting Stigma on Marshall," *Atlanta Journal*, 7 July 1986; Dunnavant, June 1986; R. Schwartz to Lucas, 6 June 1986, Dr. Lucas' Book of Letters, MSFC Center Archives.

52 Jay Reeves, "Testimony Correct, Mulloy Insists," *Huntsville Times*, 10 June 1986; Marshall Ingwerson, "Recovering from the Challenger Disaster," *Christian Science Monitor*, 8 July 1986; Lucas OHI, 3 November 1992, pp. 18, 22.

53 Jay Reeves, "MSFC Engineers Dispute Claims of Rogers Report," *Huntsville Times*, 12 June 1986; "Report Unfair, Says MSFC," *Huntsville Times*, 11 June 1986; "Commission Treated Center Fairly in Report, Says Lucas," *Huntsville Times*, 11 June 1986.

54 "Lucas and Mulloy are Targets at Hearing," *Huntsville Times*, 11 June 1986; Randy Quarles, "NASA Officials: Close Booster Study Should have been Ordered Earlier," *Huntsville Times*, 13 June 1986; Howard Benedict, "Senators Blast Commission Report," *Huntsville Times*, 18 June 1986; Randy Quarles, "Adm. Truly Defends Marshall at Hearing," *Huntsville Times*, 11 June 1986; "Hollings Wants Blame Fixed For Shuttle Accident," *Defense Daily*, 18 June 1986, p. 265.

55 Charles Fishman, "NASA's Mulloy Acknowledges Preflight Errors," *Washington Post*, 12 June 1986; Philip M. Boffey, "Official Explains How He Failed to Prevent Space Shuttle Flights," 12 June 1986, Fletcher quote; Laurie McGinley, "NASA Headquarters Staff is Criticized at House Hearing on Shuttle Accident," *Wall Street Journal*, 13 June 1986; John Noble Wilford, "Congress to Press Shuttle Inquiry," *New York Times*, 15 June 1986.

56 Kraft OHI, 28 June 1991, p. 22.

57 Kingsbury quoted in Dunnavant, June 1986; Stuhlinger to Lucas, 4 June 1986; Kraft to Lucas, 11 June 1986, Dr. Lucas' Book of Letters, MSFC Archives.

58 Kingsbury quoted in Dunnavant, June 1986; Stuhlinger to Lucas, 4 June 1986; Kraft to Lucas, 11 June 1986, Dr. Lucas' Book of Letters, MSFC Archives.

59 R. Truly, Associate Administrator for Space Flight, "Strategy for Safely Returning the Space Shuttle to Flight Status," 24 March 1986, JSC, 51–L Mission Documents, JSC History Office.

60 Myers quoted in "Shuttle Management Revamped to Resemble that of Apollo," *Aviation Week and Space Technology*, 10 November 1986, pp. 30–31; R. Truly, "Organization and Operation of the NSTS Program," 5 November 1986 in NASA, "Response to the Recommendations of the House of Representatives Committee on Science and Technology Report of the Investigation of the Challenger Accident," 15 February 1987, Enclosure 2; Marshall quoted in "Shuttle Management Changes Announced," *Marshall Star*, 12 November 1986, pp. 1–2.

61 Kathy Sawyer, "Old Hand Takes up a Mandate to Break Space-Center Tradition," *Washington Post*, 30 September 1986.

62 Thompson OHI, 26 August 1994, pp. 5–7.

63 Thompson to Cohen, "Data from the SRM Tests," 16 June 1986, SRM Design Team 1987 folder, MSFC Archives.

64 "Thompson Tells Subcommittee Marshall 'Deeply Committed to Success,'" *Marshall Star*, 28 January 1987, p. 5; R.J. Schwinghamer, "Improved Internal Communications—Branch Chief and Team Leader Attendance at Lab Staff Meetings," July 1987, S&E folder; "1986 MSFC Productivity Accomplishments Report," 1986 and "NASA Productivity Improvement Quality Enhancement Program: 1986 Accomplishments Report," July 1987, Productivity folder, MSFC Archives.

65 McCool OHI, 7 March 1990, p. 33.

66 MSFC management training comments, 8 April 1987, Management Development Program folder.

67 Robert Schwinghamer, OHI by Jesse Whalen, 30 August 1989, pp. 17–18, 21.

68 "Thompson," *Marshall Star*, 28 January 1987, p. 4.

69 McCool OHI, 7 March 1990, pp. 29–30.

70 NASA, "Response to House," pp. 7–8; NASA PR 87–91, "NASA Established Safety Reporting System," 4 June 1987; George A. Rodney, Associate Administrator for SRM&QA, HQ, "Problem Reporting, Corrective Action and Trend Analysis," 6 July 1987; Rodney, "SRM&QA Criteria for the NSTS Award Fee Contracts," 1 June 1987; Bryan O'Connor, JSC, "Recommendations of Space Flight Safety Panel," 16 June 1987, Safety folder, MSFC Archives; Sneed OHI, 8 April 1988, pp. 19–21.

71 "Thompson," *Marshall Star*, 28 January 1987, p. 4.

72 NASA HQ, SRM&QA, "Final Report of the STS Safety Risk Assessment Ad Hoc Committee," 18 August 1987, pp. 7–11; Committee on Shuttle Criticality of the Aeronautics and Space Engineering Board of the National Research Council, "Post-Challenger Evaluation of Space Shuttle Risk Assessment and Management," January 1988, pp. 1–9, quoted p. 6, Safety folder, MFSC Archives.

73 NASA, "Response to House," pp. 1–4, 6–7; Bob Marshall, "Shuttle CIL Board," 22 July 1986; Arnold Aldrich, "Assignment of Level II Personnel to Support the FMEA, CIL, and HA Reevaluation Task," 2 October 1986, Shuttle 1986 General folder, MSFC Archives; Marshall quoted in Martin Burkey, "'Criticality 1' List Expanded for Shuttle, Advisers Told," *Huntsville Times*, 5 May 1987.

74 "Changes also Made to External Tank," *Marshall Star,* 21 September 1988, p. 11; Porter Bridwell and Mike Pessin, OHI by Jessie Whalen and Sarah McKinley, 18 December 1987, pp. 14–18.

75 Bob Marshall, MSFC to R. Kohrs, JSC, "House Science and Technology Comm. STS 51–L Recommendation," SRM 1987 Design Team folder, MSFC Archives; "Reports Indicate Engine Test 'Successful,'" *Marshall Star*, 23 July 1986, p. 1; "Teamwork, Harmony Sum Up Engine Improvement Efforts" and "Shuttle Main Engine Improvements," *Marshall Star,* 21 September 1988, p. 10; "Technology Test Bed Receives First SSME," *Marshall Star,* 30 March 1988, p. 1.

76 Gerald Smith, OHI by Jesse Whalen, 30 June 1989, pp. 7–8, 17–18; "SRM Redesign Effort Paved Way to Launch," *Marshall Star*, 21 September 1988, p. 12; "One Hill, Two Paths to Booster Redesign," *Marshall Star*, 21 September 1988, p. 8; Tepool OHI, 31 October 1988, p. 15.

77 NRC, *Collected Reports of the Panel on Technical Evaluation of NASA's Redesign of the Space Shuttle Solid Rocket Booster* (Washington, DC: National Academy Press, 1988).

78 G. Smith OHI, 30 June 1989, pp. 17, 22–24.

79 NASA PR–87–19, "NASA and MTI Reach Preliminary Understanding," 24 February 1987; "NASA, Morton Thiokol Agree on SRM Contract Modification," *Marshall Star,* 18 May 1988, p. 2; Vaughn, pp. 248–250.

80 Craig Covault, "Congress, Booster Manufacturers Criticize Joint Redesign Program," *Aviation Week and Space Technology*, 9 February 1987, pp. 116–17.

81 Major General Bernard L. Weiss, USAF to T. J. Lee, MSFC, "Contractor Operations Review Conducted at MTI," 4 August 1986, Safety folder; G. Smith to A. McCool, "Assignment of Co-Located Safety Office Personnel to the SRB Project Office," 19 February 1987, SRB 1987 folder; W.C. Bunn, "Review of MT System for Process Control," 24 March 1987, SRM 1987 Design Team folder; T.A. Lewis, MSFC to R.R. Bowman, "NASA SRM and QA Concerns," 14 August 1987, SRB 1987 folder; J. G. Thibodaux to Chief Propulsion and Power Division, JSC, "Visit to Solid Propellant Rocket Plants," 4 April 1987, Safety folder, MSFC Archives.

82 Bob Marshall, MSFC to R. Kohrs, JSC, "House Science and Technology Committee STS 51–L Recommendation," SRB 1987 Redesign folder, MSFC Archives.

83 "New SRM Design Announced," *Marshall Star,* 20 August 1986, pp. 1–2; "New Motor Design Changes," *Marshall Star,* 21 September 1988, p. 3; Martin Burkey, "MSFC Says Redesign of Booster Joints Cures Problem with Sealing in Cold," *Huntsville Times*, 13 August 1986.

84 "New SRM Design Announced," *Marshall Star*, 20 August 1986, pp. 1, 2; "New Motor Design Changes," *Marshall Star*, 21 September 1988, p. 3.

85 "NRC Commends Marshall Effort on Redesign," *Marshall Star,* 13 August 1986, pp. 1, 4; "Thomas 'Confident' with SRM Design," *Marshall Star,* 29 October 1986, pp. 1, 3; J. Thomas, MSFC to L. Wear, MSFC, "Solid Rocket Propulsion Industry Participation in Shuttle SRM Recovery Program," 26 January 1987, SRM Design Team 1987 folder, MSFC Archives; Couvault, 9 February 1987, pp. 116–17.

86 Chuck Vibbart, MSFC manager of TPTA, OHI by Jesse Whalen, 23 May 1988, pp. 1–11; "Full-Segment SRM Tests Begin," *Marshall Star*, 27 August 1986, p. 1–2; "Deliberate

Flaws Employed as part of SRM Test, *Marshall Star*, 4 March 1987, pp. 1–2; "First Transient Pressure Test Article Fired," *Marshall Star*, 25 November 1987, p. 1; "STA–3 Test Series Completed, Successful," *Marshall Star*, 13 April 1988, p. 1–2; "New Motor Design Changes," *Marshall Star*, 21 September 1988, p. 3.

87 Craig Covault, "Shuttle Booster Redesign, Tests Raise Schedule Delay Concerns," *Aviation Week and Space Technology*, 26 January 1987, pp. 26–27; John Young, JSC to Aaron Cohen, JSC, "Redesigned SRM Program Review," SRM Design Team 1987 folder, MSFC Archives; Smith OHI, 30 June 1989, p. 21; "Full Duration Motor Flaw Testing Philosophy," 1987, SRM Design Team 1987 folder, MSFC Archives; "PVM–1 Fired Thursday; 'Booster is Ready to Fly,'" *Marshall Star*, 24 August 1988, p. 1.

88 "One Hill," *Marshall Star*, 21 September 1988, p. 8; "Thompson," *Marshall Star*, 28 January 1987, p. 5.

89 John Thomas to Don Bean [?], [manuscript comments], 29 March 1993, MSFC History Office.

90 Edward H. Kolcum, "NASA Overhauls Shuttle Launch Decision Process," *Aviation Week and Space Technology*, 23 May 1988, pp. 20–21; J.R. Thompson, "Ground Thermal Environment," 1987 SRB folder, MSFC Archives; NASA, "Response to House," pp. 9–10, 11–12.

91 *Huntsville Times*, 30 September 1988; "The Magic Is Back!," *Time*, 10 October 1988, p. 21.

92 Thompson interview in Richard Lewis, "Space Shuttle: The Recovery" video, *Aviation Week and Space Technology*, (McGraw-Hill, 1988).

Chapter XI

Spacelab: International Cooperation in Orbit

Spacelab, one of Marshall's longest and most successful programs, is a Shuttle-based habitat that allows scientists to work in shirt-sleeves. Spacelab enabled NASA to accomplish several objectives. Commissioned in the aftermath of the 1972 decision to forego development of a large Space Station, Spacelab provided the Agency an interim means to conduct the types of space science experiments suited for a Space Station. Developed by European interests, Spacelab allowed the Agency to fulfill a mandate to foster international cooperation. With Congress pressing NASA to privatize, Spacelab gave the Agency a means by which American businesses and universities could conduct space science at a relatively modest cost.

The program also perfectly suited Marshall's needs. Any new start was welcome in the post-Apollo era, and Spacelab helped revitalize the Center. Spacelab also offered new opportunities, allowing the Center to pursue its goal of diversification into space science, systems integration, and orbital operations. By moving into new areas, Marshall created new alliances with scientists and engineers, and became the NASA installation with the greatest experience in international space ventures.

Sortie Can and the Spacelab Concept

Spacelab emerged from NASA's scramble to find successors to Apollo between 1969 and 1971. NASA planners had discussed transporting modules to space for some time, and had incorporated the concept into early Space Station studies in the late 1960s. In 1969, Associate Administrator for Manned Space Flight George Mueller proposed that NASA construct a semi-permanent Space Station by the mid-1970s by assembling a series of modules, each with its own function.[1] Marshall and Houston's Manned Spacecraft Center (MSC) planned for such modules in their early Space Station studies.

Over the course of the next three years, the plan for Spacelab emerged. Three key developments define Spacelab's early history: the assignment of Lead Center responsibility to Marshall; the decision to continue the module concept as a part of the Shuttle program after the deferral of a large Space Station; and the agreement to build Spacelab with the Europeans.

Marshall's designation as Lead Center for a manned module for space science seemed unlikely in 1969, when Huntsville still had a reputation as principally a propulsion center. That Marshall won the assignment owed both to efforts at Headquarters to divide tasks equitably between its major manned Space Centers and to aggressive efforts at the Center to obtain new business. Mueller was Marshall's most forceful advocate at Headquarters in the immediate aftermath of Apollo, and when discussing prospects for launching a Space Station by the mid-1970s, he suggested that Marshall would likely become the Lead Center if the project won approval.[2] When Houston became Lead Center for the Shuttle, Marshall was in line for compensation, and Spacelab offered some solace.

But compensatory awards alone would not have been enough had Marshall not demonstrated the capacity to manage such a program. *Skylab*, a program similar in many respects to Spacelab, provided just such a demonstration. Moreover, Marshall's expertise in propulsion gave the Center experience that could be applied to the laboratory. "It was in fact a pressurized structure," explained Marshall Spacelab Program Manager Thomas J. (Jack) Lee, and the Center's work with propellant tanks gave it knowledge about the operation of pressurized systems. Marshall knew "how to design, develop, qualify and have the in-house expertise to ensure that a pressurized structure in orbit was sound. In other words we had that technical capability. I think that's the reason that we got it."[3]

Concurrently the new Program Development Directorate began to seek more work for the Center, and payload development, management, and operations offered a fruitful new field. "We'd been into payloads even before we became a part of NASA," remembered William Lucas. "We began searching and looking in the field. What is there that needs to be done that we at Marshall can do? Where do we have the talents? What do our talents match?"[4]

O.C. Jean was one of those in Program Development who believed that payloads might offer the answer to Lucas's question. "Marshall Space Flight

Center needed an activity that would sustain its base without being slave to a development project," Jean recalled. Jean headed a group that included Bill Sneed and Bob Marshall. "We worked that problem for three months and came up with a recommendation of what Marshall should do." Their recommendations included work in payloads and development of a Marshall operations Center. Spacelab enabled the Center to pursue both goals.[5]

Pressure from another source pushed the Center in the same direction. Ernst Stuhlinger, the Center's associate director for science, advocated a Marshall specialty in payloads. He reported that scientists from around the country wanted to work with NASA, and expressed "a considerable willingness . . . to discuss space projects, and to develop plans for participation." The opportunity suited Marshall's needs and experience. "We did have a science component that was small but significant, and they had had an interest in payloads," Lucas continued. "Utilizing the science component of the Center . . . supported by the science community and universities" would allow Marshall to begin developing payloads. "We did not compete for small payloads. We thought that our expertise would lend itself to large systems."[6]

Program Development initiated a payload planning study that examined possible concepts for the Shuttle. On one level, the goal was to establish criteria for categorizing experiments by weight, size, mission duration, and orbital requirements in order to determine payload groupings and vehicle assignments. But Program Development also sought to ensure that Marshall would have a continuing role in payload management. A 1971 internal report established goals that would place Marshall in control of Shuttle payloads from inception through operational supervision:

- Establish MSFC's role in the *development* and *operation* [emphasis in original] of Shuttle payloads such as: RAM [Research Applications Module], Tug, and Space Station.
- Develop an operational concept for Shuttle utilization that establishes MSFC as the Center that:
  - Plans the mission
  - Aids and coordinates the experiment P.I.'s [Principal Investigators]
  - Has hard mock-up facilities to verify systems compatibility to actual flight hardware
  - Trains the P.I.'s that will make the flights
  - Recycles mission hardware

Thus while the primary goal was to establish policies for payload planning, Program Development wanted Marshall at the focus of that activity; the document twice emphasized that *"The concept must put MSFC between the Shuttle and the experiment P.I."* (emphasis in original).[7]

During 1971 it became clear that budget constraints would prevent NASA from developing both, Space Station and the Space Shuttle. As the Nixon Administration and NASA moved toward deferring Station and developing the Shuttle, the module concept offered a means to use the Shuttle cargo bay to house an orbiting laboratory. Although Shuttle flights would be of short duration, Research and Applications Modules (RAM), as they were now called, might provide opportunity for space science investigations in the years before a Space Station. NASA envisioned short-duration Shuttle flights, or sortie missions, employing RAMs for experimental work in astronomy, materials science, and manufacturing in space.

The Agency expected to develop manufacturing techniques for projects in crystal growth, metallurgical and glass processes, biological preparations, and physical and chemical processes in fluids. The Shuttle could accommodate a variety of payloads, but increasingly NASA began to focus on a pressurized payload carrier called the sortie can, which Headquarters considered "the least expensive and simplest member of the family of research and applications modules."[8]

In September 1971, Headquarters asked Marshall to conduct a design study of the Sortie Can. NASA envisioned the Sortie Can as a bare-bones pressurized module, and as a possible candidate for in-house development and manufacture. The Sortie Can would be suitable for short-duration missions of five to seven days, and could be extended from the Shuttle bay to enhance viewing capabilities for astronomy or Earth observations. Headquarters suggested Ames Research Center's high altitude test program as a model. Ames had used a converted Convair 990 for a variety of experiments, short lead-time between selection and flight, and an opportunity for investigators to assume direct responsibility for their experiments—all goals for the Sortie Can.[9] Marshall's assignment was comprehensive: the Center would have to design the module and develop plans for manufacture, test and inspection, and funding. At the time, NASA conceived Sortie Lab as an in-house project since the Agency could not expect additional funds for the coming fiscal year.[10] A small in-house team in the Preliminary Design Office worked from September 1971 to January 1972, when it recommended that the Sortie Can should be a cylinder 15 feet in diameter and 25 feet long. The study had substantial impact on the evolution of Spacelab design—it was "perhaps the most important" of the early studies, according to Douglas Lord, NASA's Spacelab director in Washington.[11]

By late 1971 the Program Development strategy for Marshall to move into manned science payloads began to bear fruit. The Sortie Can was but one of several payload studies assigned to the Center, and when NASA divided the $10.5 million allocated for experiment definition, nearly $7 million went to Huntsville. During the next several months, the Center conducted payload studies of many possible Shuttle cargoes, including the Sortie Can and other more complex RAMs, the High Energy Astronomical Observatory (HEAO), and an orbit-to-orbit vehicle called the Space Tug. Marshall moved into the forefront of NASA payload planning, conducting in-house studies while contractors worked on parallel investigations.[12] The Center's Sortie Can studies examined ways to use off-the-shelf laboratory equipment and investigated guidelines for temperature, acoustic, and pressure environments.[13]

Since NASA traditionally assigned development responsibility to the Center that managed definition studies, the payload studies carried with them the potential for substantial prolonged projects. With so much at stake, other Centers vied for a share, and Marshall once again found itself competing with Houston. Internal rivalries became endemic during the era of scarce resources that characterized NASA's post-Apollo years. Intercenter disputes were intense during the program definition phase when the Agency divided responsibilities; working relationships improved after Headquarters assigned tasks. But even after Headquarters divided the pie, competition continued in areas where responsibility was not clearly defined.

"I am sure that MSC will not be happy about their portion," Program Development Director James T. Murphy told Center Director Eberhard Rees after learning of Marshall's allocation for payload studies.[14] Similarly Rees worried that Houston might capitalize on its position as Shuttle Lead Center to seize other Shuttle-related programs. Coincident with early Space Station studies, Marshall developed a Concept Verification Test (CVT) project designed to use simulators to evaluate space activities proposed during station definition studies.[15] Rees worried his Space Station team was missing an opportunity to use CVT to support early Shuttle payloads. "If we don't change this attitude drastically," he cautioned, "we will find ourselves pretty soon out in the cold and MSC does the Sortie Can."[16]

Early in 1972, Headquarters directed Marshall to prepare for the Sortie Can definition phase, acknowledging the Center's work on the Sortie Can, RAM, and concept verification as "the hard core of our manned payload opportunities utilizing the Shuttle."[17] In April the Center established a Sortie Can Task Team headed by Fred Vruels of Program Development.[18]

A hazily defined area that opened an arena for Center rivalry was the question of which Center should work with customers who wanted to fly experiments on the Shuttle. With Marshall assuming responsibility for payloads and payload carriers, and Houston serving as Lead Center for the Shuttle orbiter, someone had to satisfy user demands and minimize impact on the Shuttle. Rees, following the strategy of always keeping the Center between the Shuttle and experimenters, suggested that since Marshall already had contact with the user community, it should coordinate. Users would "see" the Sortie Can or the tug, not the orbiter. MSC Director Chris Kraft countered that "the Shuttle/payloads interface is fundamental to the Shuttle development task," and insisted that Houston should reconcile user requests through an MSC Payloads Coordination Office.[19]

"Houston at that time seemed to want to control every interface with the Shuttle," recalled Lucas. "Ultimately it came out to be the logical thing that if Marshall's going to control the Spacelab, they need to control the people directly and then meet the interface with the Shuttle. You don't need to speak to someone in Houston to speak to your customer. . . . The logic is that as long as the Spacelab meets the established interface with the Shuttle, then why should the people responsible for Spacelab go through Shuttle management to get to Spacelab? That's the way it turned out to be. I like to think logic prevailed."[20]

The disagreement over user coordination was typical of the intercenter disputes that arose as Marshall diversified. The Center guarded its flanks to prevent other Centers from closing potential avenues of expansion. When MSC opposed initiation of a Shuttle Payload Data Bank study that would have enhanced Marshall's interface with Shuttle payload customers, Murphy acknowledged that "the objections to this study stem from the fact that MSFC has been posturing itself to play a key role in the Shuttle payloads business, and other

Centers are viewing MSFC's growing payload activities with some concern."[21] Rees also worried that Houston might encroach on Marshall's other emerging specializations. "I have been having a certain feeling for quite some time that MSC wants to wedge themselves into the Shuttle Payload business," he told his technical deputy, Lucas, in the fall of 1972. Rees believed that Houston would "try anything to get on the Payload and Tug Bandwagon," and cautioned that "we should be constantly aware of this tendency of MSC and fight it wherever we can."[22]

International Partnership

NASA had since its inception wanted international partners, an imperative that became more pressing after the 1969 Space Task Group included such a recommendation in its report.[23] NASA's tight budget made international participation more attractive. European interest in a cooperative venture also increased in the early 1970s. The European Launch Development Organization (ELDO) and the European Space Research Organization (ESRO)—already engaged in negotiations that would lead to the formation of an all-encompassing European Space Agency (ESA) in 1975—both explored the possibility of a joint venture with the Americans. By 1971, when it became clear that NASA's next major project would be the Shuttle, Europe's options narrowed to development of a specific part of the Shuttle (such as the payload doors), the Space Tug, or the Sortie Can.[24]

The European consortium spent $20 million on studies of the three alternatives, and in the process began working with Marshall.[25] During 1971 and 1972, ELDO conducted design studies of the tug under Marshall supervision. By February 1972, ELDO informed Marshall representatives that the Europeans were very interested in developing the Tug.[26]

Space Tug was "a natural" for Marshall, Lucas recalled, since it entailed a propulsion system and a Shuttle interface. In addition to its work with ELDO, the Center monitored Tug studies by American contractors McDonnell Douglas and North American Rockwell. Other NASA Centers and the Department of Defense helped develop design and interface requirements.[27]

Department of Defense participation doomed the hope that Space Tug might be an international program even before budget pressure forced NASA to abandon the concept. In June 1972 the Agency decided that the Europeans would not develop the Tug. "There was no way that was going to happen, not from NASA's standpoint but from the military's standpoint," explained Lucas. "That Tug was to serve both NASA's interest and the military's payload interests. The military certainly would not have been willing to have a foreign entity that they had no control over to be in the loop as far as their payloads were concerned."[28]

NASA also decided not to accept European participation in the development of the Shuttle. One assessment suggested that the Europeans lacked the organization, experience, knowledge, and laboratory depth needed to make much of a contribution to the Shuttle.[29] The Agency worried about dependence on foreign sources for critical items, and feared that it would lose more than it could gain.[30] The only alternative remaining for international participation was the Sortie Can, which Lucas said went to the Europeans as "sort of a consolation prize."[31] The Sortie Can required less advanced technology, and if it lagged in schedule or ran over budget, it would not affect the Shuttle.[32]

The Europeans hesitated to participate in development of the Sortie Can, however—and for good reason. Many Europeans questioned whether they had much to gain with Sortie Lab. Douglas Lord, who as director of NASA's Space Station Task Force negotiated with ESRO regarding participation on the Sortie Can, acknowledged NASA's advantages. "We are dealing with a potential supplier who is seriously considering investing $250 million of his own funds in the development of a spacecraft to be used primarily by the U.S.," Lord told Marshall. "This is not a typical joint venture since the direct benefits are heavily in our direction."[33]

NASA pressed the Europeans for a decision by September 1972, requesting a "start-to-completion" agreement.[34] The Europeans were not in a strong position to bargain, and would later admit that in 1972 they lacked confidence in their capabilities and believed they needed American assistance to establish their own manned program. Political scientist John Logsdon concluded that at least some of the Europeans were "willing to pursue cooperation on almost any terms, no matter how one-sided."[35] The Europeans could not be pushed into a hasty accord, however, and deliberations dragged past NASA's September target.

While the Agency conducted negotiations with the Europeans, Marshall continued its in-house definition studies of the research and applications carrier, which now bore the more elegant name "Sortie Lab." Lee succeeded Jack Trott as Phase B director of the Marshall task team. "I had a small staff of people in PD [Program Deployment], and then I drew on the whole of the engineering capability of the Center to put down the details of the design," Lee remembered. The in-house work preserved NASA's options in case the Europeans decided not to join. "We were pretty far along on the completion of that Phase B," Lee explained, "so that we could either try to build it in-house or go to contracting out."[36]

Center management followed the European negotiations with interest, since Headquarters told them that Marshall should expect a substantial role if the Europeans decided to participate. Rees told Program Development to begin planning Marshall's managerial approach if the Europeans accepted, since project management would be "somewhat different from our usual Phase C/D project management with American contractors."[37]

Marshall's role in the development of the Sortie Lab could not be defined until the Europeans decided whether to participate. The logjam began to break late in 1972 when the Europeans approved involvement by ESRO member states. At a European Space Conference in November, ministers removed obstacles blocking member nations from contributing to Phase B studies and endorsed formation of a single European space organization to supersede ESRO and ELDO.[38] In January ESRO voted to work on the lab. During the next four months, representatives of ESRO and NASA worked out the details that led to a Memorandum of Understanding. The Europeans agreed to develop a pressurized manned laboratory and an unpressurized instrument platform, or pallet. ESRO accepted responsibility for the "definition, design, development, manufacturing, qualification, acceptance testing and delivery" of an engineering model and a flight unit to NASA. They also agreed to provide ground support equipment and engineering support through the first two flights. ESRO agreed to deliver the flight unit one year before the first Shuttle flight, then scheduled for 1979. NASA would operate the lab and purchase additional units from the Europeans if needed, but the agreement did not guarantee additional purchases.[39]

Marshall's role evolved during the international conferences leading to the formal agreement. Headquarters insisted on "strong centralized management and coordination of all activities related to the Sortie Lab" under direction of an

Agency-level Sortie Lab Task Force. NASA Associate Administrator Dale Myers assured Rees that Marshall would be Lead Center, however, and directed the Headquarters task force to develop a plan for the eventual transfer of authority to a Marshall task team.[40]

The Center laid the groundwork for its assumption of Lead Center responsibilities. The Center reviewed the European Phase A Sortie Lab studies, and considered the results "reasonable." But reviewers lamented that the Europeans lacked understanding of orbiter interfaces and space limitations, and applied requirements so rigidly as to cause "extreme penalties on cost, weight, power, and other design factors." Marshall's reviewers determined to "prevent a similar happening during their Phase B."[41]

As in concurrent Shuttle development, cost became a major factor in Sortie Lab planning. The Europeans insisted on an escape clause that would allow them to back out if costs exceeded $300 million.[42] Lucas, technical deputy to new Marshall Center Director Rocco Petrone, advised that the Sortie Lab would have to be kept simple "to provide the greatest cost advantage," and directed Program Development to "maintain this cost consideration as a primary design driver."[43]

Selection of Marshall as Lead Center enabled the Center to resolve differences with the Johnson Space Center (JSC) over management of Sortie Lab. Huntsville requested JSC assistance on its Phase B studies, and the two Centers divided other responsibilities in meetings in the spring of 1973. Marshall would provide technical support to the Europeans related to the design and definition of the lab; Houston would provide interfaces for the lab with the Shuttle and direct overall safety, crew training and requirements, and mission operations.[44]

Marshall's efforts to define its Lead Center responsibilities for Sortie Lab provoked renewed concern in NASA over the larger issue of payloads. Late in 1972 Headquarters directed that the long-delayed Shuttle System Payload Data Study proceed, a decision that Marshall welcomed as "another step forward in enhancing MSFC's Shuttle Payload activities."[45] Marshall's role in payload management grew in the months that followed. Headquarters gave Marshall responsibility to integrate NASA's payload requirements, but also established a Payload Requirements Board staffed by representatives from Payload Program offices.[46] Even these assignments left questions unanswered and lines of authority hazy. Deputy Administrator George Low worried that the "the question of how Shuttle payloads will be handled and assigned within NASA is so important to the future of the Agency that it is not possible to address some of the lesser goals and objectives until it is resolved." He directed establishment of an Agencywide team under former Langley Deputy Director Charles J. Donlan to examine the distribution of payload responsibilities.[47]

By this time, however, Marshall's central role in payload development was well established, and in fact NASA augmented the Center's responsibilities a week after commissioning the Donlan study. Not only was Marshall to continue its current studies, but it would update the Shuttle payload model, conduct payload and mission planning, and develop payload accommodations for the Shuttle, Sortie Lab, and Tug based on comments from users.[48] Marshall's payload duties remained undiminished when the Donlan group submitted its report the following spring.[49]

Months of international negotiations culminated on September 24 in a formal ceremony in Washington when NASA Administrator James C. Fletcher and Dr. Alexander Hocker, Director General of ESRO, signed a Memorandum of Understanding. The accord established a Joint Spacelab Working Group (JSLWG)—soon dubbed "Jizzlewig'"'—to coordinate NASA and ESRO. With American and European program heads serving as co-chairs, the group could resolve technical and managerial issues, exchange information, and identify potential problems. Finally, Fletcher announced that the Sortie Lab would now be called "Spacelab," the name preferred by the Europeans.[50]

Building Spacelab

With the formalities of an international accord complete, Marshall assumed its role as Lead Center. The Center changed its internal management of Spacelab, moving it out of Program Development to a new Spacelab Program Office in December with Lee as manager.[51] Lee's first major task was to represent NASA during the European competition to select a prime contractor for Spacelab Phase C/D design and development. ESRO tried to achieve equity on its projects by seeking geographic distribution of contracts based on the financial participation of its member states. In the case of Spacelab, West Germany's 54.1-percent contribution placed it far ahead of second place Italy's 18-percent participation, virtually assuring that the prime contractor would be a German company. The leading contenders were two consortia: Messerschmitt-Bölkow-Blohm (MBB) of Munich, and ERNO of Bremen, a VFW-Fokker subsidiary. So close was the competition that the evaluation team refused to choose; an adjudication committee selected ERNO based on its management, technical concept, and design.[52]

Before the contract could be awarded, a serious problem emerged. When NASA Administrator Fletcher met with Dr. Hocker, he learned that both the MBB and ERNO proposals were overweight and would undercut payload capability. When Fletcher learned about the discrepancy, he insisted that the proposals were unacceptable, and that differences would have to be resolved before proceeding. Lee worried about holding the agreement together. "It is not black, but I have no idea how bright it will be," he reported as he anticipated another round of meetings. "We need to satisfy all parties concerned."[53]

Fletcher's reaction hit ESRO like a "bombshell," according to Lord. The new trans-Atlantic partnership entered its first crisis.[54] The European press criticized NASA. Typical was a Dutch newspaper that complained that NASA's action "took both ERNO and MBB completely by surprise." The paper blamed NASA for rejecting the design proposals "on the very moment that the Dutch space organization ESRO/ESTEC in Noordwijk wanted to place a contract with the European industry."[55] Lee helped to diffuse the tension, meeting with his counterpart Heinz Stoewer and ESRO Director General Roy Gibson and encouraging them to explain that the weight issue reflected a joint NASA/ESRO concern. Stoewer concurred, but ESRO insisted that the problem was less serious than NASA claimed, surely not so critical as to invalidate the award to ERNO.[56]

John F. Yardley, the new NASA associate administrator for Manned Space Flight, flew to Europe to help resolve the dispute. NASA and ESRO agreed to reduce weight and to develop different categories so that weights could vary from mission to mission. Fletcher and Hocker agreed that the issue was not so weighty as to force abandonment of the selection of ERNO as prime contractor. On 5 June, ESRO awarded the Bremen consortium a six-year, $226 million contract.[57]

The weight controversy demonstrated the fragility of NASA's relationship with the Europeans. In a legal sense, NASA and ESRO were partners; state agreements sanctioned the Memorandum of Understanding, and diplomats on both sides of the Atlantic celebrated the Spacelab agreement as symbolic of an international partnership. "It was very much, by necessity, a partnership relationship," Lucas insisted. "Europeans were very sensitive about that. They were supplying most of the money so you couldn't think of it as a contractor."

But in many ways NASA—and Marshall as Lead Center—found themselves acting as if ESRO was a contractor. Lucas acknowledged the dual nature of Marshall's position, explaining that the Center had to "act like we had a contractor but not let them know that. In other words, we had to give them a lot of guidance, but we had to do it in a discrete way rather than like you would work with a contractor here. . . . It's just much less direct than the contract relationship."[58]

Lee, who bore the major responsibility for Marshall's contact with the Europeans, told Lucas in 1978 that ESA "resents being treated like a contractor."[59] Lee understood ESA's concern, and years later he explained that the Europeans "made it very clear that ESA was not a contractor of NASA. We honored that. It was difficult sometimes because I found myself being the judge on the imposition of certain requirements." Lee tried to avoid dictating NASA specifications and requiring ESRO to impose them on the contractor; he sought instead to give basic requirements, inform Stoewer of the criteria that would be used to judge "whether what we were going to fly was acceptable," and allow ESRO to make major development decisions about how to proceed. Lee's approach applied what would later be called performance specifications. "I saw it better to let them have the flexibility of working against performance specification," he explained, "instead of me having to have to follow along with all the detailed specs."[60]

The weight controversy, although resolved amicably, exposed the potential for problems in this unusual relationship. And after resolution, anticipating a joint NASA and ESRO discipline-by-discipline review to ensure that ERNO's proposal matched the requirements stipulated by the NASA–ESRO agreement, Lee commented that the review would "allow a more thorough penetration on our part."[61] It was the language of a contracting officer, and although Lee did not specify whether he meant penetration of the partner or the partner's contractor, it was clear that the relationship was indeed unconventional.

The multinational character of ESRO also complicated relations with the Europeans. "Not only were we dealing with a different culture, but we were dealing with ten different cultures," Lucas recounted.[62] "Communication made it more difficult," according to Lee. Problems were not only cultural, but institutional. Lee believed that the program might have been completed sooner had it not been for the difficulties in getting agreements between ESRO's member states. "I suspect that we waited on them more than they waited on us," he said. Congress did not interfere with the relatively inexpensive Spacelab program, but ESRO operated under "more of a parliamentary process so quite often we would have to wait for a year. Ministers don't meet, and you don't call them together to deal with it."[63]

Selection of the prime contractor signified an important milestone. As the project moved into development, Marshall's role and Lee's responsibilities changed. "My role then became a little bit different. We weren't doing in-house design any more," Lee recalled. "We were more focused on what we considered a program function." Lord assumed NASA's Level 1 responsibilities at Headquarters in Washington; Lee's duties as program manager placed him at Level 2. ESRO established its development Center at the European Space Technology Center (ESTEC) at Noordwijk in the Netherlands. Lee and Stoewer, his European counterpart, met frequently and arranged for exchanges of information, means of monitoring progress, and program coordination.[64]

Marshall's relations with Houston also tested its diplomatic skills. Lucas tried for nine months to get Houston to assign an individual as "a single point of contact with authority to represent JSC on all Spacelab technical and programmatic matters." At one point he became so exasperated with Houston's failure to cooperate that he wrote on the margins of a note: "Don't want to call again. Just file as a reminder of how JSC cooperates with us."[65] Finally JSC appointed Glynn Lunney, who had been working on the Apollo-Soyuz Test Project.[66]

The liaison with Houston was critical since the two Centers had to coordinate interfaces for two projects, Shuttle and Spacelab, that were both in development; changes in one inevitably affected the other. "Spacelab ended up costing quite a bit more than the Europeans originally thought, partially because the Shuttle kept changing," according to Marshall's Stanley Reinartz. "And if you're trying to do two things in parallel, it can run up the bill, particularly if you're trying to do one thing in this country and one thing in another."[67] Both programs had to learn how to adjust. "On the front end, you sort of had the instinct that everything wasn't defined, but yet on the other hand you didn't know what it all was until you got in and started handling it," Lunney recalled. "The Marshall and ESA people would go back to the Spacelab project and get definitive [data and] we would go back to the orbiter and Shuttle program." Gradually a system evolved; by developing a series of interface control documents (ICDs), Houston and Marshall were able to coordinate the simultaneous development of the Shuttle and Spacelab.[68] With Lunney serving as liaison, coordination between Houston and Marshall improved.[69] Center rivalry diminished, James Kingsbury explained, as everyone in NASA worked hard "to show one front to ESA.'"[70]

Planning Spacelab Missions

While Marshall's program office coordinated Spacelab development, the Center's payload activities became more focused. Marshall's payload studies through the spring of 1974 concentrated on developing candidate payloads based on research at the Center and proposals submitted by users.[71] The Donlan committee report of April 1974 recommended establishing a Headquarters office with supporting activities at Marshall for payload planning and at Houston for flight planning and mission assignments. The committee also recommended that the Marshall Center handle integration and payload flight control for multipurpose Spacelab flights. Marshall would assemble and check out payloads for early Spacelab flights, then relinquish this duty to KSC. JSC would be in charge of Spacelab subsystems during flight as part of its Shuttle operations management.[72]

Marshall's Program Development office was at the Center of NASA's payload planning activities, taking a leading role on panels examining payload profiles for the first six Shuttle flights and for Spacelab. The Center chaired a NASA committee charged with defining payload requirements in light of Shuttle and Spacelab hardware design. Headquarters assigned Marshall responsibility for planning the first Spacelab mission, and the Center continued to work on a broader profile of the first 20 Shuttle missions.[73]

To coordinate its payload activities Marshall established a Payload Planning Office under Jean.[74] "O.C. Jean impressed me as a manager," remembered David Jex, who worked for him. "One philosophy that he espoused that always stayed with me was it doesn't matter who gets the credit as long as the work gets done."[75] In June 1975, Marshall shifted planning for the first Spacelab flight from Program Development to Jean's office.

Headquarters assigned Marshall management of payloads for the first three Spacelab missions, including responsibility to plan, develop, integrate and operate the payloads.[76] In one sense the assignment was a logical extension of the Center's development of Spacelab, particularly since the first two missions would verify Spacelab systems. NASA Chief Scientist John E. Naugle commented that "Introduction of another Center into the Shuttle/Spacelab/NASA/ESA operation would have converted a very complex barely manageable problem into a completely unmanageable one."[77] But the assignment also signaled the maturity of the Center's diversification into payloads, and gave Marshall the opportunity to broaden its experience in space science and operations.

While the principal task of the first two missions was to evaluate Spacelab systems, NASA believed there would be enough space, resources, and time available to conduct additional space science experiments. Marshall intended to incorporate several disciplines and experiments from European and American investigators to demonstrate the range of Spacelab capabilities for research.

The Marshall Center's payload work opened scores of opportunities, but like other diversification projects of the 1970s it also placed Marshall in competition with other Centers. Goddard Space Flight Center (GSFC) worried that Marshall's work in space science payloads might infringe on its specialties. Johnson Space Center found reason for concern in Marshall's involvement in operations, payload specialist selection and training, and life sciences.

Spacelab gave Marshall a chance to broaden the operations experience it acquired during *Skylab*, and although JSC preferred to manage all Shuttle-related operations, it accepted a role for Marshall. Early Spacelab missions required a dual structure for operations; JSC would have responsibility for the orbiter, Marshall for Spacelab payloads. Marshall's mission management team would work out of a Payload Operations Control Center (POCC) located in Building 30 at JSC, while orbiter operations would be conducted from Houston's Mission Control. In the POCC, Marshall's team could work side by side with ESA representatives and principal investigators whose experiments were aboard the orbiting laboratory.[78]

Although Houston accepted a Marshall role in operations, it was less conciliatory in relinquishing its monopoly over astronauts. Spacelab introduced a new category of astronauts, the payload specialists. Their selection process was different from that of traditional astronauts, and therein lay the basis for Houston's objections. An Investigators Working Group (IWG) comprised of the principal investigator (or chief scientist) for each experiment on the mission selected scientists or engineers as payload specialists. Not only would Marshall have influence in the selection process by virtue of its role in payload integration, but the Center would provide mission-specific training in Marshall's Payload Crew Training Complex (PCTC), thereby infringing on Houston territory (of astronaut training).[79] Houston Center Director Kraft objected vigorously to this process, arguing that Spacelab payload specialists ought to be "selected from the present corps of mission specialists residing in Houston" since they were suited by training, experience, and involvement in Spacelab design and development.[80]

Kraft's proposal made no headway against an approach already accepted by Headquarters, and Marshall relished its victory. "Dr. Kraft is going to fight the payload specialist philosophy that NASA has developed and we are implementing on Missions 1, 2, and 3," Jean informed Lucas. "His whole supremacy collapses if a non-JSC man flies in space. I believe we have the whip and can do the driving."[81]

More important was Marshall's expansion of its involvement in space science. Marshall and its predecessor organization, the Army Ballistic Missile Agency (ABMA), had long worked with outside scientists, but Spacelab offered the Center opportunity to expand this activity. In planning for Spacelab 1, for example, the Center selected experiments in 1976, and the following year brought all chosen principal investigators to Marshall to form the Investigators Working Group.[82] Spacelab also afforded a chance for Marshall to develop its own experiments and to attract scientists to work at the Center.

When Marshall began developing payloads in life sciences, eyebrows raised at JSC and Ames. Marshall had begun investigations in life sciences as part of its early payload studies, but the other Centers saw this as their prerogative, and Marshall "got our arms broken," in the words of the Marshall Center's John Hilchey. Marshall found ways to remain active in life sciences nonetheless, concentrating on non-human subjects and accommodating Ames and JSC experiments in ways that enabled Center Director Lucas to justify the Center's work to Headquarters. Hilchey remembered that "Lucas would say to Herm Gierow, 'Hey Herm, how's that life science's program going?' The standard answer was, 'Dr. Lucas, we don't have a life sciences program; we have a payloads program of Space Station accommodations for payloads; and that's what we're doing, and life sciences is just one of the disciplines we deal with.' Lucas would just grin at him."[83]

A Troubled Partnership

More troubling than Center rivalry were emerging problems with ESA, for NASA was becoming more concerned about the performance of its European partner. The peculiar relationship between NASA and ESA led to unexpected difficulties. Marshall often learned of emerging problems before NASA Headquarters through on-site visits. In one such report—on the Spacelab thermal control systems in 1975—the Center learned how internal communications deficiencies, poor systems integration, customary exclusion of working level people from meetings, lack of experience, and limited facilities caused delays. Subcontractor Aeritalia, for example, had come to rely on McDonnell Douglas engineers, and had to replace them with Italian engineers, none of whom had spacecraft experience. Marshall's Kingsbury worried that similar shortcomings were "widespread in all subsystems."[84]

Such difficulties had serious implications, leading to delays, misunderstandings, and uncertainties. It was difficult to implement changes, in part because ESA's contractors operated under the assumption that systems were defined at the time of the proposal, and that they were to "design to cost." Unless contracts were completely definitive, contractors disclaimed responsibility. ESA thus found itself in the unusual position of having to persuade its contractors to make changes. Contractors, operating under fixed price contracts and working with limited engineering manpower, were seldom receptive.[85]

Although many of ESA's problems were the sorts of difficulties that customarily occur in large development programs, the Europeans became increasingly sensitive to NASA supervision. "The Europeans are a proud group. They didn't want us telling them how to do something," explained Kingsbury, whose contact with Spacelab came because of his position as head of Marshall's Science and Engineering Directorate. "When we would go and say to them, 'What you're doing isn't going to work,' they would say, 'Thank you very much,' and do it anyway. Then they would never tell us it didn't work. The next thing we knew they'd changed to something else. Of course we knew what happened to it. We had to get them around to a system that would work and that we could live with by trying not to offend them by telling them what they were doing was crazy. Engineers are not by their very nature very tactful people usually. We had some people who tried very, very hard to be tactful. We had some who couldn't stand it any more and lost all tact."[86]

In part problems arose because of the number of people working on Spacelab for each Agency. Marshall had 180 of its own people and 115 support contractor employees assigned to Spacelab in FY 1975.[87] Lee assigned 6 percent of them to monitoring ESRO, and 14 percent to assisting ESRO. The other 80 percent were divided equally between those assigned to systems engineering and the development of NASA-provided software and equipment, and those engaged in operations planning and experiment integration. In contrast, ESRO had only 80 people assigned to Spacelab, and Headquarters worried that either Marshall had too many on the project or ESRO had too few. Lee defended his manpower as the minimum required and expected to increase by about one-third over the next two years.[88] Some NASA administrators worried about the imbalance, and particularly about assigning too many NASA personnel to posts in Europe. "This approach might even be . . . harmful if it appears that NASA is 'taking over' the program," suggested one internal NASA assessment.[89] On the other hand, NASA's concern about the sensitivities of its European partner obscured a basic issue; as Marshall's Lowell Zoller, who was on duty in Europe, suggested, "ESA is facing about a 40 percent increase in manpower requirements to get the job done."[90]

NASA knew of European sensitivity to excessive penetration, but the Agency believed that ESA needed both managerial and technical advice. NASA approached ESA General Director Roy Gibson, criticizing project management and offering a "combined technical/management advisory package." Gibson admitted problems, but said he "would prefer by far not to accept an offer of NASA advisory support" below the program management level.[91] The two agencies negotiated an arrangement under which 12 NASA technical experts and 3 management advisers took assignments at ESTEC and ERNO. Although the Americans initially wanted a dual reporting system in which its experts would be responsible to both ESA and NASA chains of command, they agreed to an arrangement in which the individuals would be integrated into ESA's organization and have no responsibility for implementing NASA's requirements.[92]

ESA had its own reasons for dissatisfaction. As time passed, it became clear that Europe was going to get less than expected from Spacelab. Although the 1972 Memorandum of Understanding required delivery of only one Spacelab, the Europeans had anticipated selling NASA perhaps as many as four additional units. After all, NASA's 1973 plan for Shuttle utilization required six Spacelabs (and seven orbiters), and as recently as December 1974, NASA had projected flying as many as 25 Spacelab missions per year. A year and a half later Shuttle development lagged, and it was apparent that there would not be such high levels of use. Now NASA would commit only to one Spacelab with an option on a second. Many Europeans believed that too much money was going to Spacelab. As ESA's budget declined, less money was available for European utilization after completion, and there seemed to be little ESA could do to prevent the initiative from slipping to the Americans.[93]

By the summer of 1976, it was clear that the Spacelab program was in trouble. So concerned was NASA about both schedule slippage and ultimate performance capabilities of the European program that it took steps to initiate studies at Marshall and JSC for ways to back up ESA's work.[94]

Although the Europeans were reluctant to acknowledge the depth of their difficulties, Marshall representatives in Europe observed serious shortcomings. Zoller noticed "striking similarities" between the difficulties Marshall had experienced with the Shuttle main engine a few years earlier and the Spacelab problems. "Neither ESA or ERNO have very efficient management systems," he observed, "and the top management on both sides spends an inordinate amount of time fighting over fee and image." He worried that ERNO management was "concentrated at the top," and that ESA was "basically a one-man show," leaving inexperienced subordinates like Stoewer so cautious that they would postpone "sticky" issues until after key reviews. Another sign of excessive caution was a tendency to overdesign rather than analyze requirements that were peculiar to certain payloads or missions. Zoller nonetheless believed that both ESA and contractor were competent, but that "the biggest detriment to the program is the mistrust that is so evident among all the contractors and ESA."[95]

Spacelab began to experience the cost and schedule problems familiar to most post-Apollo space programs. Zoller cautioned Marshall about the emerging dilemma before the Europeans were willing to acknowledge it. "The technical baseline is not clear, and therefore the schedule and cost are up in the air," he warned. "Needless to say, ESA doesn't readily admit to the full implications of the programmatic problems."[96] By the end of the year, even ESA was ready to address the crisis.

Costs were particularly problematic because of ESA's multinational funding arrangements. ESA members participating in Spacelab understood the volatile nature of funding Big Science projects, and had agreed to support a 20-percent overrun. If costs exceeded 120 percent of initial support commitments, however, member states would be allowed to withdraw. As early as February 1975, ESRO began suggesting delays, descoping, or split deliveries after its contractor submitted funding requests in excess of the Agency's budget.[97] By November 1976, ESA's cost projections had already exceeded 100 percent. "We have always been very much afraid of being forced to exceed the 120 percent," reported Michel Bignier, who was soon to take Stoewer's place as Lee's ESA counterpart. Bignier also lamented that "a certain number of systems are now behind schedule and . . . it will be difficult to catch up completely." He suggested simplifications that might reduce delays and costs.[98]

The problems plaguing Spacelab strained the international partnership. NASA believed it had no alternative but to apply pressure. ESA, perturbed by its limited ability to compel changes from its contractors, fearful that budget overruns might lead to withdrawal of member states, and disappointed by diminishing returns from its large investment, reluctantly succumbed. By the mid-1970s, the NASA–ESA relationship was at best an unequal partnership. R.N. Lindley, one of NASA's representatives in Paris at the time, observed that "Far too many people, on both sides of the Atlantic (and I have been one of them) have looked upon this relationship as one which places ESA almost into the role of a contractor to NASA (with a no-cost plus no-fee contract)."[99]

Cost reductions and schedule adjustments dominated meetings in the months that followed. ESA proposed a "comprehensive overhaul of the management of the project" and a "descoping."[100] The Europeans suggested revisions in the schedule of equipment to be delivered, including the deletion of some equipment requirements, and offered to replace key personnel. They agreed to appoint a task force to review contractor management, and to include NASA representation.[101] Bignier took over as Program Director for Spacelab, and he conceded the need for "very tough Program Management, which is not exactly the image presented by Heinz Stoewer."[102]

Technical Challenges

Development of Spacelab continued while the Europeans and Americans established management for the program. The basic configuration was now set. Spacelab would have two elements: a pressurized chamber in which scientists could work in a shirt-sleeve environment, and an unpressurized pallet, or platform, for instruments requiring direct exposure to space. Modular design allowed for flexibility. The habitation module would have two segments—a core segment and an experiment segment. The core segment would contain basic support equipment and several cubic yards of experiment racks; the experiment segment would be devoted entirely to experiments. Up to five U-shaped pallet modules could be added, allowing for a variety of arrangements depending on the mission. When pallets would be flown without the core segment, supporting equipment would be protected in a small pressurized temperature-controlled chamber called an "igloo." Experimenters could also arrange modular experiment racks to suit a particular flight, and integrating experiments for the early flights would be Marshall's responsibility.

As Lead Center, Marshall had duties in addition to its supervision of the Spacelab module development. In order to ensure proper weight distribution aboard Shuttle, Spacelab would nest toward the rear of the orbiter's cargo bay, so Marshall would have to devise a crew tunnel from the orbiter flight deck to the laboratory. Much of the program's complexity centered around Spacelab's subsystems, which included structure, environmental control, electrical power, command and data management, and payload support equipment.[103] In addition to monitoring these subsystems, Marshall also bore responsibility for development of an instrument to provide precise alignment of experiment instruments.

Of the technical challenges involved in Spacelab development, the instrument pointing system (IPS) posed the most obstacles. Solar physics and astronomy experiments required a system that could align large instruments with pinpoint accuracy and stabilize them for long periods. It was "new and different and proposed requirements that we hadn't done before," Lee recalled in describing the IPS as the "most difficult" of all Spacelab projects.[104] Indeed the IPS was extraordinarily complex, requiring drive motor systems for movement in three axes, mechanisms to secure the gimbals for loading and unloading experiments, an optical sensing system for alignment in relation to stars and the Sun, a system for directional control and stabilization, support structures, a clamping system to secure the delicate instrument during ascent, and a means of temperature control—all of which had to ensure precise accuracy and stability. Lord concurred with Lee that "in terms of technical complexity, organizational responsibilities, schedule difficulties, and cost escalation," the IPS was the most challenging part of Spacelab.[105]

Perhaps no Spacelab subsystem demonstrates as well as IPS how Marshall carried out its Lead Center responsibilities, since the Center's presence was apparent throughout development of that system. Marshall's previous experience with the Apollo Telescope Mount and *Skylab* gave the Center unmatched experience in instrument pointing systems for manned space flight, and the Europeans turned first to Marshall for guidance.[106]

ESRO, its hands tied by tight budgets, proceeded with a single development approach, an imaginative concept called the inside-out gimbal that differed from conventional ring gimbals. That same principle had been used for gyrostabilized platforms on recent rockets, such as the Saturn V, in contrast to older systems that used ring gimbals. Marshall had no objections to the method, but its approach differed from that of the Europeans. ESRO sought to satisfy the requirements set forth in the Spacelab Memorandum of Understanding and its requirements document, while Marshall wanted to satisfy the broader demands of experimenters. By early 1975 Lee worried that "no one IPS design will satisfy all the users' pointing requirements."[107] Lord conceded that "it was very difficult to get designers to agree on a statement of specifications."[108]

Marshall launched a "total effort" on the IPS, monitoring ESRO progress, briefing customers on IPS capabilities and limitations, and examining alternative approaches for a small IPS under study at Marshall and Goddard. Finally ESA, hemmed in by rising costs, suggested less restrictive specifications, and then abandoned the inside-out gimbal approach altogether in favor of a less expensive alternative. Marshall developed simulations to test the new ESA proposals, and in March 1976 NASA concurred with the Marshall recommendation to proceed to Phase C/D development in spite of the resulting schedule slippage.[109]

NASA hoped to have IPS ready for the second Spacelab mission, but development through the end of the decade was plagued by continuing cost, schedule, and technological problems that prompted contention between NASA and its European partner. NASA maintained that ESA was failing to provide adequate documentation, and ESA complained that NASA was continuing to develop competitive IPS systems. In 1977 ESA suggested removing the system from its Spacelab program in order to find another means of development.[110] ESA frustration boiled over in 1978, when member states refused to approve additional funding for IPS.[111] Increased load requirements rendered earlier specifications insufficient, and forced IPS contractor Dornier to make modifications and slip the schedule. Both ESA and NASA complained of a lack of cooperation from Dornier, and suspicions rose that the company was trying to use legitimate redesign demands resulting from load changes to hide other problems.[112]

Reviews conducted by ESA and Marshall in 1979 and 1980 raised questions as to whether the IPS as currently designed would meet requirements. Technical, management, and safety concerns dominated review reports. In April 1981 Dornier submitted a proposal for a redesigned IPS, and NASA accepted the proposal in July. ESA restructured its IPS contract, and Marshall assigned Gene Compton as a full-time liaison at Dornier. Although the redesigned IPS also encountered development difficulties, they were less onerous than those of the late 1970s. ESA delivered the first flight unit in November 1984, and delays elsewhere in the Shuttle program made it possible for the IPS to fly on Spacelab 2 as originally intended.[113]

Challenging as they were, the technical problems posed by the instrument pointing system were restricted to a single subsystem, and development of other subsystems and the Spacelab modules proceeded apace. In March 1977 Marshall awarded a systems analysis and integration contract to McDonnell Douglas Technical Services Company (MDTSC). The most significant Spacelab contract to go to an American company, it called for systems engineering, experiment integration, software development, and the design, development and fabrication of most of the Spacelab hardware under Marshall's purview, including the crew transfer tunnel.

Marshall also continued its monitoring of ESA's progress in Spacelab development. Beginning in 1974, Marshall conducted a series of periodic reviews of all major Spacelab subsystems. Reviews served first set baseline requirements, then monitored design modifications. At each step, these sessions helped to bring both technical and managerial problems to the attention of Marshall and NASA management, and to ensure that Spacelab-Shuttle interfaces were protected. Marshall also participated in reviews conducted by ESA's contractors. The most important step in the long review process was the Spacelab critical design review (CDR), initiated by ESA and NASA in March 1978 and completed in December. The CDR was particularly important for the Americans, since it was their last opportunity to make major changes in Spacelab design.[114]

Even after completion of the CDR, a final technical problem interrupted preparations for the first Spacelab mission. Weight status reports early in 1979 indicated that each of the first three Spacelab missions exceeded acceptable criteria as a result of orbiter-supplied equipment. ESA was near its weight limits, and could not be expected to make adjustments, so NASA's Spacelab Program Office suggested that upgrading landing capability was the most acceptable solution, and that reduction in payload weight should be considered only as a last resort. The issue demonstrated the importance of coordination between the Spacelab and Shuttle programs, and also the fact that resolution of problems required the cooperation of both Marshall's Spacelab and Payload Offices, as well as representatives of Headquarters and the JSC Shuttle Office. Headquarters believed that manipulating landing capability would set a bad precedent, and NASA found ways to absorb the difference for each mission without modifying the Spacelab module or significantly impacting payloads.[115]

Recession and Realignment

Costs, schedule, and technical challenges continued to be the three problems that defined Spacelab, but by the late 1970s the issue of money dominated. Simply put, the European dilemma was that costs rose inexorably while expected benefits dropped. Besieged by the oil crisis, the economies of the United States and Western Europe declined during the late 1970s, and the space programs of both were not immune to economic contraction.

ESA worried that design changes, additions to the program, development difficulties, and schedule slippage had increased "cost-to-completion" estimates to the point that member states began to question whether the commitment to Spacelab had been worthwhile. "Our biggest problem is cost," reported a senior ESA official.[116]

Not only were Spacelab costs rising, but returns in terms of technology for manned spaceflight, cooperative flights, and opportunities to support Spacelab integration seemed to be crashing. Even worse for ESA, the operating cost per mission—estimated at $60 million in 1979—meant that the returns from Spacelab operations would likely be less than originally expected. ESA had entered the program expecting that Spacelab would be less expensive to experimenters, but by 1979 the estimated cost per mission had tripled. Meanwhile, the Europeans had been counting on "follow-on procurement"—the sale of additional Spacelabs and support equipment to NASA after the completion of the initial program—as a means of recouping part of their investment. Now the Americans seemed unlikely to buy more than required under the narrow terms of the Memorandum of Understanding.[117] Complicating the reduced likelihood of follow-on procurement was an ESA concern that NASA and American contractors were violating the Memorandum of Understanding by duplicating Spacelab equipment, producing their own versions of the instrument pointing system and pallets.[118]

NASA tried to accommodate ESA's concerns, but only to a point. The Agency suggested that if ESA was not getting what it expected out of Spacelab, it might be their own fault; officials expressed "amazement" that ESA was not "utilizing Spacelab commensurate with their development investment."[119] NASA agreed to "descoping," cutting back some of the originally agreed upon ancillary equipment. On matters unrelated to cost, NASA tried to meet the Europeans more than halfway, conceding to most ESA technology requests, encouraging cooperative flight proposals, giving ESA the same data rights as U.S. Government civil agencies, and forming a joint Duplication Avoidance Working Group. When money was at stake—and it was the root of most of ESA's problems—NASA was less forthcoming. The Carter years were lean for NASA, and the Agency could ill afford to loosen its purse strings. Marshall even sought legal opinion to ensure that NASA would not be compelled to purchase follow-on equipment the Agency no longer desired.[120]

Before NASA and ESA could resolve their differences, the European member states had to decide how much money they were willing to commit to Spacelab. Participating members already had pledged up to 120 percent of their original commitment, and that money would last only until September 1979. After long deliberations, only Italy refused to increase its contribution beyond the 120-percent level, and ESA agreed to present a proposal of 140 percent to the Spacelab Program Board.

By the end of the 1970s, with the first Spacelab flight still years in the future, both Europeans and Americans had reason for disappointment with the partnership. The new ESA funding arrangement left both parties dissatisfied. ESA nations resented that they were going to have to pay far more than they expected; NASA resented that ESA was unwilling to take risks normal to the space business by honoring their commitment to bear responsibility for Spacelab design through the first two flights. "The concern here is ESA's inability to anticipate operational changes and fund for them," Lee told Lucas. "NASA has essentially the same problem in planning for unforeseen changes."[121] Furthermore, the initial agreement between the Europeans and NASA had no cost ceiling.[122] Evaluating their Spacelab experience shortly after working out the new funding relationship, ESA's Spacelab Programme Board concluded that "At the present time, Spacelab is the only possible base from which Europe could make significant progress and thus be able to play a role towards the end of the 1980s."[123] Lucas admitted that this was "shaky" support. "The only thing to drive the Board in the direction of support will be that there is no other choice," he wrote.[124] Ultimately, ESA delivered two pressurized module assemblies to NASA, the first under the original Memorandum of Understanding, and the second as part of follow-on procurement.

For Marshall, there was another reason for disappointment. One of the attractive aspects of Spacelab was the opportunity to further diversify. But while Marshall's Spacelab work gives indication of the success of the Center's diversification, NASA had no intention of making Marshall the Agency's sole payload integration Center. The Center remained in a precarious position, and Center administrators and the Huntsville community watched NASA decisions for indications of Marshall's fate. Thus it was not surprising that when NASA transferred some of Marshall's projects including important Spacelab work (sending Spacelab sustaining engineering to KSC) and gave managerial authority over six Spacelab missions to Goddard, alarms went off in Huntsville. Congressman Ronnie Flippo, who represented the Alabama Fifth District (including Huntsville), alleged a trend of moving projects out of Marshall, and asked NASA administrators if they had plans to backfill the losses. He questioned the wisdom of moving Spacelab activities out of Huntsville since Marshall had developed both the expertise and the facilities to manage the program.[125]

NASA's response was barely reassuring. Headquarters told Flippo that there was no conscious effort to erode Marshall on a project-by-project basis.[126] John Naugle insisted that it was reasonable to have JSC and Goddard involved in

Spacelab mission management. "It is essential that there be Centers other than MSFC involved," he insisted. "A monopoly by MSFC would seriously inhibit the kind of innovation and competition that is required to develop Spacelab into a cheap, effective research laboratory in space."[127]

Budget problems continued to plague NASA in the early 1980s as the Agency had to absorb reductions similar to those experienced by other federal organizations. Cuts in NASA's Space Science budget for Fiscal Year 1981 forced one-year schedule slips for Spacelabs 4, 5, and 6.[128] James C. Harrington, who had succeeded Lord as the director of the Spacelab Program at Headquarters, lamented the impact of these reductions: "Over the past four years the planned SL-1 launch date has slipped three years. Worse yet, over the past 13 months we have slipped this milestone 17 months. Additionally, the manifest of SL flights has been reduced from 4–5 flights per year to the current 2 flights per year through 1986."[129]

Harrington presented an insightful analysis of the impact of budget reductions on Spacelab that by extension demonstrated the plight of all NASA programs in hard times. Preparing the budget for Fiscal Year 1982, Headquarters first slashed field Center Spacelab budget requests by 20 percent, then subtracted another 8.5 percent before submitting the NASA budget request to the Office of Management and Budget. Then the Reagan Administration amended its budget, reducing NASA's line by another $30 million. Harrington argued that NASA had no alternative but to slip the schedule for early Spacelab missions, which was costly in terms of user interest and support, ESA confidence, and overall program costs. Delay never saved money; runout would add costs to maintain program readiness, increase expenses for users or force them to abandon experiments, and "will not aid in relieving our budget difficulties, but only compound them." Although few in NASA would have disputed Harrington's persuasive argument, the Agency had little choice but to implement cutbacks. Harrington proved prescient.

The Early Missions

While Marshall's administrators worried about budgets and transferred projects, technicians continued their preparations for the first three Spacelab missions. In addition to checking out Spacelab systems, Marshall wanted to incorporate a wide variety of experiments into the first two missions in order to demonstrate what the new spacecraft could do. The payload plan for the first mission was to use experiments to demonstrate the capability to investigate a wide variety of phenomena in a microgravity environment. Marshall's Harry Craft explained that "the emphasis was on microgravity, life sciences, materials processing, although we flew an array of experiments in just about every discipline." Included were research on Earth's atmosphere, crystal growth, cloud microphysics, observations to monitor Earth's surface for environmental quality and for the development of remote sensing methods, investigations of ultraviolet and infrared radiation, and life science experiments involving humans, animals, plants, cells and tissues. The second mission also intended to be multidisciplinary, emphasizing "astronomy, solar physics, and high energy astrophysics," according to Craft. Spacelab 3 would be the first mission dedicated entirely to applications and science, and would emphasize processing in space.[130]

The mission plan was indeed ambitious, and Houston's Kraft believed the schedule for Spacelab 1 was overly so. In Kraft's view, Marshall was "structuring the 7-day first flight of Spacelab to be as complex and ambitious as *Skylab*."[131] Lucas insisted that the wide variety of experiments was important to maintain the interest of potential users, and that less than half of the experiments selected would place moderate to heavy demands on the crew. Most experiments required no crew activity, or merely the flipping of a switch.[132] Reviews conducted at Marshall and in Europe in the fall of 1979, however, confirmed Kraft's worries, and NASA simplified the first mission.[133] Budget problems also forced reevaluation of the schedule for early missions, and compelled NASA to delay experiments.[134]

Installation of OSTA–1 in the orbiter Columbia *before the second Shuttle mission in November 1981.*

While administrators debated the budget, payload, and schedule, preparations continued for the first mission. Two early Shuttle flights prior to the first Spacelab mission served to validate Spacelab hardware. The OSTA–1 (for the Office of Space and Terrestrial Applications) mission in November 1981 used an engineering model of the Spacelab pallet. The following March the OSS–1 (for the Office of Space Science) used an engineering model of the pallet to mount eight of its nine experiments.[135]

Spacelab offered an opportunity to merge NASA's two primary activities, space science and manned space flight. As one of the Agency's manned space flight Centers, Marshall was under the umbrella of the Office of Manned Space Flight. But post-Apollo diversification had established expertise in science at the Center that prepared it to lead a merging of the two ventures, and Marshall would work closely with the Office of Space Science. "Manned spaceflight and science came together really for the first time in *Skylab*," explained Rick Chappell, mission scientist for Spacelab 1 and later Marshall's director of science. "But that was a one shot deal. It was with the Shuttle [that] we're going to take these two major pieces of what NASA did, science and manned spaceflight, and merge them."[136]

Marshall conducted training in a Spacelab mission simulator at the Center. "We have a fullscale Spacelab pressurized module and pallets as a part of our training capability," explained Ralph Hoodless, a manager for the development of Spacelab. "We configured that for Spacelab I and II and actually used that to train hands-on."[137]

Payload specialists for Spacelab 1 train in mock-up at Marshall in June 1981.

Marshall began assembling the hardware for Spacelab 1 at Kennedy Space Center late in 1981. Equipment for experiments began arriving in October, and the Spacelab module and pallet followed in December. In February 1982 Vice President George Bush attended the unveiling of Spacelab, and NASA formally accepted flight hardware for Spacelab 1 (SL–1).

Over the next several months, engineers tested components and began integrating experiments. Marshall technicians installed the life sciences mini-lab and its flight rack in the module in February, and in May began placing equipment on the pallet. Integration of major assemblies, including a platform of 12 European experiments, continued through the summer. In December the team moved the pallet into position behind the module, and completed integration by installing experiment racks in the module. Mission sequence tests during the early months of 1983 culminated in July with remote operation of experiments from the POCC in Houston. The orbiter *Columbia* arrived at Kennedy in November 1982, and the integration team began modifications necessary to place Spacelab in its cargo bay.

"The experiments were brought in by their various scientific teams," recalled Mission Manager Craft. "We would let them check the experiment out initially in an off-line capability and then we'd bring them into a room and just make sure the instrument had met the transportation environment and still worked. They would do some checkout and they'd turn it over to us." Then the Marshall-Kennedy team integrated the experiments "into a Spacelab rack if it was inside the module or integrated onto a pallet if it was outside."[138]

While preparations proceeded in Florida, all Spacelab systems and Shuttle interfaces underwent reviews. The design certification review in January 1983 followed months of preparation during which Marshall and MDTSC reviewed performance and design requirements and examined all Spacelab subsystems in collaboration with representatives of ESA and ERNO; Headquarters lauded the team's careful preparation and considered the session "exemplary." Other reviews examined flight operations and all aspects of integration.[139]

On 15 August 1983, technicians moved Spacelab to KSC's Orbiter Processing Facility, and the next day placed the module and pallet in *Columbia's* cargo bay. On 23 September *Columbia* moved to KSC's mammoth Vertical Assembly Building, and five days later to the pad in preparation for a scheduled launch on

30 September. Unfortunately a problem with the solid rocket booster nozzle delayed the mission, soon rescheduled for 28 November.

The 28 November launch date of Spacelab 1 culminated years of preparation. With Spacelab nestled in the cargo bay of the orbiter *Columbia*, Marshall representatives in Huntsville, Houston, and at the Cape took their stations to support the mission. Experimenters huddled with the Marshall team in JSC's POCC, where a large Marshall Center banner hung on the wall, signifying a Marshall beachhead in what former Program Manager Douglas Lord called "intercenter warfare."[140] The Huntsville Operations Support Center (HOSC) operated much as it had during the *Skylab* mission, supplying technical advice. The composition of the six-man crew—a commander, pilot, two mission specialists, and two payload specialists—signaled the beginning of a new era in space science.[141] As the crew settled into its routine for the 10-day mission, Marshall's central role soon became apparent to those monitoring the flight. The communications call, "Marshall operations, this is Spacelab 1," registered more often than calls to JSC's mission control. The crew divided into two teams for 12-hour work rotations, and by the end of the first day they had already initiated 25 experiments. Instrumentation problems slowed progress as the mission went on, but NASA believed that the success of the crew in repairing balky equipment demonstrated the value of humans to space science.[142]

The mission experiments required 40 instruments, 18 on the pallet, 19 in the module, and 3 with components in both locations. In order to demonstrate Spacelab's capabilities, the crew conducted 72 experiments ranging across five disciplines: atmospheric physics and Earth observations, space plasma physics, material sciences and technology, astronomy and solar physics, and life sciences.[143]

Mission Payload Operations.

The variety of experiments aboard Spacelab 1 makes that mission a useful measure of the range of activity that attracted Marshall scientists and mission managers. Spacelab provided an exciting environment from which to study the chemical composition of the atmosphere and the effect of Earth-based human activity on the upper atmosphere. The Imaging Spectrometric Observatory (ISO) could measure multiple constituents in a slice of Earth's atmosphere, and proved its value by providing for the "first comprehensive spectral atlas of the upper atmosphere." The Grille Spectrometer aboard Spacelab 1, designed to measure constituents in the atmosphere between altitudes from 10 to 95 miles, found methane (produced by biological decay and the burning of fossil fuels) at the surprisingly high altitude of 30 miles above the surface of Earth. Two cameras aboard Spacelab 1 recorded aerial photographs of Earth's surface in three days that would have taken 10 years to accumulate using conventional methods, providing data for agriculture, archaeology, and cartography.[144]

Space plasma physics experiments studied the ionosphere, Earth's uppermost atmospheric envelope which extends from 40 to 60 miles above Earth's surface. Using both passive and active instruments, Spacelab scientists examined the behavior of the ionosphere's electrically charged gasses, or plasmas. Among experiments employing active instruments, the Space Experiments with Particle Accelerators (SEPAC) and the Phenomena Induced by Charged Particle Beams (PICPAB) were ambitious attempts to inject particle beams into the ionosphere to examine changes in electric and magnetic fields. In both cases, passive instruments measured the effect of particle injection on theories of particle acceleration. Among the surprising results of these experiments was the discovery that neutral gas injections could quickly neutralize induced charges on the spacecraft.[145]

Because NASA hoped that the private sector might demonstrate interest in manufacturing in space, experiments in materials processing aboard Spacelab 1 were particularly important. Crystal growth experiments have been among the most successful on several Spacelab flights, and Spacelab 1 set the tone. The mission demonstrated the practicality of reducing defects by growing crystals in microgravity. Crystal experiments in the Mirror Heating Facility demonstrated that certain defects in silicon crystals grown on Earth were not gravity-induced, but rather were caused by surface tension. Other materials processing experiments proved that microgravity was an ideal environment for determining the diffusion coefficient of liquid metals—a measure of how metals diffuse through each other.[146]

Spacelab 1 carried instruments to make observations in the ultraviolet and X-radiation portions of the electromagnetic spectrum and contributed to knowledge of astronomy and solar physics. The Very Wide Field camera, which could survey a 60-degree field of view, made 48 photographic images of 10 astronomical objects, and returned excellent images of stellar clouds in the ultraviolet range. The Far Ultraviolet Space Telescope (FAUST) experienced problems with fogged film and overexposure, but scientists expected that equipment modifications would yield promising results on future flights. Because the background level of cosmic ray activity in space was lower than anticipated, X-radiation data collection surpassed expectations; the astrophysics experiments aboard Spacelab 1 included 200 hours of accumulated X-ray data. NASA and ESA instruments designed to measure energy output of the Sun also yielded promising results.[147]

The scientist-astronauts aboard Spacelab 1 served as subjects for life science experiments, several of which sought to evaluate the response of the human vestibular system, vision, and reflexes to microgravity. The vestibular system, which is located in the inner ear, controls balance and orientation. Experiments found a relationship between balance and eye movements, and provided data on the effect of head movements on motion sickness. These and other experiments helped evaluate the adjustment of the sensory motor system to microgravity, the ability of people to estimate mass in space, and the effect of microgravity on muscle mass and blood.[148]

The flow of data from Spacelab generated excitement on the ground even before the Shuttle returned to Earth. By the time the mission ended when the Shuttle landed at Edwards Air Force Base in California, Mission Manager Craft could report that the mission had accumulated 20 million pictures and 2 trillion bits of data.[149]

The mission achieved most of its goals, and Samuel Keller, deputy associate administrator for Space Science and Applications, deemed Spacelab "an unqualified success." Chappell considered the flight a "very successful merger of manned space flight and space science." The crew accomplished all systems verification objectives, with only minor anomalies. Several months later, Chappell and his ESA counterpart Karl Knott reported that the mission had achieved 80 percent or more of its objectives in all but atmospheric physics and Earth observations (which achieved 65 percent). Spacelab proved its viability for research in all five disciplines investigated.[150]

In the months following the first Spacelab mission, NASA realigned the Shuttle payload schedule. Because of changes in Defense Department Shuttle requirements, redesign of the instrument pointing system required for Spacelab 2, problems related to the satellite tracking system, and Shuttle-related delays, Headquarters moved the Spacelab 3 mission ahead of Spacelab 2.

Spacelab 3 rack/floor installation at KSC in May 1984.

Spacelab 3, NASA's first dedicated mission, concentrated on the acquisition of scientific data with a focus on microgravity rather than a wide range of disciplines. Again, Marshall provided management of mission development and operation; J. W. Cremin served as mission manager, George H. Fichtl as mission scientist. The mission, which flew in the orbiter *Challenger* from 29 April to 6 May 1985, included experiments in materials science, life sciences, and fluid mechanics, and carried out atmospheric and astronomical observations. A Marshall ground control team managed the mission from JSC's POCC, and scientists stationed in rooms adjacent to the POCC had opportunities to confer directly with mission and payload specialists aboard Spacelab.

By maintaining a stable attitude for the six-day experimentation period, the crew established an ideal setting for microgravity research and developed methods "to provide the best low-gravity environment achievable from this system." Materials science experiments focused on crystal growth, testing ways to grow more homogeneous crystals by reducing the effect of gravity as a means to produce crystals that might be used for applications such as Earth resources surveys, medical diagnostics, and infrared astronomy. The fluid dynamics experiments were the first controlled experiments on free-floating drops, thus providing an opportunity to test theoretical predictions without the acoustic forces that impact such experiments on Earth. Two monkeys and 24 rats accompanied the crew into space to assist in life science experiments, and on return to Earth the rats demonstrated loss of muscle, bone loss mass, and other data that researchers said "may well be the most significant contribution on biological systems in space ever gained from a single mission."

The mission's atmospheric observations gathered more data on trace elements in the upper atmosphere. Spacelab 3 recorded the first observations of the Southern Hemisphere aurora from a lateral perspective; previous observations had been only from Earth or from satellites in a higher orbit. The mission's most successful astrophysical experiments focused on low-energy cosmic-ray observations.[151]

After repair of a problem with the Shuttle main engine, Spacelab 2 launched on 29 July, three months after Spacelab 3. The delay in launch provided opportunity for one of the experimenters to rework hardware, and showed the range of Marshall mission support. The Jet Propulsion Laboratory planned an experiment to test the behavior in space of superfluid helium, which they hoped could be used as a coolant for infrared telescopes. Mission Manager Roy Lester and other Marshall resident personnel at the Cape facilitated repairs that enabled the experiment to fly successfully. They helped rebuild and test an essential vacuum pump and coordinated trouble-shooting between the Marshall Center and KSC. Personnel responsible for the Infrared Telescope took time from servicing their own equipment to assist JPL's experimenters.[152]

Shutdown of a main engine late in ascent forced the Shuttle to orbit lower than planned, but did not interrupt the experiment schedule. The troubled instrument pointing system performed erratically, working best when relying on one of the independent telescopes for alignment rather than its own optical sensor package. Astronauts, directed by experts from Marshall's Huntsville Operations Support Center, attempted to make repairs. At times the IPS worked perfectly, demonstrating the capability of the system once repairs could be made, and the mission succeeded in gathering invaluable data about the Sun.[153]

Marshall's work on Spacelab 2 won praise from one of its experimenters, who suggested that the records set by the mission "will stand until the era of the Space Station because no payload now under consideration matches the complexity of SL–2, which tested the limits—of hardware, software, and people—

everywhere in the system. The success we experimenters are now enjoying was made possible because all of the people at MSFC associated with SL–2 did their utmost to make it so."[154] The eight-day mission officially marked the completion of the Spacelab development program. A German mission, Spacelab D–1, flew in October; it was the last Spacelab mission before the *Challenger* accident.

The hiatus on Shuttle flights following the *Challenger* accident interrupted Spacelab as it did all NASA's manned space flight programs. The Marshall-managed Astro–1 mission in December 1990 was the first Spacelab mission to fly after the return to flight. The Astro–1 payload featured four telescopes—three Marshall ultraviolet instruments and Goddard's Broad Band X-Ray Telescope. Spacelab's pallet-borne instrument pointing system aligned the telescopes for observations of distant galaxy clusters, white dwarfs, binary stars, and active galactic nuclei.

For Huntsville, the Astro–1 mission marked another milestone, the first use of the new Spacelab Mission Operations Control facility. No longer did the Marshall team have to travel to Houston's POCC to direct payload operations. In the early morning hours shortly after launch on December 2, Mission Specialist Robert Parker opened his communication lines saying, "Huntsville, this is Astro," marking the first time that there had been direct communications between Huntsville and astronauts in space.[155]

Marshall Spacelab Mission Operations Control facility during Astro–1 mission in December 1990.

Like most of NASA's post-Apollo Programs, Spacelab was plagued by budget problems from its inception, and forced the Agency and its European partner to confront the question of whether space development programs can be designed to cost. Inevitably tight money led to delays, but concurrent delays in the Shuttle program lessened their impact. The problems that plagued Spacelab development traced back to tight budgets, and successful completion of the system testified to Marshall's accomplishment under trying circumstances. Program Manager Lord praised Marshall as "an effective and responsible Lead Center."[156] Ultimately Spacelab proved to be one of NASA's workhorses, and Lee's successful management of the program at Marshall paved the way for his selection as Center Director after J.R. Thompson.

Spacelab anticipated Space Station. Delays in Space Station development made Spacelab all the more valuable as a platform for space science research into the 1990s. Like Spacelab, station would be undertaken as an international partnership, and both ESA and NASA entered the latter program having learned their own lessons from Spacelab, determined not to repeat the same mistakes.[157]

If Spacelab benefited from Marshall's contributions as Lead Center, the Center also gained from its management of the project. The Marshall Center emerged from Spacelab development more diversified in terms of technical capabilities, and with experience in science operations, international relations, systems integration, systems management, payload integration, and space science. By the mid 1980s, Spacelab helped expand the Center's expertise to the point that no other NASA field Center could match the range of Marshall's experience.

---

1 H. J. Richey, "Space Laboratory Plan is Outlined by Mueller," *Huntsville Times*, 11 March 1969.
2 *Ibid.*
3 Thomas J. Lee, Oral History Interview by by Andrew J. Dunar and Stephen P. Waring (hereinafter OHI by AJD and SPW), 9 April 1993, Huntsville, Alabama, p. 2.
4 William R. Lucas, OHI by AJD and SPW, 1 March 1993, Huntsville, Alabama, p. 1.
5 O. C. Jean, OHI by Jessie Emerson/MSI, Huntsville, Alabama, pp. 13–15.
6 William R. Lucas, OHI by AJD and SPW, 1 March 1993, Huntsville, Alabama, p. 1; Stuhlinger to Rees, 1 June 1971, Sortie Lab 1972 folder, MSFC History Archives.
7 Gierow Weekly Notes, 29 March 1971, Sortie Lab 1972 folder, MSFC History Archives.
8 Terry Sharpe, report to Murphy, "Payload Planning Activity: MSFC's Concept for Shuttle Utilization," 3 May 1971, Box 62, MSFC History Archives.
9 Philip E. Culbertson, statement, U.S. Congress, House, Subcommittee on Manned Space Flight of the Committee on Science and Astronautics, *1973 NASA Authorization: Hearings*, 92nd Cong., 2nd sess, Part 2, 24 February 1972, pp. 233–38.
10 Douglas R. Lord, *Spacelab: An International Success Story* (Washington, DC: National Aeronautics and Space Administration Special Publication–487, 1987), pp. 2–5;

James A. Downey III, "Shuttle Sortie Mission Modes," 23 April 1971, Sortie Lab 1972 folder, MSFC History Archives; Douglas R. Lord, Director, Headquarters Space Station Task Force, to James T. Murphy, Director, MSFC Program Development, 10 September 1971, 10s Space Shuttle 1971 folder, MSFC History Archives.

11 Rees to Lucas, 22 March 1972, Space Shuttle Payload Studies 1970–1973 folder, box 55, MSFC Directors Files.

12 Lord, p. 46.

13 James T. Murphy to Eberhard Rees, 16 November 1971; Eberhard Rees to Christopher Kraft, 13 March 1972, Space Shuttle Payload Studies 1970–1973 folder, box 55, MSFC Directors Files.

14 Roger E. Bilstein, "International Aerospace Engineering: NASA Shuttle and European Spacelab," August 1981, pp. 7–13.

15 Murphy to Rees, 16 November 1971.

16 MSFC Release 74–27, 27 September 1974.

17 Rees marginal notations, Heimburg Notes, Weekly Notes, 24 January 1972, Weekly Notes file, MSFC Directors Files.

18 Dale D. Myers to Rees, 16 February 1972, Sortie Lab 1972 folder, MSFC Directors Files.

19 MSFC Release 72–41, 4 April 1972, Shuttle RAM 1972–73 folder, NASA Headquarters History Office (hereafter NHHO) Vertical File.

20 Rees to Dale D. Myers, 25 February 1972; Kraft to Myers, 27 March 1972; Rees to Myers, 19 April 1972, Space Shuttle Payload Studies 1970–1973 folder, box 55, MSFC Directors Files.

21 William R. Lucas, OHI by AJD and SPW, 1 March 1993, Huntsville, Alabama, pp. 9–10.

22 Murphy to Rees, 28 September 1972, Space Shuttle Payload Studies 1970–1973 folder, box 55, MSFC Directors Files.

23 Rees, memo to Lucas, 30 September 1972, Space Shuttle Payload Studies 1970–1973 folder, box 55, MSFC Directors Files.

24 The 1958 Space Act established international cooperation as one of NASA's objectives.

25 John M. Logsdon, "Together in Orbit: The Origins of International Participation in Space Station Freedom," unpublished paper prepared under NASA contract NASW–4237, December 1991, pp. 7–9, NHHO; Roger E. Bilstein, "International Aerospace Engineering: NASA Shuttle and European Spacelab," August 1981, pp. 7–10.

26 Logsdon, p. 9.

27 Erich Goerner Notes, MSFC Weekly Notes, 17 January 1972, 21 February 1972, Weekly Notes file, MSFC Directors Files.

28 William R. Lucas, OHI, 1 March 1993, p. 2; Erich Goerner Notes, MSFC Weekly Notes, 31 January 1972, Weekly Notes file, MSFC Directors Files.

29 Lucas, OHI, 1 March 1993, p. 2; Huber Notes, MSFC Weekly Notes, 26 June 1972, Weekly Notes file, MSFC Directors Files.

30 Dale D. Myers to James C. Fletcher, 1 February 1972, Box 1, Bilstein Spacelab Collection, JSC History Office.

31 Logsdon, p. 10.

32 Lucas, OHI, 1 March 1993, p. 2.

33 Myers to Fletcher, 1 February 1972.

34 Lord to Fred Vruels, Manager, MSFC Sortie Lab Task Team, 15 October 1972, Sortie Lab 1972 folder, MSFC Directors Files.

35 James A. Downey III, Memorandum for the Record, 17 July 1972, Sortie Lab 1972 folder, MSFC Directors Files.

36 Logsdon, p. 13. See also Bilstein, pp. 8–9.

37 Lee, OHI, 9 April 1993, p. 3.

38 PD Task Team Notes, Vruels Notes and Rees marginal comments, 17 July 1972, Weekly Notes file, MSFC Directors Files.

39 Richard J. H. Barnes, file memorandum, 9 November 1972, Sortie Lab 1972 folder, MSFC Directors Files.

40 Lord, pp. 18–27; NASA Release 73–12, 19 January 1973, Shuttle RAM 1972–73 folder, MSFC Directors Files; "Spacelab: Guidelines and Constraints for Program Definition," Level I, NASA MF–73–1, issued jointly by NASA and ESRO, 23 March 1973, Spacelab 1973 folder, MSFC Directors Files.

41 Myers to Rees, 19 December 1972, Spacelab 1973 folder, MSFC Directors Files.

42 Trott notes, Program Development Weekly Notes, 20 November 1972, Weekly Notes file, MSFC Directors Files.

43 "UK/France Join Sortie Lab Definition," *Space Daily*, 2 March 1973, p. 15.

44 Lucas to Petrone, 5 March 1973, Spacelab 1973 folder, MSFC Directors Files.

45 Jack Trott to Jack Herberlig, 13 March 1973, Spacelab Chronological Files, Box 3, JSC History Office; Charles F. Beers, "Agreements, Conclusions, and Actions from Special JSC/MSC Sortie Lab Meeting on March 30, 1975," 4 April 1973, Spacelab Chronological Files, Box 3, JSC History Office; "Management Instruction: Sortie Lab Program Management," 3 May 1973, Bilstein Spacelab Collection, box 1, JSC History Office.

46 Gierow notes, Program Development Weekly Notes, 20 November 1972, Weekly Notes file, MSFC Directors Files.

47 Philip E. Culbertson, Director, NASA Mission & Payload Integration, to distribution, 21 June 1973, DM01 Files: Shuttle 1973, MSFC Directors Files.

48 Low to Program Associate Administrators, 15 August 1973, DM01 Files: Shuttle 1973, MSFC Directors Files; Lord, pp. 64–65. Donlan was Deputy Associate Administrator for Manned Space Flight (Technical); the committee bore the title Payload Activities Ad Hoc Team.

49 Myers to Petrone, 22 August 1973, DM01 Files: Shuttle 1973, MSFC Directors Files.

50 C. J. Donlan, "Shuttle Payload Activities Ad Hoc Team Report to the Deputy Administrator," 17 April 1974, MSFC Shuttle History Historical Documents Collection (hereinafter SHHDC) No. 3892.

51 Lord, pp. 27–29, 59; Bilstein, pp. 19–21.

52 Petrone, internal MSFC memo, 18 December 1973; "Spacelab Program Office Charter," MM 1142.2, Charter Number 89, 18 December 1973, Spacelab 1973 folder, MSFC Directors Files.

53 Lord, pp. 72–74.

54 Lee, phone message to Lucas via Gertrude Conard, 31 May 1974, Spacelab 1974 folder, MSFC Directors Files.

55 Lord, p. 74.

56 "NASA Disapproves Spacelab Design Plan," De Telegraaf, 29 May 1974, Spacelab 1974 folder, MSFC Directors Files.

57 Lee to Lucas, 7 June 1974, Spacelab 1974 folder, MSFC Directors Files.

58 Lord, pp. 74–75.

59 Lucas, OHI, 1 March 1993, pp. 3, 4.

60 Lee, note to Lucas, 14 July 1978, Spacelab 1978–79 folder, MSFC Directors Files.

61 Lee, OHI, 9 April 1993, pp. 5, 8, 22.

62 Lee notes, Weekly Notes, 17 June 1974, Weekly Notes file, MSFC Directors Files.

63 Lucas, OHI, 1 March 1993, p. 3.

64 Lee, OHI, 9 April 1993, p. 9.

65 Lord, p. 61; Lee, OHI, 9 April 1993, p. 9.

66 Lucas handwritten note, 24 February 1975, Spacelab 1975 folder, MSFC History Archives.

67 Kraft to Lucas, 27 August 1975, Spacelab 1975 folder, MSFC History Archives.

68 Stan Reinartz, OHI by Jessie Emerson/MSI, 22 February 1990, Huntsville, Alabama, p. 28.

69 Glynn Lunney, OHI by AJD, 1 July 1991, Houston, Texas, p. 3. The development of Interface Control Documents came out of the efforts in 1975 of working groups in three areas (Structural/Mechanical, Environmental Control Support System, and Avionics) to establish joint control over interfaces. JSC and MSFC later extended the use of ICDs to other subsystems, and NASA and ESA applied them to interfaces ground support equipment and checkout areas at Kennedy Space Center. Lee, statement before the Subcommittee on Space Science and Applications of the Committee on Science and Technology, undated (late 1976 or early 1977), SHHDC No. 3709.

70 Lunney to Kraft, note attached to Yardley to Space Flight Management Council, "Spacelab Visibility at JSC and KSC," 30 July 1976, Box 1, Bilstein Spacelab Collection, JSC History Office.

71 James E. Kingsbury, OHI by AJD and SPW, 3 March 1993, Huntsville, Alabama, p. 18.

72 The Shuttle System Payload Data study, under Harry Craft, culminated with a report describing 44 possible Spacelab payloads. Another study early in 1974 examined possible payloads from recommendations submitted by the Joint User Requirements Group. Lord, p. 63.

73 The report also recommended that MSFC should be responsible for configuration management during the initial operational period. Donlan, "Shuttle Payload Activities Ad Hoc Team Report to the Deputy Administrator," 17 April 1974, MSFC SHHDC No. 3892.

74 Program Development Weekly Notes, 4 March 1974, 15 April 1974: Lee Weekly Notes, 15 April 1974, 22 April 1974; Otha C. Jean to Lucas, 17 October 1975, Weekly Notes File, MSFC History Archives.

75 Lord, pp. 65–67.

76 David Jex, OHI by Jessie Emerson/MSI, 27 February 1991, Huntsville, Alabama, pp. 5–6.

77 Payload management of the first two Spacelab flights fell under the Solar Terrestrial Programs Division of the Office of Space Science, and the third was under the Office of Applications (renamed the Office of Space Science and Applications by the time of the

mission). Charles W. Mathews to Lucas, 9 March 1976, Spacelab Payloads Projects folder, MSFC History Archives; John E. Naugle to Distribution, 7 April 1976, Spacelab folder, NHHO.

78 Naugle to Deputy Administrator, 21 April 1978, Spacelab folder, NHHO.

79 "STS–9 to Carry European-Built Spacelab on Its First Mission," MSFC Release No. 83–75, 10 November 1983, p. 47.

80 *Ibid.*, pp. 60, 61.

81 Kraft to Lucas, 20 June 1977, Spacelab Payload Projects 1977 folder, MSFC Center Directors File.

82 Jean to Lucas, 8 July 1977, Spacelab Payload Projects 1977 folder, MSFC Center Directors File.

83 Charles R. Chappell and Karl Knott, "The Spacelab Experience: A Synopsis," *Science* 225 (13 July 1984), 163.

84 Dr. John D. Hilchey, OHI by Thomas Gates, MSI, 6 November 1987, Huntsville, Alabama, pp. 24–25, 28–29, 42–43, 47–49.

85 G.D. Hopson to Lee, 22 October 1975; Kingsbury to Lucas, 31 October 1975, Spacelab 1975 folder, MSFC Center Directors File.

86 Hopson to Lee, 22 October 1975; R.H. Gray, Manager KSC Shuttle Projects Office, to Director, KSC, 13 July 1976, Spacelab 1976 folder, MSFC Center Directors File.

87 James E. Kingsbury, OHI by AJD and SPW, 3 March 1993, Huntsville, Alabama, p. 12.

88 132 MSFC civil servants and 49 support contractor employees assigned to the Concept Verification Test program (CVT) also performed Spacelab-related work.

89 Lee to Lord, 13 February 1975; Lord to Associate Administrator for Manned Space Flight, 16 January 1975; Lee, "MSFC Spacelab Program Office Manpower Position Statement," undated (probably January 1975), Spacelab 1975 folder, MSFC History Archives.

90 R.H. Gray to Director, KSC, 13 July 1976, Spacelab 1976 folder, MSFC History Archives.

91 Lowell Zoller to R.G. Smith, 20 October 1976, Spacelab 1976 folder, MSFC History Archives.

92 A.W. Frutkin, NASA Assistant Administrator for International Affairs, rewrite of "Spacelab Program," 23 July 1976; Gibson to Frutkin, 26 July 1976, Spacelab 1976 folder, MSFC History Archives.

93 NASA Sensitive Working Paper, "NASA Conditions and Recommendations," 21 July 1976; Yardley to distribution, "Roles and Responsibilities for NASA Advisors and Technical Specialists Assigned in Europe in Support of Spacelab," 20 October 1976, Spacelab 1976 folder, MSFC History Archives.

94 "Spacelab Director Resigns," *Flight* 110 (July 1976[?]) p. 106. The 25 flight per year figure is from Lucas to Yardley, 2 December 1974, Spacelab 1975 folder, MSFC History Archives. The Shuttle utilization figure of six Spacelabs is from Program Development Notes, Jean Notes, 1 July 1974, Weekly Notes file, MSFC Directors Files.

95 Lee To W. R. Marshall, 28 June 1976; John F. Yardley, draft memorandum to Directors of Space Shuttle and Spacelab Programs, 28 June 1976, Spacelab 1976 folder, MSFC Directors Files.

96 Zoller to R. G. Smith, 20 October 1976, Spacelab 1976 folder, MSFC History Archives.
97 *Ibid.*
98 Lee Notes, 3 February 1975, Weekly Notes file, MSFC Directors Files.
99 Bignier to Yardley, 22 November 1976, Spacelab 1976 folder, MSFC History Archives.
100 R. N. Lindley to D/SL, June 1977 (date obscured), Spacelab 1977 folder, MSFC History Archives.
101 Bignier to Yardley, 22 December 1976, Spacelab 1976 folder, MSFC Center Directors File.
102 Office of Space Flight Management Council, Summary of Agreements and Actions, 14 January 1977, MSF Management Council—HQ 14 January 1977 folder, MSFC Directors Files.
103 Bignier to Spacelab Project Team, 17 December 1976, Spacelab 1976 folder, MSFC Directors Files.
104 For general descriptions of Spacelab configuration and subsystems, see Spacelab Fact Sheet, June 1977, SHHDC No. 3713; and Spacelab One Fact Sheet, MSFC Release 79–5, 15 January 1979, SHHDC No. 3705. More detailed information is in Lee, statement before the Subcommittee on Space Science and Applications of the Committee on Science and Technology, undated (late 1976 or early 1977), SHHDC No. 3709.
105 Thomas J. Lee, OHI by AJD and SPW, 9 April 1993, Huntsville, Alabama, p. 14.
106 Lord, pp. 227, 231.
107 Program Development Notes (Gierow), 15 July 1974, Weekly Notes File, MSFC Directors Files.
108 Ernst Stuhlinger to Mike Wright, comments on draft of this manuscript, 21 November 1994, MSFC History Office; Lee Notes, 13 January 1975, Weekly Notes File, MSFC Directors Files.
109 Lord, p. 228.
110 Program Development and Lee Weekly Notes, 6 May 1974, 29 July 1974, 13 January 1975, 21 April 1975, Weekly Notes File, MSFC Directors Files. Luther E. Powell to Lucas, 24 March 1976, Spacelab 1976 folder, MSFC Directors Files.
111 Office of Space Flight Management Council, "Summary of Agreements and Actions," 14 January 1977, MSF Management Council—HQ 14 January 1977 folder, MSFC Directors Files. Lee to Distribution, 19 January 1977, Spacelab 1977 folder, MSFC Directors Files. IPS development was so complex that ESA contractor Dornier included a statement in its proposal that would later seem ironic: that realization of its goals "in some cases may be precluded by cost, schedule, or technological considerations." NASA properly found the statement "completely unacceptable," but indeed all three concerns became realities. Thomason, in Kingsbury Notes, 26 January 1976, Weekly Notes File, MSFC Directors Files.
112 Lee to Lucas, 28 February 1978, Spacelab 1978–79 folder, MSFC Directors Files.
113 Lee to Lucas, 20 March 1979, Spacelab 1978–79 folder, MSFC Directors Files.
114 Lord, pp. 232–39.
115 "MSFC Spacelab Program," 10 January 1983, MSFC History Office Fiche #1781.
116 Lord to Associate Administrator for Space Transportation Systems, 18 June 1979; Andrew J. Stofan to Associate Administrator for Space Transportation Systems,

21 June 1979; Lord and Chester Lee, Director HQ STS Operations to Manager, Shuttle Payload Integration and Development Program Office, JSC, 28 June 1979; Chester M. Lee to Yardley, 9 July 1979, Spacelab General 1979–1984 folder, NHHO.

117 Dr. David Shapland quoted in Dick Baumbach, "Cost of Spacelab Europe's Top Fret," *Today* (27 April 1979).

118 Viewgraphs, "NASA Administrator-ESA Director General Spacelab Programme Meeting," 14 January 1977; Lee to Distribution, "NASA Administrator/ESA Director General Spacelab Program Review Summary," 19 January 1977, Spacelab 1977 folder; Lee to Lucas, 14 July 1978, Spacelab 1978–79 folder, MSFC Directors Files. "ESA's Spacelab Director Assesses Reality Vs. Expectations," *Aerospace Daily* (13 April 1979), pp. 221–222. These documents raise the issues cited; the issues are further elaborated in various documents in the Spacelab files for the years from 1977 to 1980 in the MSFC Directors Files.

119 "ESA Cautions U.S. About Duplicating Spacelab Elements," *Defense/Space Daily* (25 February 1980), p. 283.

120 John W. Thomas to Lee, 7 October 1980, Spacelab 1980–81 folder, MSFC History Archives.

121 Lee to Lucas, 14 July 1978. William L. Walsh, Jr. to Lee, 5 October 1979, Spacelab 1978–79 folder; "Spacelab Duplication Working Group Report," 11 February 1980, Spacelab 1980–81 folder, MSFC Directors Files.

122 Lee to Lucas, 2 April 1979; Lee to Lucas, 27 February 1979, Spacelab 1978–79 folder, MSFC Directors Files.

123 "ESA's Spacelab Director Assess Reality vs. Expectations," *Aerospace Daily* (13 April 1979), p. 221.

124 ESA Spacelab Programme Board, "Desirability of continuing the European effort in the manned-systems area beyond Spacelab," 14 May 1979, MSFC Directors Files.

125 Lucas to Lee, 7 June 1979, Spacelab 1978–79 folder, MSFC Directors Files.

126 Flippo to Robert A. Frosch, 11 April 1978, Spacelab folder, NHHO; Judith A. Cole, Memorandum for the Record of Lovelace meeting with Congressman Flippo, 18 April 1978, Flippo, Rep. Ronnie (Alabama) folder, NHHO.

127 Cole, Memorandum for the Record, 18 April 1978.

128 Naugle to Deputy Administrator, 21 April 1978, Spacelab folder, NHHO.

129 "NASA Charts Spacelab Science Cutbacks," Aerospace Daily (11 June 1980), p. 230.

130 Harrington to Deputy Comptroller, 6 May 1981, Spacelab 1980–81 folder, MSFC History Archives.

131 Harry Craft, OHI by AJD, 14 July 1994; "Spacelab 1 & 2 Payload Project Plan," August 1977, Spacelab Payload Projects 1977 folder; "Spacelab Mission 3 Payload Project Plan," May 1976, Spacelab Payloads Projects 1976 folder, MSFC Directors Files.

132 Kraft to Lucas, 20 June 1977, Spacelab Chronological Files, Box 17, JSC History Office.

133 Lucas to Kraft, 12 July 1977, Spacelab Chronological Files, Box 17, JSC History Office.

134 Robert A. Kennedy, NASA HQ Manager, space Science Missions to Bignier, 15 January 1980, Spacelab General 1979–1984 folder, NASA HQ History Office.

135 Alan M. Lovelace, NASA Deputy Administrator, statement before the Subcommittee on Space Science and Applications Committee on Science and Technology, House of

Representatives, 31 March 1980, Spacelab General 1979–1984 folder, NASA HQ History Office.

136 "OSTA–1 Mission" and "OSS–1 Mission," Mission Summary Sheets, June 1983, Fiche No. 697.

137 Charles R. Chappell, OHI Thomas Gates, 24 September 1987, p. 2.

138 Ralph M. Hoodless, OHI by AJD, 25 July 1994.

139 Craft, OHI.

140 Lord, pp. 330–35.

141 Lord, p. 340.

142 Commander John Young, Pilot Brewster Shaw, Mission Specialists Dr. Robert Parker and Dr. Owen Garriott, and Payload Specialists Dr. Ulf Merbold of ESA and Dr. Byron Lichtenberg of NASA.

143 Chappell and Knott, p. 164.

144 Marshall's planning for the payloads had been directed by a team including O.C. Jean, Rick Chappell (the mission's chief scientist), Jim Downey, Bob Pace, and Harry Craft. Lord, p. 346.

145 NASA MSFC, *Science in Orbit: The Shuttle & Spacelab Experience, 1981–1986* (Washington: U.S. Government Printing Office, 1988), pp. 71–77, 84–89; David Shapland and Michael Rycroft, *Spacelab: Research in Earth Orbit* (Cambridge: Cambridge University Press, 1984), p. 139.

146 *Science in Orbit: The Shuttle & Spacelab Experience, 1981–1986*, pp. 55–59; Shapland and Rycroft, pp. 137–139.

147 *Science in Orbit: The Shuttle & Spacelab Experience, 1981–1986*, pp. 27–33; Shapland and Rycroft, pp. 143–147.

148 *Science in Orbit: The Shuttle & Spacelab Experience, 1981–1986*, pp. 97–107; Shapland and Rycroft, p. 136.

149 *Science in Orbit: The Shuttle & Spacelab Experience, 1981–1986*, pp. 6–15; Shapland and Rycroft, pp. 148–152.

150 Lord, pp. 352–60; Craig Covault, "Spacelab Returns Broad Scientific Data," *Aviation Week & Space Technology* (5 December 1983), 18–20.

151 Lord, pp. 363–64; *Marshall Star*, 14 December 1983; Chappell and Knott, pp. 163–65.

152 George H. Fichtl, John S. Theon, Charles K. Hill and Otha H. Vaughan, editors, *Spacelab 3 Mission Science Review*, NASA Conference Publication 2429 (Washington: NASA Scientific and Technical Information Branch, 1987), 1–10.

153 Lew Allen to W.R. Lucas, 8 August 1985, Spacelab Payloads 1985 folder, MSFC History Archive.

154 Lord, pp. 374–88.

155 Dianne K. Prinz to William R. Lucas, 25 September 1985, Spacelab Payloads 1985 folder, MSFC History Archive.

156 *Marshall Star*, 29 August 1990, 25 November 1990, 5 December 1990.

157 Lord, p. 401.

Chapter XII

The Hubble Space Telescope

The Hubble Space Telescope was the most costly and challenging science technology project managed by the Marshall Space Flight Center (MSFC). Because of Hubble's $2 billion expense, Marshall became a leading member of a complex coalition that moved Congress to continue support. Because of the project's complexity, Marshall constantly struggled to balance scientific and technical requirements with financial resources. Whenever the project fell out of equilibrium, the Center worked with the coalition to find a new alignment. Like any middle manager, Marshall often got more blame for problems than it got credit for achievements. But the Center's efforts to overcome management and engineering troubles helped ensure that the Space Telescope became a scientific success.

Conception and Coordination

Scientists and space pioneers had long recognized that a telescope in space would escape many conditions distorting observations from the ground. Some of these early conceptions had a Marshall connection. Wernher von Braun, in *Collier's* in 1952, had envisioned space observatories tended by a Space Station. In 1965 and 1967, Marshall had let contracts for studies of Space Telescopes.[1]

Professional astronomers associated with universities and research institutes, however, first lobbied for the Large Space Telescope (LST), which eventually became the Hubble Space Telescope. Most prominent among them was Lyman Spitzer at Princeton. In the late 1960s Spitzer and other astronomers urged NASA Headquarters to support an optical telescope with a three-meter primary mirror. Headquarters responded by sponsoring a scientific and engineering conference for the telescope organized by the Marshall Center in Huntsville in the spring of 1969. Later that year some Headquarters officials in the Office of

Advanced Research and Technology and in the Office of Space Science and Applications (OSSA) began urging von Braun and Marshall to push the project. The most active and determined promoter of this effort was Jesse Mitchell, director of physics and astronomy programs at OSSA.[2]

In an era of budget cuts and personnel reductions, Marshall needed new work. Von Braun told one member of the space telescope coalition, "That's the project I would like to see Marshall do." Ernst Stuhlinger, the Center's leading scientist, noted that the project was gaining support just when the Center was phasing out of the Saturn project, and argued that "We can hardly afford not to consider it as a very promising future MSFC project." He believed that "the LST Project would utilize many of the skills existing at MSFC, including technical, scientific, test, quality assurance, and project management types. It would help us retain and even strengthen our in-house capabilities." While von Braun had not wished to initiate a major new project until the Saturn project ended, his successor as Center Director, Eberhard Rees, decided to submit a formal proposal for a space telescope. Accordingly, late in 1970 when Headquarters approved preliminary management and engineering studies, Marshall established a telescope team within the Program Development Directorate.[3]

The team, with James A. Downey III as manager and with Jean Olivier as chief engineer, followed Program Development's entrepreneurial routines. Downey, who, as a member of the Space Sciences Lab, had also helped bring the High Energy Astronomy Observatories (HEAO) project to Marshall, guided planning studies for the space telescope. Later he recalled that the space telescope team followed formal procedures while HEAO had been "catch-as-catch-can." The telescope team had regular channels for working with Center laboratories and communicating with outside groups. Downey recollected that "within Program Development it was certainly the major scientific activity, far and away the major scientific activity we were studying." Team members drew ideas and information from scientists like Spitzer, Herbert Friedman, Robert O'Dell, and Riccardo Giacconi. Also useful were previous studies by Langley Research Center (LaRC) that suggested including a space telescope in plans for a Space Station. Indeed some of Marshall's early plans for the telescope were similar to the design of the *Skylab*-Apollo Telescope Mount; like *Skylab* the telescope would be joined to a Space Station or a Research and Applications Module (later called Spacelab) and would record data on photographic film that astronauts would regularly change. Costs, and lack of support for a Space Station, quickly drove the team toward an untended satellite concept.[4]

Marshall faced competition from Goddard Space Flight Center (GSFC) for management of the project. A strong rival, Goddard had numerous professional astronomers and superior experience with astronomy satellites. In contrast Marshall had less experience in optics and astronomy, and no astronomers with doctoral degrees. Moreover some officials in the Headquarters Office of Space Science and Applications preferred working with Goddard, a Center they had worked with frequently. This preference showed in November 1970 when OSSA personnel described the assets of both Centers in a management meeting at Headquarters. Some present wondered whether Marshall was up to the task of managing such an ambitious science project. Dale Myers, associate administrator for Manned Space Flight, was blunt, saying that "MSFC could not do the large space telescope program."[5]

Nevertheless Marshall had advantages. Goddard had too many commitments and too few people and so its director did not support the new project. Marshall, in contrast, had too many people and too few commitments; a Center manpower study argued that "MSFC could accept and successfully pursue the lead role assignment for the LST and our assigned Shuttle responsibilities, in addition to continuing with our on-going and other anticipated programs." Moreover, Marshall leadership had become enthusiastic about the Space Telescope. Before one planning conference, Stuhlinger urged Program Development to show Headquarters' officials that Marshall was "willing to put its full strength behind the LST project." Whenever NASA had a telescope meeting in Huntsville, recalled one astronomer, Marshall practically welcomed the visitors with "a brass band and red carpet." Beyond the style, the substance of Marshall's plans was often impressive; in January 1971 Jesse Mitchell, NASA director for physics and astronomy programs, praised the Center's Program Development team for giving an "excellent" presentation.[6]

Behind the scenes, some NASA officials, like Administrator James Fletcher, feared that Marshall's personnel surpluses could lead the Office of Management and Budget to close the Center. Fletcher told his successor that Marshall had to be kept open to preserve its expertise for the Shuttle program. These circumstances led Hubble historian Robert Smith to the charge that "the manpower argument" was "decisive" in determining the assignment of Lead Center and that Marshall became the manager for reasons other than technical competence.[7] This contention seemed doubtful to Downey. Looking back years later, he thought that Headquarters officials had worried more about the success of the large Space Telescope than about the needs of Marshall; "they will go to the Center where they think it [a project] can best be done."[8]

Headquarters recognized Marshall's technical and managerial strengths. The Center had experience with previous scientific satellites and was Lead Center for *Skylab* and its experiments. Marshall's early designs included a pressurized cabin to facilitate repairs in space, and Goddard could not match the proficiency on manned projects that Marshall had accumulated on *Skylab*. Most importantly, Marshall had far more expertise than Goddard did in managing big engineering projects, coordinating numerous organizations, and integrating diverse hardware. William Lucas remembered that "those people [at Headquarters] who saw or grasped the significance of the systems engineering involved saw it as a Marshall program."[9]

Even so, as early as mid-1971 Headquarters proposed a division of labor between Huntsville and Greenbelt. Jesse Mitchell, the key Headquarters official who promoted the telescope, expressed his conviction that Marshall was better prepared to do the project than Goddard, but he insisted that the two Centers cooperate. He said Washington expected Marshall to answer the question, "How can MSFC work with GSFC in a gainful way?" Marshall suggested that Goddard provide scientific specifications for the spacecraft, direct development of the scientific instruments, and manage orbital operations; Marshall could develop the overall spacecraft and the optical apparatus. By early 1972 Marshall's plans called for a large Space Telescope with three hardware modules, with the optical telescope assembly (OTA) and the support systems module (SSM) for itself, and the scientific instrument package (SIP) for Goddard. Under this scheme "the Scientific Instrument Package [would] be 'sub-contracted' to GSFC for development along with the Flight Operations." Under these terms, the Agency made Marshall Lead Center for the LST in April 1972.[10]

Although this plan would use the strengths of both Centers, it left many questions unanswered. Could the Centers work out a clear division of labor on a complex project that lacked clear borders between science, management, and engineering? Would the engineering development Center be able to direct the science and operations Center? Which Center would coordinate communications with the telescope's customers, the astronomers? How would the Agency settle conflicts? NASA would spend years answering these questions.

The Centers began working on solutions and, by late summer 1972, had agreed that Marshall would select the project scientist with Goddard's consent. Marshall did not want a Goddard scientist in the post; James Murphy, MSFC's director of Program Development, feared that GSFC would use their person to "run the LST project." Accordingly the Centers agreed on an outsider, Dr. C. R. "Bob" O'Dell, the former director of the Yerkes Observatory at the University of Chicago. Members of the Marshall Center first believed that Lyman Spitzer should be the project scientist; in fact, Center personnel had read the acronym LST (Large Space Telescope) as Lyman Spitzer Telescope. However, Spitzer suggested O'Dell, who agreed to accept the position. O'Dell recognized the scientific and political prestige of scientists outside NASA, and wanted external astronomers to control the science aspects of the Space Telescope. His ideas coincided with Marshall's traditional use of contractor scientists and with its efforts to avoid Goddard's control. Following O'Dell's advice, Marshall proposed creating an LST science steering group to provide scientific support to the project and to facilitate communications with external astronomers. The Center argued that "NASA does not now have sufficient astronomical expertise to internally provide all necessary scientific judgment." The new advisory group would be composed of the project scientist, science officials from Headquarters, Marshall, Goddard, and eight outside astronomers.[11]

Goddard accepted the advisory group, but the two Centers disagreed about the project's science organization. Goddard and Marshall disputed which Center should manage the contracts for the scientific instruments and communications with the scientists. In November Murphy reported that the Centers were "in a state of serious disagreement" such that Marshall's "ability to effectively interface with GSFC on a daily basis at the working level has been seriously impaired," and his counterpart at Goddard agreed that "our positions on the issues . . . are fairly far apart." Murphy complained that Goddard wanted "to assume practically total science responsibility and authority, including interfacing with the scientific community" and had "prematurely assumed a design configuration and integration philosophy for LST which would maximize their management and integration role in the scientific instrument development without regard for other program considerations." To ensure effective project management, Marshall Director Rees insisted that "the main contact with the scientists had to be through the Project Scientist who is assigned to Marshall."[12]

The Centers tried to resolve their disagreements using a typical NASA matrix. Marshall would provide a contracting officer to monitor the finances of each scientific instrument contract, allowing the Center to penetrate Goddard's activities. The Centers would coordinate technical issues through overlapping science teams. Each scientific instrument would have a team of external specialists who would report to Goddard. Above this would be the LST science steering group composed of the project scientist, the leaders of the instrument teams, and some Goddard experts; this group would report to Marshall. In theory the instrument teams and steering group would assume responsibility for the project rather than for the parochial interests of the Centers.[13]

In practice, however, relations between the Centers remained problematic, and many people associated with the project would blame later problems on the troubled marriage of Marshall and Goddard. Although the Centers struggled to define overlapping responsibilities by dividing technical tasks,[14] their agreements left many problems unresolved. As early as 1976, Goddard's LST manager remarked that the difficult relationship had led to "a tremendous amount of wasted effort and dollars."[15]

Nonetheless by late 1972 Marshall had organized the project and begun preparations for Phase B activities. In December NASA issued a request for proposals inviting astronomers to join the LST teams that would help define the scientific instruments and preliminary designs. To share information about NASA's plans, O'Dell, officials from Program Development, and Goddard experts addressed scientists at Cal Tech, the University of Chicago, and Harvard; Headquarters officials presented the same material at Frascatti, Italy. In addition to their technical purpose, O'Dell said these "dog and pony shows" tended to help in "drumming up business" for the telescope. Marshall Director Rocco Petrone told Headquarters that the scientists' response had been "extremely enthusiastic" and had "justified MSFC's development of this plan and will serve to guarantee future science community support for the LST Observatory Program."[16] Marshall and the rest of the Space Telescope coalition would need this support in the trying days ahead.

Money and Machinery

While NASA worked on management, it also struggled with money. The cost of the Space Telescope would be a constant concern and create a political and technical conundrum. To get congressional support, NASA had to minimize costs; but to keep scientific support, it had to ensure the telescope's performance.[17] As Lead Center, Marshall had to balance conflicting goals and build support for the telescope. In the process the Center functioned as an engineering organization and a behind-the-scenes political machine.

Financial pressure pushed the Center's design activities and often forced it to relinquish conservative engineering principles. The Center's March 1972 project plan called for three telescopes, an engineering model, a "precursor" flight unit, and the final LST. Design and development would cost between $570 and $715 million. Headquarters believed this was too expensive. In a December 1972 meeting, NASA Administrator Fletcher "emphasized that the current NASA fiscal climate was not conducive to initiation of large projects" and suggested $300 million as a cost target.

By April 1973 Marshall had proposed three ways to cut costs. A "protoflight" approach would eliminate the engineering and precursor units; a single spacecraft would serve as test model and flight unit. The protoflight approach had been successfully tried for Department of Defense projects, and the Center expected it to reduce costs—which would please Congress—and speed progress to operations—which would please the astronomers. The telescope maintenance strategy also changed. Rather than designing for extensive repair in orbit inside a pressurized cabin, Marshall suggested a design that would eliminate the cabin and minimize repairs in orbit. The new design assumed the Space Shuttle could return the telescope to Earth for major repairs. These changes simplified the overall LST design and development scheme.

More problematic was a contracting method that used two associate contractors rather than a prime contractor for the support systems module and a subcontractor for the optical telescope assembly. NASA would pair large aerospace contractors working on the SSM with optical companies working on the OTA. Several motives determined NASA's decision. Downey recalled that NASA recognized the complexity of the optical systems and wanted two contractors to proceed with preliminary design. The Agency could then judge proposals for the OTA separate from those for the SSM and match the best contractors. In the development phase, the associate approach would allow the Center to penetrate the OTA contractor directly rather than having to go through an SSM prime contractor. Finally planners expected to save costs by making Marshall, rather than a prime contractor, responsible for systems engineering and integration activities. All these changes lowered the projected cost to between $290 and $345 million.[18]

The associate contractor approach, however, complicated an already complex management structure. In December 1972 a Headquarters report observed that the scheme was "rife with interfaces" because "MSFC itself plays several roles; study manager, project synthesizer, (dual) development contractor, integrator, with GSFC in the wings as instrument developer and ultimate systems operator, all this without mentioning the role of the astronomical profession." The report worried that the resulting management problems would drive up costs.

Looking back years later, Center officials wondered about the associate approach. Downey believed that Headquarters had at first only wanted the associate approach for the design phase; after the Agency had gained confidence in its designs and after the project received approval, they had expected to turn to a prime contractor. But Downey said management turnover at Headquarters led to a loss of memory and to perpetuation of the initial scheme. James Kingsbury, the director of Marshall's science and engineering labs, believed the associate approach was a mistake. "We were not telescope manufacturing people," he said, and since "neither one could tell the other one what to do, and it was exceedingly difficult for somebody like us to be in sufficient position to be sure what the right thing was if the two were at odds. We had to make some decisions that were made with the best knowledge and intelligence that we had and in a few cases months later we had to reverse them because they were wrong."[19]

Throughout the last half of 1973 and the first half of 1974, NASA continued to elaborate the LST design and prepare for Phase B. The telescope astronomer teams met and refined the science requirements. Their advice led to the decision to use new detectors for the telescope. Innovative electronic detectors would replace film cameras, because the astronomers worried that film would reduce data quality and increase risk, especially when astronauts replaced film canisters. Moreover, the scientists, following O'Dell's lead, also simplified the telescope's optical structure. O'Dell defined standard modular science instrument (SI) envelopes, each with identical mechanical and electrical interfaces with the telescope. This standard interface greatly simplified development and made orbital replacement of SI's practical.

Marshall's Information and Electronic Systems Lab and its Systems Dynamics Lab helped contractors with the pointing and control system. Preliminary design studies by the labs and by Martin Marietta investigated whether moving the spacecraft or moving its mirrors best met the pointing requirements. After determining that accurate pointing of the entire spacecraft was possible, they chose reaction wheels over control moment gyros to guide the spacecraft, deciding that reaction wheels were more stable, reliable, and cost effective. Contractors, while preparing proposals for Phase B, also studied how to reduce weight by using new materials and designs for the spacecraft structure. In mid-1973 the Center awarded two identical $800,000 contracts for preliminary design and program definition of the OTA to the Optical Systems Division of the Itek Corporation and the Perkin-Elmer (PE) Corporation.[20]

NASA intended to ask Congress for a new start for the telescope in FY 1976, but decided to list the project's Phase B funds as a separate item in the FY 1975 budget request. The strategy intended to alert Congress of the need for future money, but in effect the telescope faced the double jeopardy of two new start decisions. The plan backfired in June 1974 when the House Appropriations Committee reasoned that the LST was too ambitious and lacked support from the National Academy of Sciences. Based on this recommendation, the House deleted the project's $6.2 million from NASA's budget.[21]

The telescope coalition, including space astronomers, aerospace contractors, and optics firms quickly began lobbying to restore the money. Marshall, largely through O'Dell, facilitated the efforts from behind the scenes. From his first days as project scientist, O'Dell had mixed technical and political activity. Deputy Center Director Lucas wrote that O'Dell "is fully aware that the project may not move out as rapidly as we would like, and he considers one of his important responsibilities to be of assistance in selling this project to the scientific community." O'Dell had tried to sell the large Space Telescope through articles in popular science journals like *Sky and Telescope* and in his dog and pony shows at professional meetings. These presentations fell short of formal lobbying but blended promotional appeals in technical information, in much the same way that von Braun had publicized previous plans.[22]

Immediately after the House deleted telescope funding, Headquarters' Offices of Space Science and Legislative Affairs told O'Dell "not to communicate with the scientific community." Marshall managers believed this was a mistake because the project scientist had NASA's closest contacts with astronomers. Evidently Headquarters officials agreed because they soon removed "the gag" from Dr. O'Dell. Although he could not work openly, O'Dell led part of what Space Telescope historian Smith has called the "Princeton-Huntsville axis" that fought to restore funding. O'Dell helped transform the scientists on Marshall's LST teams into lobbyists, and furnished "scientists specific names and addresses of Congressmen and their staff members that the scientists may wish to contact." He also channeled information between Agency groups, contractors, and astronomers. The coalition argued that, contrary to the House interpretation, the National Academy of Science actually supported the Space Telescope. By August, claims like this convinced the Senate and the conference committee to restore funding. After this success, new Marshall Director Lucas congratulated O'Dell for "your very substantial effort and the catalytic effect you had on the others."[23]

The coalition had won the battle, but the struggle transformed the telescope project. While approving funds, Congress cut the budget from $6 million to $2 million, thus forcing NASA to extend Phase B planning and delay the new start. Congress also wanted a less expensive telescope, and in August 1974 directed NASA to scale down the project and to get international help. Headquarters therefore told Marshall to define "a minimum 'LST class' observatory" with a total cost of $300 million and plan for a new start in FY 1977. Once again politics required that Marshall's telescope task team and science groups design to cost.[24]

In the fall, NASA decided to seek European assistance for the project. NASA expected that foreign participation would not only reduce the charges to Congress, but also raise the project's chances in Congress. Marshall's director of Program Development explained that "If we can get the UK. and/or ESRO to support a non-critical part of the LST with dollars then our chances are improved for a final 'go-ahead.'" The Center prepared for the negotiations by looking for hardware modules with clean interfaces that the Europeans could develop. Then Headquarters and Marshall project officials traveled abroad, beginning discussions of European development of various scientific instruments or parts of the spacecraft structure. These negotiations culminated in 1977 in an agreement in which the European Space Agency (or ESA, the successor of ESRO) would develop a faint object camera to observe the ultraviolet, visual, and near-infrared spectrum, build the solar energy arrays for the spacecraft, and support scientific research and orbital operations.[25]

By December 1974 the Program Development task team had downsized the telescope. As before the team had to balance cost and performance and devise a design pleasing to Congress and the astronomers. Team leader Downey said the Agency wanted "to procure the lowest cost system that will provide acceptable performance" and would "be willing to trade performance for cost." Working with the LST science groups and contractors, the team reduced the telescope's primary mirror from a 3-meter aperture to 2.4 meters. This major change mainly resulted from new NASA estimates of the Space Shuttle's payload delivery capability; the Shuttle could not lift a 3-meter telescope to the required orbit. In addition, changing to a 2.4-meter mirror would lessen fabrication costs by using manufacturing technologies developed for military spy satellites. The smaller mirror would also abbreviate polishing time from 3.5 years to 2.5 years. The redesign also reduced the mass of the support systems module from 24,000 pounds to 17,000 pounds; the SSM moved from the aft of the spacecraft to one-third of the way forward and became a doughnut around the primary mirror. These changes diminished inertia and facilitated steering of the spacecraft, thus permitting a smaller pointing control system. The astronomers chose to reduce the number of scientific instruments from seven to four. Finally, the Marshall team believed that designing for repair would allow for lower quality standards. Together the changes lowered the telescope's cost to $273 million. Alois W. Schardt, the director of physics and astronomy programs at Headquarters, praised the team for doing "an outstanding job" of planning with "design to cost" criteria.[26]

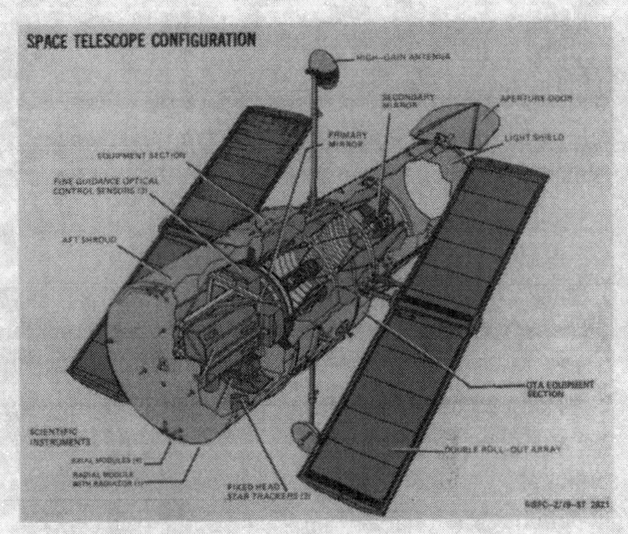

Space Telescope configuration.

The following year and a half was very trying for Marshall and the Space Telescope coalition. NASA's top management postponed the request to Congress for a new start from FY 1977 until FY 1978, fearing that more money for the telescope would mean less for the Shuttle.[27] The telescope thus became caught in the Catch-22 of the budget priorities of the Shuttle program: Agency managers justified the Shuttle by its capability to carry scientific payloads like the telescope but also justified sacrifices from science projects by the needs of the Shuttle.

During the waiting period Marshall walked a tightrope, balancing the telescope project's terrible twin needs for money and cost containment. In the fall of 1974 the Center pressed Headquarters to begin the Phase B industry study contracts. Murphy contended that the project needed the studies to learn about costs and technical problems. Moreover, delaying the contracts could disrupt the coalition and force industry to disband its telescope teams and withdraw its political backing. He told Headquarters that "we need strong industrial support at our congressional hearings" and another delay "could greatly impact all supporters of the LST." For these reasons and the need to accomplish a more thorough definition of the complete spacecraft, NASA issued a competitive solicitation to industry. This led in November 1974 to the award of preliminary design and program definition contracts for the SSM to Boeing Aerospace, Lockheed Missiles and Space, and Martin Marietta Corporation. In January 1975 the Agency extended the study contracts of Itek and Perkin-Elmer for the optical telescope assembly.[28]

At the same time the Center was spending money, however, it had to contain costs to please Congress. Balancing realism and salability was especially problematic when telescope officials tried to devise a project budget. They recognized that technical challenges would make the project expensive, especially during a period of high inflation: if they underestimated costs, they would eventually have to beg Congress for more money. On the other hand, too large a contingency would be self-defeating and make the project's budget "too high to be sold" in the first place.[29]

Given that a project without a new start was not a project, Headquarters in mid-1975 emphasized salability and directed Marshall to minimize cost estimates. Noel Hinners, associate administrator for Space Science, informed Center Director Lucas "to continue to explore every avenue towards realistically reducing the LST cost and to actively look at ways to keep early year funding as low as possible" because "our chances of obtaining a . . . new start hinge on this." Deputy Administrator George Low informed the telescope contractors that "the costs now being projected would be impossible to include in any NASA budget in the foreseeable future and the project therefore might well be canceled." He advised them to try cutting costs in half by "relaxing the requirements." Low's efforts to convince Congress that the project had purged extravagance led him to change its name from the Large Space Telescope to the Space Telescope.[30]

Lobbying for the project continued throughout this period. Because the astronomers and contractors had improved their organization since 1974, Marshall participated less. Downey, the task team manager, recalled that the Center "did a lot of kind of wringing of our hands in that period [of lobbying], because we'd done about what we could do." Even so, Project Scientist O'Dell continued to make public presentations and contribute to Headquarters' campaigns. He described the telescope's benefits in nontechnical terms, calling it a time machine that could study the history of distant stars and the origins of the universe. In addition, Marshall officials drafted a letter for North Alabama Democratic Congressmen Ronnie Flippo's signature, trying to get support from the chair of the House Appropriations Committee.[31]

Finally in 1976, Congress approved a new start for the Space Telescope. This approval owed much to Marshall's efforts to define a salable program. The search for support, however, had led to major changes, including reduction in the size and capability of the spacecraft, addition of the European Space Agency, adoption of an associate contractor approach, and, most importantly, degradation of realistic cost projections. According to historian Smith, the "price" of political support was a project that was "both oversold and underfunded," making the telescope "a program trapped by its own history."[32] Eventually the trap would squeeze tightly on Marshall and its contractors.

Design and Delay

In the late 1970s, Marshall clarified the project's organization, selected contractors, and elaborated final designs. Again the Center encountered problems squaring science and engineering, especially when working with Goddard. And even as hardware design and development progressed, the Space Telescope project showed early symptoms of organizational and financial ills.

In this period, the project's greatest controversy was a struggle to control the orbital operations, and ultimately the science, of the Space Telescope. The struggle emerged from differences among astronomers, Goddard, and Marshall. Many astronomers believed NASA should establish an independent institute, much like the institutes for ground telescopes, to manage the Space Telescope's science operations and data dissemination. Since this proposal threatened Goddard's position as NASA's space science Center, Goddard opposed it. When Marshall backed the academic astronomers, Headquarters stepped in to find a solution pleasing to both its scientific customers and its Centers.[33]

Initially Marshall's support for the telescope institute came from O'Dell. As project scientist he served as spokesman for the Science Working Group and as early as 1974 began presenting its wishes to the Agency. In a letter in 1975, O'Dell expressed the group's fears to John Naugle, associate administrator for space science. Goddard's plans for operations, he argued, were based on "Center parochialism" rather than the needs of the scientific community. "GSFC has a large body of resident astronomers, feels it must carve out a meaningful role for these people, and is unwilling to commit substantial resources to LST." Worse yet, Goddard's astronomers lacked the expertise of academic scientists but refused to accept advice. In contrast, "MSFC does not have a large body of resident astronomers, has no reservations to looking outside for guidance, has been substantially reduced in size, is looking for more business, and is willing to commit significant resources to LST." O'Dell got support from other Marshall officials. Stuhlinger, Marshall's associate director for Science, thought the telescope institute could be anywhere, raising Huntsville as a possibility. He also suggested to Headquarters that "Mission Operations should be at the Center where design, development, fabrication, integration, testing, launching, checkout, and initial operation of LST has been managed, i.e., at MSFC."[34]

Marshall's support for an institute for science operations and quest to become Lead Center for spacecraft operations put Goddard on the defensive and worsened the Centers' already troubled relationship. Goddard officials believed that they were the science Center for the telescope, but that Marshall and the academic astronomers wanted to reduce Goddard to the status of a contractor. William Keathley, who became the Marshall telescope task team manager in 1976, described "GSFC's working level attitude" as "distrustful, uncooperative and even hostile at times." By late 1976, the conflict had impaired negotiations on the intercenter agreement for the telescope. From Keathley's perspective,

Goddard wanted an "associate role" in project development, and sought responsibility for the science institute, the principal investigators and all operations planning, and equal authority with Marshall for all contractual and engineering matters affecting the scientific instruments and spacecraft control.

Such proposals, Marshall officials believed, would complicate management, and thereby raise costs and reduce quality. Keathley thought that the Goddard plan "limits the authority of the Project Manager and degrades the position of the Project Scientist" and risked "jeopardizing MSFC's ability to fulfill our commitments for overall management of the project."[35] O'Dell agreed, believing that "having all responsibility for operations turned over to GSFC would make the Project Scientist directly responsible to GSFC" and thus "make his role ambiguous." In discussions with Headquarters, Center Director Lucas argued that the Marshall-Goddard relationship for the scientific instruments was no different from the Johnson-Marshall relationship for the Space Shuttle main engines and that success of the project required "Marshall penetration" of Goddard. But rather than accepting subordinate status during development, Lucas believed Goddard wanted its "head of the Mission Operations Office to have veto power over the whole program." After one meeting in which each Center explained its perspective to Headquarters, he wrote that "I can't recall having participated in a meeting dealing with such an unreasonable position." Marshall not only resisted Goddard's co-management, but proposed that NASA remove Goddard from the project and give MSFC complete responsibility.[36]

Finally Headquarters arbitrated the dispute. By December 1976, Headquarters science officials, including Hinners and Warren Keller, who was the defacto program manager, had accepted the idea of an independent science institute and wanted to avoid making project development any more complicated than it already was. They informed Goddard that it had no authority over engineering details and threatened to assign the entire telescope to Marshall if Goddard refused to back down. Consequently Goddard capitulated and Headquarters revised the intercenter management agreement in order to "make it acceptable to all parties."[37] Once the Centers settled on an organization, their relationship improved. Keathley informed Lucas that the arrangement had "worked well" and that Goddard personnel in Huntsville had "established good working relationships in S&E [labs]."[38]

Another two years passed before NASA resolved the orbital operations issue. Goddard sought control of the science institute, and Marshall and the astronomers continued to resist. Lucas recalled having lunch with NASA Administrator Fletcher. Fletcher asked, "Why should this be Marshall's? Goddard is right there in the middle of Johns Hopkins and all the other universities around the Washington area. Who does Marshall have?" Lucas replied, "We have UAH [the University of Alabama in Huntsville]."[39] Fletcher was not impressed, and after he left the Agency in spring 1978, new NASA Administrator Robert Frosch decided that the astronomers would get an independent institute for science, and Goddard would control spacecraft operations and direct the institute contract. To address Marshall's concern about divided authority, Goddard's mission operations manager would co-locate in Greenbelt and Huntsville and work under the Marshall project manager. Following this decision, university consortia competed for the site of the telescope institute, and in January 1981, NASA chose Johns Hopkins University.[40]

If Headquarters resolved the basic conflicts between Marshall and Goddard, their disputes left their mark on the project. Principal investigators complained about working with two Centers, each with a unique culture, management pattern, and testing philosophy, and they believed this created waste. They also thought that rivalry contributed to poor communication between the Centers and that Goddard remained so resentful of Marshall's intrusions that it failed to assign its top talent to the project.[41] Hinners, who had helped initiate the project at Headquarters and then became Goddard director in 1982, agreed that when he took over, GSFC had "an attitude problem." He said that "the Space Telescope team here at Goddard had not really gotten the Center's support" because its leaders decided "we'll do the minimum—screw it."[42] In 1984 Dr. Nancy Roman, the chief astronomer at Headquarters in the early seventies, said that "I think an awful lot of the problems that Space Telescope has had are because of the Marshall-Goddard split."[43]

Marshall officials had similar complaints. Fred Speer, Marshall's telescope project manager from 1979 to 1983, found communications between the Centers difficult. Budget austerity restricted travel, forcing the project to rely on teleconferences, and created competition for resources, leading to "a tendency to shift responsibility to the other side." Speer thought that working with ESA was easier than with Goddard and discovered that "you can't tell another Center what to do. It tells you what it will do." Marshall's Director for Science and

Engineering Kingsbury believed that the friction arose because the Centers' early relationship was one of "competition" and Goddard felt threatened by Marshall's reliance on outsiders for scientific expertise. Lucas thought the project would have been better off if one Center had received complete management of the project. Still Marshall officials thought the relationship with Goddard improved as the project progressed and that whatever problems existed were slight compared to those with the contractors.[44]

Meanwhile Marshall helped form the contract team for the telescope and sought an organization suited to the complexity of the project. In the fall of 1977, NASA chose 18 scientists as principal investigators and members of the science working group who would advise Marshall during the project's C/D phase. They would design the scientific instruments and help NASA with the fine guidance system, optical hardware and instrumentation, and control and data systems for the telescope. In addition, in 1978 the Center established a special project review committee, an advisory panel of scientists and engineers who were not on the project or from Marshall or Goddard.[45]

In January 1977, Marshall and the Agency solicited bids for the associate contracts. In July they chose the aerospace firm Lockheed Missiles and Space for the Support Systems Module and the optics house Perkin-Elmer Corporation for the Optical Telescope Assembly.[46] Although Lockheed had little expertise on astronomy satellites, both firms were very experienced with military satellites and had worked together on the KH–9 reconnaissance satellite.[47]

Years later, because of budget overruns and technical failures, the selection of Perkin-Elmer would become controversial, and in 1977 Marshall personnel also had some reservations about the firm. The Source Evaluation Board said that "our only concern about the Perkin-Elmer approach Centers around their plan to utilize an as yet unverified computer controlled mirror polishing technique." The company compounded risks because it had no plans for an end-to-end ground test for the OTA. In contrast, Eastman Kodak, had planned to use traditional polishing technology and end-to-end tests. On management issues, the Agency also fretted that Perkin-Elmer showed "a lack of understanding of interface configuration, documentation [used in] sustaining engineering and hazard analysis requirements" and had "a performance management system that did not meet the intent of the cost and schedule performance criteria."

Such doubts would prove prescient, but at the time the board thought Perkin-Elmer's bid was superior. The board believed that the "single most significant technical discriminator involved the different approaches to the development of the fine guidance sensor" (FGS) because without an effective sensor, the telescope would be unable to lock on its targets. Based on this criterion, the board decided that the Perkin-Elmer design for the FGS was the most simple, flexible, and inexpensive. Moreover the firm's matrix organization allowed for flexible staffing, and its overall projected costs were lower than those of its competitors.[48]

Unfortunately at the beginning of the telescope's detailed design and development phase, the Marshall Space Flight Center had restrictions on its traditional systems of contractor penetration and automatic responsibility. These limitations, which would soon contribute to problems, originated in a personnel cap imposed by NASA Headquarters. Under the cap, Marshall could only assign 90 employees to the telescope. In part the limitation stemmed from an Agency agreement with the Department of Defense; Lockheed and Perkin-Elmer were working on military contracts and the Pentagon wanted to restrain NASA penetration and reduce risks of exposing secret technology. In addition, Headquarters officials believed that in the past, Marshall had over-penetrated some contractors, leading to excessive demands, gold-plated hardware, and high costs. The personnel cap obliged Marshall to assign small staffs to its project offices in Huntsville and at the contractor plants and to restrict engineering support from its laboratories.[49] In retrospect Lucas recalled that "I never thought that we had enough penetration at Perkin-Elmer" and indeed "we never had enough penetration that we had in most any other project we ever did. We had as much penetration as we were allowed to have given the resources that we could devote to it."[50]

The limitation proved unfortunate, because the Marshall team soon discovered that the design and development of the telescope was more complex and costly than anticipated. The project faced formidable, often unprecedented, technical challenges. Jean Olivier, the Center's chief engineer, recalled at the beginning that people had incorrectly believed that "this is just spitting out something using technology that we already fully understand." Experience proved, he said, that "technologies were much, much more demanding across the board than we ever realized when we got into it. We were naive." At times during the project Olivier wondered if "this whole Hubble Telescope was made out of Unobtainium!"[51]

Probably the greatest challenge was the pointing and control system. The telescope would be the largest astronomical instrument in space; the size of a semi-truck, it would measure 43 feet long and 14 feet in diameter, and weigh over 12 tons. Yet this huge spacecraft would have pointing requirements more stringent than any previous satellite. To make images from faint objects, the scientific instruments needed long exposures, demanding a pointing accuracy of 0.01 arc second and holding onto a target within accuracy of 0.007 arc second. In other words if the telescope were in Washington, DC, it could focus on a dime in Boston and not stray from the width of the coin.

Early in the project, engineers had chosen reaction wheels to move the spacecraft, but had to resolve the mechanical, dynamic, and structural problems of pointing control. The Center's labs helped Perkin-Elmer with the fine guidance system, working on sensors, actuators, and control systems that would find and lock on guide stars. Lab engineers, working with Lockheed, devised requirements to prevent the communications antennas and the solar arrays from moving in ways that affected the image stability. Lockheed and the labs became concerned that the spinning of the reaction wheels could produce enough vibrations to jiggle the spacecraft off target or blur the images. Working with Sperry, the contractor for the reaction wheels, they improved the bearings and balance.[52]

The complexity of telescope development showed when Marshall's team began designing for orbital repair and replacement. The telescope was the first scientific satellite designed for maintenance in orbit and for an operational life of 15 years, a very long time for space technology. NASA had justified a repairable design as means of using the Shuttle to solve potentially calamitous problems and of containing development costs. Beginning in 1979 Marshall contributed extensively to these efforts, drawing lessons from how *Skylab* ground crews and astronauts had improvised repairs of the jammed solar array and failing gyroscopes. For the telescope the Center's labs studied reliability data from components and subsystems and identified which were most likely to fail. They designed these items, mainly the scientific instruments and communications and control systems, as replaceable modular technology with standard connectors and bolts and with latches which doubled as thermal controls and hardware mounts. Working with astronauts from the Johnson Space Center (JSC), they helped design special tools and support equipment to accommodate the limitations of astronauts. The design included 31 foot restraints for freeing the astronauts' hands, and 225 feet of handrails for crawling around the telescope without damaging it. The Marshall team confirmed their ideas using models and trial runs both in the laboratories and in the Center's Neutral Buoyancy Simulator. In the simulator's huge tank, engineers and astronauts used full-size mockups to test equipment and procedures. Finally the repair and refurbishment team planned how to store replacement units on the ground and retrieve technical information for future use.[53] Although justified at the time as a means to save development dollars by reducing hardware tests, participants in the program later argued that design-for-repair drove up costs while reducing operational risks.[54]

The incompatibility of solving complex problems and staying within cost and schedule projections showed first in work on the optical telescope assembly. This hardware had to be completed first because it would be transported from the Perkin-Elmer plant in Danbury, Connecticut, to the Lockheed facility in California to be joined to the support systems module.

One of the first challenges was thermal control and material structure. Expansion and contraction caused by passage from direct sun to complete shade could warp the OTA and distort optical images. Part of the solution came from minimizing hardware linkages and using "kinematic joints" that isolated parts from one another and allowed independent movement.

After studying several materials, Marshall's Structures and Propulsion Lab recommended graphite epoxy for the OTA metering truss and focal-plane structures. These systems precisely aligned the mirrors, scientific instruments, and fine guidance system. Graphite epoxy was a new composite that was lightweight, low in thermal expansion characteristics, and nonmagnetic. The material was relatively untried for space hardware, and Marshall and Boeing, Perkin-Elmer's subcontractor for the metering truss, conducted more tests than originally planned to prove its proficiency.[55]

Marshall's Materials and Processes Lab worked on other materials problems. The designers became concerned that particulate contamination from dust and lubricants and molecular emissions from nonmetallic materials could foul the optical systems. Contamination of the primary mirror could scatter ultraviolet light and reduce the telescope's capability to see faint objects. Consequently the lab tested and qualified for flight all nonmetallic materials on the spacecraft. Later, engineers on the project learned how atomic oxygen in Earth orbit caused many materials to decompose. The lab retested materials for the impact of atomic oxygen and selected a clear polymer as a protective coating for exposed surfaces. Protecting the telescope from contamination became a major cost, requiring not only careful selection of materials, but also sophisticated cleanrooms and transportation systems.[56]

Another major challenge for Perkin-Elmer was the primary mirror. The 2.4-meter primary mirror would be the largest in space, yet it had to be lightweight and provide a precise reflecting surface. The company's subcontractor, Corning Glass, made the mirror blank from ultra-low expansion glass. The mirror would have a 94-inch (2.4-meter) aperture and would be a foot thick with a Center hole two feet in diameter. To save weight, the mirror's solid, one-inch-thick top and bottom plates would sandwich a lattice with open cells much like a honeycomb. From the beginning Marshall officials recognized that "the telescope will never be better than its mirrors" and that "telescope image quality begins with the mirror figures [curvature]." A Center report noted that a flaw in the mirror figure could result from "manufacturing error due to polishing limitations" or "measuring limitations."[57] To protect the program schedule in case Perkin-Elmer ran into problems polishing the primary mirror, Marshall had Eastman-Kodak develop a back-up mirror using conventional grinding technology and required that Perkin-Elmer try its new computer controlled polishing technique on a smaller 1.5-meter mirror.[58]

Troubles plagued the polishing of the 1.5-meter mirror in 1978 and 1979. Perkin-Elmer initially had difficulties calibrating an interferometer, which checked the mirror's figure, and later had problems with the polisher, which damaged the mirror. Following the polishing incident, a center engineer reported in the Weekly Notes that "the history of problems with computer controlled polishing coupled with the criticality of this process call for unusual penetration by NASA to ensure that safeguards are adequate." He observed that Perkin-Elmer's quality inspectors were dependent on the firm's OTA manager and so recommended that Marshall undertake "substantial participation" in all technical reviews. The company eventually completed the 1.5-meter mirror, and this success made project officials confident about the subsequent polishing of the larger flight model.[59]

Initial polishing of space telescope primary mirror blank at Perkin-Elmer, Danbury, Connecticut, May 1979.

Not withstanding Perkin-Elmer's technical progress, by spring 1979 Marshall officials began worrying that the firm lacked the management systems necessary for a project as complex as the telescope. One Center manager noted the "continued concern on Perkin-Elmer planning" and worried that the company's delays were generating "so much bad news." But Marshall believed that its pressure was making the firm become more systematic. By summer the official argued that Perkin-Elmer "was making considerable progress in improving their schedule control" although it was over budget.[60]

Unfortunately by fall 1979, adjustments to unforeseen problems had subverted the project plan and the Center could no longer meet milestones with fixed resources. In October 1979 NASA Headquarters led a cost review and participants discussed the merit of either adding money to maintain the schedule and performance or debasing performance to maintain the schedule and budget. Marshall helped convince the Agency to draw on the project's reserve to stay on schedule for a December 1983 launch, perhaps fearing that a delay would encourage contractor laxity. This proved only a stopgap measure, however, because the Center ran out of reserve money by spring 1980. When Lucas informed Headquarters that the reserve "will not be adequate," Thomas Mutch, the associate administrator for Space Science, expressed reluctance to provide more money and warned that "specific actions must be taken to control the rate of reserve usage that had been experienced to date." Marshall reassured Headquarters that "we will continue a very tight budget policy in all project elements."[61]

Even so by summer 1980, Marshall realized that NASA had underestimated the cost of meeting the telescope's technical requirements. Perkin-Elmer needed more personnel; its mirror polishing was behind schedule. Lockheed was over budget. Some of the scientific instruments were overweight, and the project had added several costly orbital replacement units. In July, Marshall established an assessment team and its report was bleak. The "engineering budget for total program [was] approximately 2/3 spent, approximately 1/3 work accomplished" and the "manufacturing budget [was] approximately 1/2 spent, approximately 1/4 work accomplished." The project was 4 to 6 months behind schedule. The team attributed the problems to "unrecognized hardware and management interface complexity" and "unrecognized tasks recently discovered." Lucas, in a handwritten notation on the report, believed that Perkin-Elmer had a "good tech[nical] understanding of job—not a good understanding of cost." Lockheed had similar problems and could not plan properly because Goddard and Perkin-Elmer communicated changing requirements ineffectively. The assessment team recommended improvements in systems engineering and planning, elimination of unnecessary tests, transfer of tests from contractors to Marshall, elimination of some back-up systems and orbital replacement units, and reduction of technical requirements.[62]

The Center's proposal to reduce technical requirements, or in the parlance of space engineers, "descope," revealed how it was walking a tightrope. Marshall needed to contain costs because continued overruns could lead to cancellation of the telescope and threaten the Center's reputation. Moreover Headquarters instructed the Center to stay within budget because deficits would hamper the Agency's ability to get future funding. Simultaneously, however, Marshall had to preserve scientific performance, because scientists would reject a gutted instrument. Speer, who left Marshall's HEAO project to become telescope manager in February 1979, had saved HEAO by descoping. He proposed to do the same for the telescope and suggested elimination of two scientific instruments. In project meetings in late July, Headquarters, Goddard, and the science working group opposed the plan, but Speer forced Headquarters to acknowledge that the program lacked resources. Accordingly Marshall received permission to exceed the personnel cap and plan a later launch.[63]

By the end of 1980, the Agency had restructured the telescope program without removing any scientific instruments. The new plan would free money for present problems by delaying work and pushing higher costs into later fiscal years.

Marshall would implement most of the assessment team's recommendations, which included using contract incentives to curb cost growth, assigning 40 more people to the project, limiting technical changes, reducing the orbital replacement units from 124 to less than 20, and stopping work on the Kodak back-up mirror. The new plan pushed back the launch date 10 months to October 1984 and would raise the overall cost from $575 million to $645 million. In December the science working group congratulated Speer for his ability to "balance the conflicting needs of the Project to produce a viable plan which we can all enthusiastically support."[64]

The studies by Headquarters and Marshall showed that systems engineering remained uncertain. Marshall attributed the problem to Lockheed's having "a 'prime's' responsibility with associate contractor's authority and accountability." The Center's solutions included appointing a NASA co-chair for all technical teams, setting up more teams, requiring that Lockheed assign a chief systems engineer to the project, and establishing a Space Telescope Systems Engineering Branch within Marshall's Science and Engineering labs.[65]

Despite the changes, the reforms had not addressed some problems that had been raised during the reassessment. A Goddard report lamented that the program had "almost no spare hardware and was already down to an absolute minimum level of testing" and that "there is no provision for new unanticipated problems." William Lilly, NASA's associate administrator and comptroller, also worried that the project still had a "success oriented" schedule and questioned the Marshall review process since "the team did not see indices of the problems that occurred this year."[66]

Toil and Trouble

In the next two years Marshall oversaw progress in several technical areas. By late 1982, however, a crisis developed within the telescope project, mainly as a result of politically expedient decisions made during program design. Congress and NASA Headquarters conducted thorough investigations but sometimes unfairly blamed the problems on Marshall.

Marshall helped the project pass several milestones in 1981 and 1982. In May 1981 Perkin-Elmer completed the shaping of the primary mirror. The company proclaimed that the mirror was "within microinches of perfection" and NASA bragged about "the finest mirror of its size anywhere in the world." By year's end, the firm had applied a reflective coating of aluminum three millionths of an inch thick and a protective coating of one millionth of an inch. In mid-1981 ESA's contractor for the solar wings began deployment tests, and in early 1982 Marshall tested the solar power cells and began work to improve their interconnects. By the end of 1982 fabrication of the scientific instruments neared completion, Perkin-Elmer had begun final construction of the OTA, and Lockheed had held major design reviews and started fabrication of all major parts of the support systems module.[67]

Again, however, progress came at a slower pace and a higher cost than NASA had predicted, and again Marshall attributed most of the problems to management failings at Perkin-Elmer. Indeed the Center experienced constant frustration with the contractor. Kingsbury, director of MSFC's Science and Engineering labs, remembered getting a phone call from a distraught Center engineer in Danbury who reported that Perkin-Elmer intended to support the primary mirror with two cloth straps and move it with a ceiling crane, thereby risking months of polishing.[68] In October 1981 Marshall Director Lucas told the firm that it had put the telescope in "serious jeopardy" because of "lack of sound planning, insufficient schedule discipline, many instances of engineering deficiencies, and inadequate subcontractor support." Consequently in one quarter of FY 1982 the firm's cost projections had increased 35 percent over its recently rebaselined budget. In reply the vice president in charge of the corporation's optical division admitted that "a viable plan for implementing the OTA Program for Space Telescope does not exist." After one meeting between Perkin-Elmer and Marshall, a software consultant from JSC recorded amazement that the firm admitted they had left a "problem open after 1 1/2 years of work!" and that corporate officials gave "a very unsatisfactory response to Dr. Lucas' question 'How can this be'?" A Marshall report on the company in February 1982 summarized the problems: "schedules always too optimistic, funding and manpower estimates always too low, analyses frequently lag design and fabrication, hardware rework extensive."[69]

Marshall tried numerous methods to control Perkin-Elmer. The Center increased the number of personnel devoted to the project from 150 to more than 200 and expanded the resident office staff. But attempts at deeper penetration did not lead to significant improvement. After Perkin-Elmer used improper test procedures and damaged orbital replacement latches, Lucas asked, "Do we need more QC [Quality Control] penetration? We must get this situation under control." Kingsbury replied that "we have provided more support than one usually expects for a problem such as this one; however, as you note, we haven't found the formula for success." Perkin-Elmer responded that the shortage of funds would necessitate personnel layoffs and cause more delays. Marshall pressured the firm to implement scheduling systems, which it did in April 1981, and change project managers, which it did in October 1981, but problems only worsened.[70]

The Center also tried strong-arm tactics, insisting that the firm stay on schedule and within budget and applied the financial clauses in Perkin-Elmer's contract. But this was also ineffective because penalties for cost overruns and schedule slips were less than awards for technical excellence, and so the firm lacked incentive to assign its best people and overhaul project organization.[71] Lucas believed that Perkin-Elmer was "probably, from the corporate level, the least responsive contractor we've ever dealt with. Their top management really didn't give a lot of attention, it appeared to us, to this program." He attributed their lack of responsiveness to the fact that the OTA "didn't constitute a sufficiently significant part of their total business base" and they were not worried about NASA moving the project, because the Agency had nowhere else to take it. Kingsbury agreed and considered the telescope as "absolutely the most frustrating program I've ever worked in." He remembered that Marshall's people in Danbury "were almost out of their minds" trying to get action.[72]

In August 1981, NASA Administrator James Beggs requested a special briefing on the telescope, and Marshall began special investigations of Perkin-Elmer. The next month four lab directors and the assistant Center Director for policy and review studied the firm's management. The Marshall Program Assessment team found "Perkin-Elmer seems very proficient on optical testing" but had skills in nothing else. Perkin-Elmer's managers thought their problems stemmed from lack of money and manpower. The Marshall team believed, however, that "past schedule performance, current hardware status, and planning do not support PE's position." Perkin-Elmer's project organization suffered from "lack of management discipline across the board" with "schedules not in place, ability to meet schedules highly uncertain, manpower and budget requirements unknown." The "schedule is very unsettled and changing daily." Consequently "PE will likely need both additional dollars and time" with perhaps a 6-month launch delay. In addition, the team believed, Marshall would have to "increase surveillance and control" and "day-to-day interaction between MSFC and PE responsible engineers." Most importantly, the Center would have to teach Perkin-Elmer how to plan. Lucas's notes described the situation at the contractor as "disorganized, no discipline, sloppy habits, attitude problem, no systems, lack of exp[erience] on big systems job;" the firm's plans had "no credibility" because there was "nobody steering ship."[73]

Unfortunately the Marshall briefing to Beggs on 3 November 1982 did not include this account of Perkin-Elmer's organizational failings. The briefing, presented by Marshall's telescope project office, acknowledged the firm's hardware development problems, especially with orbital replacement latches, but assumed that the existing organization could solve these problems with modest amounts of extra time and money. Lockheed's problems also resulted from a shortage of $11.2 million. The remainder of the briefing was upbeat, emphasizing progress on the solar arrays and scientific instruments. With infusions of cash and a launch delay to April 1985, the office said, the telescope would soon be on target.[74]

Meanwhile Marshall had sent the deputy project manager and a team of planning experts to the contractor plant. Their goal, according to Lucas, was to "enforce schedule discipline at PE." Lucas himself took a special trip to Danbury. His preparatory notes for discussions with the contractor reveal his consternation. Despite "at least 2 major rebaselinings," he wrote, "OTA project has never been comfortably under control." The "schedule had been slipping about 1 wk/mo up to rebaselining on Jan. 1, 1982," but afterwards "slip continued at approximately mo quarterly" and "now we seem to have gone critical—current rate of slip greatest of any time in the program." The Center Director believed that the company had an "attitude problem" and its pride in its technical excellence contributed to managerial complacency. All in all there was "very little progress evident in overcoming a lack of experience on big systems." After the trip Lucas demanded that the project office penetrate the contractor more; "it is time to get some of our experts deeply involved."[75]

Only in late December 1982 could Marshall appreciate the scope of the crisis. Former Goddard Manager Dr. Donald Fordyce, now the new Perkin-Elmer telescope manager, opened the company to Marshall for perhaps the first time. The Center's team helped the contractor install a scheduling system and for "the first time" tried "to assure that all jobs are identified and accounted for." During the Christmas holidays, they discovered, in Fordyce's words, "we didn't have a program."[76]

On 14 January 1983 Marshall broke the bad news to NASA Administrator Beggs. Describing the firm's technical problems, the Center said that the orbital replacement latches could not align the instruments precisely, the fine guidance system could not meet pointing requirements, and the primary mirror had a layer of dust. Perkin-Elmer's poor scheduling and planning systems and poor communications between engineering and manufacturing groups had stymied progress. The firm needed additional test equipment, manpower, and engineering analyses, but had not planned for them. Technical teams had learned by costly experience that the protoflight concept required step-by-step rehearsal of any work in order to avoid damage to flight hardware. At times Perkin-Elmer groups had fallen behind schedule milestones by a day or more for each day of work. The delays on the optical telescope assembly would slow progress and hence impose costs on the support systems module and on the scientific instruments. Perkin-Elmer needed another 8 months delay to a launch date in March 1986 and "significant funding increases"—perhaps as much as $100 million.[77]

The news upset Headquarters officials. After Marshall's report, Samuel Keller, the NASA deputy assistant administrator for Space Sciences, said that the telescope program was "out of control." Administrator Beggs was angry; he had told Congress after Marshall's November briefing that the project was on track, but now he would have to beg for more money. Witnesses said that he told Lucas, "you have done dirt to this Agency."[78]

Not surprisingly the program underwent a new round of inquiries by Headquarters officials, by a NASA team led by James Welch, by the House Surveys and Investigations Staff, and by the House Subcommittee on Space Science and Applications. The investigations confirmed that Perkin-Elmer had major management problems; in an ironic moment at these reviews, the contractor verified its weaknesses in scheduling by failing to reserve a meeting room for the NASA committee.[79] But the contractor's crisis also led to discussion of Agency management and why NASA had been unable to understand and solve the problems.

Participants believed that communication broke down between Marshall and Headquarters. The House study quoted an unnamed senior NASA official who said that communications between Marshall and Headquarters were "at best 'horrible.'" Beggs told Congress that the information flow was "poor." In part

Beggs blamed administrative turnover for rupturing continuity at Headquarters; the Office of Space Sciences (OSS) had four associate administrators and four telescope program managers after 1977.[80] OSS had never managed two field Centers and two associate contractors on such a technically complex program. Its small staff, Lucas recalled, never penetrated the project like the Office of Manned Space Flight routinely did and so never fully understood the Center's problems.[81]

In part the poor communications was Marshall's responsibility. Astronomers and Headquarters officials believed that the formal reviews emphasized good news. Dr. Robert Bless, one of the principal investigators, said that "Quarterly reviews in some instances became jokes." Reviews "were often designed to give the impression that everything was going well, that any problems were understood and being solved, and that schedules were being met. However, conversations among participants in the hallway or over a beer often revealed drastically different pictures."[82] In an interview with the *Huntsville Times,* Sam Keller said "I don't think they lied to us. It's not that sort of thing. All engineers think they're going to find the answer tomorrow. But I think they should've told us earlier that you can't get from here to there. I think they were very optimistic and 'had their head in the sand.'"[83] A memo from 1983 reveals the Center's desire to avoid damaging publicity. In June a senior Marshall official complained that the telescope scientists had shown the project's dirty laundry to congressional investigators. He was "extremely disappointed in the large number of negative comments attributed to members of the science community" and wanted project scientist O'Dell to "let his colleagues know what their irresponsible comments are doing to their project."[84]

At the time Marshall disputed criticism about miscommunication with Headquarters. Project manager Speer believed that "Sam Keller is starting with an incorrect premise" that information was "hidden." Actually "there is no lack of communications on any level within the ST program." The Center had communicated the bad news when it was available in late December 1982. Center Director Lucas agreed, believing that the formal reviews and reports "provided an effective means for communicating the very best information available."[85] In 1990, however, Speer acknowledged clogged communications. Marshall was so worried that overruns could lead to project cancellation, he said, that "we were very concerned about the wrong message getting out. The press couldn't be told anything, Headquarters couldn't be told anything, the other Centers shouldn't be told anything."[86]

Even so, the greatest failure of communication occurred between Perkin-Elmer and Marshall. Part of the problem rested with the two parties. Beggs believed the firm had deliberately hidden its problems, and he told Congress that "the contractor was not coming clean . . . to Marshall" and was "covering over what were problems."[87] Likewise Marshall managers admitted to House investigators that they had overestimated Perkin-Elmer's abilities and had underpenetrated. "Marshall 'assumed' that Perkin-Elmer Corporation was capable of doing contracted work with the same level of NASA supervision as large aerospace firms—this proved to be a grievous and costly error."[88] A March 1983 Marshall review of its reports to Headquarters revealed that the Center had typically neglected to report managerial problems at the contractor. The Marshall review found that "there were little or no references to management or systems engineering difficulties. Instead, technical problems, underestimation of complexity, underestimation of subcontractor costs, and growth in engineering and manufacturing were provided as reasons for schedule slips and cost increases."[89]

Structural problems, however, were more important in slowing information and retarding Marshall's responses. Center officials and the House and Welch reports blamed Agency procurement policy and the agreement with the Department of Defense. Marshall had no prime contractor to compensate for Perkin-Elmer's weaknesses. Center officials lamented the limitations of a "procurement strategy that required use of an optics house to do a major systems job."[90] The Center's personnel cap initially limited it to 35 project officials and 65 support engineers, less than half the normal staff of similar programs. Although the Agency removed the cap in 1979, the limitations had hampered management planning and engineering analysis and an increase to 250 people was too little, too late.[91] Speer said that "on a complex program of the magnitude of Hubble, you just need almost a comparable number to Apollo, to really look at everything in depth and to stand up and say, 'Yes, this will work.'"

Likewise the defense agreement and the "black world" of military secrecy had restricted the Center's access to Perkin-Elmer work sites and information. Speer recalled that when his people went to Danbury they continually encountered "locked doors" and closed books. Consequently Marshall had little choice but to accept the firm's word.[92] Moreover, early in the project Headquarters had believed that autonomous contractors would contain costs and had therefore directed the Center to change its traditional practices and minimize penetration.

Lucas thought that "we were somewhat victimized in this by the thought that 'Hey, we've got to learn new ways of doing things to lower costs and let the contractor do it.'" But Perkin-Elmer had learned bad habits working on defense contracts and preferred to solve problems by spending money.[93] All in all, according to a report prepared for the Welch team, the "level of detail needed to see deficiencies [was] not [the] normal level at which MSFC manages."[94]

Short schedules and tight budgets also confined Marshall. Robert Smith, the historian of the project, has suggested that the problems mainly stemmed from how NASA had oversold and underfunded the project. The Center tried to work within unrealistic program plans, mainly because both Headquarters and Marshall managers wanted to avoid surfacing problems until necessary. Headquarters wanted to keep its promises to Congress. Marshall believed that failure to follow plans could result in canceling the project or closing the Center. This reluctance to confront reality not only led to misinformation about progress, but contributed to engineering difficulties.[95] The House staff report argued that "the applied 'design to cost' theory precluded engineering test models and resulted in a 'rush to hardware.'" The Welch group questioned Marshall's emphasis "on technical problems as opposed to management difficulties" and its "commitment to fiscal year constraints ([which] forced deferred work [and] increased 'bow-wave' effect)."[96]

Looking back, project manager Speer believed that the Center was trapped by "a system that I was totally unable to change." He said that "you can really put it on a nice, simple denominator: the program was underfunded. You cannot get something like that for the money that was set aside." Consequently "almost every month we found a gap. Every gap we found meant additional money was to be spent." Money shortages created a crisis atmosphere and "you are always with the overtone of 'who is responsible for this?'" rather than "how do we solve the problem?" Speer thought the Space Telescope was "a good case history for how not to run a big program."[97] Lucas agreed, arguing that the telescope proved "there is no low cost way of doing a job half way. This is just a costly business to do a new, first time invention."[98]

In a letter to Beggs, Lucas summarized how the crisis had occurred. He believed Marshall was "not able to fully recover from the inherent problems introduced into the program as a result of those early decisions" about protoflight and procurement. Nonetheless, he wrote, "I believe we have made considerable technical progress on the development of the Space Telescope. The extreme complexity and demanding requirements, coupled with the inherent problems associated with some early decisions, have made it extremely difficult to assess schedule progress or accurately predict cost requirements in a timely and effective manner. The inability to do this and the perceived necessity to remain under annual and budgetary commitments caused us to continuously understate our budgetary needs. This understatement of budgetary needs resulted in certain critical program decisions being made that, in retrospect, would be judged to have introduced too much risk into a project of such complexity and importance. They were, however, made with full knowledge of all parties at the time they were made. While I do not offer the above as an excuse, or justification, for the problems now confronting the Space Telescope, I do believe that appropriate consideration must be given them in assessing what went wrong, if for no other reason than to preclude similar decisions being made on future projects."[99]

Reorganization and Realization

Even before the completion of the investigations in March 1983, Marshall had started reorienting the telescope project and helping the coalition reorganize. New infusions of talent and cash enabled development to proceed without the previous crisis atmosphere. The born-again project received a new name in October 1983, when NASA renamed it the Edwin P. Hubble Space Telescope in honor of one of America's foremost astronomers.[100]

Headquarters assigned the Space Telescope project a higher priority within the Agency and gave it resources to match. Beggs wrote Lucas that "the Large Space Telescope is the second most important program you have at Marshall, coming only a little behind your activities on the Space Shuttle, and I therefore believe that we should apply as much of the best talent available at Marshall without, of course, sacrificing any attention from the Shuttle." The Agency delayed the launch to the fall of 1986 to give ample time for development and testing. NASA also received forgiveness and money from Congress, amounting to a total budget of $1,175 million, far above the original 1977 projected cost of $475 million. The telescope program thus transcended its origins and its buy-in, design-to-cost strategy and for the first time had resources consistent with its technical difficulty.[101]

The Agency also reorganized the program, with NASA Headquarters assuming greater responsibility and authority. The goal, Keller wrote, was to prevent "a management situation such as had existed at Perkin-Elmer to surprise us" and to "ensure a much higher level of knowledge regarding this project." Without this information the Agency could not rationally distribute resources and maintain a favorable relationship with Congress and the Office of Management and Budget. Keller tried to reassure the Centers that his goal was "penetration rather than management." He said that he was "concerned that we do not bring the project management function into Headquarters and that Washington 'micromanage' the project." Nevertheless, Marshall officials worried that Headquarters would get too involved in details. During a conversation in which Administrator Beggs vented displeasure with the "massive problem" of Perkin-Elmer and Marshall, Center Director Lucas wrote "micro-manage" on his notepad and underlined it 10 times.[102]

The reforms transferred power from the field Centers to Washington. Headquarters expanded its telescope staff from 4 people to 15, created a new Space Telescope Development Division, and hired a systems engineering contractor. Welch, who had managed development of military satellites and conducted the program review, became the new program manager. Welch took responsibility for Level I engineering decisions, which reduced Marshall's authority. Moreover Keller insisted that the Marshall project office immediately report any departures from the program plan and provide monthly briefings in addition to the formal reviews. Headquarters also supported the principal investigators' efforts to reassert their influence. The scientists had found that the science working group was too large and met too infrequently to affect development decisions. Accordingly the astronomers created a smaller executive committee called the Space Telescope Observatory Performance and Assessment Team that reported to Headquarters rather than MSFC.[103]

These resources allowed the project to reduce risks and restore engineering conservatism. "Penny-wise, pound-foolish judgments," Welch believed, had been forced on Marshall by years of cost-cutting. NASA, goaded by the scientists, increased funding, added time for more tests, and increased the number of spares and back-ups (notably one for the Wide-Field Planetary Camera, arguably the most important instrument on the telescope). Marshall also reduced risk by increasing the number of orbital replacement units to 49; it had fallen to 20 after having been as high as 120.[104]

Rather than being demoralized by the crisis, criticism, and changes, Marshall redoubled its efforts. Director Lucas, explaining the telescope's reorientation to the project staff, expressed renewed determination. "As usual," he said, "the press has amplified bad news," but "when you get into the situation we are in, no amount of talking will help—performance is the only answer—so we'll just have to 'hang-in' and deliver the Agency's and the world's most outstanding telescope."[105] Already Marshall had implemented changes in personnel. Speer became associate Center Director for science and would advise Lucas on the project. Jim Odom, who had proven effective in the development of the Shuttle's external tank, became the new telescope project manager. One of the astronomers said that "Odom more than any one individual, at least at Marshall, deserves a heck of a lot of credit for turning around what was almost a disaster in '83, into perhaps not a smoothly running project but certainly, considering the complexity of this one, [a] well done project." Another suggested that "the whole flavor of the program changed. You could discuss problems in a open way and nobody would think less of you." Odom observed that discussing problems was much easier after 1983 because the Agency had the money to fix them.[106]

Marshall made several improvements in the project. To facilitate penetration of the contractors, the project office created separate OTA and SSM offices. To maintain control over interfaces, Marshall improved its systems engineering. Odom and Fred Wojtalik, who became deputy project manager for systems engineering, recalled that before 1983 the Center had lacked resources to fund both hardware development and integration activities, and so had concentrated on development. Although engineers on the project did not get much credit, Odom said, they had done excellent work on design of the pieces and on interface control documentation. After 1983, Wojtalik said the pieces and subsystems were largely built, and Marshall had to provide the money and staff to integrate them. The Center created a new systems engineering office for the project and expanded the telescope systems engineering branch in the Center's Systems Analysis and Integration Lab to a division. Marshall also established interdisciplinary panels in a dozen functional areas and assigned responsibility for ensuring technical support to high-ranking lab personnel. Lockheed also received more responsibility and resources for systems engineering.[107]

The Center also penetrated Perkin-Elmer more deeply. Marshall sent its OTA project office to Danbury, thus increasing the size of its resident office from 4 to more than 25. Lucas said that "I don't recall any case where the deficiencies of project management were equivalent to what we encountered at Perkin-Elmer" and so the team had to help the firm apply new planning systems.[108] Marshall also pressured the firm to select new managers. NASA, goaded by publicity about the delays, also obliged the firm to pay back $1.4 million in previously awarded fees and revised the OTA contract so that subsequent overruns would be "non-fee bearing" and Perkin-Elmer would not profit from its incompetence.

Penetration soon showed results. By May 1984, Jerry Richardson, Marshall's OTA project manager, reported that although the firm still missed 40 to 45 percent of its production deadlines, this showed "some improvement" and corporate management had assumed a "take charge—can do" attitude. Still progress mainly came because extra money allowed Perkin-Elmer to add 100 more people to the project, and in December Marshall was still complaining about the firm's mismanagement.[109]

The Center also helped its contractor overcome several technical challenges. Fordyce, the Perkin-Elmer project manager, said Marshall's team was "probably the finest team that I've seen NASA yield—a good technical team. They're not continuously yelling at us for why don't we do this, why don't we do that. They're trying to help us solve problems."[110]

Marshall's labs contributed to the latches for the orbital replacement units and scientific instruments. NASA and Perkin-Elmer had underestimated the difficulty of designing the 20 different latches. Project manager Odom said that "to call those devices latches is a tremendous understatement and misnomer. You are literally taking devices that are thermal insulators and that have to hold phone booth size objects within one or two ten thousandths of an inch through a thermal gradient that you get in each orbit, as well as handling the launch and ground handling tasks."[111] Dynamic tests showed that the latches experienced "galling," in which the outer layer of aluminum oxide rubbed off and resulted in misalignment. Early in 1983, officials identified the latches as the telescope's primary technical problem. By late in the year, however, Marshall engineers proposed a tungsten carbide coating which withstood galling tests.[112]

Although the Center still complained about its contractor's overruns and delays, the OTA project had overcome major hurdles by late 1984. The guidance system passed pointing and tracking tests in April, and in June a cleaning system, using jets of nitrogen gas, removed the layer of dust that had accumulated on the primary mirror. In May, Marshall engineers completed development of a balance beam to help ground crews install the telescope's fine guidance sensors and scientific instruments. In November, Marshall handled transportation of the OTA from Danbury, Connecticut, to Lockheed's plant in Sunnyvale, California, for mating with the SSM; a surplus *Skylab*-Apollo Telescope Mount canister protected the optical system. In 1985 the National Society of Professional Engineers recognized the optical telescope assembly as one of 1984's top 10 engineering achievements.[113]

As work on the telescope moved from fabrication of the pieces to putting them together, Marshall's attention increasingly turned to Lockheed. Now Lockheed began experiencing problems of systems management and engineering similar to those at Perkin-Elmer. Odom informed Lucas in July 1984 that "the most significant area of attention had been to try to instill in Lockheed a felt sense of systems responsibility, rather than a reactive mode of response to MSFC direction." A Marshall report that fall worried that a "team relationship between Lockheed Missile and Space Corporation (LMSC) management and MSFC, GSFC, and P-E on-site personnel does not exist." Marshall sought to help by transferring its project office to Sunnyvale. Nevertheless, the integration and testing of such complicated technology and complex organizations proved more expensive and time-consuming than anticipated. By summer 1985 Lockheed fell three months behind and went 30 percent over budget, and Center Director Lucas warned the project office that the telescope was "dangerously close to breaking the congressional ceiling on the budget."[114]

The Hubble teams received an unwelcome respite from the *Challenger* disaster. NASA had planned to launch the telescope in the second half of 1986, but the January accident grounded the Shuttle fleet. Marshall worried that the launch delay could lead key personnel to desert the project, but many stayed on. Government and contractor teams continued assembly and verification tasks, completing a major thermal-vacuum test in June 1986. After this time they reworked problem areas, adding more powerful solar panels, enhancing redundancy, improving software, installing better connectors, and labeling orbital maintenance features. Marshall and Lockheed also changed battery type, worrying that nickel-cadmium batteries had a history of failure. Although nickel-hydrogen batteries had never flown in low-Earth orbit, the Center's Astrionics Lab used the extra time and resources to build a simulator of the whole telescope power system, test the nickel-hydrogen batteries, and confirm their reliability. The lab also improved the controls for the power system to prevent overcharging the batteries. The various telescope organizations also rehearsed procedures for orbital verification and operations.

This work resolved weaknesses that had crept into the program before 1983. By the time Hubble Space Telescope was ready for launch in April 1990, it had cost over $2 billion and become the most expensive scientific instrument ever built.[115]

Mirror, Mirror

In the weeks before launch, Marshall's Hubble team felt a deep sense of accomplishment. "Many people here and at our contractors have devoted their best years in developing that system," said Wojtalik, the project manager since 1987. Everywhere expectations about Hubble were high. NASA had been promoting the telescope project for years; Administrator Beggs had liked to call the Space Telescope the "eighth wonder of the world." Marshall had contributed to the public relations campaign with releases like "The Amazing Space Telescope" which described the technical wonders of the pointing and optics systems; it promised that Hubble would yield spatial resolution 10 times better than any previous telescope and could "detect the light from a typical two-battery flashlight from a quarter of a million miles away."[116]

Unfortunately, the boosterism set up Marshall for a fall when the telescope did not perform as anticipated. MSFC located a team of engineers at the GSFC Hubble Space Telescope Operations Control Center to direct orbital verification of the Hubble for two months, until Goddard took over operations and the Lead Center role. Following the successful launch, the team encountered glitches in communications and control. Such glitches were normal for scientific satellites. MSFC's Max Rosenthal, a test support team manager, said "no matter how much testing and research you do on a piece of hardware on the ground, there are some things you just can't do" and "so you make adjustments." The controllers struggled with drifty star trackers, and signal disruptions caused by unexpectedly high radiation over the South Atlantic Anomaly where the Van Allen belts dip close to Earth. A communications antennae kept snagging on its coaxial cable loop, and until controllers compensated for it, the spacecraft periodically shut down in safe mode.

Initially Marshall had the most difficulties with vibrations in the solar panel booms. Dr. Gerald Nurre, Marshall's chief scientist for pointing control systems, recalled noticing the problem almost immediately. As the telescope traveled in and out of Earth's shadow, temperature changes bent materials in the booms. Project engineers had anticipated minor deformations, but ESA had predicted no serious problems would result. What they had not expected was the array's deployment and orientation mechanisms to magnify the deformations and bounce the whole telescope. The vibrations were severe enough to prevent the guidance system from locking on guide stars and to cause "jitter" in the optical images. The booms only stabilized in the last few minutes of daylight, and so the pointing system initially met its design specifications in about 10 percent of its orbit. Nurre's team in Marshall's Structures and Dynamics Lab worked with Lockheed to change the control program in the telescope's computer, directing the pointing and control system to counteract the vibrations. The new program brought the pointing system within the telescope's stringent specifications in 95 percent of the orbit.

Nurre drew lessons from the problems with the antenna and solar arrays, arguing that financial and organizational limitations had helped cause both. Noting that travel restrictions prevented pointing-and-control experts from inspecting key processes, he speculated that if they had attended integration of the Hubble in the Shuttle payload bay at Kennedy Space Center (KSC), they could have noticed the antenna cable loop, and if they had attended deployment tests of the solar booms in England, they might also have spotted their mechanical weaknesses. Moreover, the associate contractor arrangement, the agreement with ESA, and the lack of a prime contractor limited Marshall's ability to perform systems engineering and analyze the telescope's complex interfaces between power, communications, and pointing systems.[117]

The mission controllers made progress and by 21 May began receiving the first optical images from the telescope. These views of a double star in the Carina system, scientists believed, were much clearer than those from ground-based telescopes.[118] Such success left project officials surprised on the weekend of 23–24 June when the telescope failed a focus test.

The controllers had moved the telescope's secondary mirror to focus the light, but a hazy ring or "halo" encircled the best images. Subsequent tests determined that the blurry images resulted from the "spherical aberration" of the primary mirror; spherical aberration reflected light to several focal points rather than to one. It occurred because Perkin-Elmer had removed too much glass, polishing it too flat by 1/50th of the width of a human hair. This seemingly slight mistake, however, prevented the telescope from making sharp images.[119]

Disappointment and outrage characterized the initial reaction from project participants, politicians, and the press. NASA scientist Ed Weiler said "the Hubble is comparable to a very good ground telescope on a very good night, but it's not better than the best." Charles O. Jones, Marshall's deputy chief of guidance, control and optical systems remarked that "we are rather astounded at this error." Senator Barbara Mikulski (D–MD) protested about the waste of $2 billion and called the telescope a "techno-turkey." Senator Al Gore (D–TN), chair of a panel on science and space, referred to the solid rocket boosters, observing "this is the second time in five years that a major project has encountered serious disruption by a serious flaw that was built in 10 years before launch and went undetected by NASA's quality control procedures." Humorist David Letterman made a list of "Top 10 Hubble Telescope Excuses," which included "bum with squeegee smeared lens at red light." Editorialists pointed out the Marshall connection of the *Challenger* and Hubble failures. One asked "Is it coincidence that NASA's Marshall Space Flight Center was in charge of the telescope program, as well as the faulty solid rocket boosters that caused the *Challenger* accident?"[120]

Space pundits analyzed the Agency's institutional weaknesses. John Logsdon described how the problem emerged in the late seventies, "a time when the Agency was not being honest with itself or with anyone else. It was an Agency not expected to have problems or to fail, but it didn't have the resources required to assure success. In that situation, you can't say stop, and you can't say I need more money. You take risks and hope they work out." Howard McCurdy said the Agency's "whole philosophy had changed from the Sixties when they knew there would be trouble and they planned for it" and "in the Seventies, they didn't plan for trouble and prayed that it wouldn't come."[121]

NASA established an investigating committee under the chairmanship of Lew Allen, director of the Jet Propulsion Laboratory. The Allen Report attributed the technical failure to misassembly of the reflective null corrector, an optical device was used to determine the figure of the mirror. The commission found the device intact and discovered enough evidence to interpret what happened.

Technicians in the Optical Operations Division had mismeasured the location of a lens in the device, mistaking a spot on a metering rod where an end cap had worn away as valid scale, and thus erred in spacing the lens by 1.3 mm. Consequently the null corrector guided the polisher to shape a perfectly smooth mirror with the wrong curvature. Analysis of data from Hubble showed that the curvature flaw in the primary mirror exactly matched the flaw in the null corrector.

The device also tested the mirror perfectly, but verified that the mirror's curvature matched the wrong pattern. Basically the tests compared light reflected from a flat reference mirror with light reflected from the curved primary mirror, as modified by passage through the null corrector. Technicians compared light beams from the two mirrors and photographed the interference patterns. In each test, the patterns matched and hence they concluded that the mirror was perfect. The technicians had contrary evidence from similar tests using two other null correctors; their interference patterns showed the flaw in the primary mirror. But rather than interpreting discrepant data as proof of a problem, the firm's optical operations personnel dismissed the evidence as itself flawed. They believed the other two null correctors were less accurate than the reflective null corrector and so could not verify its reliability. Since they assumed the perfection of the mirror and reflective null corrector, they rejected falsifying information from independent tests, believed no problems existed, and reported only good news.[122]

The Allen Commission emphasized that the technical failures rested on managerial failures. It noted that the mistakes occurred in 1981 and 1982 when Perkin-Elmer and Marshall managers were distracted by cost and schedule problems. Nevertheless, Perkin-Elmer had serious failings in quality control and communications that Marshall did not correct. The use of a single test instrument "should have alerted NASA management to the fragility of the process, the possibility of gross error (that is, a mistake in the process), and the need for continued care and consideration of independent tests." The project required no formal certification for the reflective null corrector despite its use as the primary test device. The project had not established clear test criteria or formal documentation to assure compliance with quality procedures. Perkin-Elmer's Optical Operations Division operated "in an artisan, closed-door mode." The commission also found that "the Department of Defense project did not prohibit NASA Quality Assurance from monitoring the P-E activity." Even so the Center had concurred in the firm's decision to exclude even its own quality assurance personnel from the work area during key times. The quality people who did participate were not optical experts but "concentrated mainly on safety issues" and reported to the same managers they were monitoring. Perkin-Elmer did not use its optical scientists to monitor fabrication and testing and neither did Marshall require this.[123]

Other factors also prevented independent verification. The commission believed that "the NASA project management did not have the necessary expertise to critically monitor the optical activities." Marshall's managers did not compensate by using Eastman-Kodak, Perkin-Elmer's subcontractor that had worked on a back-up mirror, to verify the flight mirror. Instead the project office relied on its science working group, who had the necessary theoretical expertise and should have questioned the process, but lacked experience with fabrication and testing. If the contractor and Center had not made such mistakes, the commission believed, they would have caught the technical mistakes and "have been aware that communications were failing with the Optical Operations Division." Finally the Allen Commission noted that "poor communications" and the containment of problems "at the lowest possible level" also resulted from the "apparent philosophy at MSFC at the time" to "consider problems that surfaced at reviews to be indications of bad management."[124]

The mirror problem depressed Marshall people deeply. One official said that the aberration was the most disappointing part of his career and lamented that because of one bad measurement Center personnel became "goats" rather than "heroes." Even so, many sought to learn lessons that could be applied to later projects. Olivier, the chief engineer, recalled how the team had discussed end-to-end optical tests, but had ruled them out because of their cost, imprecision, or potential to contaminate the telescope. In worrying about the need for precise tests, however, he said they had overlooked the desirability of a simple "sanity check" which could detect a gross error and failed to conduct tests with independent experts using different measuring instruments. "That was a paramount lesson learned," Olivier said, "be sure to have cross-checks." He noted that Congress required that NASA prove the optical system on AXAF, the Advanced X-Ray Astrophysics Facility, before proceeding with funding for the whole satellite.[125]

Other Marshall officials pointed out the limitations imposed on them. Speer, the project manager at the time, recalled the difficulty of penetrating Perkin-

Elmer, especially given the defense regulations and Marshall's resource shortages.[126] Kingsbury, director of the Center's labs, explained how Marshall had trusted the contractor's expertise. He said that "The Marshall Space Flight Center is not now, nor was it ever, the optics Center of the world. We employed a contractor who was one of the three recognized and accepted optics Centers of the world. All we could do was assure that which we knew he should do . . . he did properly. But in the particular scheme of polishing, nobody [at MSFC] knew anything about polishing mirrors. We are propulsion people. We had a very, very marginal contractor. I used to say, 'If you want a piece of glass, a perfect mirror, get Perkin-Elmer. Don't ask them to do anything else, but they can polish glass.' Now I'm not sure."[127]

Center officials also blamed the contractor for not surfacing bad news. Project manager Speer, chief engineer Olivier, and chief scientist O'Dell denied receiving any information about the problem and the commission found no evidence that any NASA official saw the discrepant data. Marshall personnel also denied that their Center had a history of suppressing bad news. Wojtalik said, "I don't know of any time in any project I've been in where people were told 'don't bring me something that's a problem.' I've been here 33 years."[128] Downey argued that after *Challenger,* Marshall had become the Agency's "whipping boy" and "scapegoat." "If there was anything that Bill Lucas drilled into us," Downey said, it "was 'If you have a problem, I don't want to be surprised. Please, please communicate it to me.'"[129] Kingsbury said he had never had a contractor hide something, but "this one hid it." To discover the secret, Marshall would have needed a one-for-one match of contractor personnel with civil servants. Kingsbury wondered if resource starvation had not stifled contractor officials; they may have avoided reporting problems because "they were always behind schedule and over budget. We did beat on them mercilessly to get on top of this thing."[130] Basically accepting the idea that Perkin-Elmer had been at fault, Congress in 1991 considered changing government regulations to make contractors liable for their mistakes.[131] Nonetheless, in retrospect, it was a mistake not to have NASA experts, supported by specialists in optical testing, present during the crucial tests of the main mirror; Marshall's suspicions about Perkin-Elmer's competence during this time certainly justified such a presence and the Department of Defense did not prohibit such monitors.[132]

Despite its flaws, Hubble remained a powerful scientific instrument, in large part because its operators found ways to compensate. Not only did NASA engineers compensate for the vibrations of the solar arrays, but they also made similar adjustments to the mirror flaw. Luckily NASA had intended to use computer processing to improve the images and the aberration was so perfectly symmetrical that software could eliminate some of the blurry halo and sharpen the images.[133]

Spherical aberration limited Hubble's performance in some areas more than others. It most affected the telescope's wide field/planetary camera, faint object camera, and the use of fine guidance sensors for making astronomic measurements. The flaw hampered spatial resolution and faint object imaging because the halo effect blurred fine details and wiped out dim images, making Hubble performance similar to the best ground-based telescopes. The computer processing, however, could remove much of the aberration for bright, high contrast objects and for these bodies Hubble was much superior to ground-based instruments.

Spectroscopy, the analysis of radiation wavelengths, could still be done because the instruments required less focused light. By increasing exposure time, scientists could still perform most of their tasks. The faint object spectrograph performed well in imaging bright objects and determining a target's physical and chemical properties. Users of the high resolution spectrograph, which studied ultraviolet radiation, found that the aberration flawed "crowded field" observations of overlapping images. But their images were unmatched because ultraviolet radiation could not be studied by earthbound instruments. Unfortunately scientists found that the jitter from the solar arrays rendered the high-speed photometer, designed to measure light intensity and fluctuations, almost useless. The small aperture of the device could not focus because of the tremors.

Nonetheless, in the first 18 months of operation, the telescope carried out more than 1,900 observations of nearly 900 objects. The information attained was high in quantity and quality; at the January 1992 meeting of the American Astronomical Society, 25 percent of all papers on space observations described Hubble results. Exciting images included Pluto's satellite Charon, storms on Saturn, star generation in 47 Tucanae, planetary formation around Beta Pictoris, and remnants of Supernova 1987A in the Large Magellanic Cloud.[134]

Almost as soon as NASA became aware of the telescope's problems, the Agency began planning repair missions. It had planned maintenance missions for every

3 years of the 15-year lifespan of the telescope. Although the primary mirror was not one of the replaceable units, its aberration could be corrected, much like the way an eye doctor corrects poor vision with spectacles, by modifications to "second generation" scientific instruments. COSTAR, the corrective optics Space Telescope axial replacement, would replace the high speed photometer and use relay mirrors mounted on movable arms to focus the scattered light.

Marshall's contributions would be part technical support and part training. The Center characterized the spherical aberration, measuring the error in the null corrector, correlating it with results from the telescope, and thus providing information for the corrective optics. Marshall operated a simulator of the Hubble battery and power system to help Goddard understand flight problems. In addition, the Center upgraded its Neutral Buoyancy Simulator to support the long training sessions required for the six-hour-long spacewalks.

NASA's repair of Hubble in December 1993 was a spectacular success. The astronauts successfully conducted a series of spacewalks of several hours each, using the tools, modular technology, and space support equipment that Marshall had helped design years before. The astronauts installed new optics, changed failing gyroscopes, and replaced shaky solar arrays. Goddard found, however, that Marshall's modified control software was still needed to compensate for array jitter. Within a few weeks, Hubble's performance was much closer to the Agency's expectations and had the potential to accomplish most of its scientific goals. The telescope began making images of faint objects never seen before. Images of Galaxy M87 confirmed theories that predicted the existence of black holes. Crowded starfields, which before the fix appeared as clouds of lights, afterwards became visible in detail and revealed stellar collisions and rejuvenation. Other images included the formation of planetary systems in the Orion sector, the bending of light by gravity, and the effects of comet impact on Jupiter.[135]

For years after launch, Marshall continued to support the Hubble Space Telescope. Indeed Marshall's history and the project's coincided and shaped one another for more than two decades. The project suffered from some of Marshall's own ills, experiencing the troubles created by diversification, reliance on contractors, management, and communication of complex technological projects, and technological invention in an era of scarce resources. Both Marshall and the telescope often got more publicity from failures than from successes.

Finally the Center, more than any other organization, made the Space Telescope what it was, designing its systems, shaping its team, managing its resources, fixing its problems, more than once saving it from crisis, even oblivion. The Hubble became Marshall's greatest contribution to science, embodying its dream of forging instruments for exploring the heavens.

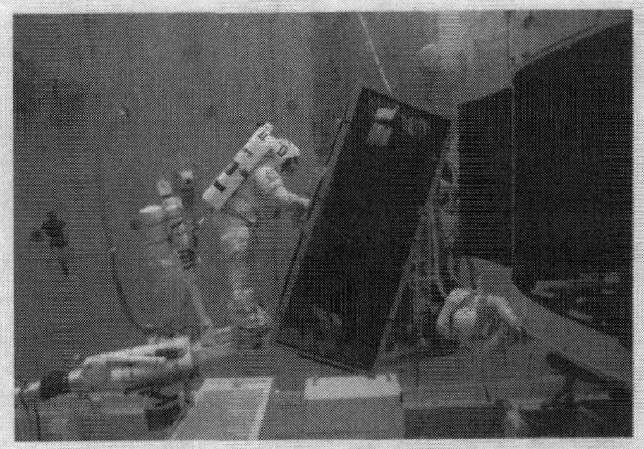

Astronauts practice installing the corrective optics module into the Hubble Space Telescope mock-up in MSFC's Neutral Buoyancy Simulator, June 1992.

---

1 Robert W. Smith, with contributions by Paul A. Hanle, Robert H. Kargon, and Joseph N. Tatarewicz, *The Space Telescope: A Study of NASA, Science, Technology, and Politics* (Cambridge: Cambridge University Press, 1989), p. 64. Smith's book is one of the best in the field of space history and we have relied heavily on it; nonetheless, it sometimes reflects the perspective of NASA Headquarters and the Hubble scientists.

2 Smith, pp. 68–71, 76–79.

3 Von Braun quoted in Smith, p. 84; E. Stuhlinger to E. Rees, 29 August, and 31 August 1970, LST 1971 folder, MSFC Archives; E. Stuhlinger, "Chapter Comments," 21 November 1994.

4 James A. Downey, III, Oral History Interview by Stephen P. Waring (hereinafter OHI by SPW), Madison, AL, 14 December 1990, pp. 25–27, quoted p. 25; James A. Downey, III, OHI by Robert Smith, National Air and Space Museum (NASM) Oral History Program, 18 January 1984, pp. 1–8, quoted p. 15; Jean Olivier, OHI by Robert Smith, NASM Oral History Program, 18 January 1984, pp. 1–4; Jean Olivier, Personal Communication to Robert Smith, NASA Oral History Program, no date, pp. 1–2; Program Development, "LST Systems Definition Study (RAM) Concept," May 1971, LST 1971 folder, MSFC Archives.

5 Myers quoted in W. Lucas to E. Rees, note attached to "OSSA 'LST Program' presentation to Dr. Low, NASA HQ," 9 November 1970, LST 1971 folder, MSFC Archives; Smith, pp. 82–84.

6 Smith, pp. 82–84; J. A. Bethay, MSFC, to H. Gorman, HQ, 31 March 1972, LST 1972 folder, MSFC Archives; E. Stuhlinger to J. Murphy, 10 January 1972, LST 1972 folder, MSFC Archives; Robert Bless, OHI by Robert Smith, 3 November 1983, quoted p. 23,

NASM Oral History Project; J. Mitchell to E. Rees, "LST," 17 January 1972, LST 1972 folder, MSFC Archives.

7 Smith, pp. 82–84, quoted p. 84.

8 Downey OHI, 1990, p. 32.

9 Smith, pp. 82–84; OSSA, "LST Program," 9 November 1970, LST 1971 folder, MSFC Archives; Lucas OHI, 1 March 1993, p. 25.

10 "J. Mitchell's Remarks Concerning Dr. Naugle's Visit (1/27/72)," 27 January 1972, LST 1972 folder; E. Stuhlinger to E. Rees, 19 July 1971; E. Rees to W. Lucas and J. Murphy, "LST," 14 September 1971; J. Murphy to E. Rees, 22 September 1971, LST 1971 folder; E. Rees to J. Naugle, "LST Program Data," 21 March 1972; MSFC, "LST" n.d., LST 1972 folder, MSFC Archives. In this last document, Marshall also informed Headquarters that "MSFC has the desire, capability and capacity to handle the SIP, if such would be desired by OSS[A]."

11 J. Murphy, PD to E. Rees, 27 June 1972; MSFC–GSFC, "Memorandum of Agreement on LST Project," 19 July 1972; W. Lucas to E. Rees, 12 September 1972; "LST Project Review to Dr. John Naugle," 25 October 1972, LST 1972 folder, MSFC Archives; Smith, pp. 93–95; E. Stuhlinger, "Chapter Comments."

12 J. T. Murphy to E. Rees, 3 November 1972; J. T. Murphy to G. Pieper, GSFC, "Clarification of Memorandum of Agreement on LST Project," 9 November 1972; E. Rees, "LST Cooperation with GSFC and Delineation of Work, Memorandum for Record," 13 November 1972; G. F. Pieper, GSFC, "Clarification of Memorandum of Agreement on the LST Project," 17 November 1972, LST 1972 folder, MSFC Archives.

13 C. R. O'Dell through J. Murphy to E. Rees, "Clarification of Points of Debate Between MSFC and GSFC in the Implementation of the Center Directors' Agreement on LST," LST 1972 folder, MSFC Archives.

14 "MSFC/GSFC Agreement for LST Definition Effort," 9 July 1974, LST 1974–75 folder, MSFC Archives.

15 Smith, pp. 94–96, quoted 96.

16 MSFC PR–72–162, "Scientists to Help Plan Space Telescope," 20 December 1972, MSFC History Office; O'Dell quoted in Smith, p. 96; R. Petrone to J. Mitchell, HQ, "LST Instrumentation," 6 February 1972, LST 1973 folder, MSFC Archives.

17 Smith, p. 87.

18 W. Lucas to J. Murphy, 30 November 1972, LST 1972 folder; J. Downey, "Memo for Record, LST Meetings, NASA Headquarters, 21 December 1972," 9 January 1973, LST 1973 folder; "LST Planning Strategy," 13 April 1973, Meeting with Dr. Low, Washington, DC, 4/13/73 folder, MSFC Archives; Downey OHI, 18 January 1984, pp. 12–13; Jean Olivier, OHI by Robert Smith, 18 January 1984, p. 15, NASM Oral History Program; Jim Downey to Mike Wright, 28 September 1993, MSFC History Office; Smith, pp. 86–91, 99–100.

19 A. O. Tischler, Director, Space Cost Evaluation to G. Low, Deputy Administrator, "Large Space Telescope (LST)," Space Telescope Documentation folder, NHDD; J. Olivier, "Personal Communication" to Robert Smith, 1984, NASM Oral History Program; Downey to Wright, 28 September 1993, MSFC History Office; James Kingsbury, Oral History Interview by Andrew J. Dunar and Stephen P. Waring (hereinafter OHI by AJD and SPW), 3 March 1993, p. 43.

20 Smith, pp. 94, 104–110; Olivier OHI, 18 January 1984, pp. 10, 13; Olivier, "Personal Communication," p. 9; Gerald Nurre, OHI by SPW, 10 August 1994, MSFC; Downey OHI, 18 January 1984, pp. 11–13; "HST Highlights," n.d., Space Telescope (A) folder, MSFC History Office 4491; Jean Olivier, "Comments on Chapter Draft," June 1996.

21 House Appropriations Committee Report, 21 June 1974, LST 1974–75 folder, MSFC Archives; Smith, pp. 125–127.

22 W. Lucas to E. Rees, 12 September 1972, LST 1972 folder, MSFC Archives; C. R. O'Dell, "The Large Space Telescope Program," *Sky and Telescope* (December 1972), pp. 369–71.

23 J. Downey, "LST Congressional Status," 25 June 1974; J. Downey, "LST Congressional Problem," 26 June 1974; J. Downey, "Status of LST Congressional Problem," 28 June 1974; G. Low, NASA HQ to Senator W. Proxmire, 3 July 1974; C. R. O'Dell to W. Lucas, "Report on LST Congressional Situation," 1 August 1974; Lucas to O'Dell, handwritten note, n.d., LST 1974–75 folder, MSFC Archives; Smith, pp. 125–35.

24 N. Hinners, Assoc. Admin. for Space Sciences, to W. Lucas, "Lower Cost Options for the LST," 4 October 1974, LST 1974–75 folder, MSFC Archives; Smith, pp. 136–38, 144.

25 J. Murphy to W. Lucas, 18 October 1974; C. R. O'Dell, "Discussions with U.K. and ERSO Officials," 16 October 1974, LST 1974–75 folder; MSFC PR 77–188, "ESA Signs Agreement with NASA on Space Telescope," 12 October 1977, LST 1977 folder, MSFC Archives; Smith, pp. 136–38, 179.

26 Smith, pp. 144–53, Downey quoted p. 153; MSFC Public Affairs Office, "2.4 Meter Aperture Chosen for Large Space Telescope," n.d., probably 1975, Space Telescope Folder, MSFC History Office 4491; Craig Covault, "Space Telescope Size Cut Eases Costs," *Aviation Week and Space Technology* (24 February 1975); Downey to Wright, 28 September 1993, MSFC History Office; A. W. Schardt, HQ to J. T. Murphy, MSFC, "LST," 18 December 1978, LST 1974–75 folder, MSFC Archives.

27 Smith, chap. 5, esp. pp. 164, 169–71.

28 J. Murphy to N. Hinners, HQ, "Request for LST Information," 4 September 1974; MSFC PR 74–308, "NASA Selects Three Firms for First Competition," 15 November 1974; "HST Highlights," n.d., Space Telescope (A) folder, MSFC History Office 4491.

29 J. Downey, "OSS Physics and Astronomy Program Managers Review," 15 May 1975, LST 1974–75 folder, MSFC Archives.

30 N. Hinners, HQ to W. Lucas, MSFC, "LST Aperture Decision," 22 May 1975; G. Low, HQ, "LST," 17 June 1975; Downey to Lucas, 15 October 1975; G. Low, "Space Telescope," 28 October 1975, LST 1974–75 folder, MSFC Archives.

31 Downey OHI, 18 January 1984, p. 23; Smith, pp. 166–67, 180.

32 Smith, pp. 185–86.

33 Smith, pp. 187–91, 196–98, 201–202. See MSFC, "Project Negotiations of Intercenter Agreement," 3 November 1976, Space Telescope 1976 folder, MSFC Archives.

34 C. R. O'Dell, MSFC to J. E. Naugle, HQ, 30 May 1975, NHDD, 1974–75 folder; E. Stuhlinger, MSFC, to H. Smith, HQ, 7 July 1975, LST 1974–75 folder, MSFC Director's Files; Smith, pp. 200–202.

35 W. C. Keathley to W. Lucas, "MSFC/GSFC Intercenter Agreement for Space Telescope," 12 October 1976, Space Telescope 1976 folder, MSFC Archives.

36 O'Dell to Lucas, "Scientific Aspects of MSFC/GSFC Discussions," 22 October 1976, Space Telescope 1976 folder; W. Lucas, "Telephone Call from Dr. Noel Hinners," 13 October 1976, Woods Hole Report & ST folder; Lucas to R. G. Smith, "Space Telescope Meeting," 22 December 1976, Space Telescope 1976 folder; Keathley to Lucas, "Briefing Notes for Meeting on ST MSFC/GSFC Intercenter Agreement," 26 October 1976, Space Telescope 1976 folder, MSFC Archives.

37 A. J. Calio, Dep. Assoc. Admin. for Space Science, HQ, "MSFC/GSFC Memorandum of Agreement on Space Telescope," 11 January 1977; Lucas to R. G. Cooper, GSFC, "Memorandum of Agreement for the Space Telescope," 17 February 1977, Space Telescope 1977 folder, MSFC Archives; Smith, pp. 208–209, 228–30.

38 Keathley to Lucas, "Co-Location of GSFC Mission Operations Manager at MSFC," 31 December 1979, Space Telescope 1978–79 folder, MSFC Archives.

39 Lucas OHI, 1 March 1993, p. 29

40 Keathley to Lucas, "ST Science Institute Contracting Method," 18 April 1978; Keathley to Lucas, "ST Science Institute," 24 April 1978; N. Hinners, HQ to Lucas, "Organizational Interface for Contracting Space Telescope Science Institute," 16 May 1978; Lucas to Hinners, "Organizational Interface for Contracting Space Telescope Science Institute," 22 May 1978; Hinners to Lucas, "Space Telescope Science Institute Planning," 30 May 1978, Space Telescope 1978–79 folder, MSFC Archives; "Group from Goddard Part of MSFC Family," *Huntsville Times*, 21 June 1978; Smith, pp. 210–18.

41 Bless OHI, 3 November 1983, pp. 24–26, 31; Bless, "What's Wrong with NASA," pp. 39–40.

42 Noel Hinners, OHI by Robert Smith and Joseph Tatarewicz, 4 December 1984, pp. 17–18, NASM.

43 Dr. Nancy Roman, OHI by Robert Smith, 3 February 1984, p. 14, NASM Oral History Program.

44 Fred Speer, OHI by SPW, 9 October 1990, pp. 27, 30; Kingsbury OHI, 3 March 1993, pp. 44–45; Lucas OHI, 1 March 1993, p. 28.

45 "NASA Selects 18 Scientists for Space Telescope," MSFC PR 77–217, 21 November 1977; MSFC, "Special Projects Review Committee," 4 January 1978, Space Telescope Assessment Group folder, MSFC Archives; Smith, pp. 240–58.

46 "Space Telescope Proposals Sought from Industry," NASA HQ PR 77–19, 1 February 1977; "Two Firms Chosen for Space Telescope Contracts," MSFC PR 77–137, 22 July 1977.

47 Smith, pp. 222–25.

48 "Selection of Contractor for an Optical Telescope Assembly, Marshall Space Flight Center," 26 August 1977, Space Telescope 1977 folder, MSFC Archives.

49 Smith, pp. 232–34.

50 William Lucas, OHI by AJD and SPW, 1 March 1993, p. 39.

51 Jean Olivier, OHI by SPW, 20 July 1994, MSFC, pp. 57, 36. For technical requirements, see MSFC Public Affairs, "Space Telescope Fact Sheet," February 1978, ST folder, MSFC 4491; Lockheed Missiles and Space Co (LMSC), "NASA Specifications for the Hubble Space Telescope," *HST Media Reference Guide*, n.d., Appendix D, MSFC History Office.

52 Olivier OHI, 20 July 1994, pp. 7–12, 16–19; Gerald Nurre, OHI by SPW, 10 August 1994, MSFC.

53 "Space Telescope Servicing in Orbit Studied," MSFC PR 79–79, 31 July 1979, Space Telescope folder, MSFC 4491; Elmer L. Field, "Space Telescope, A Long-Life Free Flyer," 29 October 1979, Space Telescope B folder, MSFC 4491; Craig Covault, "Simulation Verifies Telescope Servicing," *Aviation Week and Space Technology* (3 May 1982): 47–55; MSFC Space Telescope Project Office, "Designing an Observatory for Maintenance in Orbit: The Hubble Space Telescope Experience," 1987, 1987 TA01—HST folder, MSFC Archives; Olivier, 20 July 1994, pp. 39–41.

54 Smith, pp. 262, 265; Robert Bless, "What's Wrong with NASA?" *Issues in NASA Program and Project Management* (Spring, 1991), pp. 35–42; Lucas OHI, 1 March 1993, p. 53.

55 Olivier OHI, 20 July 1994, pp. 2–4; Moore, "Space Telescope Sensors," 16 April 1979, Weekly Notes, MSFC History Archives; Smith, pp. 234–36, 261–63; F. Speer to W. Lucas, "Dr. Lovelace Visit to Boeing," 14 August 1980, Space Telescope July–September 1980 folder, MSFC Archives.

56 Olivier OHI, 20 July 1994, pp. 27–29, 38–39.

57 "Work on Primary Mirror Blank for Space Telescope Begins," MSFC PR 77–233, 23 December 1977; Charles O. Jones, MSFC, "Control of Optical Performance on the Space Telescope," 10 January 1977, p. 9; Space Telescope F folder, MSFC 4491.

58 Smith, pp. 236–40.

59 See Weekly Notes, 22 May 1978, 30 May 1978, 5 June 1978, 23 October 1978, 27 August 1979, quotations from 24 September 1979, MSFC History Archive.

60 J. S. Potate, MSFC to Lucas, 6 April 1979; Keathley to Lucas, "OTA and SSM/ST Quarterly Reviews," 14 May 1979; J. Potate to Lucas, 16 July 1979, Space Telescope, 1978–79 folder, MSFC Archives.

61 Smith, pp. 264, 267–71; Lucas to Mutch, HQ, "OSS R&D Program Operating Plan 80–1," 10 March 1980; Mutch to Lucas, 11 April 1980; F. Speer, MSFC to Lucas, "ST OSS Budget Review," 25 April 1980, Space Telescope January–June 1980 folder, MSFC Archive.

62 Lucas to Speer, 1 July 1980; "ST Assessment Team," viewgraphs, 11 July 1980, 17 July 1980, 18 July 1980, Space Telescope July–September 1980 folder, MSFC Archives; "Space Observatory Costs Increasing," *Aviation Week and Space Technology* 29 September 1980.

63 Smith, pp. 271–81.

64 Smith, pp. 281–89; Speer to Lucas, "ST POP 80–2," 4 August 1980; MSFC, "Response to T. Mutch Letter (ST)," September 1980, Space Telescope July–September folder; Speer to Lucas, "Space Telescope Mirror Polishing at Kodak," 27 October 1980; MSFC, "Space Telescope Rebaseline Review, 30 October 1980; Speer, "Rebaselining Charts for the Frosch Presentation," 5 November 1980; O'Dell to Speer, "Project Response to Space Telescope Science Working Group, Resolutions of October 24," 8 December 1980, Space Telescope October–December folder, MSFC Archive.

65 Handwritten viewgraphs, December 1980, Space Telescope October–December 1980 folder; W. Lucas to J. Beggs, HQ, "Space Telescope," ST (General) April–June 1983 folder, MSFC Archive.

66 R. S. Kraemer, GSFC, to Director, GSFC, "Independent Assessment of Space Telescope Payload Status," 31 July 1980, Space Telescope July–September 1980 folder; Jerry Allen, "Memorandum to Dr. Lucas and Mr. Potate," 9 October 1980, Space Telescope October–December 1980 folder, MSFC Archive.

67 See *Marshall Star* articles, 13 May 1981, July 1981, 16 December 1981, 5 January 1982, 13 January 1982, 5 January 1983; quotation from "Optical Telescope Assembly: Technical Challenges," 1982, Keller Papers, ST Reports No. 2, NHDD; quotation from Perkin-Elmer, "Space Telescope Fact Sheet," n.d., Space Telescope (F) folder, MSFC 4491.

68 Kingsbury OHI, 3 March 1993, pp. 38–39.

69 Lucas to E. F. Ronan, PE, 30 October 1982; Ronan to Lucas, 10 November 1982; B. Tindall, "Space Telescope Quarterly Review [notes]," 1 September 1981, Space Telescope July–December 1981 folder; McCulloch, MSFC, "Observations on Perkin-Elmer," 9 February 1982, Space Telescope January–June 1982 folder, MSFC Archive.

70 Speer to Lucas, "ST Civil Service Manpower," 19 March 1981, Space Telescope January–June 1981 folder; Lucas notes on Speer, "Scientific Instrument Latches," 4 January 1982; Kingsbury to Lucas, 22 January 1982, Space Telescope January–June 1982 folder; Speer to Lucas, "Perkin-Elmer Budget Planning," 6 October 1981; "OTA Resident Office Weekly Notes," 8 October 1981, July–December 1981 folder; J. D. Rehnberg, PE, to Speer, 3 April 1981, January–June 1981 folder; Speer to Lucas, "OTA Organization Changes at PE," 31 October 1981, July–December 1981 folder, MSFC Archive.

71 J. S. Potate, "Award Fee Findings and Determination," 16 March 1981, January–June 1981 folder, MSFC Archive; "Space Telescope Development: Award Fee Payments to Perkin-Elmer Corporation and Lockheed Missiles and Space Company," 30 April 1983, Space Telescope Documentation, NHDD.

72 Lucas OHI, 1 March 1993, pp. 37–38; Kingsbury OHI, 3 March 1993, pp. 40–43.

73 MSFC, "Significant Problems—PMA," 11 October 1982; Speer to Lucas, 23 August 1982, Space Telescope July–October 1982 folder, MSFC Archive; Lucas to NASA HQ, Attn: A/James M. Beggs, "Space Telescope," 8 April 1983, Keller Papers, ST Reports No. 2, NHDD; B. H. Sneed and J. C. McCulloch, "Program Assessment of Optical Telescope Assembly," 25–28 October 1982; Lucas notes, "Bill Sneed—re PE," 29 October 1982, "ST," 22 October 1982, July–October 1982 folder, MSFC Archive.

74 F. Speer, "Space Telescope Project Assessment," 3 November 1982, Space Telescope November–December 1982 folder, MSFC Archive.

75 Lucas notes, 2 November 1982; Lucas, "For trip to PE," 17 November 1982, Lucas notes on memos, Speer to Lucas, "ST—DF-224 Flight Computer Spare," 19 November 1982 and Speer to Lucas, "ST Scientific Instrument Dynamic Test—Latches," 22 December 1982, Space Telescope July–October 1982 folder, MSFC Archive.

76 B. Taylor, MSFC to J. Kingsbury, MSFC, "Space Telescope PE Assessment," 8 November 1982, ST July–October 1982 folder, MSFC Archive; Michael Ollove and Gelareh Asayesh, "Anatomy of a Failure," *Baltimore Sun*, 29 July 1990, p.17.

77 "Space Telescope," 14 January 1983, ST (General) January–February 1983 folder, MSFC Archive.

78 Smith, p. 304; Ollove and Asayesh, p. 15.

79 J. Bush to B. Sneed, "A Trip Report on Perkin-Elmer Visit March 1 & 2 and Telecon

Discussion with Jim Welch on March 3 & 4," March 1983, Space Telescope (General) March 1983 folder, MSFC Archive.

80 House Surveys and Investigations Staff, "A Report to the Committee on Appropriations…on the NASA Space Telescope Program," 16 March 1983, pp. ii-iii; J. Beggs Testimony to House Subcomm. on Space Science and Applications, 22 February 1983, pp. 167, ST (General) January–February 1983 folder, MSFC Archive.

81 R. Quarles, "Keller: Telescope's Trouble 'Inevitable,'" *Huntsville Times*, 3 April 1983; Lucas OHI, 1 March 1993, pp. 28–30.

82 Dr. Bob Bless, OHI by Robert Smith, 3 November 1983, NASM, p. 30; R. Bless, "Space Science: What's Wrong at NASA," *Issues in NASA Program and Project Management* (Spring, 1991), p. 40.

83 R. Quarles, "Keller: Telescope's Trouble 'Inevitable,'" *Huntsville Times*, 3 April 1983.

84 B. Sneed to W. Lucas, 8 July 1983, ST (General) July–September folder, MSFC Archive.

85 Speer to Lucas, "ST Development Review Staff," 7 January 1983, ST (General) January–February 1983 folder; Lucas to B. I. Edelson, HQ, "Space Telescope Review and Reporting System," ST (General) April–June 1983 folder, MSFC Archive.

86 Speer OHI, 9 October 1990, pp. 22, 32.

87 Beggs testimony, p. 168.

88 House Report on ST, p. 10.

89 J. Bush to B. Sneed, "ST Key Milestone Changes," 31 March 1983, ST (General) March 1983 folder, MSFC Archive.

90 B. Sneed, "Discussion points for Dr. Lucas' Use in Executive Session with Mr. Beggs on March 16," ST (General) March 1983 folder, MSFC Archive.

91 House Report on ST, p. 9; J. Welch, "Space Telescope Development Review Staff, Briefing to the NASA Administrator," ST (General) March 1983 folder, MSFC Archive.

92 Speer OHI, 9 October 1990, pp. 23–24.

93 Lucas OHI, 1 March 1993, p. 39. See also Speer OHI, 9 October 1991, pp. 23–24; House Report on ST, p. 9.

94 Colleen Hartman to Ron Konkel, "ST Project Review, February 9–10, 1983," 10 February 1983, Keller Papers, ST Reports No. 2, NHDD.

95 Smith, pp. 303–305, 311–13.

96 House, Report on ST, p. 10; Welch Briefing, p. 9.

97 Speer OHI, 9 October 1990, pp. 21, 26, 29–30.

98 Lucas OHI, 1 March 1993, p. 39.

99 W. Lucas to J. Beggs, "Space Telescope," 8 April 1983, ST (General) April–June 1983 folder, MSFC Archive.

100 *Marshall Star*, 5 October 1983.

101 Beggs to Lucas, "Large Space Telescope," 24 March 1983, ST Reports No. 2, Keller Papers, NHDD; Smith, pp. 314–15.

102 S. Keller, HQ to Burt Edelson, HQ, "Space Telescope," 4 February 1983; Keller to Lucas, MSFC, and N. Hinners, GSFC, "Space Telescope Development Review Staff," 4 February 1983; Lucas, handwritten notes of conversation with Beggs, 24 February 1983, ST (General), January–February 1983 folder, MSFC Archive.

103 J. Welch, HQ, "Management of Space Telescope Development," 5 May 1983, ST (General) April–June 1983 folder; Keller to Lucas, "Actions Resulting from the Space

Telescope Rebaselining Activity," 31 March 1983, ST (General) March 1983 folder, MSFC Archives; "Project Management Report," 1983, ST Reports No. 2, Keller Papers, NHDD.

104 Smith, pp. 322–23; James Welch, OHI by Robert Smith and Paul Hanley, 2 August 1984, p. 12.

105 Lucas, handwritten speech to MSFC ST staff, 11 April 1983, ST (General) April–June 1983 folder, MSFC Archive.

106 J. Bush to Lucas, 18 March 1983, ST (General) March 1983 folder, MSFC Archive; Dr. E. Weiler, OHI by Robert Smith, 17 March 1983, quoted p. 9; Ollove and Asayesh, quoted p. 18; Jim Odom, OHI by AJD and SPW, 26 August 1993, pp. 10–11; Smith, pp. 314, 320–21.

107 "NASA Management Reorganization Relative to Space Telescope," 11 April 1984, Hubble Documentation, NHDD; MSFC, "Organization Charts for Center and ST systems engineering," March 1983, ST (General) March 1983 folder, MSFC Archive; Jim Odom, OHI by AJD and SPW, 26 August 1993, Huntsville, pp. 6–7; Fred Wojtalik, OHI by SPW, 20 July 1994, MSFC, pp. 1–4, 8–9, 11.

108 Lucas, OHI, 1 March 1993, p. 40.

109 J. Odom to J. Richardson, MSFC, "Potential P–E Personnel Changes," 30 April 1983; J. N. Foster to Lucas, "Rebaseline of OTA Contract with Perkin-Elmer Corporation," 19 August 1983; J. Richardson, "Project Manager's Assessment: Perkin-Elmer's Performance," 16 May 1983; W. Lucas, "Telephone Conversation with Bill Chorske, Senior VP and GM, Optical Group, PE," 9 December 1983, ST (OTA) 1983 folder, MSFC Archives.

110 Fordyce OHI, 31 October 1983, p. 11.

111 James Odom, OHI by Robert Smith, February 1985, p. 12, NASM.

112 H. R. Coldwater to Lucas, "ST SI Radial Latch Modal Test," 15 June 1984; Smith, pp. 327–29.

113 *Marshall Star,* 25 April 1984, 9 May 1984, 27 June 1984, 7 November 1984; Olivier, 20 July 1994, pp. 46–47; MSFC PR 85–5, 1 February 1985, Space Telescope, NHDD.

114 Odom to Lucas, "Assessment of the Lockheed ST Systems Engineering Effort," 20 July 1984; W. C. Bradford, MSFC, "On-Site Review of LMSC–A&V Operations," 5 November 1984, SSM (1984)—Lockheed folder; J. Odom, "HST Verification Review," 1985, ST 1985 folder; MSFC, "Assessment of LMSC Performance," 15 July 1985; Lucas, "Notes for my discussion w/our team at LMSC," 5 June 1985, SSM 1985 folder, MSFC Archive; Smith, 356–58.

115 J. Odom, "HST Project Planning," 5 February 1986, ST 1986 folder; Olivier OHI, 20 July 1994, pp. 29–31, 35; Smith, pp. 369–72.

116 Martin Burkey, "Marshall bids fond farewell," *Huntsville Times* (9 April 1990); Bob Davis, "Hubble Space Telescope," *Wall Street Journal* (4 May 1990); MSFC, "The Amazing Space Telescope," January 1990, MSFC Spacelink.

117 Nurre OHI, 10 August 1994, MSFC; *Marshall Star*, 8 May 1991; Douglas Isbell, "New Software Readied for Hubble to Counter Solar Panel Shudder," *Space News*, 15–21 October 1990; Luther Young, "Vibrations from Power Supply mean more trouble for Hubble," *Baltimore Sun,* 7 November 1990.

118 John N. Wilford, "Troubles Continue to Plague Orbiting Hubble Telescope," *New York Times* (15 June 1990); *Marshall Star,* 9 May 1990; MSFC, "Hubble Space Telescope's First 125 Days," 1990, MSFC Spacelink.

119 Lew Allen, Chair, "The Hubble Space Telescope Optical Systems Failure Report," 27 November 1990, p. iii, C.1–6 (hereinafter the Allen Report); "HST First 125 Days."

120 "NASA Begins Probe of Hubble Telescope," *Washington Post* (29 June 1990); R. Jeffrey Smith and William Booth, "Pentagon Tests Could have found Hubble Defects Before Launch, Officials Say," *Washington Post* (30 June 1990); Kathy Sawyer and William Booth, "Hubble Probe Opens as Scientists Reassess Mission," *Washington Post* (29 June 1990); Warren E. Leary, "NASA is Assailed on Quality Control," *New York Times* (30 June 1990); "Maybe NASA Should Check This," *Space News,* 21–27 May 1990; "More to Fix than Hubble," *Space News* (2 July 1990), p. 14.

121 Michael Ollove and Gelareh Asayesh, "Anatomy of a Failure," *Baltimore Sun,* 29 July 1990, pp. 15, 18–19.

122 Allen Report, Sections 4, 6, 7.

123 Allen Report, especially Section 8 and 9, quotations pp. 10–1, 10–2, 8–2, 9–4.

124 Allen Report, quoted pp. 9–2, 10–2.

125 Nurre, 10 August 1994, MSFC; Dr. Joseph Randall, 3 August 1994, MSFC; Olivier OHI, 20 July 1994, pp. 58–62.

126 Speer OHI, 9 October 1990, pp. 21–23.

127 James Kingsbury, OHI by SPW, 22 August 1990, p. 28.

128 Martin Burkey, "MSFC Disagrees with criticism in Hubble Report," *Huntsville Times* (26 November 1990); William Booth, "Hubble Flaw was Found in '81," *Washington Post* (6 August 1990).

129 Downey OHI, 14 December 1990, pp. 43–44.

130 Kingsbury OHI, 3 March 1993, pp. 42–43.

131 Douglas Isbell, "Congress Gets Tough on NASA Contractors," *Space News* (5–18 August 1991).

132 Based on E. Stuhlinger, "Chapter Comments."

133 Michael Alexander, "Computers Polish Hubble Images," *Computerworld* 26 (20 January 1992), p. 20.

134 "HST Performance Report," NASA Fact Sheet, 25 March 1991; MSFC, "HST's First 18 Months in Orbit Report," 1992; MSFC Spacelink; "Astronomers Blast Hubble," *Space News*, 2 February–3 March 1991.

135 J. N. Wilford, "Question for Telescope Is, How Much to Repair," *New York Times*, 23 October 1991; NASA HQ, PR 93–76, "Hubble Telescope Servicing Mission," 23 April 1993; NASA HQ, PR 93–96, "Hubble Servicing Mission Study Completed," 25 May 1993, MSFC Spacelink; "Hubble Daily Reports," January 1994, MSFC Spacelink; Wojtalik OHI, 20 July 1994; Nurre OHI, 10 August 1994; Dr. Frank Six, OHI by SPW, MSFC, 15 August 1994.

Chapter XIII

Space Station: A Visionary Program in a Pragmatic Era

"A major attribute of the Space Station program is the flexibility to adapt to changes in funding."

Space Station Phase A Report, November 1968

From the time when people began to dream of vehicles to escape Earth's gravity, two images dominated their thoughts: rockets and space stations. Marshall Space Flight Center (MSFC) has played a central role in the realization of both dreams, building Apollo's Saturn rockets and using the S–IVB stage as the basis for *Skylab*.

Progress toward a permanent Station in orbit was slow, but Huntsville's space team was at the Center of American dreaming, planning, and development. Perhaps no program shows as well the tortuous path from creative imagination to hardware. Marshall's involvement with Space Station encompasses von Braun's visionary sketches of the 1950s; conceptual studies in the 1960s; management of *Skylab*, America's first Space Station; development of payloads suitable for Space Station experimentation; management of major portions of NASA's *Space Station Freedom* program; and the political, budgetary, and organizational struggles of the 1980s and 1990s.

Space Station has been NASA's most visionary and frustrating program. The program had the misfortune of maturing at a time when the nation was not seeking visionary quests, but rather trying to trim federal expenditures and evaluating programs on the basis of cost effectiveness. Space exploration and the Space Station were hard to justify with quantifiable standards. Bob Marshall, who directed MSFC's Program Development directorate, explained the dilemma: "The main reason we're building the Space Station is not because of what I can tell you we're going to do with it, which I can't. The main reason is because

I can't tell you what we're going to do with it. And if you don't ever do it, you'll never find out."[1]

As in most post-Apollo programs, costs determined what NASA could do. Limited budgets, constantly under revision, forced the Agency to follow a "design to cost" approach for Space Station. This philosophy affected every aspect of the program including the configuration, division of labor, management approach, contracting, and schedule.

Design to cost led to programmatic complexity, bureaucratic infighting, and unprecedented political intrusion. Unlike the straightforward division of labor between Marshall and Houston under Apollo, NASA divided Space Station work among several Centers, and made the split on the basis of overlapping systems rather than separate hardware. This made systems integration difficult, and spawned debates between Centers, and between the Centers and Headquarters and led to political controversies that by the early 1990s threatened to kill the program.

Many NASA veterans insisted that the programmatic challenges of Space Station were greater than the technological barriers. This was a great source of difficulty for Marshall; the Center was accustomed to meeting technological challenges, but programmatic issues were often beyond its control. Initially, Marshall was at the center of the Space Station program, sharing the largest development role in a roughly equal split with Johnson Space Center (JSC). Nonetheless, because of managerial, political, and budgetary problems, the Center often found itself buffeted by winds from Washington.

Early Visions

Although fanciful notions of Space Stations appeared in fiction in the 19th century, it was not until the early 20th century that people with scientific training speculated about platforms to establish a permanent human presence in space. Pioneers in rocketry who speculated about space stations included the Russian Konstantin Tsiolkovsky in 1903, the American Robert Goddard in 1918, and the German Hermann Oberth in 1923.[2]

In speeches beginning in January 1947 and in his illustrated article in *Collier's* in 1952, Wernher von Braun advocated a space station for exploration, meteorology, navigation, and as "a terribly effective atomic bomb carrier." The *Collier's* conception, a 250-foot wheel in an orbit 1,075 miles above Earth, became the dominant public image of what a space station should look like. Herman H. Koelle, later a von Braun colleague at the Army Ballistic Missile Agency (ABMA) and Marshall, worked with von Braun on investigations of the feasibility of Mars exploration. Koelle proposed a space station design in 1951, a combination observation post, scientific laboratory, and engineering test site.[3]

The von Braun team began working on space station designs while still part of ABMA. Koelle headed the Future Projects Design Branch, which became the Future Projects Office after Marshall joined NASA. "We were one of Dr. von Braun's favorite little groups down in the bowels of the ABMA," recalled Frank Williams, who later succeeded Koelle.[4] Most of Koelle's young recruits were engineers, but others brought skills in disciplines like life sciences. One of these was John Hilchey, a physiologist who arrived in 1959, and who claimed that his only qualification was that for 25 years "I had read science fiction and dreamed and schemed it."[5]

The von Braun Space Station wheel in Collier's, 1952.

John Massey, author of one of the early ABMA space station studies, arrived at ABMA two years before the establishment of Marshall Space Flight Center. "Ever since I first came here in April of '58," he remembered, the group discussed "various programs of space-based, lunar-based, or space station-type of programs."

Von Braun and Koelle told the group to start with the premise "let's envision a space station and what [it] is made up of, what it can perform and not worry too much about how we would get it up there." Massey remembered that the group had free rein, and considered "early designs which encompassed everything from von Braun's wheel on down to virtually every concept you can come up with: globes, a disk, long arms, just everything."[6]

When the National Advisory Committee for Aeronautics (NACA) asked von Braun to take part in a committee devoted to long-range planning for the national space program, he turned to Koelle's group. "Several of us from that organization got to work directly with Dr. von Braun to help him put together thoughts and concepts and proposals and reports to take forward," Williams remembered. "We'd go back and rap among ourselves and come up with ideas and designs and concepts and do performance trades."[7]

One of the results of such brainstorming was Project Horizon. Koelle's group brought in representatives of the Army early in 1959 for a 90-day study conducted in a three-story cinder block building that later became Marshall's Structures Lab. "We went at it night and day," Williams remembered. "We laid out building a transportation system which did in fact require the use of a space station or transportation node in orbit. It was a filling station in orbit." The report envisioned operating a 12-man station by 1966.[8]

The report reflected modifications in von Braun's ideas about a space station that evolved in the 1960s in response to technological changes. The development of intercontinental ballistic missiles rendered the possibility of using a space station as a weapons platform obsolete, and advances in computer and electronic technology meant that people would not be needed for orbital Earth observations. Von Braun believed that a space station might best serve as a "house trailer" for astronauts on their way to the Moon or Mars, or for other activities in space such as the assembly of large spacecraft from components. Other uses would undoubtedly emerge over time.[9]

After President Kennedy committed NASA to a lunar landing program, plans for a station contributed to the Earth-orbit rendezvous (EOR) mode proposal advocated by von Braun, now the director of the Marshall Space Flight Center. Although EOR would not have required a space station, the orbital maneuver necessary to transfer propellant from one Saturn to another would have anticipated the type of activity for which a space station would be suited.[10] "In the very beginning it was envisioned by most people around here that we'd probably go to a space station as a stepping stone to a lunar exploration program,"

Massey remembered. Koelle's group proposed an orbital launch facility (OLF), a permanently manned space station with capabilities that would be useful long after a lunar landing, insisting that no purpose would be served if the lunar mission were to be an end in itself.[11]

NASA selected the lunar-orbit rendezvous approach advocated by Houston's Manned Spacecraft Center (MSC) in June 1962, however, and the Agency subordinated space stations to lunar exploration. Many of those involved in Marshall's early space station planning regretted the decision. "Technically and from an evolutionary point of view, the Earth Orbital Rendezvous mode was the correct way to go," Hilchey insisted years later. Others agreed. "The decision to go to a lunar base rather than an orbital build-up was purely political," Massey argued. "The concept that won out didn't require orbital build-up, just lunar landing which I think was to the ultimate detriment of NASA because it left us with, 'What are we going to do next, now fellows?'"[12]

Although a space station was no longer high priority after the mode decision, the studies of the late 1950s and early 1960s proved valuable to NASA, and forced the Agency to ask important questions. Should a space station be a closed-loop system, or should it rely on resupply from Earth? If resupply were to be necessary, what kind of a system could be used for frequent, dependable, low-cost visits? Should a space station have a zero-g[ravity] environment, artificial gravity, or a combination? And in light of the mode dispute between Houston and Huntsville, how could such a project be divided between NASA Centers?[13]

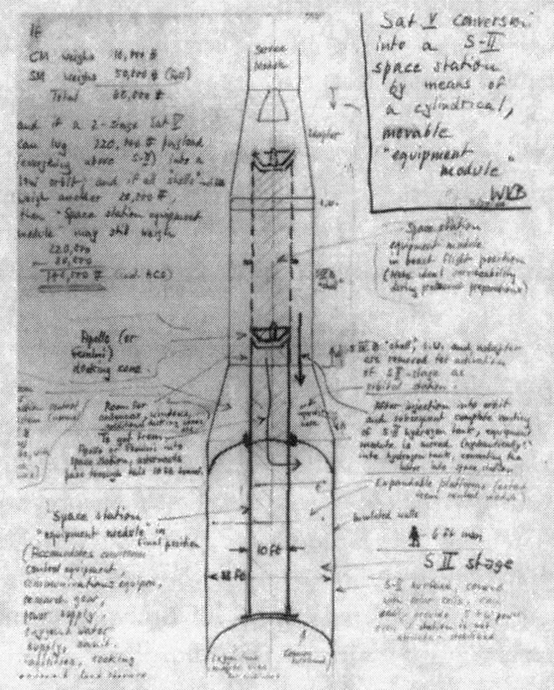

Early sketch of space station concept by Wernher von Braun, 1964.

Space Station in the Shadow of Apollo

The mode decision forced NASA's hopes for a space station to the periphery. The space station vision clashed with reality, as low priority, sparse funding, and competition from the Air Force limited planning. Rather than abandon plans, the Agency resorted to protracted studies, incremental planning, and Apollo technology to keep space station plans alive.

Marshall, the Manned Spacecraft Center, and Langley Research Center all directed contractor studies, but in light of the "understandable preoccupation with the Apollo mission," funding was meager. NASA decided to split planning into small segments in order to spread spending over a longer period. "That's what I had expected," von Braun remarked. "OMSF just hasn't got the doe [sic]!"[14] Marshall received the smallest portion of study funds allocated by Headquarters—only $300,000 for contractor work in 1963, less than 10 percent of the money distributed among the three Centers.[15]

Furthermore, the program lacked direction. Joseph F. Shea, who coordinated Space Station planning for the Office of Manned Space Flight (OMSF), found only diffuse support from other Headquarters offices, and even his deputies termed the justification and requirements for station "nebulous."[16]

Prospects for a NASA Station suffered not only from poverty and malaise, but from competition with the Air Force. NASA and the Department of Defense agreed that there should be only one space station to meet both defense and civilian requirements. But they had not agreed who should build it, what form it should take, and who would control it, so the Air Force proceeded with studies for a manned orbital laboratory (MOL). Early in 1963, NASA Associate Administrator Robert C. Seamans, Jr. appointed a special task team to evaluate NASA's plans for a manned Earth orbiting laboratory (EOL), and appointed Marshall's James Carter to the committee. By June, however, it was clear that NASA would not be able to initiate a major new program. Seamans was noncommittal when the group presented its report. "NASA HQ is simply very cautious with respect to any new starts in view of Apollo overruns [and] Congress[ional] sentiments," von Braun commented when he received Carter's report. "We must lie low for awhile!"[17]

Budget constraints forced NASA to set priorities, and by 1965 the Agency had to acknowledge that "approved programs are making heavy demands on limited financial and human resources."[18] The Agency shelved ambitious plans for large space stations.

The new fiscal environment posed unprecedented challenges to Marshall, but ironically thrust the Center into a leading role in space station planning. MSFC had to contend with declining resources for the decade after 1965.[19] NASA's need to capitalize on existing programs rather than initiate large new missions offered opportunity, however. It gave birth to the Apollo Applications Program (AAP), under which Marshall developed *Skylab*, and thereby became the only Center to manage a space station program. When NASA revived studies for a large station, Headquarters would not be likely to assign Marshall only a marginal role.

Skylab was the major AAP program for both NASA and Marshall, but neither the Agency nor the Center abandoned hopes of building a large manned space station superseding Apollo technology. Von Braun insisted that a large manned space station should be the "next major objective in the manned space flight program." Not surprisingly, he suggested that the AAP program would be "a logical first step for the generation of the necessary operational experience, knowledge and techniques that are required for the establishment and useful operation of a space station," an assumption that would place Marshall in the forefront of the next major NASA goal.[20]

NASA continued to refine plans for Station, looking for ways to reduce costs, defining experiments, and adjusting the concept to the expectations of experimenters.[21] Station plans, however, showed the impact of conflicting pressures. Headquarters, caught between Centers that were demanding more and a Bureau of the Budget that delivered less, sent contradictory signals.

For the next two years, Space Station planning reflected the new environment of fiscal austerity. In 1966 a committee headed by Charles Donlan advocated a station manned by 8 to 12 people capable of operating for up to five years, and serviced by vehicles already in NASA's inventory.[22] NASA requested $100 million in its FY 1967 budget for Phase B definition studies based on the Donlan report. When the Bureau of the Budget refused to approve funding, NASA continued Phase A conceptual studies out of advanced mission funds during 1967 and 1968.[23] The Phase A study concluded that one of the attributes of Space Station was its "flexibility to adapt to changes in funding," and showed what it meant by slashing its intended operational life to two years and reducing its crew to six with a provision that it could be operable with a crew of only three.[24] In six years budget constraints had forced NASA to lower its sights from a 21-man station to one that could be operated by a crew no larger than that of an Apollo capsule.

NASA managers, including Marshall's von Braun, were not accustomed to thinking small, however. In December 1968, Acting Administrator Thomas O. Paine showed his dissatisfaction with the Phase A report by querying Center Directors about the goals, configuration, size, and uses of Space Station. The Center Directors cheered Paine's instinct to seek a bolder concept. Von Braun assured Paine of his support for a "truly forward-looking program."[25]

Marshall wanted to play a central role in the planning for a larger space station. When von Braun assigned William R. Lucas to head the Program Development Directorate in December 1968, he made clear that a major duty of the new entrepreneurial organization was to "'harden' complete package plans for promising new programs, such as the Space Station."[26] Over the Christmas holidays Lucas visited William Brooksbank, who had experience with the orbital workshop, and convinced him to leave the Structures and Mechanics Laboratory to head Space Station work in Program Development.[27]

One of Lucas's first tasks was to assist the Center's executive staff in the preparation of a five-year institutional plan, an exercise mandated by NASA's Office of Manned Space Flight. For MSFC, the key issue was the "determination of Marshall's desired roles in the new programs (space station and lunar exploration)."[28] Lucas and the executive staff decided to make a bid for substantial Space Station work, including provision of Saturn launch vehicles; Station design, development and production; experiments in astronomy, technology, and manufacturing; integration of all experiments; and assistance work on a reusable logistic vehicle.[29] OMSF wanted a Station by 1975, and Marshall proposed that it could deliver with a budget peaking at $199 million and manpower peaking at 1,000 Civil Service and 7,300 contractor employees in FY 1973.[30]

Before NASA could allocate Space Station assignments and move into Phase B program definition, a fundamental issue had to be resolved: should a Space Station provide artificial gravity? The issue divided MSC and Marshall. Von Braun and George Mueller, associate administrator for manned space flight, agreed that artificial gravity was unnecessary and inordinately expensive. Apollo manager George Low suggested that a Station ought to include both artificial gravity and zero gravity, but warned that "it would be extremely difficult, expensive and time-consuming to re-invent all that we have learned during the past century to obtain measurement instruments that would work in zero-G." MSC Center Director Robert Gilruth, however, argued forcefully in favor of artificial gravity, and refused to accept a "zero 'g'" station. Furthermore, Gilruth was reluctant to accept a compromise in which Phase B would consider both zero gravity and artificial gravity since he believed the strong advocacy of Mueller and von Braun would mean that artificial gravity would not receive fair consideration.[31] Von Braun retorted that while he was not opposed to artificial gravity, he was not in favor of making a major commitment to it "until we understand the phenomenon and its implications [including] technology, design, operational considerations, schedule, cost, and attraction of potential users."[32]

Charles Mathews found a compromise that addressed Gilruth's reservations. The 1975 station was to be the first step toward assembly of an enormous craft of assembled modules. If Paine wanted a bold plan, Gilruth offered him one in the form of a 100-man space base. NASA agreed to accept a space base (reduced to a 50-man crew) as a long-term goal, and agreed that it would have a classic wheel form with artificial gravity in the perimeter, and zero gravity in the hub. This concession allowed for the construction of an interim 12-man Space Station targeted for a 1975 launch.

Mathews's compromise was so technologically complex, politically naive, and financially extravagant that it helped to kill Station prospects. It satisfied no one in the NASA community, and led to acrimonious meetings at Headquarters in January and February 1969. Marshall argued that the module should be integrated into the Station; Houston wanted it to be a prototype. Marshall still believed that the 1975 station should not require artificial gravity since experimenters wanted zero gravity, and suggested that Mathews was ignoring potential users. The Center in fact disagreed so strongly with Mathews that it presented an alternative plan a week later, but Gilruth and Mathews rejected the MSFC approach as having "too many pieces." Gilruth and Lewis Director Abe Silverstein wanted to move directly to a large Station without an interim step.[33] Ultimately, politics rendered Mathews's compromise unfeasible. The Nixon Administration told NASA to expect cuts.[34]

Before Mathews adjourned his series of Headquarters-Center meetings, he directed the Centers to study module designs for the 1975 launch. Each Center would direct a contractor design study for a "common" module, so called because it could both serve as a building block for a space station and operate independently. By late April, Headquarters set base requirements: the module would have to be 33 feet in diameter, carry a crew of 12, and serve either a zero- or artificial-gravity space base. MSFC would then investigate zero gravity, MSC artificial gravity.[35]

While Mathews and the Centers were fashioning hubbed pie-in-the-sky plans, budget realities forced Mueller to make a choice between Shuttle and Station. But even while Mueller and NASA brass struggled to find a way to build both a Space Station and a Shuttle, the Centers continued their station planning.

Von Braun named Brooksbank to head Marshall's Space Station task team, and Brooksbank established rapport with his Houston counterpart. Cooperation between the two teams showed not only that MSC and Marshall could work together, but that there were immediate advantages to doing so. "Rene Berglund and I were quite compatible, which was somewhat unusual between the two Centers," Brooksbank recalled. "Both of us were mature, and we managed to get along very well." Cooperation strengthened their hands at their respective Centers. "If we reached agreement fairly soon on most major issues, we were able to make our point of view stick within our own Centers which eliminated a great deal of friction."[36]

Planning now began in earnest, as Marshall and Houston each directed $2.9 million Space Station program definition studies. Working from identical statements of work, McDonnell Douglas conducted the Marshall study while North American worked for Houston. These Phase B studies aimed to design a 12-man Station to be launched in 1975, examine concepts for a 50-man space base to be operational in the late 1970s or early 1980s, and plan logistics systems to support the station and base.[37]

One of the conundrums facing NASA in its post-Apollo planning was to find a managerial approach that would preserve the strengths of the semi-autonomous field Centers and impose the centralized control needed for large national space programs. When Mathews assigned Frank Borman to the new post of field director and instructed him to chair a Space Station review group that would "integrate" the Phase B studies, von Braun feared intrusion on traditional

Center authority. He worried that the review group might undermine Center management and interfere with Center-contractor relations. "I would want to be assured that the review group does not provide direction to the Field Centers and especially not their contractors," he insisted.[38]

The field director's office never became as intrusive as von Braun feared, but Marshall worried about Headquarters micromanagement.[39] Program Development Director Lucas noted that "an inordinate amount of time has been spent in reporting," and added that "most of the extra reporting requirements have been generated by Headquarters."[40] When Washington warned new Marshall Center Director Eberhard Rees to give contractors maximum latitude in their Phase B Shuttle studies (see Chapter VIII), the warning had implications for Station. Brooksbank insisted that close contact with McDonnell Douglas was essential to the success of the Station, telling Rees that "MDAC and Marshall have established a total Space Station team to the mutual advantage of MSFC and NASA, and a Phase B study would be sterile within the written guidelines without this personal interplay."[41] Rees insisted that "our scheme of using working groups staffed by senior MSFC personnel allows efficient penetration without interference."[42] Cooperation between Brooksbank and JSC's Space Station task team leader Rene Berglund also prevented intrusion from Washington. "We found that Headquarters could not stand if the two of us agreed on something beforehand," Brooksbank recalled. "They always acquiesced to the approach we would take."[43]

In the Shadow of Shuttle

Redefining the relationship between Headquarters and the Centers would be a continuing issue as the Space Station program evolved, but by 1970 it became a peripheral matter as NASA, industry, and the Nixon Administration entertained doubts as to whether Space Station was realistic. In the months following the Apollo moon landing, altered circumstances placed the program in jeopardy. Tight budgets, suspension of Saturn V production, the reluctance of Congress and the administration to endorse a plan encompassing both Shuttle and Station, and the realization that early plans had been too optimistic forced NASA to reconsider plans for a Space Station.[44]

In March 1970 President Nixon selected the Shuttle and Station as national goals, but deferred Space Station until after development of the Shuttle. During the next two years Marshall, MSC, and Headquarters struggled to redefine the

Space Station program, first seeking to salvage as much as possible from the original Phase B studies in a new modular design, then trying to find ways simply to keep the program alive, and finally incorporating portions of the Space Station concept in other NASA programs.

The new environment forced NASA to adopt a fresh perspective on the Station, and four concepts drove design studies. The Station would use the Shuttle; early studies had relied on the Saturn. Station plans applied a conservative engineering approach; the Agency would build on Apollo and Apollo-derived technology (such as *Skylab*) rather than attempt to break new engineering barriers. The Station design would be evolutionary; most designs for the next decade planned to start simple and grow. Finally, the Space Station would involve international partners.

Grandiose plans for a space base thus gave way to in-house studies of a less expensive, more flexible modular Station with more flexibility. "When it became clear that the next program was going to be Shuttle," William Huber of Marshall's Program Development office remembered, "the first thing we did was a study activity of how we could modularize the space station into modules which would fit inside the Shuttle." Studies out of Huber's office examined ways to use the 15- by 60-foot modules "to accomplish the same objectives as the big one, but doing it in modules." Clusters of modules could approximate the capability of Phase B plans, but also give NASA a fallback position in which a limited one-module facility could be launched by a single Shuttle. Modules offered other advantages: reduction of initial and total costs, ease of replacement, and the opportunity to return them to Earth for refurbishment. In June 1970, MSC and Marshall began 90-day in-house studies evaluating module options.[45]

JSC and Marshall Station plans diverged as the Centers sought ways to salvage the Station. The planning staff in Houston urged cancellation of the launch of a first Station element, now scheduled as part of a 1976 Bicentennial extravaganza, since the Station might damage NASA's reputation either by delivering less than *Skylab* or by costing more than Congress could support. Houston considered more extensive revisions of earlier plans than Huntsville.[46]

The Space Station needed more than a new design if it was going to survive, however, and NASA tried to bolster public confidence. In September 1970, the Agency tried to create a Station constituency by sponsoring a meeting at Ames

Research Center of engineers, scientists, aerospace corporation executives, academics, and government representatives from the United States and foreign nations. Even those who supported the concept of a space station doubted whether sufficient funding would be available. Others questioned the wisdom of proceeding since most work projected for a space station could be done on a Shuttle, and scientists questioned the need for another manned vehicle. Ernst Stuhlinger, one of Marshall's representatives at the meeting, concluded that scientists, engineers, and corporate leaders alike were "acutely aware of the discrepancy between our total program (station, shuttle, tug, nuclear stage, Viking, Grand Tour, astronomy, exploration of the moon, exploration of the solar system) and our dwindling resources."[47] If potential space station users doubted NASA's dreams of two new major programs, Congress, the administration, and the general public were even less supportive.

Uncertainty pervaded NASA's Station redesign efforts. After the Centers initiated in-house modular studies in the summer of 1970, they requested their contractors to examine modular concepts. After Marshall's Phase B contract with McDonnell Douglas and Houston's with North American Rockwell concluded early in 1971, the two Centers initiated new studies with their contractors (termed Phase B Extended) for a modular station that would be compatible with the Shuttle, acknowledging "the funding constraints imposed by current budget estimates."[48]

The new studies were barely underway before a new threat loomed. The Office of Management and Budget, reasoning that "the current and anticipated pace of the space program clearly indicates that space station activity would follow the shuttle by at least several years," directed that Space Station funding would be "constrained," and that current station funds be expended more for Shuttle-related programs (such as the Sortie Can) than for long-range Shuttle planning.[49] Now began a complex dance in which Marshall and MSC competed for management of NASA's major manned space flight programs of the next two decades, and in which Headquarters struggled to find appropriate managerial tools to direct the Agency in a dramatically altered post-Apollo environment. Each of the three parties—Marshall, MSC, and Headquarters—had much at stake. Each took many uncertain steps, and in the process raised questions that NASA would wrestle with for more than two decades.

Indications were that Houston would be Lead Center for the Shuttle. But that left numerous projects up for grabs, including Sortie Can, Space Station, nuclear propulsion studies, payload studies, and Space Tug, as well as major elements of the Shuttle itself. It appeared that Space Station would be the next plum assignment. Competition was clouded by increasing awareness that the Agency would not be able to buy everything on the menu—or would at least have to order smaller portions, as was already the case in Space Station.

For Marshall, being decimated at the time by post-Apollo reductions-in-force, management of new projects offered opportunity to diversify. If Marshall was aggressive in pursuit of new projects, MSC was on the defensive. In May 1971, Associate Administrator Dale Myers recommended that MSC be assigned Lead Center on Shuttle.[50] With control of Shuttle within its grasp, MSC looked for ways to prevent Marshall from encroaching on its authority for operations, astronauts, and manned vehicles. But *Skylab* was clouding Center roles and missions, giving Marshall experience in all Houston specialties. Houston thus argued that its management of shuttle necessitated control of key interfaces, some of which would have precluded Marshall expansion.

Headquarters also found itself on uncertain terrain. In the aftermath of Apollo, Headquarters had to tread carefully between often-contradictory alternatives. Headquarters wanted to ensure that the Agency would have ample funds to support NASA programs, and could do so only by avoiding political problems and developing constituencies among aerospace contractors, researchers, and the public. Headquarters wanted to control Huntsville and Houston; but the engineering talent rested in the Centers and a Washington-based bureaucracy might destroy NASA's technical culture.

Part of the Headquarters' management approach was to balance Huntsville and Houston. When Myers recommended that MSC manage Shuttle, he suggested that any future work on RAMs (Research and Applications Modules, the forerunners of Spacelab) should be assigned to Marshall. Furthermore, Marshall would be designated Lead Center for Space Station at the conclusion of the Phase B studies. In July, a week after assigning Shuttle to Houston, Myers formally awarded Marshall integration responsibilities for RAM and Space Station, a task that entailed "definition, design, and verification of design concepts."[51] The last word in Station management decisions had not been said; in fact Myers had rendered only the initial paragraph of a long treatise.

Whether Marshall's assignment meant anything remained to be seen, since Space Station seemed to be performing a disappearing act. Congress would fund only one major space program, and Space Station became a dream deferred. Marshall's Space Station task team finished its contractual modular station studies in December 1971 and disbanded the following June.[52]

Marshall continued to conduct station-related studies under the auspices of a new Concept Verification Test (CVT) program, established to simulate environmental control and life support systems applicable to future manned systems. Brooksbank, Marshall's Space Station task team manager, directed CVT on the assumption that the limited funding available to Station in the mid-1970s could be applied in select critical areas, cutting costs and accelerating Space Station into Phase C/D.[53]

Lucas, now serving as Rees's technical deputy and thus the second-ranking administrator at Marshall, recognized the long-term benefits to the Center: "The attractive thing about all the elements of the prospective program is that, in addition to supporting a Space Station sometime in the distant future, the technical development will be very important to what lies between now and the Space Station, for example: RAM and Shuttle Cargo Bay. All the work we do will determine whether we obtain a Space Station or not." Support for CVT offered both technical and political advantages. "In some respects, we will be competing with MSC again," Lucas continued, "but I think we must do this to offer the strong capability in Spacecraft subsystems and systems design that we have developed in the *Skylab* program."[54]

CVT enabled Marshall to win Lead Center responsibility in June 1971 for an integrated Earth orbital systems effort in which the Agency kept Space Station planning alive, but it also led to contention with Houston. "After space station studies themselves were over [and] CVT was underway, we ran into some very, very confrontational politics between the two Centers," Brooksbank recalled.[55] Once again Marshall and MSC were moving on parallel paths, since Houston was developing a Space Station prototype (SSP) in a project contracted to Hamilton Standard. Both projects required the development of pressurized enclosures as preliminary steps toward Space Station development, and NASA could not afford duplication. Headquarters reduced Houston's funding and directed that Marshall provide the containers for testing, and instructed the Centers to coordinate their projects to ensure compatibility.[56] Cooperation between the Centers did not come easily, and on occasion Marshall had to request Headquarters give direction to Houston rather than work directly with MSC. Gilruth complained to Associate Administrator Myers about the incorporation of Houston's SSP and Environmental Control and Life Support System (ECLSS) into Marshall's CVT program, claiming that "planning has proceeded with a minimum of consultation with MSC" and with a "significant lack of understanding of the intended use of the hardware."[57]

"I don't believe the issue on our lead role in the CVT is now open," James Murphy, Marshall's director of Program Development, worried in November when Houston delayed delivery of SSP equipment to Huntsville. "I would not want to embarrass the Center by requesting delivery early just to enforce our lead Center role."[58] Indeed in late November 1971, Myers reaffirmed Marshall's role, insisting at the same time on closer cooperation between the Centers. "In terms of your role in CVT," he told Gilruth, "I envision MSC as a prime subcontractor for ECLSS, just as MSFC serves as a prime subcontractor to MSC for the Shuttle Booster."[59]

Development of life support systems was at the heart of the dispute and its resolution would affect later Space Station decisions. George Hopson, who had years of experience in the field, explained that it was clear very early that "probably the pacing technology for a space station would be the environmental control and life support systems." Other systems drew on earlier technology, "but on space station where there's several people living there for extended periods of time, everything that they use has to be resupplied. You don't have to do much calculation to see that one of the biggest problems is water and oxygen and the atmosphere that they breathe. . . . Most people, including myself, think that's the toughest job on the Space Station."[60]

Rees and Gilruth worked out an agreement which Headquarters accepted with slight modifications. The final decision retained some ambiguity; Marshall would control ECLSS, but Myers said he would "look to MSC as the lead Center in life support development" to recommend test objectives.[61] The solution took care of the short-term problem by giving both Centers jobs, but was no resolution; indeed it was the birth of a long running controversy over which Center should manage ECLSS.

In spite of intercenter competition, CVT kept Space Station studies going during shuttle development. "Every test we did in CVT for the first two years," Brooksbank insisted, was "directed and aimed at space-station problems." CVT examined some of the more challenging technological problems the Agency expected to encounter when the Space Station program could be revived. "We took those technologies that were long tent poles in designing the stations," Brooksbank explained, "and tried to implement them through the technology route." High-density solar arrays, the Astromast used to deploy the arrays, and a high data-rate system were all incorporated into the CVT study.[62]

Marshall could not afford to devote much of its scarce resources to a distant dream, however. Rees worried that the CVT team was so involved in Space Station that it might jeopardize the Center's efforts to secure related projects with a more immediate payback, and directed the group to broaden its focus.[63] Space Station consumed a declining portion of Center attention. Task team members found other assignments; Brooksbank became the deputy manager of Spacelab. For the time being, Marshall's and NASA's interest in building a Space Station remained alive mainly in related programs such as *Skylab*, Spacelab, and Shuttle.

New Strategies: Evolution Versus Revolution

Although Space Station was but a footnote in NASA's activities during the decade beginning in 1974, Marshall and JSC continued planning. The two Centers applied different philosophies as they worked on Station plans, with Marshall proposing evolutionary development of a station that could grow incrementally, and Houston urging commitment to a larger concept that could win program approval up front, an approach that NASA planners deemed "revolutionary." Each Center pursued its plans demonstrating how intercenter competition could generate creativity.

NASA clung to the belief that Space Station would be the next logical step, the major new start after Shuttle. The Agency also had a general idea of what it wanted: a modular station that could be positioned in either geosynchronous, low inclination, or low-Earth orbit, and could serve both as an orbiting laboratory and a space construction base, service facility, or Shuttle depot.[64] The new baseline station of the mid-1970s was more modest than its predecessors: a four-person Station capable of being placed in orbit by two Shuttle flights, one of which would carry a subsystems module and a habitability module, the other a logistics module and a payload module. The arrangement would allow for expansion.[65]

In 1974 the Agency began a series of Space Station studies, most of which were either managed by MSFC or parallel studies under Marshall and the Johnson Space Center. In August 1974, Marshall contracted a $274,000 study for a nine month McDonnell Douglas study of a Manned Orbital Systems Concept (MOSC), a permanent orbital station. The MOSC study was "probably the most fundamental study of that period in the '70s," according to Robert A. Freitag, NASA's deputy director of Advanced Programs, since "it really got us into the serious Space Station activity."[66] The study concluded that a MOSC facility could deliver more man-hours of space study at a lower cost than comparable Shuttle-launched Spacelab missions could provide.

The following summer Marshall, Johnson, and Kennedy formed a joint action group to devise an option for a geosynchronous space station.[67] In March 1976 Marshall and JSC negotiated $750,000 contracts for Space Station systems analysis with Grumman and McDonnell Douglas, respectively.[68]

With space station planning accelerating, Marshall reestablished a Space Station task team within the Program Development Directorate in the same month that the Center initiated the Grumman contract. Lucas named Huber as manager, and directed the team to analyze Station systems and configuration options.[69]

While the mid-1970s studies helped NASA refine the type of station it wanted, the Agency also sought convincing arguments to explain why it wanted to build a station. NASA was committed to a space station, but Congress, the public, and the White House had to be convinced that the expenditures for another major space program in a new "era of limits" was worthwhile. At a management meeting in March 1976, Frietag asked representatives of the Centers and Headquarters to list 20 reasons for a station in "compact, pithy language." Everyone could compile a list, but Jerry Craig, manager of one of JSC's Station studies, summarized NASA's promotion problem: "I think we must recognize that in virtually every objective considered singly, you cannot present an absolute argument for a permanent space station as opposed to multiple Shuttle flights."[70]

Recommendations for potential uses of a space station posed a dilemma. Bob Marshall remembered that the three basic proposals for using station were not compatible:

"First, science for viewing the universe and studying earth are generally compatible except for the direction for viewing. Second, materials science has been a user and desires maximum zero gravity conditions. Thus, any movement of men or repositioning interferes with processes requirements. Third, a refueling station for vehicles planned for deep space and planetary exploration would require frequent traffic with attendant disturbances and very hazardous operations."[71]

Freitag, however, had his own idea of the purpose of a space station, and during 1976 began to promote "space industrialization" as a goal, sparking a shift from the traditional concept of a station as an orbiting scientific laboratory. Freitag suggested material processing, construction of communications antennae, use of solar energy, and Earth observations as worthy topics for space station studies, and advocated employing a space station as a space construction base.[72]

With the new MSFC task team beginning operation, Freitag's approach provided grist for Marshall's mill. In 1976 alone, the Center solicited proposals for space industrialization studies, managed a Grumman Space Construction Base study, and included space construction and processing scenarios in a July in-house station definition. Marshall's Program Development office proposed that early shuttle flights include demonstrations in assembly of large space structures.[73]

Problems in winning support for a new Space Station program influenced NASA's development approach. The Agency debated whether to build the Station incrementally, or seek approval of a large program comparable to Apollo or shuttle. "Our thought was we get to Space Station by a series of well-planned steps, a few steps at a time," Huber explained. "The other theory is that NASA progresses in these momentous presidential decisions—Apollo, Shuttle, Space Station. Multi-billion-dollar steps."[74]

"The Marshall approach back in the seventies and the early eighties was build something that the country can afford," said Cecil Gregg, who worked on several of Marshall's concepts during the period. "Then expand from that."[75] The Center was convinced that "smaller is better," and pushed the idea of modular stations launched by the Shuttle. "Bill Lucas referred to the MSFC approach as a colony of stations in orbit," Bob Marshall remembered. "Through a modularization of elements, three or more separate stations could be built at an equal or lower cost."[76]

Once again Huntsville and Houston were on opposite sides of the question. "The folks at JSC said they felt they would like to have permission to take a look at doing [something] really big," remembered William Snoddy of Program Development. "Wham. Here it is, all in one chunk. It was referred to by some of us as the revolutionary space station. It didn't evolve; it was white-paper brand new. . . . We were trying to be more cautious, and they were proposing the big thing."[77]

Unlike the CVT dispute in which Marshall and JSC wrestled for control of a study project, the debate over the Space Station development approach showed how NASA intended to employ intercenter competition to unleash the creativity of both Centers. Each Center developed plans independently, giving NASA a chance to evaluate two viable options. JSC proposed a Space Operations Center (SOC) that Center Director Chris Kraft described as "a permanent manned facility in low earth orbit, dedicated to the development and use of space construction techniques, and to the servicing of space vehicles including assembly, launch servicing, refueling, and re-use."[78] It would employ two each of three different types of modules—service, cargo, and habitability—positioned along solar arrays that would span 433 feet. The SOC thus would be devoted primarily to operations, while most station proposals had concentrated on scientific purposes.[79] "We really never believed that was the way we wanted to go," explained Gregg, who helped develop Marshall's alternative. "We felt the science station . . . was the right way to go, not to try to move the whole mission operations and mission control function to orbit."[80]

Marshall's evolutionary approach centered on establishing a platform or module in space that could be used as a building block. Center engineers suggested in 1977 that either a Shuttle external tank or a Spacelab module could be employed in such a fashion.[81] Headquarters was more interested in another Marshall proposal, a 25-kilowatt power module designed to extend the Shuttle's time in orbit by providing additional power. The Office of Space Flight told Marshall to plan for a $90 million hardware development effort, and in March 1979 the Center established a project office under Luther Powell to direct development.[82] It was "just a big power supply in the sky," according to Snoddy. "When you went up with a Spacelab mission in the back of the orbiter you could plug into this thing, get more energy for the experiments, and also more energy for the orbiter; thus you could extend its lifetime on orbit for another week or two."[83]

Extra time in orbit was an important selling point for the power module, since the short 7-day duration of Shuttle flights fell short of the 89-day *Skylab* mission. "The science community began to realize what was there," recalled Powell. "Quite a few of them were enamored with the idea that here's a rich power supply in orbit." Scientists could "put experiments onboard and they can stay there forever and can be changed out by the astronaut crew."[84] Scientists in NASA also recognized the potential provided by the 25-kilowatt power module. Andrew J. Stofan, deputy associate administrator for space science, suggested that shuttle flight durations of 20 days might be possible by using the module, perhaps in combination with a JSC-sponsored power extension package (PEP) aboard the Shuttle. Stofan even suggested that combinations of platforms, Spacelabs, and power modules might allow flight durations of as much as 60 days.[85]

Marshall explored other platform concepts, any one of which could have provided an initial building block for a space station. In 1979 the Center initiated studies of a Science and Applications Space Platform (SASP) and a geostationary platform.[86] The Center sponsored a workshop on space platforms early in 1981, sharing its ideas with representatives of federal agencies, the aerospace industry, and space communications companies. By now engineers envisioned the 25-kilowatt power module as the foundation of an incremental manned space platform system. The addition of extension arms could transform the module into an SASP. By adding more modules later, the complex could be enhanced to host crews of eight or more astronauts.[87]

Planning From Headquarters

Soon after his inauguration, President Ronald Reagan nominated James Beggs as NASA Administrator and former Secretary of the Air Force Hans Mark as his deputy. Beggs, a NASA veteran who had been working in private industry, believed that a space station was "the next logical step" for the Agency.

Indeed the change of leadership in the White House and at NASA Headquarters offered opportunity to reinvigorate the Space Station program. The Carter Administration had not been enthusiastic about space programs, and never considered a major new start for a space station. Administrator Robert Frosch had all he could handle trying to keep shuttle development apace. Many in NASA, and particularly those involved in space station studies, viewed Beggs's arrival as an opportunity for a fresh start. After years of trying to "keep the system alive," according to Powell, "we felt like all that we had done to keep that embryo breathing paid off for us."[88]

The change also gave Headquarters opportunity to assert control over Space Station. From the early studies of the 1960s into the 1990s, NASA wrestled with the question of whether Space Station should be managed by Headquarters or by its development centers. Indeed Apollo and Shuttle witnessed experiments in organization, but Space Station demonstrated the Agency's ambivalence in unusual ways; for the first time the Agency vacillated between Headquarters management and relative center autonomy within one program.

At the time of Beggs's confirmation in June 1981, Marshall and JSC station studies offered options ranging from the JSC Space Operations Center to the MSFC evolutionary platforms based on the 25-kilowatt power system. Marshall tried to convince the incoming NASA leadership of the viability of its approach, and seemed to win support. Bob Marshall presented Huntsville's evolutionary approach to major contractors and to Headquarters, and received a favorable response. Headquarters directed JSC to assess using the MSFC power system and Spacelab as the foundation for an initial station.[89] MSFC Center Director Lucas explained the Marshall position to Mark shortly before Mark's confirmation, insisting that the Center still believed it was the best way to go. "That is the only way to go," Mark responded.[90]

Beggs agreed, and often insisted that he wanted to buy the space station "by the yard." What that meant would become clearer as Beggs sought presidential approval for a space station in the two and a half years that followed, but it implied both the evolutionary development approach favored by Marshall and the process of winning approval described by political scientist Howard McCurdy as "incremental politics."[91] In November, Beggs appointed Philip E. Culbertson as associate deputy administrator and directed him to manage planning for Station. John Hodge, another NASA veteran who had left the Agency, and Freitag joined Culbertson's staff.[92]

Freitag drew up a charter for a Space Station task group to coordinate Station planning out of Headquarters. "The reason I did this," Freitag explained, was that "when we had set up the competition between Marshall and Houston to look at both sides of it we were overly successful and we had set up a dichotomy that was disastrous. They were absolutely destroying each other." Freitag hoped to "wipe out all vestiges of the inter-center rivalry," even if it would take six months or a year. He believed that the only way to proceed was to cancel out Center projects like Marshall's platforms and power modules and Houston's Space Operations Center, and "bring everything into Headquarters."[93]

General James Abrahamson, associate administrator for Space Transportation Systems, who was organizing NASA's Space Station definition effort for Beggs, pulled funds from the Center Station study budgets to initiate contractor mission studies and "waived off" JSC and MSFC objections.[94]

Marshall objected to commissioning more contractor studies.[95] The Center wanted NASA to begin development of a space platform and conduct Phase B studies of a habitable module, an approach consistent with the Center's commitment to evolutionary development of station. Jack Lee received assurance from Headquarters that Beggs still favored Marshall's platform approach, and that he would seek approval for a start in 1984.[96] MSFC Program Development Director Bob Marshall argued that hardware under development would mean more to the Agency than more requirements studies, since once development began and metal was bent programs are seldom canceled.[97] Abrahamson was adamant, however, and soon announced plans to proceed with several contractor studies.[98] Furthermore, politics made an evolutionary station unlikely. Hans Mark was convinced that station would be a decision made at the top; there would be no "tolerant or permissive" attitude that might permit a low-cost evolutionary approach.[99]

Conceding that the mission studies (comparable to Phase A) would be directed out of Washington, JSC and Marshall positioned themselves for pieces of the development pie. The opening round of negotiations offered a split similar to the Shuttle/Spacelab division of responsibilities. Bob Marshall suggested to his Houston counterpart Joe Loftus that they begin program negotiations. He planned to seek MSFC management of the platform, platform orbital operations, payload modules, and payload interfaces, and conceded the habitability module, airlock, Station operations, Shuttle interfaces, and crew training to Houston.

This would leave Level II (Lead Center) responsibilities, the logistics module and the multiple docking assembly open for negotiations.[100] Unfortunately the discussion did not result in an agreement; by the time the two Centers would meet again to divide responsibilities, politics had intervened and a simple division of labor was no longer possible. Moreover, Headquarters was not about to turn responsibility over to the Centers at this point, and friction between the Centers and Headquarters was apparent. At one meeting, Houston's Loftus noted that "there were numerous references to 'the conservative Centers' (MSFC and JSC) and generally a negative attitude toward Center capabilities."[101]

Beggs announced establishment of the Space Station task group under Hodge's direction on 20 May 1982. The task group was to build a constituency for a Space Station and define a concept that might win approval for a new start for NASA. To do so, it would have to determine mission requirements, architectural options, and approaches for advanced development, systems engineering, management, and procurement. A loosely structured committee, the task group conducted most of its work through working groups whose conclusions would be reviewed by a program review committee chaired by Freitag.[102]

Hodge and Freitag had accomplished two goals even before the working groups began meeting. First, the establishment of the task group transferred Space Station impetus from the Centers to Headquarters. Second, by careful selection of the membership and leaders of the working groups, they spread Station work among the Centers to ensure that no one Center would dominate deliberations. The balanced workload minimized NASA's internal disputes at a time when the Agency needed to speak with one voice in order to combat external opposition to Space Station. It also fostered long-term problems, however, since the Centers insisted on a favorable division of the development spoils.

Headquarters did not establish all working groups at the same time it announced formation of the Space Station task group, and in fact it took nearly a year before all working groups were in place. Rumors circulated during the interim as the Centers worried about their stake in the station. As early as September 1982, members of the task force believed that Headquarters had decided to award Lead Center responsibilities to JSC, but Terry Finn of the Headquarters staff warned that Marshall should not be cut out or NASA could lose the support of the Alabama congressional delegation.[103]

Headquarters encouraged JSC and Marshall to submit proposals for Station management, and each Center made a pitch for Lead Center duties. JSC cited Apollo and Shuttle spacecraft experience. Marshall pointed to Saturn, *Skylab*, and Spacelab. The Marshall document argued that the Center was "characterized by total systems management of hardware development, high program visibility, effective program control, technical penetration, fast response, organization flexibility, and established interface with the User Community" (emphasis in original), and that the Center had a "sound success record in complex hardware performance management."[104]

Still, rumors of JSC's selection persisted, and Marshall managers worried early in 1983 that Hans Mark and JSC Director Jerry Griffin had struck a deal that would designate Houston Lead Center. "The tone and discussion in the halls of Washington is that MSFC is going to be eliminated from the space station competition," Bob Marshall, MSFC Director of Program Development, cautioned Lucas. "It is frequently stated that it is Johnson's position that they want to eliminate all competition," he continued, "and in attaining the assignment would totally operate the program from JSC."[105]

Bob Marshall also worried that Powell had been eliminated from consideration for a post in Washington, but Hodge chose Powell to head the Concept Development Group (CDG). The CDG, formed in April 1983, was one of the two most important working groups—the other being the Program Planning Working Group (PPWG), created in September 1982, and chaired by Craig at JSC.[106] NASA's planning under the task force aimed to win support for Space Station from broad constituencies. Concurrent with the establishment of the CDG, FY 1984 budget decisions curtailed further industry participation in Space Station planning. Beggs shifted NASA's effort to "an in-house effort concentrating on technology and systems engineering."[107] To close out contractor studies then underway, he ordered a series of briefings in which the companies explained their Station studies to the Agency and to the Defense Department, which had been reluctant to commit its support to a space station.[108] The briefings, held at Marshall in April 1983, gave the CDG a base on which to build its concept studies.[109]

Powell went to Washington in April 1983 on loan from Marshall and set up shop below the cafeteria in a warehouse built in the 1930s, the only quarters NASA could find in the capital. The building leaked so badly that a 50-gallon barrel filled with water each day, and in the winter frozen pipes burst. The NASA inspector general ordered the team out after discovering a sewer leak, but no other quarters could be found and the group continued to work out of the same location.[110]

By June the CDG had a full staff. When Beggs told Powell that he wanted to buy the Station "by the yard," Powell replied, "I want to first show you what the bolt's got to look like that you buy the first yard from." Describing the bolt became the CDG's task. To do so, Powell's group drew on trade studies, and sought input from interested agencies including the Department of Defense and the State Department. Powell had a small budget, but found a way to get aerospace firms to contribute without letting expensive contracts. Several firms wanted to work with the CDG. Powell offered them a deal: they could take part in discussions and receive copies of the reports of other participants if they would contribute reports of their own. Many agreed, and review meetings of the CDG often had more than 100 people in attendance.[111]

The CDG also helped set NASA's initial budget proposal, the figure on which President Reagan based his decision to support the Space Station. Shortly after taking office, Beggs asked former Administrator Fletcher to chair a panel that would estimate the development cost of an initial Station. Fletcher doubted that Congress would approve more than $1.5 to $2 billion, and decided to recommend a minimum figure in that range. Beggs was more confident that he could sell the program, and worried that the estimate might be unrealistically low. He asked Powell and the CDG for an independent estimate. Powell and his team knew the $2 billion figure was far too low. They suggested that costs could be kept down by using a common module that would eliminate duplication costs that would accrue with independent design. Powell drew a wide curve with an upper limit of $9 billion and a lower limit of $7 billion.

"I took it to Beggs, and he sat there at his table and looked at it for the longest time and grunted three or four times, and I walked him through the whole thing," Powell remembered. "I could see he was making up his mind. And finally, he just pointed to one and said, 'I'll take that one right there.' It was the $8 billion one, which was right in the middle between the seven and nine. So, I said, 'Fine.' He said, 'Go get me some more details, and go work that out and come back and tell me.'"[112]

The $8 billion figure caused problems. Beggs used it in an effort to propose a station that would be able to win presidential and congressional approval, but it was developed at a time when the Agency had insufficient information on which to base a realistic estimate and left the Agency committed to a baseline price that it could not deliver. NASA had lived on cost overruns before, but times had changed since the development of Apollo and Shuttle: Washington was more cost-conscious, the public no longer considered NASA's programs above review, and the changing international climate and tepid Defense Department support for Station diminished NASA's ability to justify the program as essential for national security.

Organizing Management

During the summer and early fall of 1983, NASA held a series of internal meetings that increased the involvement of the Centers in Station planning. Three management decisions were at stake: Would Headquarters or a Lead Center manage Space Station? Would the Centers or contractors handle systems engineering and integration? How would the Centers divide development work?[113] Answers to these questions determined the contours of the Space Station program, establishing relationships among the Centers and between the Centers and Headquarters that triggered problems.

In July the Space Station task force briefed the Center Directors on its progress. The group had defined a space station design employing a cluster concept, with a manned base comprised of habitat, utility, and operations modules, with provision for the addition of growth elements (such as experiment and logistics modules), unmanned platforms, and an orbital transfer vehicle.

NASA now turned to management issues. In August and September NASA held a two-session Space Station Management Colloquium at which the highest levels of Center and Headquarters administration confronted Station management issues. Headquarters intended the first meeting, held at Wallops Flight Facility from 29 August through 1 September, to assess program management. By now years of planning had taken place, and Space Station had yet to win approval; Center representatives showed frustration at the endless tedium of meetings with no certainty that they would ever bend metal. One Marshall manager who took extensive notes revealed his frustration, writing: "I cannot understand the position of the government. They are all powerful to be impotent, resolved to be irresolute, rabid for fluidity and adamant for drift. All the while the locusts eat."[114]

The undertone of rebellion suggested in the above comments affected discussions. Level B program management emerged as a dominant issue, and the Centers agreed that it should be at a field Center, not at Headquarters. The Centers also differed with Headquarters over who should manage systems engineering and integration (SE&I) during design and development. Headquarters, and especially Hodge, believed contractors should do it; the Centers believed the work should be done in-house. Marshall had long advocated in-house systems work, and wanted the job.[115]

Having experienced the problems associated with management of NASA programs throughout their careers, the participants enumerated the dangers to avoid. Handwritten notes from one of the task meetings documented dangers in an insightful, even hauntingly prescient listing:

1) Lack of program definition early in program
2) Lack of clear assignment of responsibilities between Centers and between Centers and Headquarters (HQ)
3) Low balling by contractors and by NASA
4) Incompetent staffing particularly in the program M[anager]
5) Complex interfaces, hardware and organizational
6) Lack of attention to details by NASA during development (contractor penetration)
7) Contractor selection
8) Lack of understanding between field Centers and HQ on the Center commitment
9) Establishing program cost as the most significant driver.[116]

The conclusions of the Wallops meeting influenced the agenda when Center Directors, the Space Station task force, and other management personnel met at Langley on 22 and 23 September. The Lead Center issue dominated discussions. Headquarters had reservations about using a Lead Center; on other programs the approach had caused problems regarding control of resources, diffusion of responsibility, and intercenter rivalry. The Center Directors, however, were united in favor of using a Lead Center on Station, and reminded Headquarters that "Centers can, and do today, 'work for' another Center." They also agreed that the Level B (lead) Center ought to have control of the money distributed to Level C Centers. The message was clear: the Center directors were so opposed to Headquarters program management that they were willing to take a vow of intercenter cooperation.

As a consensus formed in favor of adopting the Lead Center concept, discussion focused on which Center should assume the responsibility. Langley received consideration from those who believed Level B should not be located at one of the development Centers, but soon dropped out of the picture. Lewis and Goddard chose not to seek the assignment, and KSC and Ames never considered it. That left Marshall and Johnson to compete once again. As NASA's most diverse Center, Marshall was competing with several Centers on other programs: with Goddard on space science and astronomy, and with Lewis on space station power. This worked to Houston Center Director Griffin's advantage when he lobbied to form a coalition in favor of JSC. At the Langley meeting, General Abrahamson called for an informal nonbinding straw vote on which Center should take the lead. With Lucas abstaining, Marshall received only one vote. Not everyone at Marshall wanted the Lead Center role. Bob Marshall, director of Program Development, believed the Center should try to get it, but both Powell and James Kingsbury had reservations. "I quite frankly think that the Center has been a hardware Center since day one and that's our forte, and we ought to stay with that," Powell remembered telling Lucas. "The only thing we have to recognize in lead Center is that you're going to do everybody else's dirty laundry. . . . Everything that goes wrong, it's going to be your problem."[117]

Before the actual division of program assignments took place, Center directors agreed on certain management principles. They insisted that clarity was crucial for the program to succeed: clarity of definition, purpose, schedule, and money. "Don't even suggest a purpose is 'save NASA as an institution,'" they recommended. They suggested that systems engineering and integration should properly be the role of the Government.

The Langley meeting addressed NASA's major Space Station management issues but did not resolve them. In the aftermath of Langley, managers at the Centers worried about the disagreement between Marshall and JSC. Operating on the premise that agreement could come if both Centers had a meaningful part of Station and other Centers received a responsibility that fit their role, they weighed options for ways to divide major elements (habitat, air lock, support module, logistics).

By the end of 1983, the Centers and Headquarters had come to agree on three assumptions that would guide planning. Systems engineering and integration would be done in-house. The Agency would avoid committing station to one prime contractor over the life of the program. And development would be spread among several Centers to help revive the engineering capability of the Agency.[118]

Presidential Approval

The Space Station faced a critical juncture in the fall of 1983. NASA had devoted years to in-house and contractor requirements studies, conducted configuration and preliminary design reviews, and debated management options, but had yet to win presidential or congressional approval. President Reagan seemed supportive, but had backed off before when NASA thought it had won his blessing. Now Beggs and Hans Mark lobbied hard, and NASA gave a key presentation to the President during the closing days of the successful Spacelab 1 mission. But the Agency faced strong opposition from Congress and from within the administration. Budget Director David Stockman and Secretary of Defense Caspar Weinberger were vocal opponents. Beggs canvassed the Center directors to ensure that no hidden obstacles might undermine his campaign. Marshall's Lucas pinpointed NASA's conundrum: the Agency understood the technical issues, but could not demonstrate "an indisputable need and/or economical benefit." NASA needed political backing from the White House to proceed.[119]

Despite vigorous lobbying by opponents, the executive decision came in the State of the Union address on 25 January 1984, when President Reagan announced: "Tonight, I am directing NASA to develop a permanently-manned Space Station and to do it within a decade." Lucas welcomed the announcement of "an exciting new venture to which we in the Marshall Space Flight Center have looked for many years."[120]

Dividing the Pie

NASA had been planning for a space station for years, and now had presidential backing. The Agency now took on its most difficult managerial task: dividing space station work between the Centers. Two choices made in the six months following the presidential blessing created problems that plagued the program for the next decade. For political reasons NASA assigned work packages to four Centers rather than to the two major development Centers. Then NASA divided work by functional systems rather than hardware elements. These decisions multiplied interfaces into a maze of interrelated overlapping responsibilities.

Three weeks after Reagan's dramatic announcement, Headquarters decreed that JSC would be the Lead Center for Space Station. With Level B authority, Houston had responsibility for systems engineering and integration, business management, operations, integration, customer integration, and Level A (Headquarters program office) support.[121]

Although not unexpected, the announcement was a great disappointment to Huntsville. Bob Marshall was blunt: "We're not very pleased with not being named as lead Center." Hans Mark did little to cushion the blow when he said that Marshall had never been in the running, although he added that the Center would be "deeply involved" in Station work. Alabama Senator Howell Heflin demanded to know what Marshall's role would be.[122] It was a question that would take months of bitter wrangling to answer.

Center rivalry affected how NASA divided tasks on Space Station. Marshall was in the middle of the controversy, competing with Lewis Research Center and JSC. The first division concerned what NASA called the Space Station Advanced Development/Test Bed assignments, which involved the development by intercenter teams of technologies for specific space station applications. Theoretically, the advanced development tasks provided a means for research Centers (Langley, Lewis, and Ames) to contribute to space station technology development by working on teams with the development Centers (JSC and Marshall). NASA identified seven areas for advanced technology research, and in February assigned teams and Lead Centers. Three lead assignments went to Marshall (Attitude Control and Stabilization System, Auxiliary Propulsion System, and Space Operations Mechanism) and three to Houston (Data Management System, Environmental Control and Life Support System, and Thermal Management System). For the seventh discipline, Electrical Power, Headquarters assigned Marshall, JSC, and Lewis to the team, but deferred designation of a Lead Center. In each case, a team of personnel from other Centers supported the lead, so most Centers had a role in several advanced development tasks.[123]

Both Marshall and the Lewis Research Center in CLeveland wanted the lead in electrical power, and Marshall's Lucas and Lewis's Stofan lobbied to win the assignment. The Ohio congressional delegation swung its weight behind the Lewis bid. Some congressmen threatened to withhold support for Station unless Lewis won an acceptable portion of work. Deferral of the decision on the lead for the seventh advanced development task complicated negotiations for work packages in the months that followed.[124]

For the Centers, division of work packages was one of the most critical of all Space Station decisions, for it would determine their share of work on NASA's major program for the next decade, perhaps longer. During management meetings in August and September 1983, NASA had decided to divide Station assignments on the basis of work packages that would structure Phase B procurement and determine Center responsibilities for Phase C/D development. Negotiations would be driven by both political and technical considerations, and both were complicated. Politically, NASA had made broad promises to diverse constituencies in order to win approval for Space Station, and not the least of these was a pledge to involve all eight Centers. Guidelines dictated that no one Center would "own 'it' all," and that no one Center would be overloaded. But beyond that, NASA had to determine the number of work packages, the level of participation by each Center, and the types of work packages.[125] Such vague guidelines allowed for endless permutations. Everyone assumed that JSC and Marshall would have major portions, and that Goddard would have responsibility in some way for unmanned systems. Culbertson was worried that too many work packages would unnecessarily complicate an already complex system, but contention over the electrical power advanced development task brought Lewis into the picture, and Stofan insisted that the Cleveland Center ought to have one of the work packages.[126]

Technical considerations were no less complex. The station configuration was not yet set; a skunk works at JSC would develop a reference configuration concurrent with work package negotiations, but it had not even met when the Agency began to consider the division of labor. NASA had decided to keep systems engineering and integration in-house, but had yet to determine whether it should be done by Level B or delegated to the Level C work package Centers.[127] The Agency hoped to keep work package assignments consistent with Center strengths, but even this criteria was ambiguous. Houston established expertise in habitation modules during Apollo and Shuttle, for example, but Marshall's work in *Skylab* and Spacelab gave MSFC an equal claim to expertise.

At a meeting in Houston on 23 March, Headquarters assigned JSC Director Griffin the task of recommending a work package split.[128] Over the next two months Griffin engaged in what he later called "shuttle diplomacy" in an effort to reach agreement with other Center directors.

Unfortunately Headquarters had made a key decision that made Griffin's task formidable. Headquarters decisions dictated four work packages; the decision to give Lewis the electrical power advanced development assignment virtually guaranteed Lewis a work package in the same discipline, and Goddard's role in unmanned elements (platforms, free flyers and associated hardware) also fell into place. "Once that decision was made it forced us into splitting up the Station to the point where now it was difficult to have system control," Lee explained. Assignments for JSC and Marshall became much more complex as a result of the Lewis work package. Referring to the meetings in August and September 1983, Lee argued that "Some of us thought that we'd already had an arrangement between us and JSC on how that was going to be split, and we were ready to go with it." The Lewis assignment, however, "destroyed our little plan."[129]

The decision to grant Lewis a work package was political, a concession to the Ohio congressional delegation. The decision had inestimable consequences. It changed NASA's traditional modus operandi by having research Centers do development on major manned space projects. It cast into doubt the division of work between the Centers, destroying an understanding between JSC and Marshall, fostering greater (and unnecessary) Center rivalry. It led indirectly to Culbertson's decision to assign work packages to Marshall and Houston that reversed traditional Center strengths. It added complexity to an already complicated program. It made communications more difficult by adding additional prime contractors. It made distributive systems more difficult to manage by adding additional parties that had to be informed and agree to changes. In short, it may even have been the single greatest mistake in the program.

Now the split between Marshall and JSC would be more difficult, in part because of overlapping expertise, in part because of a tacit understanding that the workload should be equitably divided between the Centers. At a meeting of

Center directors late in March, Hodge suggested that it was time for "a bunch of good old boys to sit around the table and split up the pie," according to the notes of one of the participants.[130] Griffin, Lucas, JSC's newly appointed Space Station Program Manager Neil Hutchinson, and other key personnel from each Center met several times in April and May. At the first meeting in Huntsville in April, they attempted to divide work based on equal money, but the approach proved unworkable. Powell remembered one Griffin visit to Huntsville when the two Centers came tantalizingly close to agreement:

"That time, that night, to give you an example of how it shifted, Marshall was going to take on the systems integration responsibility. JSC agreed to it. . . . They were going to have the ECLSS system, and they were going to have the crew system. We were going to have the structures and propulsion. They were going to have communications. We had it all pretty well worked out. As we walked away that night, everybody was extremely happy. They thought we got this thing made. And so next morning about 9 o'clock Neil Hutchinson called me and says, 'Boy Luther, I really feel good about this thing—we've really made a tremendous accomplishment.' And about noon Jerry Griffin called Lucas and said, 'I'm sorry, I can't agree to that—all bets are off.' Then Neil Hutchinson called me and told me, 'Yeah, they couldn't agree with it.' I never understood why."[131]

Ultimately Griffin was unable to find a split satisfactory to both Centers, and at the end of May he reported to Headquarters that "Our areas of disagreement are significant and, I believe, are based on honest differences of opinion as to how the program should be structured." He explained that discussions "lacked a crispness" because they proceeded parallel to the evolution of the program, a fact that "added considerable difficulty" to negotiations.[132]

It remained for Headquarters to arbitrate. The aspect of Griffin's proposal that most troubled Hodge, now the acting deputy director of the Space Station program, was that the systems engineering and integration function would not be conducted by Level B in Houston, but rather distributed to the Level C Centers.[133] Indeed the means to handle systems integration would prove a formidable challenge.

In June, Culbertson, acting director of the Interim Space Station Program Office, asked Langley's Director Don Hearth to assist in working out a solution. Hearth and Culbertson met with Marshall officials on 11 June, and Hearth laid out principles to guide the split: strong Level B management, simple interfaces between Level B and the Level C Centers, commonality should be carefully contained and not foul up Center assignments, and an admonition that money should not be the driver in work package divisions. Lucas concurred with Hearth's suggestions.[134]

Culbertson then presented the Center directors two options; both had identical packages for Goddard and Lewis, and differed only in the JSC and MSFC assignments. The two options differed in that "Alternate A" assigned the assembly structure to Marshall's Work Package 1 (WP–1) and the common module to JSC's Work Package 2 (WP–2), and "Alternate B" reversed them.[135] After examining the proposal, Marshall argued that Alternate B provided "the worst mismatch of Center strengths and tasks," and that it threatened "such a profound impact on the total Agency, the contractors, and the development phase" and that as such "it should be rejected by all."[136]

The work package Center directors met with Culbertson and Hutchinson on 22 June. Noel Hinners of Goddard and Stofan favored Alternate B. The two JSC representatives, Hutchinson and Griffin, "waffled" according to Lucas's notes, but leaned toward Alternate A. Lucas said that he believed Alternate B "made no sense," but that Marshall "could do all or any part."[137]

Despite Lucas's reservations, Culbertson made the split similar to his Alternate B proposal; the most important deviation was that Marshall, rather than JSC, would be responsible for ECLSS. Although most in the Agency looked to Houston for expertise in life support systems, Marshall could make a strong claim. "JSC had never built an environmental control life support system that was closed-loop," Powell pointed out. "The only thing they had ever built and flown was the lithium-hydroxide canisters as filters; but we built and flew *Skylab*, which had the mol[ecular] sieve, which has the nearest thing to a closed-loop that you can get."[138] "We were very pleased that we got the ECLSS responsibility at this Center," said Randy Humphries, who had worked on ECLSS in Spacelab. But he admitted that the decision "really surprised us. . . . The way they wanted to manage this thing drove what kind of discipline responsibility they assigned to the Centers."[139]

The distinction between Marshall and Johnson roles and missions was now indeed muddy. Marshall's work package included ECLSS, but Houston had the ECLSS advanced development task; JSC's work package included the Attitude Control and Stabilization System, for which MSFC had advanced development lead. Culbertson's reasoning was that JSC, as Lead Center, ought to be responsible for the Station's structure, even though this was an MSFC strength. His work package division flowed from this logic, and thus deviated from the assumption shared by Hearth, Griffin, and Lucas that each Center ought to receive tasks most closely related to its traditional strengths. Culbertson said that since each Center would need "considerable subsystem support" from other Centers, it would not be necessary to adjust the earlier advanced development assignments.[140]

Marshall's Work Package 1 also included the "common" module, propulsion, and the orbital maneuvering vehicle. Marshall's responsibility for the module involved not only the module structure, but responsibility for provisions for its data management, power, environmental and thermal control, and communications. JSC's Work Package 2 included the structural framework, Shuttle interfaces, attitude control, communications, and data management. Lewis received the electrical power system, and Goddard the platforms and responsibility to define provisions for instruments and payloads.[141] The Marshall-Johnson split was relatively even; estimates for program costs for each Center were close, and MSFC expected about 40 percent of the total Station work.[142]

Configuration and International Partners

During the protracted negotiations leading to work package assignments, the Space Station configuration evolved at skunk works in Houston. People from other Centers joined JSC personnel under the direction of Hutchinson to elaborate the work begun by Powell's concept development group. The concept of a "power tower," a long boom with modules clustered at one end, best met user requirements, allowed for viewing and construction, and gave NASA the maximum capacity for Space Station growth. The Agency now had a reference configuration on which to base Phase B contracts.[143]

A reference configuration was not the only product of the skunk works. Level B management also developed during the four months the intercenter group met in Houston. Senior staff meetings evolved into the Space Station Control

Board (SSCB), the Level B clearinghouse for integration decisions. Hutchinson used the skunk works to organize a staff that would carry the program into Phase B. He staffed most of the key positions with JSC personnel, and as people began to depart from Houston to return to their Centers, Level B took on an even more pronounced Houston cast. JSC was of course the Lead Center, but the domination of its people at interCenter meetings had exacerbated Center rivalry as Phase B got underway.

Marshall and the other Level C Centers also organized their space station teams. Lucas commissioned a Space Station Projects Office, and moved it out of the Program Development Directorate. Project Manager Powell would now report directly to Lucas. Cecil Gregg became Powell's deputy.[144] In April the four work package Centers awarded contracts to industry teams to conduct 21-month definition and preliminary design studies. Marshall's contracts, with Boeing Aerospace Company and Martin Marietta Aerospace, were valued at $24 million, 36 percent of the total value of the contracts awarded. By the end of the summer, both contractors had established offices in Huntsville, and Boeing had announced plans to build an $8 million building near the city's airport to support its Space Station work and other contracts with Marshall.[145]

While NASA was establishing its reference configuration, organization, and procurement approach, the Agency was also seeking to fulfill another aspect of its Space Station mandate: the involvement of international partners. The Agency courted ESA for months, and in February 1985 the Europeans agreed to advance a $2 billion Italian-German project called Columbus as a means of ESA participation. In March, President Reagan and Canadian Prime Minister Brian Mulroney met in Quebec for what the press called the Shamrock Summit, and Mulroney announced that his nation would accept the American invitation to participate in the Space Station program. The next month Japan agreed to take part in the preliminary design phase, pledging a two-year commitment, and indications were that the Asian nation would likely continue beyond that date and design a laboratory for the Station.[146]

With the international partners on board, NASA worked to develop a baseline configuration. Finally the Agency adopted a baseline design first proposed by Marshall in the summer of 1985. The new configuration, a derivative of the power tower, used parallel twin booms in an arrangement NASA called the dual keel. Compared to the power tower, it had more mounting surface, greater potential for growth, and an improved pattern for microgravity experiments. Marshall and Houston "went through with a lot of analysis and determined with the modules down at the lower end of the boom, where they were located on the Power Tower, we didn't get exactly the right microgravity level," according to Gregg. With the dual keel "we moved the modules up to the center of gravity of the Station."[147]

The Perils of Complexity

The fledgling program was experiencing problems by the summer of 1985, some of which were normal growing pains, some more serious. The most troubling difficulties were hinged either to the complex work package arrangement or to budget constraints. The Space Station program was so complicated that management guru Peter Drucker said its organization chart looked more like a maze than a matrix.[148] "We created an almost impossible management and engineering job," explained James Odom, who witnessed Station development both from Marshall and from Headquarters. "I came from the school that the fewer interfaces you can have in a hardware program, between Centers, between contractors, the more straightforward, the easier it can be. Space Station doesn't limit itself to doing it that simplistically. There's hardly any way you can divide that thing up and not have numerous interfaces, but you don't need thousands. I think that's something that we did early on in the program that significantly complicated the design, the contracting, and the management."[149]

The complicated ECLSS split, with JSC managing advanced development and MSFC managing the work package that included ECLSS, was one example. Marshall complained that the two tasks were not synchronized and that JSC was not responsive to Marshall direction. Culbertson, whose split had created the problem, insisted that MSFC had system responsibility, but directed Houston to continue its advanced development project. A similar problem existed on the attitude control system, with Center roles reversed.[150] "Centers compete rather than coordinate for work," one Agency assessment concluded. Interfaces between work packages were difficult, and sometimes nonexistent; some contractors claimed that their Centers had directed them not to deal with contractors from other work packages. Neither Level B management nor the SE&I system appeared capable of holding the program together, and NASA began to worry that it was buying four "indigestible" products—work packages that would not mesh.[151]

Money had been a constraint in every NASA program since Apollo, but with Space Station the problem became particularly acute. By 1985 it was already clear that the Reagan commitment to build a space station within a decade was unlike the Kennedy vow to reach the Moon in a decade, and money was a fundamental difference. The Beggs pledge that NASA could build an $8 billion space station left NASA hedged in. Nineteen eighty-four was the only year in which NASA received its full space station budget request, in part because the Agency had limited itself to a modest $150 million, barely enough to cover start-up expenses. The decision forced NASA to design to cost, and now a year later costs had already begun to rise. Some in NASA claimed the Agency was costing the design rather than designing to cost. Problems external to the Agency exacerbated NASA's budget squeeze; federal deficits prompted Congress to trim all discretionary programs, and NASA suffered with other independent agencies.[152]

The budget crunch forced Culbertson to reexamine the Space Station program with an eye to "reducing or deferring development costs." On 14 August he directed Hutchinson to initiate a review involving both Level B and Level C, and to examine both cost reductions and changes that might affect system capability. The review, or "scrub," soon became known as "scrub mother," the first of several such exercises compelled by budget ceilings.[153]

Program reviews increased the already palpable tension between the Centers, especially since it focused attention on perceived shortcomings at Level B. Powell complained to Lucas that JSC was not delegating responsibility, and was micromanaging even tasks in the $50,000 range. He claimed that JSC failed to communicate; rather Level B was "in charge," and acted as if "We will tell you what we want you to know, what to do, and when."[154] Gregg remembered being "completely overpowered" in meetings at JSC. "You'd get down there in the conference room that would hold a hundred people, and it would be completely full of people coming in from all the [JSC] engineering and development divisions and offices. . . . It was a pretty difficult environment to work in." Disputes "pervaded the whole activity."[155] Powell remembered a meeting of the Configuration Control Board at which Marshall, Lewis, Goddard, and Headquarters each had 1 representative, and JSC had 16—and each individual had one vote.[156]

Matters came to a head at a space station management council meeting at Marshall on 24 October 1985. Hearth presented the findings of his investigation of systems integration problems. He pointed first to problems at the top: people perceived Level A to be weak, and "not in charge," and everyone was uncertain as to exactly what the Level A role was to be. Problems at Levels B and C were manifest. Key people at Level B were inexperienced, and the program manager was tired, frustrated, and "up-tight." It was unclear whether JSC was lending sufficient institutional support, and whether Level C accepted Level B authority. The Centers were plagued by excessive interfaces, Hearth said. Work packages had been driven too much by trying to preserve equality between JSC and Marshall. The Centers were too protective of turf, and were wary of international participation since foreign partners might absorb parts of their work packages.

What could be done? Some problems could be addressed relatively easily; JSC could assign more experienced people, and responsibilities at each managerial level could be defined. But the problem ran too deep for cosmetic solutions. The work packages would have to be redefined in order to simplify interfaces, allow for efficient integration, and facilitate international participation. Realignment should concentrate on Center technical capabilities, not on the relative size of the work packages or the dictate to provide "something for all Centers."[157] Hearth's report carried weight in Headquarters, where Culbertson was perturbed with continued intercenter rivalry.[158] A consensus emerged within the Agency that a change in work packages was necessary, although no one could yet define it.

The next several months encompassed the most chaotic period in NASA's history. Beggs took an indefinite leave of absence from the Agency in December as a result of fraud charges dating to his tenure at General Dynamics. Although the charges later proved groundless, Beggs's departure brought William Graham to the NASA helm as acting administrator. Graham, however, had been in the Agency for only eight days, so Culbertson became NASA general manager in charge of day-to-day activities.[159] Then JSC Center Director Griffin and Space Station Program Manager Hutchinson resigned, to be succeeded by Jesse Moore and John Aaron. Budget pressure also continued, and on 23–24 January 1986, Space Station planners discussed ways in which the "scrub mother" exercise might reconfigure Station to the \$6.5 to \$7.5 billion range.[160] The *Challenger* tragedy on 28 January thus caught NASA and the Space Station program in transition.

The *Challenger* accident was devastating to all of NASA, and the Space Station program was no exception. Station depended on Shuttle, and the grounding of the Shuttle fleet guaranteed further delays to a program already plagued by budget and management problems. Most immediately, the accident meant delays in thermal and materials experiments deemed to be "of critical importance to Space Station design."[161] Culbertson directed that the Space Station Office consider "lifeboat" rescue capability for the Space Station.[162]

Reorganization

The six months following the *Challenger* accident witnessed a wholesale reexamination of the Space Station program that resulted in a realignment of work packages, abandonment of the Lead Center concept, and establishment of a new Headquarters program office to manage Station. Marshall, buffeted by the repercussions of *Challenger* and preoccupied by the investigations that followed the accident, offered comments on the proposals floated by Headquarters and JSC, but for the most part Headquarters directed the reorganization. New leadership took charge in Houston, Washington, and Huntsville, and sought answers to an old problem: how to find the delicate balance between Center strengths and Headquarters' managerial responsibility.

The path to these tumultuous changes followed two tracks. With Culbertson stepping up to serve as NASA's general manager, Hodge took over as acting associate administrator in the Space Station Office and initiated a review from within the Space Station Program Office. He directed Marc Benisimon to lead a team dominated by Headquarters but comprised of representatives from all three levels to recommend a new work package split.[163] The other review brought back an old NASA veteran, General Sam Phillips, who had managed the Apollo program. Acting Administrator Graham asked Phillips to conduct a review of NASA management, and particularly of Space Station. Both studies had dramatic impact on the structure of the program.

Hodge's evaluation produced two alternatives. JSC and its contractor, Rockwell, advanced a plan that would have designated a single prime contractor and shifted much of Marshall's work to Houston. This "primary integration" approach, JSC argued, would provide "cost effectiveness, clear accountability, and superior flexibility."[164] The other Centers, including Marshall, preferred to stick more closely to the original structure.[165] Lucas argued that although Marshall had opposed the original split, "the present work package definition is workable," and that to make anything other than minor changes would be disruptive to the program as it neared Phase C/D.[166]

Hodge's recommendation, which he called "equal accountability," retained the four work packages of the original agreement. It made a significant modification in task definition, however, and Marshall Project Manager Powell influenced the change. NASA should "separate the inside from the outside," Powell suggested. "There's a very natural separation there," he remembered telling Headquarters. "Anything outside ought to be those people who are responsible for basic structure, and inside ought to be those people responsible for the basic module."[167]

Hodge's "inside/outside" split awarded Marshall the "inside." MSFC would develop all systems related to the "pressurized environment," which included the modules and related hardware such as tunnels, nodes, and interconnects. Houston had the "outside," or the "structure/architecture." JSC thus retained the truss and had responsibility for subsystems including attitude control, data management, and communications and tracking. The "inside/outside" split divided subsystems like thermal and communications, which had previously been assigned to one Center. The most significant implication was that each Center now had responsibility for one of the other's traditional specialties: JSC had propulsion, Marshall had ECLSS.[168]

When Graham suggested bringing in General Phillips from retirement to study space station management, Hodge told Graham, "If you give it to Sam, you can almost guess what your answer is going to be, and it is not what we've got." Hodge expected that Phillips, the former Apollo manager, would lean toward the Apollo management concept, which ran the program out of Headquarters rather than rely on a Lead Center. Phillips agreed to head the review, and accepted the task of examining station management, work package distribution, and systems integration.[169]

Phillips assembled a team that included former NASA Associate Administrators Mueller and Mathews. After discussing Station management with members of the Space Station Program Office in Washington, he visited each of the field Centers and their contractors. On 16 June, the team visited Marshall for two days of meetings with representatives of the Center and its contractors,

Martin Marietta and Boeing.[170] "Practically the whole of Marshall's Space Station role hinged on that visit," according to Powell.[171]

When Phillips returned to Washington to present his findings, James C. Fletcher had taken office as NASA administrator. Fletcher, who had headed NASA in the early 1970s, returned at the request of President Reagan to oversee the Agency's recovery from the *Challenger* accident. Fletcher was preoccupied with Shuttle, but had opinions about Station problems that predisposed him to accept recommendations for changes in management. Reviewing the flip charts of a Station review from several months earlier, Fletcher wrote on the cover: "JSC/MSFC split still an abortion," and "Bottom line: Lead Center concept would work but it depends on personalities. Level B did not have quality it deserves."[172] Phillips briefed Fletcher on 26 June. His most dramatic recommendation was that the Lead Center concept be abandoned, to be replaced by a strong program management office located near Headquarters but outside of Washington, removed from Beltway politics. The new office would have direct line authority to the field Centers. A branch office in Houston would coordinate system integration. He accepted the "inside/outside" split advocated by Hodge, modified to shift habitation module and airlock outfitting to Marshall. Within a week Fletcher announced acceptance of Phillips's recommendations and named Lewis Director Stofan associate administrator for Space Station.[173]

Marshall was the greatest beneficiary of the announced changes. The Center stood to increase its share of Space Station work from 31 to 44 percent, while JSC's would have decreased from 43 to 29 percent.[174] For Houston, the timing of the announcement could not have been worse; plunging oil prices depressed the Houston economy, and JSC Center Director Moore had just announced that he was retiring and thus would not be able to guide the transition. Houston newspapers screamed that JSC might lose 2,000 jobs, and the Texas congressional delegation enlisted Vice President George Bush to fight the decision.[175]

Fletcher retreated, announcing a 90-day cooling-off period to reexamine the changes.[176] Politics forced NASA to abandon another of its work package guidelines: that division of tasks should not be driven by traditional balance of funding between JSC and Marshall. Adjustments, including retention of the airlock at Houston, enabled NASA to give Houston and Marshall each about 36 percent of Space Station work.[177]

Cutting Costs

The *Challenger* accident guaranteed that Congress would scrutinize space station planning because it called into question NASA's technical expertise in a way that even the Apollo fire had not done. That it struck during a time of increasing concern over mounting federal deficits increased NASA's dilemma, for the Agency would now have to face criticism not only of the program's structure, but of its costs. During Apollo, NASA never had to prove that its program was cost effective. Such criticism became a factor during Shuttle development, but never overwhelmed the program. After *Challenger,* the public treated NASA as just another federal Agency competing for scarce resources. With the federal budget deficit climbing at an astonishing rate, agencies like NASA whose budgets were subject to annual review were vulnerable. Space station, a high-profile program with increasing costs and ill-defined purpose, was an easy target for cuts. Space station would have to prove itself during each budget cycle, and on difficult terms. In this environment, space station had to overcome two formidable obstacles: it was a visionary program, with returns measured in terms more related to the human spirit than cost effectiveness; and its promised material returns were far in the future and difficult to quantify.

NASA reorganized space station as part of the post-*Challenger* overhaul. Within two months in the spring and early summer of 1986, Fletcher and Stofan came aboard at Headquarters, and the Center Directors of both JSC and Marshall left the Agency. Lucas retired early in July after a 30-year career at ABMA and Marshall. On 29 September, J. R. Thompson, a 20-year NASA veteran who had managed the Shuttle main engine project at Marshall, took over as the new Center director. Fortunately project personnel remained stable at all four work package Centers; Powell continued to run Marshall's Space Station Projects Office. Managerial stability, however, was less crucial than costs. NASA had to defend the Station from cost reductions. Cuts forced delays, which increased criticism the next budget round.

In the fall of 1986, NASA conducted a review of space station design. A Configuration Critical Evaluation Task Force (CETF), under W. Ray Hook at Langley, evaluated Station design, concentrating on problems related to launching, assembly, and maintenance. "The CETF allowed us an opportunity to just stop for about a month and see where we were," explained O'Keefe Sullivan, one of Marshall's representatives. "We had had four work packages working pretty much independently during Phase B, and there had been no real coordination and compiling of what each of the elements [was doing]. All four work packages worked together with our best weights [and] power requirements, and put together a coordinated assembly sequence."[178]

Charles Cothran, another Marshall representative, worked on a reevaluation of how many shuttle flights it would take to launch and assemble the Station. Cothran's work demonstrated that early planning projecting 10 shuttle flights was overly optimistic, and gives one indication of why Congress attacked the $8 billion budget figure. "We went from 10 launches to 16 launches," Cothran explained, "and it was very obvious that we couldn't do it even in 16 launches, because we had negative margins on almost every load that we sent up. . . . And we had some hardware manifested at zero weight, which you know is unrealistic. We knew there was at least another flight or two of equipment that had to go up."[179]

The CETF review, which culminated in December, also recommended design changes that affected Marshall's participation. The team suggested enlarging Marshall's nodes and tunnels; larger "resource" nodes could be used to house equipment, thus helping reduce EVA time. Finally, the review advocated still another modification of work packages, giving Marshall responsibility for engine elements of the Station propulsion system.[180]

Upward revision in the number of shuttle flights required to build Space Station was but one of many factors increasing the estimated cost to completion. NASA had decided that an $8 billion Station was impossible, and in 1987 the Agency began to revise its estimates. The Agency informed the administration that it would cost $14.5 billion in 1984 dollars ($21 billion in 1987 dollars). An internal analysis suggested that NASA would need a $3.5 billion annual budget, while the administration had planned Station spending to peak at just over $2 billion per year.[181] Although Hodge acknowledged political, complexity, and administrative problems, he placed part of the problem at the Centers. NASA did not really "design-to-cost," Hodge believed, but rather practiced "cost avoidance" or "cost cutting." Center engineers were content to let costs rise, since this benefited their organizations. Inadequate contractor oversight caused duplication and "uncontrolled manpower loading."[182]

Myers, formerly head of manned space flight, returned to NASA late in 1986 as deputy administrator and immediately began to look for ways to cut Station costs. In doing so, he examined the roles of the Centers; his plans, had they been adopted, would have had a dramatic impact on Marshall. One possibility was to lower sights and develop an "austere" station by eliminating vertical beams and using only one cross beam, reducing the data system, and developing only one American lab/hab to be manned by a crew of five. He proposed dropping Lewis and Goddard from the work packages, suggesting that "by getting the Manned Program back in the three manned Centers, we even improve our management ability." These shifts "would reduce MSFC's workload slightly so they could take on the heavy lift launch vehicle."[183]

Myers also considered eliminating all space station work at Marshall. He believed it would be necessary to "reschedule" space station, to "half-size" the lab and hab modules, and plan for a man-tended rather than a permanently manned system. Then, since the modules would be smaller, "and since MSFC is so busy with ELV [expendable launch vehicles] and new engines, put MSFC work at JSC," he wrote. "MSFC would be out completely. Their contractor would be managed by JSC."[184]

Myers's ruminations never became Agency policy, but they reveal the character of the program early in 1987. For the second-ranking official in the Agency to consider such drastic action on the heels of a contentious work package revision demonstrates the program's instability, high-level doubts about its Station plans, and organizational problems.

Such fears were justified. The Congressional Budget Office suggested that in light of the $14.5 billion estimate, the Agency should cancel Space Station. Fletcher worried that the administration's commitment had wavered, that the international partners were getting cold feet, and that the Agency had lost control of Station and was losing its competitive edge in manned space flight. NASA delayed beginning Phase C/D for at least two years. Delays forced a schedule slip of at least two years. In March the White House agreed to a two-phase space station "stretchout" program that would result in a scaled-back station comprised of main truss, four modules (two American and one each for Japan and ESA), and a solar array power system.[185] The second phase would add two "keel" beams, provisions for more power, and a platform.[186]

Space Station was safe for the time being, but the program was now under unrelenting scrutiny. Powell insisted that the changes would not affect Marshall's work package, that there would be no reduction in the Center's hardware responsibilities. It "simply means that we will pay for the station as we go," he asserted.[187]

Moving into Phase C/D

With space station breathing new life, NASA prepared to initiate Phase C/D development. Implementation of the programmatic changes recommended by the Phillips Committee and the shift of management to the Washington area preceded publication of the call for contractor bids. Headquarters sought to control the Centers, but its new program office also introduced new managerial problems.

In the spring of 1987, Headquarters opened a new program office in Reston, Virginia. The new Level A–Prime replaced Houston's Level B.[188] Unfortunately the Reston office also introduced another level of bureaucracy, and instead of simplifying the program's interfaces, added complexity. The Centers complained about Reston micromanaging. The new office was "too involved in the next level down," according to Lee, who was Marshall's deputy director at the time of the change. "They never seemed to understand exactly what their role was." JSC's Denny Holt, who worked on systems integration, described what he called "the initial Reston fix": "Instead of taking the Level B documentation which was about the right level because it had been argued by all of us, they took it and processed 7,000 changes [and] added detail that you couldn't believe." Lee insisted that Reston never "got control of defining the program at the systems level."[189]

The frustrations prompted NASA, the White House, and the Defense Department to commission a study by the National Research Council (NRC). Seamans, a former NASA associate administrator now on the faculty of the Massachusetts Institute of Technology, headed a 13-member panel whose report contained good and bad news. The first part of the report, submitted in July, raised the frightening prospect of a $32.8 billion space station (in 1988 dollars, compared to the NASA estimate at the time of $19 billion).[190] The NRC full report in December concluded that Space Station would be a challenge "of formidable proportions," one that would stretch for two or three decades and thus could not be approached as a "one administration" program that could be built "on the cheap." The committee, however, endorsed NASA's configuration and its conception that the Block I station was only a starting point. The NRC had little to say about individual Centers, but supported developing advanced solid rocket motors for the Shuttle, which would be assigned to MSFC.[191]

Even as the NRC review proceeded, NASA finally released the RFPs for the work packages late in April 1987. Marshall's solicitation, valued at $4.5 billion, spelled out two options: one for a phased program, the other for an enhanced configuration program.[192] "We were going out with four RFPs at the same time, and we were trying to get as much common language and common items as we could, so we didn't have four completely disjointed contracts," explained Gregg, who chaired Marshall's Source Evaluation Board.[193] Marshall's Work Package, as it now stood, included two pressurized modules (one "lab" for microgravity research and one "hab" for eight crew members), three logistics support systems, four resource node structures, the ECLSS, an internal thermal management system, and internal audio and video systems.[194] In July, Boeing and Martin Marietta submitted proposals for Marshall's Work Package One.

The importance of the submissions to the contractor and the Agency, the requirement for security, and the depth of detail and sheer size of the proposals made the Source Evaluation Board's task a difficult one. Martin Marietta's two-million page proposal weighed 8,780 pounds, and filled 186 boxes. Boeing's 6,000-pound proposal filled 121 boxes.[195] Gregg set up shop in an office building on Huntsville's Memorial Parkway and posted 24-hour security. More than 200 people assisted the Board in its evaluation, some examining only small details, while others spent weeks with the group.[196]

On 1 December, Fletcher announced the successful bidders for each work package. Boeing won the competition as the prime contractor for the Marshall work package on the basis of its approach to key areas like systems engineering and integration, design and development, and program management. Boeing would have support from Grumman, Lockheed, Teledyne Brown, and TRW. NASA expected that the award might bring $800 million and 2,000 jobs to Huntsville.[197]

Development Work

While management worried about administering the Space Station program, Marshall's engineers and contractors began work on design and development. NASA had decided early that the Space Station would rely as much as possible on pre-existing technology, and most Station officials acknowledged that the programmatic challenges were greater than the technical challenges. Nonetheless NASA relied on state-of-the-art technology in some areas.

The ECLSS provided Marshall the most demanding challenge. ECLSS had seven subsystems: temperature and humidity control, atmospheric control and supply, air revitalization, the water reclamation and management system, waste management, fire protection and suppression, and EVA support.[198] It was a technological driver because other subsystems depended on ECLSS development. ECLSS relied on old technology, but Marshall sought improvements based on lessons from *Skylab* and Spacelab. "We went back and reviewed all those anomalies and made sure that . . . our design would side-step any similar type problems," according to Humphries.

In the early 1970s NASA used Marshall's powerful Saturn rockets to deliver thousands of pounds of water for *Skylab*. The Saturns were no longer available, and the shuttle's smaller lifting capacity would be used for other cargo. "The biggest difference between *Skylab* and Space Station is the fact that we didn't [have] oxygen and water loop closure," Humphries explained. For the first time, NASA would be "closing oxygen and water loops," which meant that Marshall had to design systems for recovering waste water for reuse and extracting oxygen

Space Station Freedom *mock-up at MSFC in December 1991.*

from CO_2 for rebreathing.[199] "It's imperative to have any practical space station, that you have to recycle the water," explained Hopson. To do so was essential: "If we have the right kind of system, there's no reason why you'd ever have to take water up," Hopson said.[200]

"The main source is urine, and another is condensate," Hopson continued. "The toughest is urine and there you normally use some sort of distillation process. And power is at a premium on a space station, so you've got to have some process of using heat for distilling and then later you condense the vapor. But you have to be very careful not to come up with a system that uses so much power that it's impractical."[201]

Another of Marshall's responsibilities, the habitation module, demanded fewer technological developments. "There has to be some innovative thinking of exactly how to put everything together," explained Axel Roth, who headed the project beginning in 1987. "But I don't see any pushing the state of the art." The principal problem in designing the habitation module was that "we've got a limited amount of space to do a lot of things in," Roth explained. To compensate for the crowded conditions, designers decided to separate the module into three areas: a quiet area for the crew's quarters on one side, a wardroom/galley on the other side where more activity would take place, and an intermediate area for lower-use activities, such as a health maintenance facility with its exercise machines.[202]

While the "Hab" would provide living space, the "Lab" would be the workplace of Space Station. Marshall's responsibility for the laboratory module evolved as the program changed. Originally, NASA planned to have two labs, one for life sciences to be developed by Goddard, and one for materials under Marshall. The two Centers had different ideas regarding how the labs should be structured; Goddard wanted the lab divided into floors. "We referred to [the Goddard design] as a bologna slice," recalled Marshall's Walt Wood. "We had the orientation down the longitudinal axis of the lab." The two Centers conducted studies, and Goddard agreed to use the Marshall orientation.

Budget reductions forced NASA to cut back to one lab incorporating both life science and materials research, and realignment gave responsibility to Marshall. Designers relied on racks to provide access for experimenters. "We spent a lot of time and a lot of effort trying to determine the dimensions of a rack—its depth, its size, trying to get the most volumetric efficiency we could in a rack," Wood explained. Eventually they settled on four "stand-offs," each supplied with power, fluid lines and ducts, housing a total of 44 racks.[203]

While some problems were unique to each module, each had common concerns. Contaminants posed a serious challenge in a closed-loop system. As Hopson explained, "Once you close the door you have no ventilation anymore. Some of these things you never worried about before become problems." Controlling microbes is vital, since "you're handling some pretty dirty stuff" in an environment favorable to growth. Both water and the gasses in the module atmosphere would have to be tested constantly, and the Center and its contractors had to design holding tanks and monitoring apparatus, as well as biocides and the catalytic oxidizer to eradicate contaminants.[204]

Systems Integration

Systems integration was a particularly difficult problem that had troubled Space Station plans from the beginning. Robert Crumbly of Marshall's Systems Engineering Office described the process as "making apples and oranges add up together."[205] Initially systems engineering involved defining requirements, contract specifications, and interfaces, and developing program documentation. As the program moved into Phase C/D, the job evolved into one of setting requirements to verify hardware and monitor contract performance.

Making sure that all the systems work together was anything but simple on Space Station. "The integration role and the coordination role with Level II and other Centers is probably greater than any other program we've ever had here at Marshall," according to Crumbly.[206] In order to coordinate between systems, subsystems, and work packages, NASA relied on two different types of control documents that would alert people to changes affecting their areas of responsibility. Architecture Control Documents (ACD) set forth the Station's structure, and Interface Control Documents (ICD) like those used in the Shuttle program addressed overlaps between systems. If JSC introduced a change in truss structure, for example, it might affect Marshall's modules; ACDs would alert Marshall of the alteration. An interface working group with representatives of each of the Centers resolved differences.

Integration meant close work with other Centers, particularly with JSC, and although the two Centers squabbled over division of responsibilities, people at the two Centers had worked together for years and knew how to cooperate. "There has not been acrimony," insisted JSC's Holt. "Quite frankly, at the working level, we've never had a problem of getting to an answer with Marshall." Both Centers have typically "let the technical solutions bubble and then go in at the last minute and make decisions. I think that's been almost the modus operand of Marshall-JSC operations over the whole time I've been involved."[207]

Interfaces with contractors were another matter. Because of the division into work packages, contractors under different work packages had difficulty communicating with one another, even though their responsibilities often overlapped. If Boeing had a problem that related to an interface with JSC's contractor McDonnell Douglas, Boeing could not approach McDonnell Douglas directly. Instead, they had to report to Marshall's project office, which in turn would approach JSC's project office, which would then contact McDonnell Douglas. It was a cumbersome bureaucratic system. Marshall Center Director Lee explained that "Any time you have a complex system like this and you've got to put . . . one or two government people in between two contractors to do even the simplest kind of thing, then you're inefficient."[208]

In April 1988 Odom, who had managed the Shuttle external tank and the Hubble Space Telescope for Marshall, replaced Stofan as associate administrator for Space Station. One of Odom's goals was to find a solution to the impasse in contractor-to-contractor communications. He proposed an "associate contractor" relationship. "What I wanted to do," he explained, "was put into the contracts the responsibility that if Boeing and if McDonnell Douglas had a problem, their first responsibility was to go very quickly, find the most economical way to fix it, regardless of what it would cost, which one would cost more money. Put the responsibility on them to come back to the government with one or two solutions and let the government pick the best solution."[209] Grumman, as integration contractor, would coordinate between work package contractors, but Odom believed the Grumman contract was too limited to allow them to improve communication significantly. Lee said "it's difficult to bring an outside contractor in to be systems engineer on somebody else's hardware."[210] JSC's Holt believed that Odom's plan would have succeeded in giving prime contractors incentive to work out problems, thereby bringing fewer problems to the Government. Unfortunately, however, neither the contractors nor many in NASA were accustomed to operating in such a fashion, and "as soon as Odom and [his deputy Ray] Tanner left, that went away overnight. Reston took that apart in five seconds."[211]

Hanging On

During Odom's year as associate administrator, Space Station budget battles became institutionalized. Odom and Fletcher recognized how much the struggle to justify Station had impacted the program the previous year, and tried to prevent a recurrence. "Dr. Fletcher and I very deliberately decided it was time to really decide if the nation and/or the Congress really wanted a Space Station Program," Odom remembered. Congress proposed level funding, and Odom worried that "we would have just kept going treading water and not making any real progress." Odom and Fletcher convinced Congress to fund Station at $900 million for Fiscal Year 1988.[212] They had won only a skirmish; Space Station would remain controversial well into the 1990s. When Fletcher left the Agency, he chose to emphasize funding problems in his valedictory address: "Restudy after restudy simply reinforces the conclusion that *Station Freedom* is well-conceived and well-managed, but very sparingly financed. There is simply no room for further trimming or shaping or cutting."[213]

The Space Station program became one of the most debated federal programs in the 1990s. Congress treated NASA like a spoiled child who had been given too much, and now needed to be brought up short. Congress restricted the Agency's spending, demanded rescoping, and then chastised the Agency for failing to make more progress. Costs increased, in part because of the stretchout. "You have funding instability when you have increase in cost," Lee explained. "That increase in cost gets reported, and then you get criticized for it."[214] The budget system was not designed for programs that stretched for decades. Apollo astronaut Wally Schirra highlighted the difference between the lunar program and station when he told a Huntsville audience in 1989, "We need to look at the space station as at least a 25- to 30-year program, not a quickie like going to the moon and back."[215]

Delays and stretchouts contributed to a decline in support for station. The public mood shifted, and Congress challenged Station at every turn. Even within the space community, people wearied of the lack of progress. As early as 1988, Marshall's Associate Director for Science Charles R. Chappell, worried about wavering commitment among scientists:

"This process that we have gone through with Spacelab mission cancellations has served to scare away many of the scientists who would be oriented toward doing science on the Space Station. They just hung with it, some of them for a decade, before they gave up. They wrote a proposal. It was a great idea. It got selected, and then they got money dribbled to them over the period of a decade, never coming to fruition. They just, at some point, say I can't stay with this any longer."[216]

Many factors coalesced to place Space Station in constant peril, some beyond NASA's control and others of the Agency's own making. Space Station, as Odom has said, "came about at a time when the nation didn't know what it wanted to do either nationally or internationally."[217] NASA's programs had always been political, but politics came to dominate Space Station in an unprecedented way. The driving force behind the division of space station work was an effort to ensure a geographic spread that would maintain the support of the aerospace community and Congress.

NASA, for its part, was unable to articulate its vision in a way that appealed to the public imagination. When NASA in the mid-1970s turned to industrialization in space as a justification, it started down a path that allowed the Space Station to be evaluated on the basis of what it could produce, rather than on the basis of scientific research or a visionary quest for humankind. It was a rationale without hope of short-term fulfillment, and placed Station in the wash of Shuttle's unfulfilled promise. NASA had made similar pledges for Shuttle, arguing on the basis of cost-per-pound to orbit and number of missions per year, raising expectations to levels that the Agency never came close to fulfilling.

The highly political context in which the Space Station program matured often left Marshall on the periphery. Marshall people, to be sure, played key roles throughout; the story of Space Station could not be told without reference to Wernher von Braun's and Koelle's visionary designs, the pioneering contributions of the engineers who developed *Skylab* and Spacelab, Powell's leadership of the Concept Development Group, or Odom's leadership as associate administrator. But the Center was often acted upon rather than acting. Sometimes this was by intent, since Space Station was one program over which Headquarters asserted unusual control. Lee, for example, was one of the more experienced people in the Agency in dealing with international partners, yet when asked about how much he was involved in developing the international role for

Station, he replied, "Not as much as I would have liked to have been and thought I should have been." Frequent changes in the program also had the effect of leaving Marshall to respond to the latest modification. Marshall, along with the other Centers, faced formidable external obstacles throughout Space Station development. The internal (within NASA) obstacles were primarily programmatic, since the technological challenges were less than they had been on previous projects.

Space Station has challenged Marshall in ways unlike previous programs. As an overtly political program, Space Station has drawn the Center into the political arena. "We can't lobby, but we can give information," Lee explained. "We've done more of that on Station than I ever remember we've done on any program here, and we've been asked to do that by Headquarters."[218]

In spite of the problems that plagued the program, work on Space Station displayed Marshall strengths. The Center had unusual vision; more than 30 years after Marshall engineers produced the first Space Station study, their professional heirs were working to fulfill their dreams. A culmination of more than 30 years of work in manned space systems, space station demonstrated the legacy of Apollo, *Skylab*, Shuttle, and Spacelab. Marshall engineering talent helped to solve the problems posed by ECLSS, Station's most challenging technology. And Marshall engineers and managers learned to operate a technological program under unprecedented political, budgetary, and bureaucratic pressure.

In 1993 President Bill Clinton ordered another redesign of Space Station in order to reduce costs, streamline management, and increase international involvement. The post-Cold War relationship with the former Soviet Union made possible closer ties with the Russians, who now joined the Americans, Canadians, Europeans, and Japanese as partners.

Teams at NASA Centers developed three new designs, and the administration selected the proposal designated "Alpha." Although the new design preserved 75 percent of the hardware designs of the old program, it was a fundamentally new program. Now known as the *International Space Station (ISS)*, the new design slashed projected completion costs from $25.1 billion to $17.4 billion, and cut operating costs in half. The new Station would have six laboratory modules instead of the three planned for the old design. As in the old design,

Canada would provide a remote manipulator arm and Japan and the Europeans would provide lab modules. The Russians would contribute hardware elements and employ their *Mir* Space Station in collaborative operations with the American Shuttle during the first phase of the *International Space Station* program.[219]

In August 1993 Headquarters designated JSC "host Center," meaning that the program office would operate out of Houston, but that JSC would operate only as "host," and not have the authority of a Lead Center. The change took into account earlier difficulties; there would be one prime contractor, which NASA hoped would minimize the troublesome systems integration problem. Award of the prime assignment to Boeing, Marshall's contractor, reflected well on the excellent working relationship that the company and the Center had experienced. Lee expected only minimal impact on Marshall: "I think we are still reasonably sure that there's going to be a pressurized module within our work package and that there's an environment control system that's going to be done here. We're using quite a bit of our facilities. I see us [as] not doing any less than we were doing before. The problem is the money. We know that the overall cost of the station is going to come down. That means everybody's dollars are going to come down and that means we have to again find ways to do it with less money. That would be the biggest challenge."[220]

Reorganization gave Space Station another new beginning. The new program outlined a three-phase schedule. Phase I began in 1994, employing the Shuttle and the Russian Space Station *Mir* for preliminary work and experiments. Phase II, scheduled to run from 1997 to 1999, projected assembly of the core of the *ISS* from American and Russian components and the beginning of Station research. In Phase III projected completion of the *ISS* by 2002, and initiation of 10 years of international experiments.[221]

As the new program began, Marshall remained ready to "do all or any part," as Lucas had said a decade earlier.[222] Key elements of the *ISS*, including the habitat module, underwent fabrication in MSFC's Space Station manufacturing building.[223] The Center supported Station testing, and prepared to manage payload operations and utilization. Marshall engineers worked in-house to develop the first major experiment facility for the *ISS*, the Space Station furnace facility (SSFF) for microgravity materials science research.[224] From the origins of concepts in the early 1960s to the fabrication of elements in the 1990s, and from *Skylab* to *Freedom* to the *International Space Station*, Marshall continued to be at the center of space station development.

1 Bob Marshall, OHI by Jessie Whalen and Sarah McKinley/MSI, 22 April 1988, Huntsville, Alabama, p. 46.

2 John M. Logsdon and George Butler, "Space Station and Space Platform Concepts: A Historical Review," in Ivan Bekey and Daniel Herman, editors, *Space Stations and Space Platforms: Concepts, Design, Infrastructure, and Uses* (New York: American Institute of Aeronautics and Astronautics, Inc., 1985), pp. 203–04.

3 Logsdon and Butler, p, 207; John W. Massey, "Historical Resume of Manned Space Stations," Report No. DSP–TM–9–60, 15 June 1960, ABMA, Redstone Arsenal, Alabama, pp. 19–27, Space Station Correspondence File, box 1, JSC History Office; Rip Bulkeley, *The Sputniks Crisis and Early United States Space Policy: A Critique of the Historiography of Space* (Bloomington: Indiana University Press, 1991), pp. 56–59. (Von Braun *Collier's* citation.) The 1947 date for von Braun's first speech related to space station is from Ernst Stuhlinger to Mike Wright, 21 November 1994 (MSFC History Office) in Stuhlinger's comments on a draft of this chapter.

4 Frank L. Williams, OHI by Thomas Gates/MSI, 4 June 1990, Huntsville, Alabama, p. 4.

5 Dr. John D. Hilchey/PS02, OHI by Thomas Gates/MSI, 6 November 1987, Huntsville, Alabama, p. 5.

6 John Massey/EH01, OHI by Thomas Gates/MSI, 6 October 1987, Huntsville, Alabama, pp. 1, 4.

7 Williams OHI, pp. 5–6.

8 Frederick I. Ordway, Sharpe, and Wakeford, "Project Horizon," October 1987, UAH Saturn Collection; Williams OHI, pp. 17–18.

9 Ernst Stuhlinger to Mike Wright, Comments on a draft of this chapter, 21 November 1994, MSFC History Office.

10 Stuhlinger to Mike Wright, Comments on a draft of this chapter, 21 November 1994, MSFC History Office. See Chapter 2 for an extended discussion of the LOR–EOR debate.

11 Massey OHI, p. 8; "Marshall Intensifies Rendezvous Studies," *Aviation Week and Space Technology* (19 March 1962), p. 78.

12 Hilchey OHI, p. 16; Massey OHI, p. 9.

13 See, for example, Hilchey OHI, pp. 10–12.

14 Von Braun, handwritten note on Carter to von Braun, 13 November 1962, Space Station (von Braun) 1962–1968 folder, Drawer 52, MSFC Center Directors' Files.

15 MSC received $2.2 million, Langley $800,000. William E. Stoney, Jr. to R. L. Bislinghoff, 5 October 1962, Correspondence File, box 1, Space Station Series, JSC History Office; "NASA Building Space Station Technology," *Aviation Week and Space Technology* (22 July 1963), p. 77.

16 Shea to Distribution, 17 October 1962; John E. Naugle to D.M. Shoemaker, Deputy Director for Systems, Office of Manned Space Flight, 14 November 1962; Orr E. Reynolds to Shoemaker, 10 December 1962; Deputy Director, Office of Space Sciences to Shoemaker, 21 December 1962, Correspondence File, box 1, Space Station Series, JSC History Office; Gruen, pp. 16–17.

17 Von Braun, handwritten note on Carter to von Braun, 10 June 1963, Space Station (von Braun) 1962–1968 folder, Drawer 52, MSFC Center Directors' Files.
18 "Summary Report, Future Programs Task Group," January 1965, p. 4, Space Station Series, NASA Documents, box 2, JSC History Office.
19 See Chapter 4.
20 Von Braun to Major General David M. Jones, Acting Director, Saturn/Apollo Applications, 16 December 1965, Correspondence File, box 4, Space Station Series, JSC History Office.
21 "Presentation for Dr. Paine on the Space Station," 5 December 1968, p. 9, Correspondence File, box 7, Space Station Series, JSC History Office.
22 Space Station Requirements Steering Committee, "The Needs and Requirements for a Manned Space Station," Volume 1: Summary Report, 15 November 1966, p. 82, Space Station Series, NASA Documents, box 2, JSC History Office.
23 Logsdon and Butler, p. 223.
24 NASA OMSF Advanced Manned Missions Program, "Space Station: Phase A Report," 21 November 1968, Correspondence File, box 7, Space Station Series, JSC History Office.
25 Von Braun to Paine, 24 January 1969, Space Station No. 2 folder, MSFC Center Directors' File; Logsdon and Butler, p. 224.
26 MSFC MSI Station Historical Documents Collection (STHDC) No. 0123, Entry 0646, 5 December 1968.
27 William A. Brooksbank, OHI by Thomas Gates/MSI, 2 May 1990, Huntsville, Alabama, p. 4.
28 Hans H. Maus to MSFC Distribution, 16 December 1968, 19/6 MSF Institutional Planning Study, Memo to Jay Foster from Dr. Lucas 1/20/69 folder, Terry Sharpe Files, Drawer 111, MSFC History Archive.
29 Maus Weekly Notes, 16 December 1968, 19/6 MSF Institutional Planning Study, Memo to Jay Foster from Dr. Lucas 1/20/69 folder, Terry Sharpe Files, Drawer 111, MSFC History Archive. MSFC was willing to "assume [that] other Centers will be responsible for the new logistics spacecraft and the biological, medical, earth resources and meteorological experiments."
30 Manpower for experiments and experiment integration would have peaked the following year with 1,300 civil service and 5,000 contractor employees for experiments, and 720 civil service and 2,870 contractor employees for experiment integration. "MSFC Institutional Plan," 24 February 1969, 19/6 MSF Institutional Planning Study, Memo to Jay Foster from Dr. Lucas 1/20/69 folder, Terry Sharpe Files, Drawer 111, MSFC History Archive. Not all figures were included in the bound institutional plan; they are available on draft versions of the document in this file. Management based its projections on the assumption that MSFC manpower would increase from the current level of 5,851 to 6,900 in 1974. (Terry H. Sharpe to Lucas, 14 February 1969, *ibid.* file.) This was not an unrealistic estimate at the time, since Marshall could hardly have anticipated the toll that reductions-in-force would exact on MSFC. Instead of increasing, however, manpower plunged to less that half of that estimated in 1969. See Chapter 4 for a discussion of manpower issues in the late 1960s and early 1980s.

31 Lucas to von Braun thru Jim Shepherd, 21 March 1969; Shepherd to von Braun, 25 March 1969; Lucas to Shepherd, 28 March 1969, Space Station No. 2 1969–1970 folder, Drawer 89, MSFC History Archive; George Low to J.D. Hodge, 14 December 1968, Chronological Files, Space Station Series, JSC History Office.

32 Von Braun to Paine (unsent, but hand-delivered to Charles Mathews), 16 April 1969; Lucas to Shepherd, 12 August 1969, Space Station No. 2 1969–1970 folder, Drawer 89, MSFC History Archive.

33 Mathews to Faget and Lucas, 19 February 1969; R. L. Lohman, "Notes from Mr. Mathews' Meeting Friday, 31 January 1969 on Space Station Issues," 4 February 1969; John D. Hodge to AA/Director, 20 February 1969, Correspondence File, box 8, Space Station Series, JSC History Office.

34 "Marshall Space Flight Center Space Station Program Plan," 1 August 1969, Space Station No. 2 1969–1970 folder, Drawer 89, MSFC History Archive; Gruen, pp. 46–47.

35 John D. Hodge to AA/Director, 3 March 1969, Correspondence File, box 8, Space Station Series, JSC History Office.

36 Brooksbank OHI, pp. 3, 16.

37 NASA FY 1969 Project Approval Document and Work Authorization 96–980–975 for Space Station, 24 April 1969, Correspondence Files, box 8, Space Station Series, JSC History Office; "Space Station Design Study," NASA Release 69–108, 23 July 1969, Space Station No. 2 folder, MSFC Center Directors' Files; Rene Berglund, MSC Space Station briefing, 13 April 1970, Chronological Files, box 11, Space Station Series, JSC History Office.

38 Mathews to Distribution, 21 August 1969; von Braun to Mathews, 5 September 1969, Space Station No. 2 1969–1970 folder, MSFC Center Directors' Files.

39 Borman left NASA for private industry, and the responsibilities of the field director's office passed to the Headquarters Space Station Task Force under Deputy Director Douglas R. Lord. Mathews to Distribution, 17 June 1970, Space Station No. 2 1969–1970 folder, MSFC Center Directors' Files.

40 Lucas to Rees, 24 June 1970, Space Station No. 2 1969–1970 folder, MSFC Center Directors' Files.

41 Brooksbank to Rees, 22 June 1970, Space Station No. 2 1969–1970 folder, MSFC Center Directors' Files.

42 Rees to Dale D. Myers, 29 June 1970, Space Station No. 2 1969–1970 folder, MSFC Center Directors' Files.

43 Brooksbank OHI, p. 3.

44 Philip E. Culbertson urged the centers to consider these factors in preparation for a Management Council meeting at Marshall and for a scheduled review of long-range OMSF plans. Culbertson to Distribution, 18 March 1970, Chronological Files, Space Station Series, JSC History Office.

45 Douglas R. Lord to Kraft and Lucas, 27 May 1970; MSC E&W, "Shuttle-launched Space Station Study," 25 June 1970, Chronological Files, Space Station Series, JSC History Office; Bill Huber, OHI by Thomas Gates/MSI, 24 September 1987, pp. 4–5.

46 Julian M. West to Associate Director, 2 July 1970; Donald W. Denby, KSC Executive Secretariat, Space Station Task Group, to Distribution, 10 August 1970; "Pre-Phase A

Study of a Shuttle-Launched Space Station," MSC Final Review, 16 September 1970, Chronological Files, Space Station Series, JSC History Office.

47 Paine to Rees, 21 August 1970; Stuhlinger to Rees, 12 September 1970, Space Station No. 2 1969–1970 folder, MSFC Center Directors' Files.

48 "Request for Procurement Plan Approval, Space Station Phase B Extended," 1 February 1971, Space Station No. 2 1969–1970 folder, MSFC Center Directors' Files.

49 Donald A. Derman, acting chief, Economics, Science, and Technology Division, OMB to William E. Lilly, assistant administrator, NASA, 19 April 1971, Chronological Files, Space Station Series, JSC History Office.

50 Myers to Fletcher, 20 May 1971, Shuttle Management/Lead Center folder, box 1, Shuttle Series—Mauer Notes, NASA Headquarters History Office (hereafter NHHO).

51 *Ibid.*; MSFC Release 71-118, MSI Space Station Historical Document Collection (hereinafter STHDC) 0818.

52 Rees, "Completion of Space Station Task Team Activities," 30 June 1972, Space Station No. 2 1969–1970 folder, MSFC Center Directors' Files.

53 J. Bramlet, CVT briefing, 30 March 1971, 10.n Space Station folder, Drawer 46, MSFC Center Directors' Files.

54 Lucas to Rees, 30 March 1971, Space Station No. 2 1969–1970 folder, MSFC Center Directors' Files.

55 Brooksbank OHI, pp. 14–15.

56 James T. Murphy to Rees and Distribution, 13 May 1971, Space Station No. 2 1969–1970 folder, MSFC Center Directors' Files; Douglas R. Lord to Director, Engineering and Development, MSC, 21 October 1971, 10.n Space Station folder, Drawer 46, MSFC Center Directors' Files.

57 Gilruth to Myers, 2 November 1971, Chronological Files, Space Station Series, JSC History Office.

58 Murphy to Rees, 10 November 1971, Space Station No. 2 1969–1970 folder, MSFC Center Directors' Files.

59 Myers to Gilruth, 22 November 1971, Chronological Files, Space Station Series, JSC History Office.

60 George Hopson, OHI by Thomas Gates/MSI, 12 February 1988, pp. 1–2, 6.

61 Rees to Myers, 15 March 1972; Kraft to Myers, 16 March 1972; Myers to Rees and Kraft, 7 April 1972, Chronological Files, Space Station Series, JSC History Office.

62 Brooksbank *OHI*, pp. 14, 21.

63 Rees notations, Heimburg Weekly Notes, 24 January 1972, MSFC Center Directors' Files.

64 J. W. Craig, Manager, Space Station Systems Analysis Study to Distribution, 22 April 1976, Chronological Files, Space Station Series, JSC History Office.

65 "Four-Man Station Would be Orbited by Two Shuttle Flights," *Defense/Space Daily* (16 December 1975), p. 243.

66 Freitag OHI, p. 45.

67 Gierow Weekly Notes, 9 June 1975, 7 July 1975, STHDC Entries 0943 and 0944; Gruen manuscript, pp. 51–54.

68 John H. Disher, Director Advanced Programs, NASA to Lucas, 9 March 1976, Chronological Files, Space Station Series, JSC History Office.

69 MSFC announcement, 10 March 1976, STHDC–0121.

70 Robert A. Freitag to Distribution on "Space Station Systems Analysis" Study Management Meeting, 25 March 1976; J.W. Craig to R.J. Gunkel, 10 August 1976, Chronological Files, Space Station Series, JSC History Office.

71 Bob Marshall to Mike Wright, comments on chapter draft, 23 December 1993, MSFC History Office.

72 Sylvia D. Fries, "Space Station: Evolution of a Concept," NASA HQ HHN–151, 28 February 1984, pp. 8–9.

73 MSFC Release 76–98, 28 May 1976, STHDC–0945; Charles Darwin Weekly Notes, 12 July 1976, MSI Space Station Chronological File Entry 0954; Gierow Weekly Notes, 20 September 1976, MSI Space Station Chronological File Entry 0957.

74 Bill Huber, OHI by Thomas Gates/MSI, 24 September 1987, p. 8.

75 Cecil Gregg, OHI No. 1 by Thomas Gates/MSI, 17 October 1989, p. 16.

76 Bob Marshall to Mike Wright, comments on chapter draft, 23 December 1993, MSFC History Office.

77 Bill Snoddy, OHI by Thomas Gates/MSI, 16 July 1987, p. 11.

78 Kraft to Center Directors, 30 October 1979, Chronological Files, Space Station Series, JSC History Office.

79 Dave Dooling, "Space Station May be Ready for Comeback," *Huntsville Times*, 9 December 1979; "Boeing Studies Permanent Orbiting Manned Space 'Base Camp,'" *Aerospace Daily,* 10 October 1980, 231.

80 Cecil Gregg, OHI No. 1, p. 13.

81 MSFC Release 77–36, 7 March 1977, STHDC–0820; Darwin Weekly Notes, 4 April 1977, MSI Space Station Chronological File Entry 0968.

82 Huber Weekly Notes, 28 March 1977, MSI Space Station Chronological File Entry 0967; MSFC Release 79–29, 21 March 1979, STHDC–1086.

83 Snoddy OHI, p. 7.

84 Luther E. Powell, OHI No. 1 by Thomas Gates/MSI, 21 March 1989, p. 8.

85 "NASA Sees Spacelab Leading to Space Platforms," *Defense/Space Daily,* 7 September 1979, p. 17.

86 Science and Applications Space Platform (SASP) Report, April 1979, STHDC–1103; MSFC Release 79–63, 20 June 1979, STHDC–1087.

87 MSFC Release 81–28, 10 March 1981, STHDC–1090; MSFC Release 81–68, 17 June 1981, STHDC–1091.

88 Powell OHI No. 1 by Thomas Gates/MSI, 21 March 1989, p. 10.

89 Bob Marshall to Lucas, 19 May 1981, Space Station 1981–1983 folder, Drawer 46, MSFC Center Directors' Files.

90 T. J. Lee, daily journal, Meeting with John Potate, Woody Bethay, Don Dean and Bill Walsh, 6 July 1981, Drawer 117, MSFC Center Directors' Files.

91 Howard E. McCurdy, *The Space Station Decision: Incremental Politics and Technological Choice* (Baltimore: The Johns Hopkins University Press, 1990). McCurdy's book focuses on the politics of the space station decision from the Washington perspective, as does Adam Gruen's unpublished manuscript *The Port Unknown*. Thomas J. Lewin and V. K. Narayanan's "Keeping the Dream Alive: Managing the Space Station Program, 1982–1986," (NASA Office of Management, Contractor Report 4272), 1990, balances

developments in Washington and those at the centers. The present work will treat the political dimension only insofar as it affected Marshall's role in the Space Station Program.

92 *Astronautics and Aeronautics,* 1981, MSI Space Station Chronological File Entry 0737.

93 Captain Robert Freitag, OHI by Adam L. Gruen, NHHO, 22 March 1989, pp. 40–41.

94 Charles Darwin, note to Lucas, 26 February 1982, Space Station Planning 1984 folder, Drawer 46, MSFC Center Directors' Files.

95 John Hodge would later concede that the purpose of the requirements studies was constituency building. Joseph P. Loftus, Memorandum for the Record of Fletcher Committee Meeting held 28 April 1982, 6 May 1982, Chronological Files, Space Station Series, JSC History Office.

96 Mike Weeks of headquarters told Jack Lee on 23 February 1982 that he was "a firm believer that the platform is the way to go," and that Beggs wanted it started in 1984. Lee, Memorandum for the Record, 23 February 1982, Space Station Planning 1984 folder, Drawer 46, MSFC Center Directors' Files.

97 Loftus, Memorandum for the Record, 1 March 1982, Chronological Files, Space Station Series, JSC History Office.

98 Loftus, Memorandum for the Record of Space Station Planning Meetings held 4–5 March 1982, 9 March 1982, Chronological Files, Space Station Series, JSC History Office.

99 Loftus, Memorandum for the Record of Fletcher Committee Meeting held 16 March 1982, 17 March 1982, Chronological Files, Space Station Series, JSC History Office.

100 Bob Marshall to Lucas, 2 March 1982, Space Station Planning 1984 folder, Drawer 46, MSFC Center Directors' Files.

101 Loftus, Memorandum for the Record of Fletcher Committee Meeting held 28 April 1982, 6 May 1982, Chronological Files, Space Station Series, JSC History Office.

102 *Marshall Star*, 2 June 1982, STHDC Entry 0248; Lewin and Narayanan, "Keeping the Dream Alive," pp. 15–17; "Space Station Task Force," Presentation to Center Directors, 7 July 1983, Drawer 46, MSFC Center Directors' Files.

103 Lewin and Narayanan, pp. 28–29.

104 Jerry Craig, JSC, to Robert Freitag, undated, September–December 1982 Correspondence folder, Space Station History Project, Accession number 255–93–0656, NASA HQ History Office; "Why Space Station Lead Systems Assignment to MSFC," undated, Space Station Planning 1984 folder, Drawer 46, MSFC Center Directors' Files. (Although the MSFC document is undated and appears in a 1984 folder, it is clear from internal evidence that it dates from late 1982 or early 1983. Dated documents in the folder also predate 1984.)

105 Bob Marshall to Lucas, 8 February 1983, Space Station 1981–1983 folder, Drawer 46, MSFC Center Directors' Files.

106 The remaining working groups were the Mission Requirements Working Group, the Systems Working Group, and the Operations Working Group. Lewin and Narayanan, pp. 15–16.

107 Beggs to Lucas, 11 February 1983, Space Station 1981–1983 folder, Drawer 46, MSFC Center Directors' Files.

108 Brian Pritchard, Space Station Task Force, to Distribution, 9 February 1983, Space Station 1981–1983 folder, Drawer 46, MSFC Center Directors' Files.

109 *Marshall Star*, 4 May 1983.

110 Luther E. Powell, OHI No. 2 by Thomas Gates/MSI, 30 March 1989, p. 10.

111 Powell OHI No. 1, p. 18; Powell OHI No. 2, pp. 8–9.

112 Powell, OHI No. 2, pp. 11–16.

113 McCurdy, *The Space Station Decision*, p. 204.

114 Unsigned notes, Space Station Management Colloquium, August to September 1983, Space Station Management Colloquium folder, Drawer 46, MSFC Center Directors' Files.

115 Lewin and Narayanan, pp. 37–38; "Meeting Objective" (Notes on MSFC SE&I position), 4 November 1982, Space Station Planning 1984 folder, Drawer 46, MSFC Center Directors' Files.

116 *Ibid.*, Chart labeled "Task No. 1," comparing "Negative Impact" and "Positive Impact."

117 Luther Powell, OHI No. 5 by Thomas Gates/MSI, 9 May 1989, pp. 1–3.

118 Jesse W. Moore to William R. Graham, 21 April 1986, April 1986 File, Yellow Chron, box 6 of 11, Space Station History Project, NHHO.

119 Howard E. McCurdy, The Space Station Decision: Incremental Politics and Technological Choice (Baltimore: The Johns Hopkins University Press, 1990), pp. 177–91; Charles R. Chappell to Mike Wright, comments on chapter draft, 30 November 1993, MSFC History Office.

120 *Marshall Star,* 1 February 1984, STHDC Entry 0030.

121 Beggs to Griffin, 15 February 1984, Space Station 1984 folder, Drawer 46, MSFC Center Directors' Files; *Huntsville Times*, 16 February 1984.

122 *Huntsville Times,* 16 February and 7 March 1984. Beggs reassured Heflin that Marshall's capabilities would be "fully utilized" in the space station program, and that the center was "clearly a candidate for Space Station hardware development in a number of areas." Beggs to Heflin, 5 March 1984, Space Station Documentation (1981–) folder, NHHO.

123 John D. Hodge, Statement before the Subcommittee on Science, Technology and Space, Committee on Commerce, Science and Transportation, 15 November 1983, Space Station 1981–1983 folder, Drawer 46, MSFC Center Directors' Files; Culbertson to Center Directors, 27 February 1984, Space Station 1984 folder, Drawer 46, MSFC Center Directors' Files; Culbertson, cited in Lewin and Narayanan, p. 50.

124 MSFC Release 84–11, 9 February 1984 (STHDC Entry 0209); *Marshall Star,* 7 March 1984 (STHDC Entry 0158); *Huntsville Times,* 30 March 1984; Lewin and Narayanan, pp. 49–52.

125 "Space Station Work Package Discussion with Associate Deputy Administrator," 2 March 1984, Space Station 1983 folder, Drawer 46, MSFC Center Directors' Files.

126 Lewin and Narayanan, pp. 54–57.

127 "Space Station Work Package Discussion with Associate Deputy Administrator," 2 March 1984, Space Station 1983 folder, Drawer 46, MSFC Center Directors' Files.

128 Griffin to Culbertson, 31 May 1984, December 1984 Correspondence folder, box 3 of 11, Space Station History Project Collection, NASA HQ History Office.

129 Thomas J. Lee, OHI by AJD and SPW, 1 September 1993.

130 Handwritten notes attached to agenda "Space Station Center Directors" Meeting, JSC," 23 March 1984, Space Station Planning 1984 folder, MSFC Center Directors' Files.

131 Luther E. Powell, OHI No. 3 by Thomas Gates/MSI, 6 April 1989, p. 13.

132 Griffin to Culbertson, 31 May 1984.
133 Hodge to Culbertson, "JSC Response to Work Package Division Assignment," undated but June 1984 from internal evidence, December 1984 Correspondence folder, box 3 of 11, Space Station History Project Collection, NASA HQ History Office.
134 Lee notes, 11 June 1984, Space Station Level C June 1984 folder, Drawer 46, MSFC Center Directors' Files.
135 Culbertson to Work Package Center Directors, 18 June 1984, Space Station 6/24/84 folder, Drawer 46, MSFC Center Directors' Files.
136 "Recommended Approach," undated, Space Station 6/24/84 folder, Drawer 46, MSFC Center Directors' Files.
137 Lucas notes, 22 June 1984, Space Station Level C June 1984 folder, Drawer 46, MSFC Center Directors' Files.
138 Powell, OHI No. 3, p. 17.
139 Randy Humphries, OHI by Thomas Gates/MSI, 17 March 1988, pp. 23–24.
140 Culbertson to Distribution, 12 July 1984, Space Station Management Plan and Procurement Strategy, Letter to Mr. Culbertson folder, Drawer 46, MSFC Center Directors' Files; Lewin and Narayanan, pp. 58–60.
141 NASA Release 84–85, 28 June 1984, Space Station Management Plan and Procurement Strategy, Letter to Mr. Culbertson folder, Drawer 46, MSFC Center Directors' Files.
142 NASA Release 84–85, 28 June 1984; Charles Darwin to Lucas, 13 August 1984, Space Station 1984 folder, Drawer 46, MSFC Center Directors' Files.
143 Mark Craig, "Space Station Configuration Chronology (1983–1987)," Presentation to National Research Council, 14 July 1987, NRC Space Station Study Group folder, Subject File—Space Station, Dale D. Myers Papers, NHHO.
144 STHDC Entry 0163; Marshall Star, 24 October 1984.
145 *Marshall Star,* 20 March 1985, STHDC Entry 0266; *Huntsville News,* 29 August 1985, STHDC Entry 0699; *Huntsville Times*, 29 August 1985, STHDC Entry 0700; "Space Systems Highlights: Space Station," 16 September 1985, Correspondence September 1985 folder, box 5 of 11, Space Station History Project, NASA HQ History Office.
146 STHDC Entry 0537, 4 February 1985, A&A 1985; STHDC Entry 0408, 17 March 1985, from Gowan/Morris Chronology; STHDC Entry 0539, 12 April 1985, A&A 1985.
147 Cecil Gregg, "MSFC Project Manager's Monthly Summary to the Administrator," July 1985, Space Station July–December 1985 folder, Drawer 46, MSFC Center Directors' Files; *Astronautics and Aeronautics*, 25 October 1985, STHDC Entry 0550; Cecil Gregg, OHI No. 2 by Thomas Gates/MSI, 18 December 1989, p. 1.
148 Culbertson, Memo for the Record re: Meeting with Peter Drucker, 22 January 1985, Space Station January–June 1985 folder, Drawer 46, MSFC Center Directors' Files.
149 James B. Odom, OHI by AJD and SPW, 26 August 1993.
150 Freitag, Memorandum for the Record, Space Station Management Council Notes for 13 June 1985, 18 July 1985, Space Station July–December 1985 folder, Drawer 46, MSFC Center Directors' Files.
151 Richard F. Carlisle to SE staff (based on Space Station Engineering Division personnel inputs, 29 August 1985), 23 September 1985, Correspondence August 1985 folder, box 5 of 11, Space Station History Project, NASA HQ History Office.
152 *Ibid.*; Gruen, pp. 129–131.

153 Culbertson to Hutchinson, 14 August 1985, Space Station July–December 1985 folder, Drawer 46, MSFC Center Directors' Files.
154 Powell to Lucas, 25 September 1985, Space Station (Don Hearth), 3 October and 23 October 1985 folder, Drawer 46, MSFC Center Directors' Files.
155 Cecil Gregg, OHI No. 2 by Thomas Gates/MSI, pp. 17, 18.
156 Luther Powell, OHI No. 4 by Thomas Gates/MSI, 14 April 1989, p. 13.
157 Lucas notes, Space Station Management Council meeting, 24 October 1985, Space Station Management Council 24 October 1985 folder, Drawer 106, MSFC Center Directors' Files.
158 Culbertson wrote to Beggs, asking him to remind the centers of their September 1983 agreement that the centers would work together under a lead center. STHDC Entry 0438, 14 November 1985, citing Gowan/Morris-Chronology.
159 STHDC Entry 0552, 4 December 1985, based on A&A, 1985.
160 Freitag, Minutes of Level A/B Program Strategy meeting on 23–24 January 1986, 5 February 1986, Program Office Correspondence, box 3, Space Station Series, JSC History Office.
161 STHDC Entry 0117, 28 January 1986, citing Gowan/Morris-Chronology.
162 STHDC Entry 0465, 28 March 1986, citing Gowan/Morris-Chronology.
163 Hodge to Distribution, 24 February 1986; Bensimon to Distribution, 27 February 1986, Space Station January–June 1986, Drawer 46, MSFC Center Directors' Files; Lewin and Narayanan, p. 112.
164 Jesse W. Moore to William R. Graham, 21 April 1986, April 1986 File, Yellow Chron, box 6 of 11, Space Station History Project, NHHO.
165 STHDC Entry 0565, 19 March 1986, cited from Space Station Management Council Minutes.
166 Lucas to Hodge, 14 March 1986, Space Station January–June 1986, Drawer 46, MSFC Center Directors' Files.
167 Powell, OHI No. 4 by Thomas Gates/MSI, p. 5.
168 Hodge, "Space Station Work Package Recommendation," 8 April 1986, Space Station January–June 1986, Drawer 46, MSFC Center Directors' Files.
169 Lewin and Narayanan, pp. 116–117.
170 STHDC Entry 0630, 16 June 1986, citing from "SS Management Review."
171 Powell, OHI No. 4 by Thomas Gates/MSI, p. 2.
172 Fletcher handwritten notes on D.P. Hearth, "Final Report, Space Station SE&I," 24 October 1985, Space Station folder (one of several with this title), Fletcher Papers, NASA HQ History Office.
173 Phillips, "Management Review of Space Station Program," 26 June 1986; NASA, "Space Station Management: An Analysis of Work Package Options," report to Committee on Science and Technology, U.S. House of Representatives, 3 October 1986, Space Station January–June 1986, Drawer 46, MSFC Center Directors' Files.
174 "JSC Space Station Program: Program Responsibility Changes," 12 July 1986, Space Station folder, Fletcher Papers, NASA HQ History Office.
175 "New Fears of Job Losses at Space Center," *Houston Post*, 15 July 1986; "Many Puzzled by NASA Plan to Move Jobs," *Houston Chronicle*, 15 July 1986; "Bush, Texans Asked to Find NASA Plans," *Houston Chronicle*, 23 July 1986; Lewin and Narayanan, pp. 119–120.

176 *Huntsville Times*, 31 July 1986.

177 "NASA Details Space Station Plans; Houston, Huntsville Share Work," *Washington Times*, 26 September 1986; Martin Burkey, "The Marshall-Johnson Feud Draws to an Equitable End," *Huntsville Times*, 28 September 1986.

178 O'Keefe Sullivan, OHI by Thomas Gates/MSI, 31 July 1989, pp. 11–12.

179 Charles A. Cothran, OHI by Thomas Gates/MSI, 27 October 1988, p. 20.

180 Mark Hess, "Space Station Analysis Results," NASA Release 86–81, 23 December 1986.

181 James C. Miller III, Director, OMB, Memorandum to the President, 10 February 1987, Space Station—New Cost Estimates folder, Subject File—Space Station, Dale D. Myers Papers, NASA HQ History Office.

182 Hodge to Myers, undated but apparently early 1987, Space Station Costs folder, Subject File—Space Station, Myers Papers, NASA HQ History Office.

183 Dale D. Myers, "The Austere Space Station," 25 January 1987, Space Station folder, Subject File—Space Station, Myers Papers, NASA HQ History Office.

184 Myers, handwritten notes entitled "Space Station Options," undated but accompanying (and relating to) 25 January 1987 "Austere Space Station" memo, Space Station folder, Subject File—Space Station, Myers Papers, NASA HQ History Office.

185 Theresa M. Foley, "Space Station Faces Possible Two-Year Deployment Delay," *Aviation Week & Space Technology* (9 February 1987), pp. 28–29; Foley, "White House Delays Action on New Station Cost Estimates," *Aviation Week & Space Technology* (2 March 1987), pp. 26–27; Foley, "NASA Will Proceed with Scaled-Back Space Station," *Aviation Week & Space Technology* (30 March 1987), pp. 26–27.

186 *New York Times*, 15 September 1987.

187 *Marshall Star*, 15 April 1987; STHDC Entry 0659.

188 STHDC Entry 0271, 5 May 1987, citing NASA Administrative Memo of this date.

189 Lee OHI by AJD and SPW, 1 September 1993; John D. Holt, OHI by AJD, Houston, Texas, 3 August 1993.

190 STHDC Entry 0274, based on *Huntsville Times*, 7 July 1987.

191 "Report of the committee on the Space Station of the National Research Council" (Washington, DC: National Academy Press, 1987).

192 STHDC Entry 0189, 22 April 1987, citing *Aerospace Daily*, 27 April 1987 and I&PS Chron (1987).

193 Gregg, OHI No. 2 by Thomas Gates/MSI, p. 20.

194 Susan Cloud to Tom Newman forwarding draft of Mark Hess's NASA Release 87–XX, 24 April 1987, Space Station Exercises folder, Subject File—Space Station, Myers Papers, NASA HQ History Office.

195 STHDC Entries 0287 and 0278, citing Huntsville Times, 20–21 July 1987.

196 Gregg, OHI No. 2 by Thomas Gates/MSI, pp. 20–28.

197 JSC's Work Package Two award went to McDonnell Douglas; Goddard's Work Package Three went to General Electric; and Lewis's Work Package Four went to Rocketdyne. Fletcher, "Selection of Contractors for Space Station (Phase C/D)," 15 December 1987, Space Station Documentation (1987–) folder, NASA HQ History Office; "Space Station Detailed Design and Development," from Terence T. Finn, "The Space Station," 12 February 1988, Space Station Documentation (1987–) folder, NASA HQ History Office. STHDC Entry 0272, 1 December 1987, citing *Huntsville Times*, 1 December 1987.

198 Humphries OHI, p. 7.
199 Humphries OHI, p. 20.
200 Hopson OHI, pp. 2, 11.
201 Hopson OHI, pp. 2, 3.
202 Axel Roth, OHI by Thomas Gates/MSI, 27 July 1988, pp. 9–13.
203 Walt Wood, OHI by Thomas Gates/MSI, 17 October 1989, pp. 9–13.
204 Hopson OHI, pp. 13–14.
205 Robert Crumbly, OHI by Thomas Gates/MSI, 25 October 1988, p. 13.
206 Crumbly OHI, p. 7.
207 Holt OHI.
208 Lee, OHI by AJD and SPW, 1 September 1993.
209 James B. Odom, OHI by AJD and SPW, Huntsville, Alabama, 26 August 1993.
210 Odom, OHI by AJD and SPW; Lee, OHI by AJD and SPW, 1 September 1993.
211 Holt OHI.
212 James B. Odom, OHI by Thomas Gates/MSI, 19 April 1990, pp. 2–5.
213 STHDC Entry 0837, 17 January 1989, citing Space Station News, Vol. 3, No. 3.
214 Lee, OHI by AJD and SPW, 1 September 1993.
215 STHDC Entry 0915, citing *Birmingham News*, 11 May 1989.
216 Charles R. Chappell/ES51, OHI Thomas Gates/MSI, 24 September 1987, p. 6.
217 Odom, OHI by AJD and SPW.
218 Lee, OHI by AJD and SPW, 1 September 1993.
219 NASA, "*International Space Station*: U.S. Space Station History," 14 September 1995, http://issa-www.jsc.nasa.gov/ss/prgview/iss2fact.html; NASA, Fact Book, "The *International Space Station*: A Critical Investment in America's Future," 1 June 1995.
220 Lee, OHI by AJD and SPW, 1 September 1993.
221 NASA, "*International Space Station*: Phase I–III Overview," IS–1995–08–ISS003JSC (August 1995).
222 Lucas notes, 22 June 1984, Space Station Level C June 1984 folder, Drawer 46, MSFC Center Directors' Files.
223 "Space Station This Week," 16 September 1996, NASA Spacelink at http://spacelink.msfc.nasa.gov.
224 Mike Wright, telephone conversation with AJD, 23 September 1996.

Chapter XIV

Conclusion 1960–1990

During its first 30 years, the Marshall Space Flight Center was at the center of many of NASA's most important endeavors. Marshall people helped NASA plan explorations of space, develop complex technologies, and contribute to scientific progress. At each step, they encountered uncertainties because NASA, more than any federal Agency, was charting unexplored territory. In following their dreams and in responding to opportunities and directives, Center personnel shaped their future and the future of American space exploration.

Uncertainties faced NASA from the beginning. In the late 1950s, America's political leaders and space managers debated various plans for space policy. They discussed whether space budgets should be large or small, whether the military or a civilian agency should be primary, whether spacecraft should be robotic or piloted, whether exploration should be Earth orbital or interplanetary. While still in the Army Ballistic Missile Agency, future Marshall personnel contributed to the debates by publicizing their visions of new space technology. With Wernher von Braun leading the way, the engineers and scientists devised concepts of big rockets, space stations, scientific spacecraft, orbiting telescopes, lunar rovers, and lunar outposts. Over the next decades the space team in Huntsville oversaw the conversion of many of these visions into space hardware.

The first steps from dreams to reality came after American policy makers made space exploration an arena for peaceful competition during the Cold War. They wanted a space program that could demonstrate American political, organizational, technological, and scientific superiority over the Soviet Union. The Army's missile specialists in Huntsville became a tremendous pool of talent that could help achieve these national goals. While still in the Army, the team was virtually a space agency in miniature; it developed the Jupiter–C launch vehicle and helped develop the spacecraft for Explorer I, the first American scientific satellite, and began work on the Saturn family of rockets, a new generation of large launch vehicles intended primarily for civilian payloads. In 1959 the Army agreed to transfer the missile team to the National Aeronautics and Space Administration, the civilian space organization formed the previous year. In 1960 the Marshall Space Flight Center formally became a NASA field Center.

The first decade at Marshall centered on the Apollo lunar program, and the Center successfully overcame several daunting technical, organizational, and political challenges. The political consensus supporting the Apollo mission facilitated NASA's efforts, and Marshall's engineers benefited from expandable budgets, clear technical goals, and a fixed schedule. Within this secure political environment, the Center's engineering laboratories designed, developed, tested, and operated the Saturn launch vehicles, especially the Saturn V rockets that lifted astronauts to the Moon. To cope with the enormous technical demands of Saturn, Marshall built new facilities, hired more experts, and enhanced its capabilities in systems engineering and project management. Their efforts were so successful that the Saturns never experienced a launch failure, and NASA met President Kennedy's end-of-the-decade deadline for Apollo.

Beyond the addition of personnel and capabilities, the Apollo Program helped change the Center's organizational culture and the political economy of the space program. NASA required that Marshall privatize most Saturn work, using the Apollo program to demonstrate the strengths of a public-private partnership and to spread the largesse of space spending across the political landscape. Consequently the Center moved away from the Army Arsenal system, which developed prototypes and some flight hardware in-house, and toward the Air Force system, which relied on contractors. Moreover NASA and the Johnson Administration directed the Center to pioneer new race relations, a directive Marshall carried out well enough to help remove most legal barriers to equal opportunity in Madison County. The tremendous successes of the Apollo Program convinced Center personnel and many Americans that NASA could overcome any challenge.

Even before the lunar landings ended, however, NASA began experiencing uncertainties that helped create a crisis for Marshall. Beginning in the mid-1960s, the Agency planned new missions to follow Apollo, but no new program had the political mandate that had supported NASA's lunar missions. At the same time, the Marshall Center was finishing Saturn development and its personnel were ready for new challenges. Faced with declining budgets and work, the Center experienced a long institutional crisis. From the late sixties until the mid-seventies, Marshall laid off hundreds of workers, and NASA Headquarters even discussed closing the Center.

In response to the crisis imposed from outside, and to pursue their perennial dreams of space exploration, Marshall people recognized that they had to find new tasks. Consequently Marshall reorganized to compete with other NASA field Centers for new projects and diversify outside of their propulsion specialty. In 1968 the Center created a Program Development Office which helped technical specialists from the labs devise preliminary plans and designs, and thus win new projects. In 1974, Marshall formed a more flexible laboratory organization to facilitate cooperation of specialists drawn from several different labs and to solve the complex problems of diverse projects.

With this new organization, Marshall branched beyond propulsion and successfully diversified into spacecraft engineering and space science. The most dramatic early achievement of diversification was *Skylab*, America's first space station. The Center oversaw construction of *Skylab* from a Saturn V upper stage, and built many of its subsystems, including the Apollo Telescope Mount (ATM). It also supported *Skylab*'s myriad scientific experiments, from the sophisticated solar studies of the ATM to the simple observations of spider web formation in a student experiment. Before the end of the 1970s, Marshall people oversaw development of the lunar roving vehicle, three high-energy astronomical observatory satellites, a general relativity experiment, a geophysics satellite, and solar energy and coal mining research.

In conceiving and winning new tasks, Marshall ensured its survival and became NASA's most diversified field Center. By the mid-1980s the Center had engineering expertise in launch vehicles and orbital transportation, materials and processes, structures and dynamics, automated systems, data systems, and spacecraft design. Marshall also had scientific expertise in microgravity science, astronomy and astrophysics, solar physics, magnetospheric physics, atomic physics and aeronomy, and earth science and applications.

This expertise resided in the Center's laboratories. While lab scientists and engineers had always supported major projects, such support activities were only a portion of their work. To Associate Director for Science Charles R.

Chappell, the laboratories were the heart of the Center. He compared Marshall's wide-ranging activities to an iceberg, with work on the major projects—Shuttle, Space Station, Spacelab—as the visible tip. Below the surface, spreading wide and deep, were the Center's research and technology programs. The following survey is far from exhaustive, but gives an indication of the Center's vast capabilities.

Just as Wernher von Braun's vision defined Marshall in its early years, in 1990 Marshall's vitality rested on a foundation of imagination. The Center's advanced studies helped NASA conceive future space programs, and generated innovations in research and technology. Space Station work in the 1980s comprised only a portion of the advanced studies conducted at the Center. Marshall also pursued work in transportation systems, space systems, and data systems. Development of new transportation systems to supersede the Shuttle were the most ambitious projects under consideration at Marshall. The Center conducted in-house studies and worked with other NASA Centers, government agencies, and contractors to define the next generation of launch systems and vehicles. Two propulsion projects, the space transportation main engine and the space transportation booster engine, envisioned employing liquid propellants for the next generation of launch vehicles. In related efforts, the Center conducted propulsion studies examining alternative propellants, including varieties of liquids and solids, hydrocarbon, and low-cost auxiliary booster/core systems using liquid-oxygen tank pressurization separate from the engine.

Advanced studies also sought to develop experiments and hardware to further research in space science and applications. Charles Darwin of Marshall's Program Development Directorate described the systems under investigation as "large astronomical observatories that would succeed or complement the Great Observatories, Earth and microgravity science instruments and facilities, geostationary platforms, and a variety of Space Station . . . payloads."[1] These experiments had both theoretical and practical goals. One of the theoretical projects was a spacecraft called Gravity Probe–B, an experiment in gravitational physics designed to test Einstein's theory of relativity by using four precision gyroscopes designed to detect minute changes in the structure of space and time. AXAF, the Advanced X-Ray Astrophysics Facility, was a 43-foot-long, 20,000-pound spacecraft designed to gather data on x-rays over the course of a 15-year lifetime. It included an optical system eight times as precise as that of HEAO–2, an experiment flown in 1978. Several advanced studies projects involved the use of tethers, long cables deployed between the Shuttle (or a later vehicle) and a satellite that could be used to carry electrical current, to transfer momentum from the Shuttle to the payload (and thus lift the payload to a higher orbit), or ultimately to help maintain the orbit of a Space Station.[2]

Advanced systems projects included new data systems to facilitate the collection, display, access, manipulation, and dissemination of information for various NASA efforts. Marshall supported a four-dimensional display program for the Man-computer Interactive Data Access System (McIDAS) developed by the University of Wisconsin, and explored ways to use the system for NASA's Earth Science program. Marshall also helped with another Earth sciences data system, WetNet, a system that integrated data from satellites and ground stations in order to evaluate the global moisture cycle.[3]

Marshall's involvement in research programs extended back to the pre-NASA days, when the Army Ballistic Missile Agency (ABMA) helped develop and launched Explorer I. Since then, *Skylab*'s Apollo Telescope Mount, HEAO, and the diverse payloads flown aboard Spacelab gave testimony to the Center's pathbreaking work in various emerging fields of space science.

Marshall scientists were among the principal investigators for microgravity experiments aboard Spacelab, and the Center's scientists also conducted ground-based microgravity experiments using Marshall's 105-meter Drop Tube/Drop Tower and NASA's KC–135 aircraft. They developed crystal growth experiments designed to produce new materials for technology applications and protein crystals to facilitate the development of new drugs. Other experiments investigated the effect of space processing techniques on materials in microgravity, including undercooling (cooling to below the normal temperature for solidification) and the rate of cooling, and separation techniques for proteins and other biological materials.[4] The microgravity experiments promoted progress in biology, medicine, and technology.

Marshall began managing, developing, and conducting research in the fields of astronomy and astrophysics since *Skylab* flew in the early 1970s. Development of AXAF had opened new possibilities for broader involvement, but Marshall had long been at work in infrared astronomy, relativity, and cosmic-ray research. In addition to devising experiments, Marshall worked at developing new x-ray and infrared detectors.[5]

Solar physics research at Marshall centered on examining the solar magnetic field, including studies of the properties of the field, the energy stored there, and the means by which that energy was released. Research included investigations of the solar corona, with the intent of learning more about solar flares, about the solar wind (the expansion of gasses in the corona), and about why the corona is 500 times hotter than the surface of the Sun.[6]

Magnetospheric physics investigated the volume of space influenced by Earth's magnetic field and studied how that field is influenced by the Sun. Scientists at Marshall examined the influence of the solar events and the solar wind on the magnetosphere and how they dispersed plasma outward into space. They concentrated on the "observation of low-energy or core plasma which originates in the ionosphere and has been found to supply plasma for the entire magnetosphere." They developed experiments, hardware, and software to measure plasma flow and evaluated data from previous Shuttle and satellite missions.[7]

Aeronomy examines the interaction between gasses in Earth's upper atmosphere and the Sun's electromagnetic and corpuscular radiation. By gathering data from instruments carried by balloons, on satellites, and on the Shuttle, Marshall scientists were able to learn more about the nature of Earth's atmosphere by studying photochemical and dynamical processes in the ultraviolet and infrared regions of the spectrum.[8]

One of the applied research programs at Marshall was the Center's portion of NASA's Mission to Planet Earth. The Marshall Earth Science and Applications program applied space technology to study Earth's atmosphere, land surface, and oceans. Activities included the invention of theoretical models, creation of remote sensing instruments, analysis data gathered from satellite and Shuttle flights, design of simulations, and experimentation on Earth, in the near-Earth atmosphere, and from spacecraft. Marshall's Global Hydrology and Climate Center, for example, developed sensors in support of the Earth Observing System; one of these instruments, the lightning imaging sensor, examined the global distribution of lightning. Other Earth sciences investigations involved studies of temperature variations, observations of soil and snow properties, atmospheric modeling, studies of Earth's hydrological cycle, and climate dynamics. The many direct applications of Earth science projects included predictions of weather and violent storms, and the availability of data for decisions regarding water use.[9]

If Marshall's research capability demonstrated ways in which the Center added breadth, its technology programs showed how the Center increased depth in areas of traditional strength. Marshall originated as a propulsion team, and at the beginning of the fourth decade the Center remained in the vanguard of propulsion engineering. In the late 1980s, in the aftermath of the *Challenger* accident, Marshall's Propulsion Laboratory contributed to redesign of the Shuttle solid rocket motor, but also worked on improvements in the Shuttle main engine.

The solid rocket motor redesign effort at Marshall was a high profile activity, and the Propulsion Lab contributed in many ways to returning the Shuttle to flight. The lab built a tool to measure roundness of the solid rocket motor case, and Morton-Thiokol immediately put it to use. Engineers helped develop a new material to use as a sealant to replace that previously used in the O-rings. They applied computational methods to the internal flow analysis of the booster to detect possible localized burning pockets. The Propulsion Lab was involved at every step of redesign.

Marshall's Science and Engineering laboratories also worked on improvements to the Shuttle main engine. The Materials Laboratory sought a solution for the problem of the cracking of turbine blades that continued to plague the main engine, and the Dynamics Laboratory developed a computational fluid dynamics model to study the problem. Fuel flow within the engine was always a complex problem. Engineers devised a meter to measure fuel flow in the engine, began developing a diagnostic system to measure flow at the nozzle exit, and devised means to simulate the inherently instability caused by relative motion between rotor and stator airfoils.[10]

Advanced welding techniques were among the activities pursued in Marshall's Materials and Processes Laboratory. The variable polarity plasma arc welding system was one of the significant advances in welding technology to come out of Marshall's labs, and in the late 1980s the lab developed a mathematical model to evaluate and improve the system. X-rays of welds on the Shuttle's external tank occasionally showed fine lines, and after years of investigation the lab duplicated the lines and identified their cause. Another project related to the external tank was the invention of an improved foam insulation coating. To complement the Propulsion Lab's work on turbine blade fracture, the Materials and Processes Laboratory devised a new computer code for fracture mechanics analysis. When the Shuttle main engine's high-pressure oxidizer turbopump end bearing failed to meet its design life requirements, the lab devised a bearing tester to evaluate bearing performance.[11]

The Structures and Dynamics Laboratory included facilities for testing, analyzing, and improving the dynamics and structural integrity of systems developed at Marshall. The lab evaluated the effect of such variables as load weight and distribution, temperature, vibration, fluid dynamics, strength, and durability on systems components developed at the Center. Structures and Dynamics activities ranged from designing structures and assembly techniques to developing pointing control systems, life support systems, and thermal protection systems. Its work in thermal protection systems, for example, led the lab's Productivity Enhancement Facility to create a simulation system to improve the application of spray-on foam insulation to the Shuttle external tank. The Space Station program drew on the lab's expertise to study means by which the station could contend with the threat posed by space debris and micrometoroids.[12]

Docking simulation in Marshall's Teleoperation and Robotics Research Facility.

Robotics was central to the development of new NASA systems. Marshall contributed by pioneering robotic methods of docking and remote servicing of orbital platforms. The Center's Orbital Hardware Simulator Facility gave testimony to the latest generation of sophisticated robotic technology. The facility's Dynamic Overhead Target Simulator (DOTS) operated in eight degrees of freedom, and could position a 1,000-pound load to within an accuracy of one-quarter inch. Operating in conjunction with the Air Bearing Mobility Unit, DOTS could support a variety of docking and stationkeeping operations.[13]

Marshall's Space Systems Lab performed a wide variety of tasks. Its engineers provided support for Space Station, AXAF, and Spacelab. In support of Space Station, they developed the system for the distribution of power in habitation and laboratory modules, a complex system that required the invention of computerized processors and artificial intelligence systems. They also help develop technology for the development and integration of experiments and instruments. They also helped developed methods of welding in space, created a lightweight composite intertank for the advanced launch system, and designed the technology mirror assembly for AXAF.[14]

Advanced technology solar array tested in space.

Marshall's Astrionics Laboratory had experts on electrical systems, instrumentation and control, computers and data management, software, optics, avionics simulation, and electrical power. These engineers contributed to the subsystems of virtually all of Marshall's projects. Among the lab's projects was the autonomously managed power system (AMPS), a complex apparatus designed to control spacecraft without commands from the ground. It involved subsystems for fault detection and recovery, load management, status and control, and an expert system for fault monitoring.[15]

It took three decades to build an organization of such wide-ranging capabilities. Spinning off from its propulsion specialty, the Center developed diverse skills by contributing to NASA's most ambitious and complex projects of the seventies and eighties. The Center extended its expertise in rocketry by its work on the Space Shuttle, helping produce reusable liquid-fuel engines and solid rocket motors. Marshall oversaw development of the Hubble Space Telescope with its complex interfaces and precise systems of optics and pointing and control. The Center worked with the European Space Agency on Spacelab which became the embodiment of Marshall's diversification; this experiment module for the Shuttle combined the Center's expertise in spacecraft and scientific instrument engineering, human systems, and research in multifarious scientific and technical disciplines. Marshall also drew on all its skills in its contributions to the Space Station, helping NASA conceive a configuration that could be constructed in space, carried in the Shuttle, capable of sustaining a crew and supporting experiments for decades, and salable to Congress.

Space Shuttle, Hubble Space Telescope, Spacelab, and Space Station projects had common political and technical features which produced more complicated challenges than those Marshall faced during Apollo. The technology and technical interfaces were much more complex after the 1960s. The Shuttle orbiter and propulsion system were designed as one unit while the Saturn boosters and Apollo spacecraft had been designed separately with guidance-and-control as the main interface, and Space Station designs multiplied the complexities of the Shuttle program. Even as Marshall's technical challenges grew, the Center lost the advantages of the Arsenal system and in-house manufacturing capability. Development was in the hands of contractors and measuring their performance became more difficult because Marshall could not build prototypes to use as "yardsticks." Nor could the Center address technical problems by hiring new experts because of personnel policies and austere budgets.

Moreover, Marshall personnel had to adjust to political and financial decisions that imposed severe restraints on their technical work. In the seventies and eighties, mission goals and hardware designs were more subject to external constraints and changes, mainly because Congress exercised greater scrutiny over NASA and was more willing to slash budgets. No longer did NASA have a privileged status as part of the struggle against Communism. For instance, Congress backed and funded Apollo in the sixties, but throughout the eighties kept questioning the Space Station and limiting its budget. After the 1960s, Congress would usually not give NASA the extra money needed to meet the unexpected costs typical in research and development. To cope with the budgetary shortfalls, NASA reduced tests and prototypes, stretched schedules, and restructured the project to cut costs. For example, while NASA had received sufficient funds to meet Apollo's end-of-the-decade deadline, unrealistic budgets caused the Hubble Space Telescope to fall years behind the original schedule.

In addition in the seventies and eighties, Marshall's organizational environment became more complicated than the sixties. The Center worked with other NASA Centers, multitudinous contractors and universities, other federal agencies (especially the Department of Defense), and foreign space agencies. Coordinating these complex coalitions was often difficult because each entity wanted to maintain independence, hide problems, or impose ideas on the others. Work with multinational partners introduced diplomacy as another factor in NASA decisions.

In a different way, NASA's travails with the Space Station in the eighties revealed the complex and uncertain environment in which Marshall worked. When Congress cut funding and forced redesign, international partners felt the effect. Redesigns, reorganizations, and annual congressional votes on whether to continue work and how much money to appropriate stretched schedules and forced Marshall to be flexible and resourceful.

Marshall's journeys to the heavens were further complicated by disasters and false starts in the 1980s. The *Challenger* accident and the Hubble mirror flaw demonstrated how rigorous procedures could not eliminate human error from a complex technical endeavor. Prior to each event, Marshall and its contractors had struggled with difficult questions about how to balance spending between hardware development and ground tests, devise realistic tests, interpret technical data, report complicated engineering evaluations, and extend communications. After each event, Marshall strove to learn engineering and management lessons and thus to avoid repeating the problems. The Center improved quality practices and communications and emerged stronger than it had been before.

Marshall overcame most of the challenges and constraints of the 1980s; its projects led to significant advances in space exploration and science. The Center redesigned the Shuttle propulsion system, and soon the Space Shuttle and Spacelab were again providing regular access to Earth orbit. After NASA corrected the flaws in Hubble's optics, taking advantage of how Marshall had designed the satellite for repair in space, the space telescope gave new insights into the far reaches of the universe.

Marshall and NASA in 1990 were passing through an era as uncertain as the late 1950s or the early 1970s. While using the past to predict the future is risky, the previous periods of uncertainty do provide some harbingers of events to come. In the past the engineers and scientists in Marshall's laboratories had proposed ideas for new missions, launch vehicles, experiments, and spacecraft, thus inventing new ways for NASA to fulfill its mission of space exploration. As a result of diversification, the Center in 1990 had great expertise and was ready to undertake grand endeavors. And as in earlier eras of uncertainty, decisions on the use of this resource rested outside the Center.

Aerial view of MSFC looking south in 1992.

---

1 C.R. Darwin, "Advanced Studies," in Research and Technology 1988: Annual Report of the Marshall Space Flight Center (NASA TM–100343), p. 1.
2 Research and Technology 1985 (NASA TM–86532), pp. 9–24; Research and Technology 1988, pp. 19–42.
3 Research and Technology 1989: Annual Report of the Marshall Space Flight Center (NASA TM–100369), pp. 43–57.
4 Research and Technology 1988, pp. 48–58; Research and Technology 1989, pp. 60–76.
5 Research and Technology 1989, pp. 77–81.
6 Research and Technology 1989, pp. 82–95.
7 Research and Technology 1989, p. 96.
8 Research and Technology 1989, pp. 107–110.
9 Research and Technology 1989, pp. 111–165.
10 Research and Technology 1988, pp. 124–158; Research and Technology 1989, pp. 168–207.
11 Research and Technology 1989, pp. 208–227; Research and Technology 1988, pp. 159–72.
12 Research and Technology 1989, pp. 228–234.
13 Research and Technology 1989, p. 180.
14 Research and Technology 1989, pp. 243–259.
15 Research and Technology 1989, pp. 261–62.

Appendix A:

Center Chronology

31 January 1958	Jupiter C launched Explorer I, first United States satellite
1 July 1960	Marshall Space Flight Center (MSFC) established
1961	First Mercury-Redstone launch with live chimpanzee payload
1961	First -manned Mercury-Redstone launch and suborbital flight
1961	President John F. Kennedy set goal of manned lunar landing by end of the decade
7 September 1961	NASA chose Michoud Ordnance Plant near New Orleans for production of the Saturn S-I Stage and put it under the technical direction of MSFC
October 1961	NASA created the Mississippi Test facility under direction of MSFC
27 October 1961	First Saturn I launched
1962	MSFC Launch Operations Directorate at Cape Canaveral, Florida became an independent NASA Center
July 1962	MSFC acquired Slidell Center Computer Facility in Slidell, Louisiana to service Michoud Operations
1 September 1963	MSFC reorganization established two directorates: Research and Development Operations and Industrial Operations.
1965	Huntsville Operations Support Center established

16 February 1965	A Saturn I launched the first of three Pegasus micro-meteoroid detection satellites
17 February 1966	First test firing of the S–IC–1 for 40.7 seconds
26 February 1966	AS–201, the first Saturn IB flight vehicle, successfully launched from Cape Kennedy
9 November 1967	Apollo 4, first Saturn V, SA 501 launched
1968	Neutral Buoyancy Simulator completed
1969	Major MSFC reorganization establishing directorates in Program Development, Science and Engineering, Administration, and Program Support
June 1969	MSFC assigned to develop lunar roving vehicle
16 July 1969	Apollo 11 launch for first human landing on the moon
12 January 1970	NASA announced Dr. Wernher von Braun would be transferred to NASA Headquarters, Washington
March 1970	Apollo Applications Program name changed to *Skylab*
1 March 1970	Dr. Eberhard Rees replaced Dr. von Braun as director of MSFC
May 1970	NASA selected McDonnell Douglas Astronautics Co. and North American Rockwell Corp. for definition and preliminary design studies of a reusable Space Shuttle vehicle for possible future space flight

30 September 1970	Final first S–IC–15 stage tested at MTF
30 October 1970	Final second S–II–15 stage tested at MTF
19 June 1971	MSFC assigned responsibility for the Space Shuttle booster stages and main engine
26 July 1971	During the Apollo 15 mission, first lunar roving vehicle used on the Moon
1972	Apollo 17, last lunar landing mission
1972	Space telescope assigned to MSFC
1972	Program offices established for *Skylab* and HEAO
1972	Shuttle Projects Office established
January 1972	President Nixon approved development of the Space Shuttle
1973	MSFC assigned responsiblity for design and development of the Space Shuttle main engine (SSME), external tank (ET), and the solid rocket booster (SRB)
26 January 1973	Dr. Rocco Petrone replaced retiring Dr. Eberhard Rees as Center Director
March 1973	European Space Research Organization (ESRO) announced would design, develop, and manufacture a Spacelab to be launched by the Shuttle with MSFC as Lead Center
14 May 1973	Final Saturn V placed *Skylab* space station into Earth orbit

15 May 1973	MSFC workers and engineers begin intense two-week effort to develop solution for *Skylab* solar shield problem
25 May 1973	Launch of *Skylab* rescue mission to deploy solar shield
24 September 1973	Memorandum of Understanding on international cooperation in NASA's Space Shuttle Program signed by NASA and ESRO for development of Spacelab
21 December 1973	Establishment of a Spacelab Program Office at MSFC to manage NASA's activities in the international project
1974	Science and Engineering Directorate reorganized
1974	Final *Skylab* mission of record 84 days completed
14 June 1974	NASA's Mississippi Test Facility renamed the National Space Technology Laboratories, and became an independent NASA installation
17 June 1974	Dr. William Lucas became MSFC director
24 September 1974	MSFC named Lead Center for NASA activities under the Solar Heating and Cooling Demonstration Act under the direction of NASA HQs Office of Energy Program
1975	Spacelab I and II responsibility assigned to MSFC
20 January 1975	Interagency agreement between NASA and Department of Interior to use NASA technology for mineral extraction with MSFC as Lead Center
17 July 1975	Apollo-Soyuz rendezvous

7 October 1975	Establishment of an advanced mineral-extraction task team within the Program Development directorate working with the Department of Interior's Bureau of Mines
1976	Spacelab Payload Project and Special Projects Offices established
1976	Spacelab III project management assigned to MSFC
4 February 1976	First main stage test of the SSME occurred at the NSTL in Mississippi
5 February 1976	Restoration of Mercury/Redstone test stand to original appearance as historic site at MSFC
4 May 1976	NASA launched LAGEOS
16 June 1977	Wernher von Braun died in Virginia
17 August 1977	First HEAO satellite launched
1978	Materials Processing in Space Projects Office established
1978	HOSC reactivated for Shuttle launch support
11 July 1979	*Skylab* reentered atmosphere
1980	First joint endeavor agreement between MSFC and McDonnell Douglas for materials processing in space
1981	Spacelab integration began
1981	Space telescope mirror polishing completed

12 April 1981	First Space Shuttle mission (STS–01) orbiter *Columbia* launched
1983	Tenth and final SPAR flight
28 November 1983	First launch of Spacelab
1984	Space Station Projects Office established
August 1984	Solar Array Flight Experiment OAST–1 mission
1984	Space telescope's optical telescope assembly completed and delivered
1984	Work began on Payload Operations Control Center
28 June 1984	MSFC officially assigned to a portion of Space Station responsibility
November 1985	61–B Launch—ASES (Experimental Assembly of Structures in Extravehicular Activity) and ACCESS (Assembly Concept for Construction of Erectable Space Structures)—Marshall managed payloads representing the first flight demonstration of construction of large structures in space
1985	Space telescope assembly in progress
28 January 1986	51–L *Challenger* disaster
24 March 1986	MSFC formed solid rocket motor redesign team to requalify the motor of the SRB
July 1986	Dr. William Lucas resigned as director of MSFC; Thomas J. Lee appointed as interim director

29 September 1986	James R. Thompson became Center Director
27 August 1987	First full-duration test firing of the redesigned SRM
29 September 1988	STS–26 *Discovery* Return to Flight
July 1989	James R. Thompson resigned to become NASA deputy administrator. Thomas Jack Lee became director of MSFC.

Appendix B:

Personnel

B–1

MSFC Employment as Percentage of Madison County Employment**

Source: MSFC Manpower Office and Mike Wright; Pocket Statistics; Alabama Industrial Relations

Year Employment	MSFC Permanent Employees	Madison County Employment	Percentage of Madison County
1961	5,688	43,100	13.8
1962	6,533	48,500	14.1
1963	6,821	57,200	12.8
1964	7,321	65,500	11.7
1965	7,327	72,600	10.6
1966	7,277	75,900	10.2
1967	7,177	72,200	10.3
1968	6,440	70,200	7.5
1969	6,149	70,000	9.5
1970	5,994	68,220	9.3
1971	5,760	68,770	8.8
1972	5,500	72,150	7.7
1973	5,044	74,100	7.1
1974	4,400	75,070	6.1
1975	4,081	73,300	5.9
1976	4,062	78,140	5.5
1977	3,922	81,310	4.9
1978	3,760	86,020	4.4
1979	3,598	87,100	4.2
1980	3,563	86,100	4.2
1981	3,385	88,400	4.0
1982	3,332	91,200	3.8
1983	3,350	97,000	3.7
1984	3,223	105,200	3.3
1985	3,284	111,500	3.1
1986	3,260	117,100	2.8
1987	3,385	123,400	2.9
1988	3,340	128,800	2.7
1989	3,613	131,200	2.8
1990	3,620	136,730	2.8
1991	3,789	136,630	2.8
1992	3,714	138,720	2.7
1993	3,626	140,140	2.6

*** MSFC employment is permanent full-time employment and Madison County employment is nonagricultural employment*

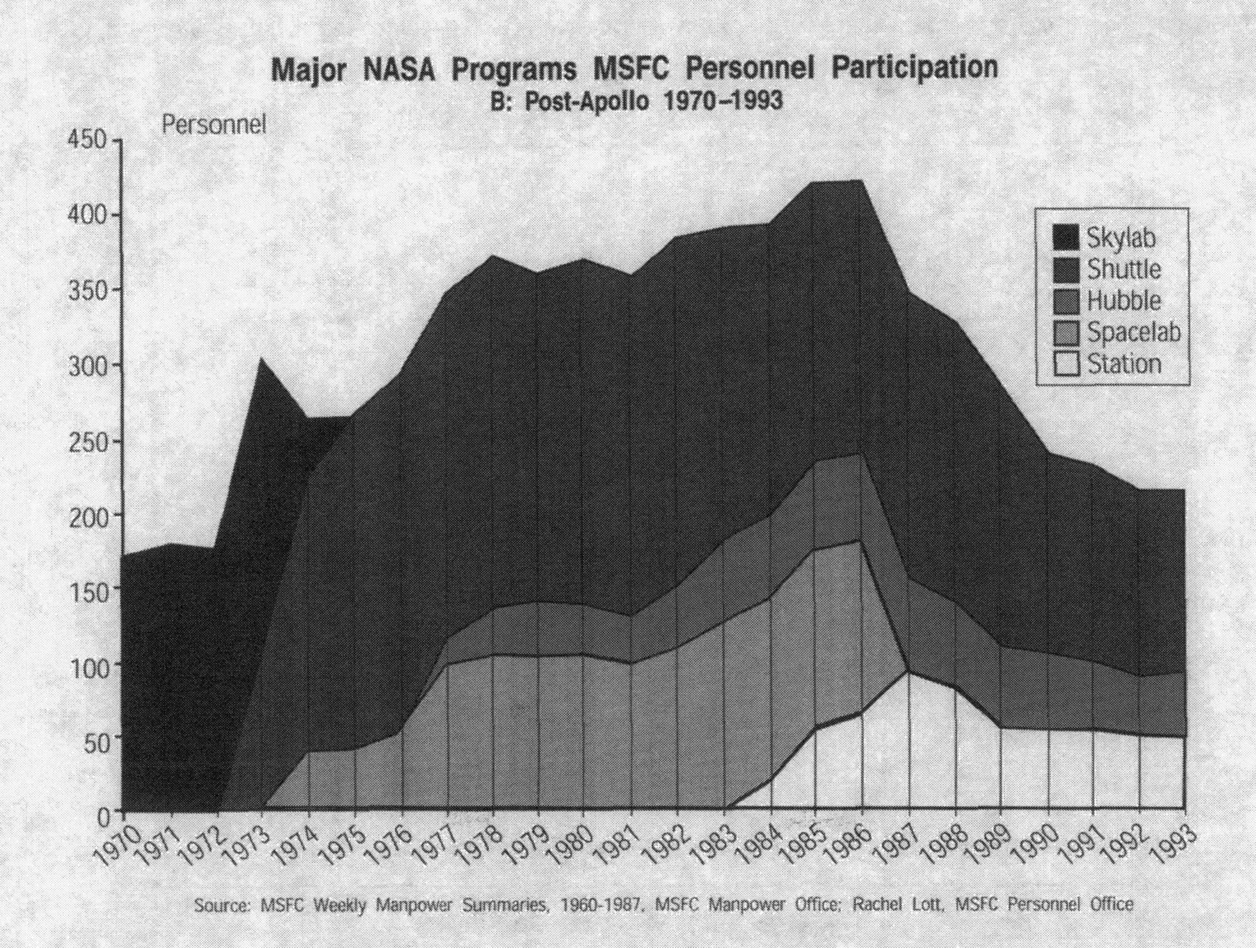
Major NASA Programs MSFC Personnel Participation
B: Post-Apollo 1970–1993
Personnel
450
400
350
300
250
200
150
100
50
0
1970
1971
1972
1973
1974
1975
1976
1977
1978
1979
1980
1981
1982
1983
1984
1985
1986
1987
1988
1989
1990
1991
1992
1993
Skylab
Shuttle
Hubble
Spacelab
Station
Source: MSFC Weekly Manpower Summaries, 1960-1987, MSFC Manpower Office; Rachel Lott, MSFC Personnel Office

B–3
Minorities as Percentage of MSFC

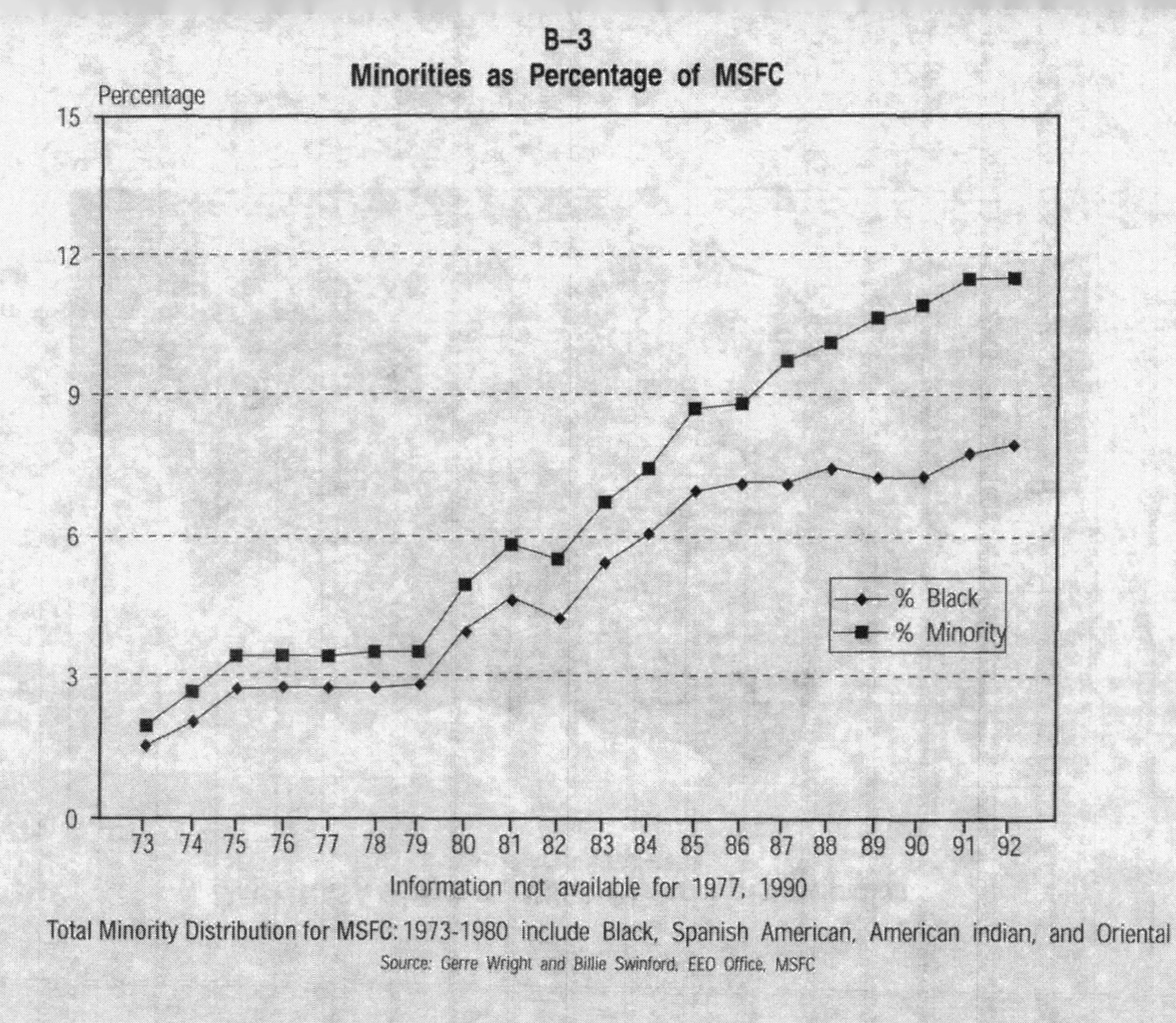

Total Minority Distribution for MSFC: 1973-1980 include Black, Spanish American, American indian, and Oriental

Source: Gerre Wright and Billie Swinford, EEO Office, MSFC

B–4
Education Levels of MSFC Employees

Gerre Wright and Billie Swinford, Equal Employment Office, Marshall Space Flight Center

Year	Total No.	Less Than Bachelor's	Bachelor's	Master's	Ph.D.
1973	5115	2397	2183	443	92
1974	4400	1860	2046	411	83
1975	4100	1667	1950	400	83
1976	4059	1621	1949	403	86
1977		Information Not Available			
1978	3760	1431	1848	392	89
1979	3598	1328	1793	383	94
1980	3563	1272	1798	396	97
1981	3385	1176	1698	413	98
1982	3332	1101	1701	431	99
1983	3412	1115	1756	437	104
1984	3264	1041	1701	417	105
1985	3352	1026	1780	431	115
1986	3260	927	1771	447	115
1987	3385	904	1888	470	123
1988	3340	890	1850	476	124
1989	3613	889	2054	529	141
1990	3620	857	2083	533	147
1991	3789	852	2212	571	154
1992	3714	826	2158	576	154
1993	3626	791	2103	573	159
1994	3311	657	1947	547	160

(As of August 6, 1994)

B–5
Women as Percentage of MSFC and NASA Civil Service

Source: Pocket Statistics and Manpower Summaries

***Women at MSFC from 1973, Source: Manpower Summaries from Gerre Wright and Billie Swinford, Equal Employment Office*

Year	Total MSFC	Women MSFC	Percentage
1973	5115	805	15.7
1974	4400	672	15.3
1975	4100	654	16.0
1976	4059	671	16.5
1977	Information Not Available		
1978	3760	619	16.5
1979	3598	628	17.5
1980	3563	692	19.4
1981	3385	710	21.0
1982	3332	695	20.9
1983	3412	748	21.9
1984	3264	752	23.0
1985	3352	842	25.1
1986	3279	853	26.0
1987	3461	983	28.4
1988	3422	1014	29.6
1989	3610	1101	30.5
1990	3619	1138	31.5
1991	3788	1226	32.4
1992	3747	1213	32.4
1993	Information Not Available		
1994	3292	1087	33.0

(As of August 6, 1994)

Appendix C:

Center Directors

Center Directors

Wernher von Braun

Eberhard Rees

Rocco A. Petrone

William R. Lucas

James R. Thompson

T. Jack Lee

Wernher von Braun

Born:	March 23, 1912
Place:	Wiersitz, Germany
Education:	Berlin Institute of Technology, Mechanical Engineering, 1932
Master's:	University of Berlin, 1933
Doctorate:	University of Berlin, Physics, 1934

Career:

1934	German Ordnance Department, Rocket Development Engineer
1937	Peenemünde Rocket Center, Director of Research
1945	White Sands, NM, White Sands Missile Range, Project Director
1945	Ft. Bliss, TX, Guided Missile Development Group, Project Director
1950	Redstone Arsenal, AL, Guided Missile Development Group, Technical Director
1955	Became a U.S. citizen
1956	Redstone Arsenal, AL, Army Ballistic Missile Agency, Director of Development Operations Division
1960	Marshall Space Flight Center, AL, Director
1970	Washington, DC, National Aeronautics and Space Administration, Deputy Associate Administrator for Planning

1972	Germantown, MD, Vice President, Engineering and Development, Fairchild Industries
1977	Retired in January, and died in Virginia on June 15, 1977

Eberhard Rees

Born:	April 28, 1908
Place:	Trossingen, Germany
Education	Stuttgart University
Master's:	Dresden Institute of Technology, Mechanical Engineering, 1934

Career:

1940	Peenemünde Rocket Center, Technical Plant Manager
1945	White Sands, NM, White Sands Missile Range, U.S. Army contract
1945	Ft. Bliss, TX, Guided Missile Development Group
1950	Redstone Arsenal, AL, Guided Missile Development Group
1954	Became a U.S. citizen
1956	Redstone Arsenal, AL, Army Ballistic Missile Agency, Deputy Director of Development Operations
1959	Winter Park, FL, Honorary degree of Doctor of Science, Rollins College
1960	Marshall Space Flight Center, AL, Deputy Director for Scientific and Technical Matters
1970	Marshall Space Flight Center, AL, Director
1973	January 26, retired

Rocco A. Petrone

Born:	March 31, 1926
Place:	Amsterdam, NY
Education:	West Point, 1946
Master's:	MIT, Mechanical Engineering, 1951
Doctorate:	MIT, Mechanical Engineering, 1952

Career:

1952 U.S. Army Redstone Arsenal, Redstone Rocket Development
1956 Pentagon, Army General Staff, missiles
1960 Cape Canaveral, Saturn Project Officer
1961 Cape Canaveral, Apollo Manned Lunar Landing Program
1966 Kennedy Space Center, Director Launch Operations
1969 Washington DC, Apollo Program Director National Aeronautics and Space Administration
1972 Washington, DC, additional assignment, Director of NASA's Apollo Soyuz Test Project
1973 Huntsville, Director, Marshall Space Flight Center
1974 Associate Administrator for Center Operations, NASA

William R. Lucas

Born: March 1, 1922
Place: Newbern, TN
Education Memphis State University, Chemistry
Master's: Vanderbilt University, Metallurgy
Doctorate: Vanderbilt University, Metallurgy

Career:

1941 Naval Officer
1952 Redstone Arsenal, AL, Guided Missile Development Group, Staff Member
1956 Redstone Arsenal, AL, Army Ballistic Missile Agency, Materials Officer
1960 Marshall Space Flight Center, AL, Program Development Directorate
1971 Marshall Space Flight Center, AL, Deputy Directory
1974 Marshall Space Flight Center, AL, Director
1986 Resigned as Marshall Space Flight Center Director
1987 University of Alabama, Huntsville, Assistant to the President for Space Initiatives Activity

James R. Thompson, Jr.

Born:	March 6, 1936
Place:	Greenville, NC
Education	Georgia Institute of Technology, Aeronautical Engineering, 1958
Master's:	University of Florida, Mechanical Engineering, 1963
Doctorate:	University of Alabama, Fluid Mechanics, course work completed

Career:

1960	West Palm Beach, FL, Pratt and Whitney Aircraft, Development Engineer
1963	Marshall Space Flight Center, AL, Saturn Launch Vehicle Project, Liquid Propulsion Engineer
1966	Marshall Space Flight Center, AL, Space Engine Section, Chief
1969	Marshall Space Flight Center, AL, Man/Systems Integration Branch, Chief
1974	Marshall Space Flight Center, AL, Main Engine Project, Manager
1982	Marshall Space Flight Center, AL, Science and Engineering Directorate, Associate Director
1983	Princeton, NJ, Plasma Physics Laboratory Deputy Director of Technical Operations
1986	Marshall Space Flight Center, AL, Director
1989	Washington, DC, NASA, Deputy Administrator

T. Jack Lee

Born:	1935
Place:	Wedowee, AL
Education:	University of Alabama, Aeronautical Engineering, 1958
Master's:	Harvard School of Business, Advanced Management Program, 1985

Career:

1958	Redstone Arsenal, AL, Army Ballistic Missile Huntsville, AL, Senior Manager, Orbital Sciences Corporation Agency, Research Engineer
1960	San Diego, CA, Centaur Resident Manager Office, Systems Engineer
1963	Blandenburg, MD, Pegasus Project, Resident Project Manager
1965	Kennedy Space Center, Saturn Program Resident Office, Chief
1969	Marshall Space Flight Center, AL, Assistant to Technical Deputy Director
1973	Marshall Space Flight Center, AL, Spacelab Program Office, Manager
1980	Marshall Space Flight Center, AL, Deputy Director
1980	Marshall Space Flight Center, AL, Director Heavy Lift Launch Vehicle Definition Office
1986	July–September, Marshall Space Flight Center, Acting Director
1989	Marshall Space Flight Center, AL, Director
1994	Washington, DC, NASA, Special Assistant for Access to Space

Appendix D:

Organization Charts

ORGANIZATION CHARTS

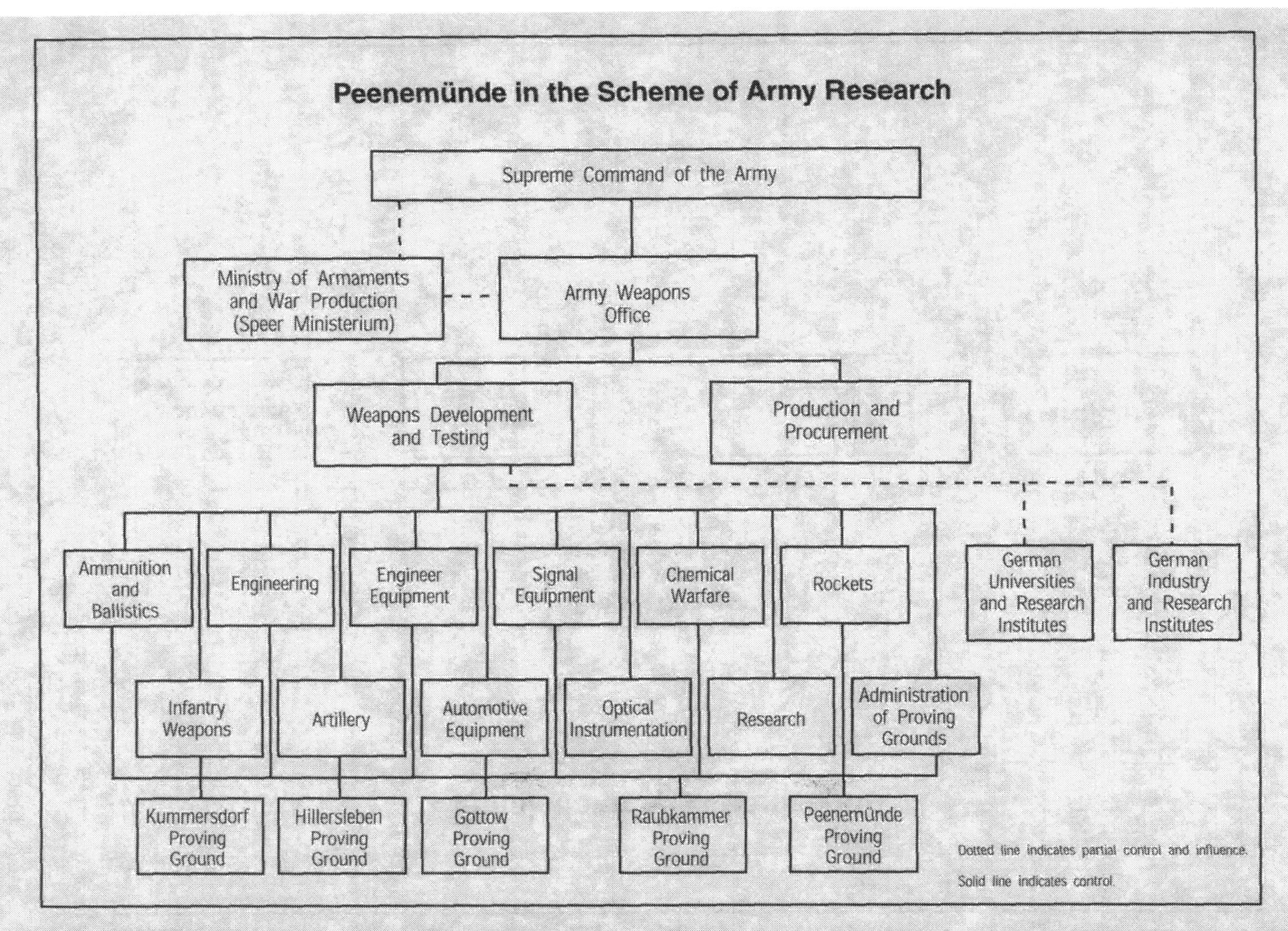
Peenemünde in the Scheme of Army Research
Supreme Command of the Army
Ministry of Armaments and War Production (Speer Ministerium)
Army Weapons Office
Weapons Development and Testing
Production and Procurement
Ammunition and Ballistics
Engineering
Engineer Equipment
Signal Equipment
Chemical Warfare
Rockets
German Universities and Research Institutes
German Industry and Research Institutes
Infantry Weapons
Artillery
Automotive Equipment
Optical Instrumentation
Research
Administration of Proving Grounds
Kummersdorf Proving Ground
Hillersleben Proving Ground
Gottow Proving Ground
Raubkammer Proving Ground
Peenemünde Proving Ground
Dotted line indicates partial control and influence.
Solid line indicates control.

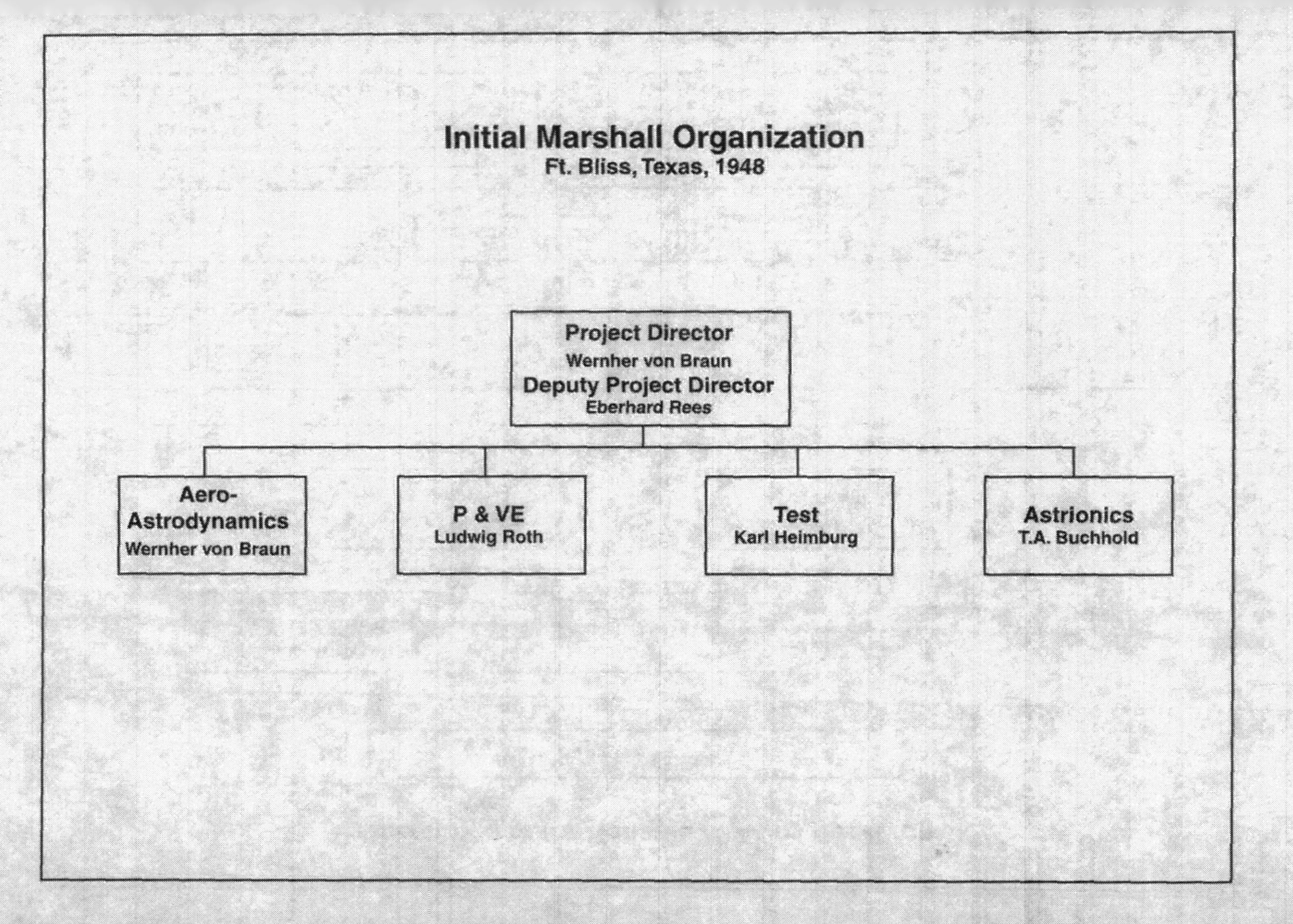
Initial Marshall Organization
Ft. Bliss, Texas, 1948
Project Director
Wernher von Braun
Deputy Project Director
Eberhard Rees
Aero-
Astrodynamics
Wernher von Braun
P & VE
Ludwig Roth
Test
Karl Heimburg
Astrionics
T.A. Buchhold

Ordnance Missile Laboratories

Director
Col. M. B. Chatfield

Asst. Dir. for Guided Missiles
Dr. J.J. Fagan

Asst. Dir. for Research & Rockets
Dr. R. C. Swann

Executive

Admin. & Management Off.

CHIEF, Mr. Edmund R. Sahag

BUDGET BRANCH
Mr. Charles R. Byerline
CENTER ADMINISTRATION BRANCH
Mr. Harold F. McMillan
OPERATIONS ANALYSIS BRANCH
Mr. Sidney Mints
TECHNICAL SERVICES BRANCH
Mr. Archie C. Bobo

Operations Research Off.

CHIEF, Mr. Emery L. Atkins

AERONAUTICAL ENG. BRANCH
Mr. J. L. Edmondson (Act. Ch.)
CENTRAL ENG. & RESEARCH BRANCH
Mr. S.C. Chambers (Act. Ch.)
MATHEMATICS BRANCH
PHYSICS BRANCH
Mr. C.C. Dalton (Act. Ch.)

Tech. Feasibility Study Off.

CHIEF, Dr. Ernst Stohlinger
DEP. CH., Mr. Kurt E. Patt

ADMINISTRATIVE BRANCH
Mrs. Eunice P. Danner
AERODYNAMICS BRANCH
Mr. R.E. Lavender (Act. Ch.)
GUIDANCE ANALYSIS BRANCH
Mr. Stephen L. Johnston
MISSILE ANALYSIS BRANCH
Mr. Fritz Kraemer
PERFORM. & FLIGHT MECH. BRANCH
MR. R.C. Callaway (Act. Ch.)
SPECIAL PROJECTS BRANCH

Rocket Dev. Division

CHIEF, Mr. Joesph F. Rush
DEP. CH., Mr. John W. Wemble

ADMINISTRATIVE BRANCH
Mr. Michael K. Foster
DESIGN BRANCH
Mr. Casper J. Keeper
ROCKET DEVELOPMENT LAB.
Mr. Frank W. James
ROCKET PROJECTS BRANCH
Mr. William C. Rotenberry
TEST & EVALUATION BRANCH
Mr. David H. Newby

Guided Missile Dev. Div.

CHIEF, Dr. Wernher von Braun
DEP. CH., CIV., Mr. Eberhard Rees
DEP. CH., MIL., Maj. P. Stebeneichen

AEROBALLISTICS LABORATORY
Dr. Ernst D. Gessler
COMPUTATION LABORATORY
Dr. Helmut Hoelser
EDITORIAL OFFICE
Mr. Herman H. Birney
ENGINEERING LIASON OFFICE
Mr. Ludwig Roth
FABRICATION LABORATORY
Mr. Hans H. Maus
FIELD TEST OFFICE
Mr. Frederick Graf Saurma
GRAPHIC ENG. & MOD. STUD. OFF.
Mr. Gerd Delteek
GUIDANCE & CONTROL LAB
Dr. Walter Haeussermann
LAUNCH & HAND. EQUIP. LAB
Mr. Hans Heuler
MISSILE FIRING LABORATORY
Dr. Kurt Debus
OPERATIONS OFFICE
Mr. Walter Wiesman
SCIENTIFIC ASSISTANT
Mr. Gerhard Heller
STRUCTURES & MECHANICS LAB
Dr. Wilbebn Pasthel
SYSTEMS ANAL. & RELIATY LAB.
Mr. Erich W. Neubert
TECHNICAL PLANNING OFFICE
Mr. Thomas C. Bodell
TEST LABORATORY
Mr. Earl L. Heimburg

Research Division

CHIEF, Dr. Eugene Miller
DEP. CH., Mr. James E. Norman

ADMINISTRATIVE BRANCH
Mr. Woodrow B. Stuart
BALLISTICS COMPUTATION LAB.

FLIGHT & AERODYNAMICS LAB.
Mr. J. Leith Potter
MATERIAL ANALYSIS LAB.

MISSILE GUIDANCE LAB.

MISSILE PROPULSION LAB.

PROPELLANTS COMBUSTION LAB.

SPECIAL PROJECTS LAB.
Mr. James E. Norman

ABMA 1955

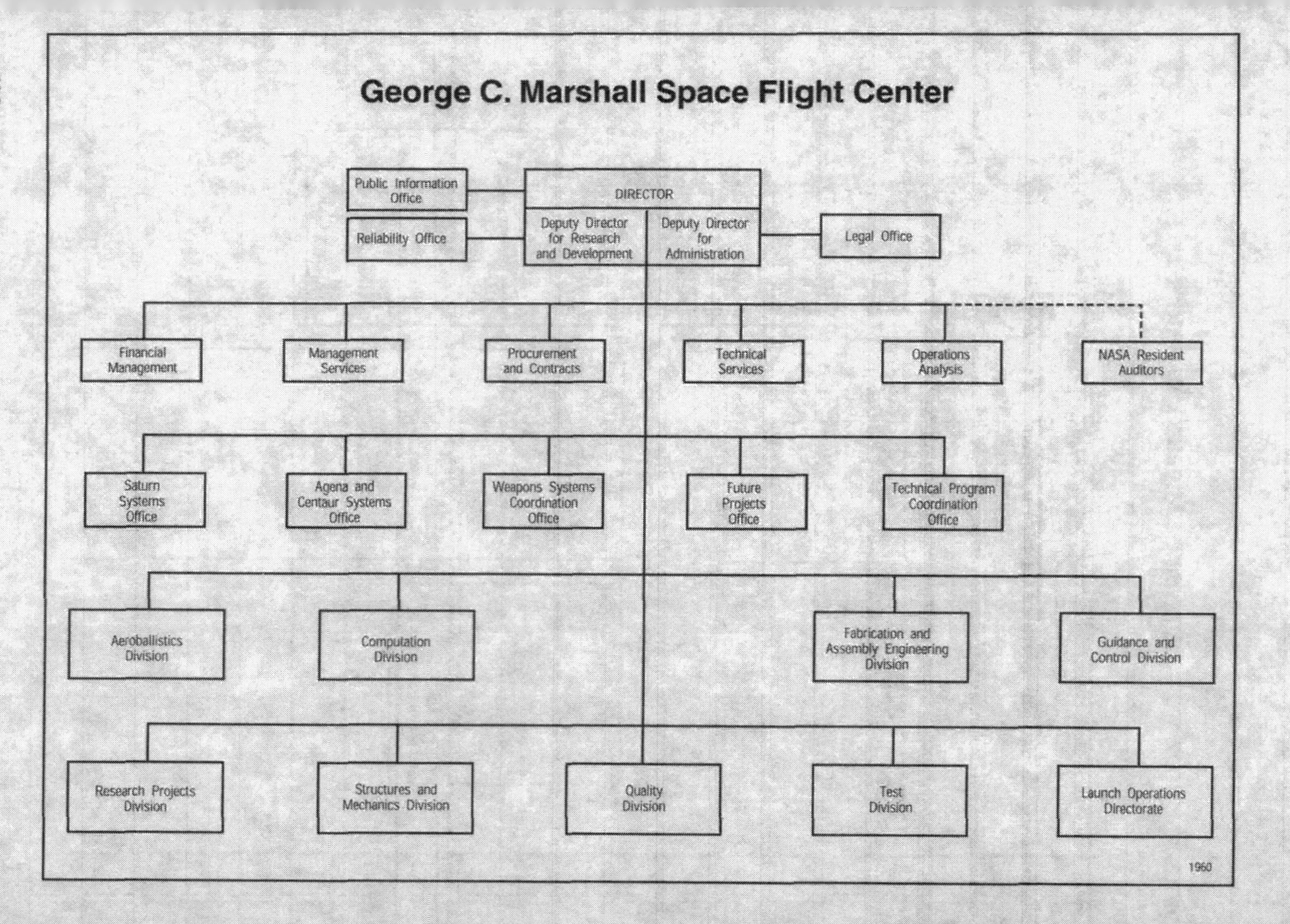
George C. Marshall Space Flight Center
Public Information Office
DIRECTOR
Reliability Office
Deputy Director for Research and Development
Deputy Director for Administration
Legal Office
Financial Management
Management Services
Procurement and Contracts
Technical Services
Operations Analysis
NASA Resident Auditors
Saturn Systems Office
Agena and Centaur Systems Office
Weapons Systems Coordination Office
Future Projects Office
Technical Program Coordination Office
Aeroballistics Division
Computation Division
Fabrication and Assembly Engineering Division
Guidance and Control Division
Research Projects Division
Structures and Mechanics Division
Quality Division
Test Division
Launch Operations Directorate
1960

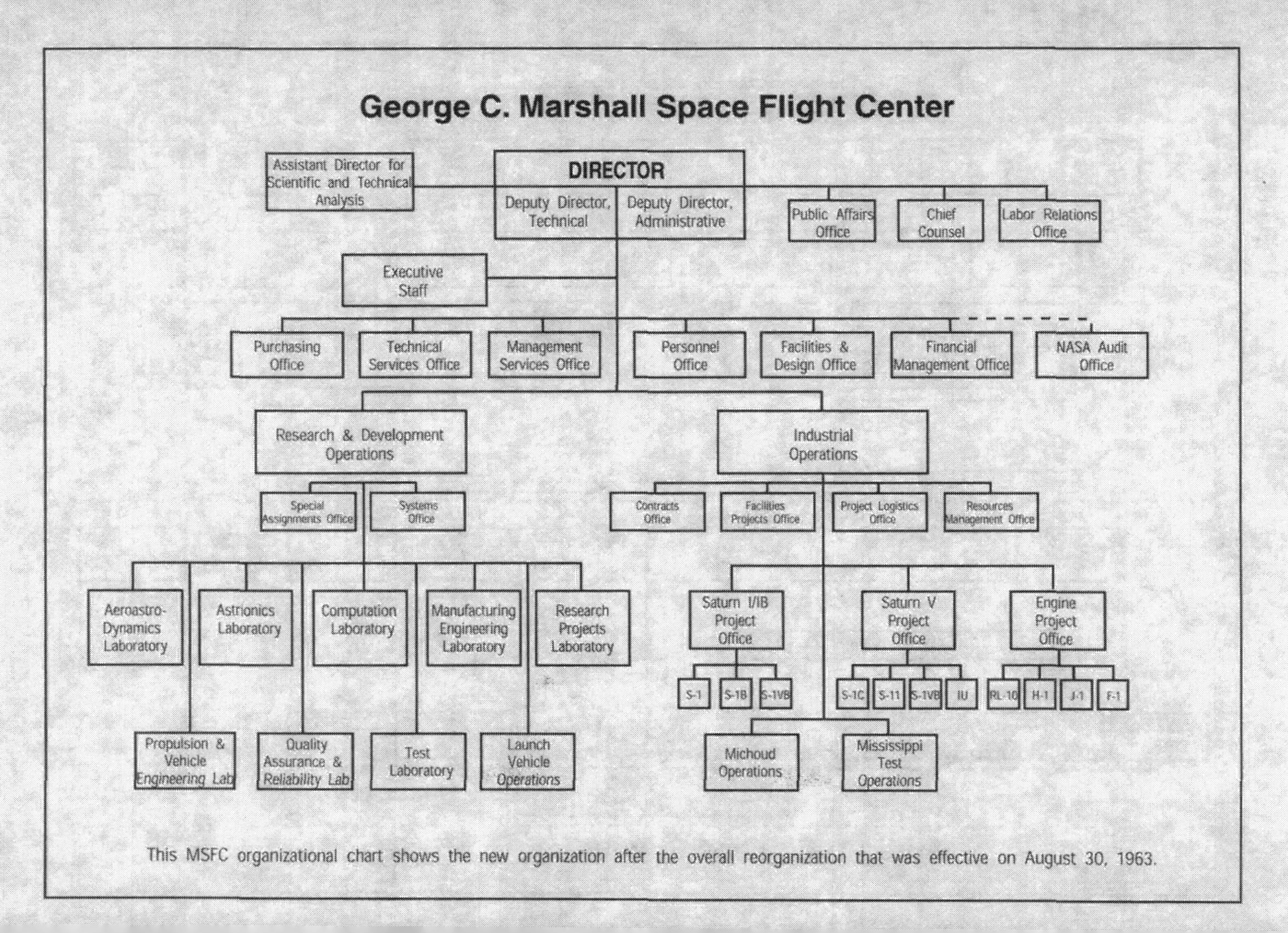

This MSFC organizational chart shows the new organization after the overall reorganization that was effective on August 30, 1963.

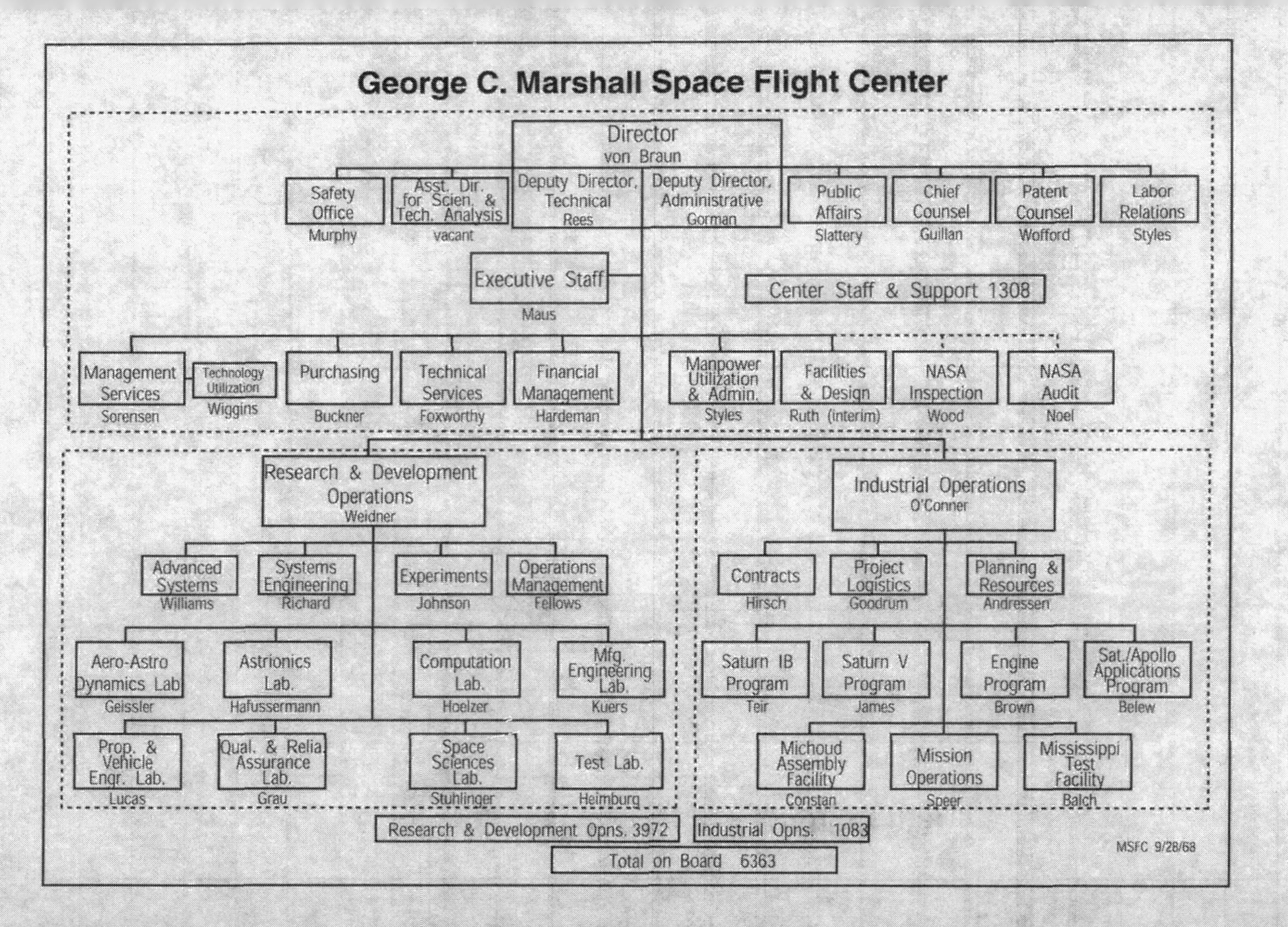
George C. Marshall Space Flight Center
Director
von Braun
Deputy Director, Technical
Rees
Deputy Director, Administrative
Gorman
Safety Office
Murphy
Asst. Dir. for Scien. & Tech. Analysis
vacant
Public Affairs
Slattery
Chief Counsel
Guillan
Patent Counsel
Wofford
Labor Relations
Styles
Executive Staff
Maus
Center Staff & Support 1308
Management Services
Sorensen
Technology Utilization
Wiggins
Purchasing
Buckner
Technical Services
Foxworthy
Financial Management
Hardeman
Manpower Utilization & Admin.
Styles
Facilities & Design
Ruth (interim)
NASA Inspection
Wood
NASA Audit
Noel
Research & Development Operations
Weidner
Advanced Systems
Williams
Systems Engineering
Richard
Experiments
Johnson
Operations Management
Fellows
Aero-Astro Dynamics Lab
Geissler
Astrionics Lab.
Hafussermann
Computation Lab.
Hoelzer
Mfg. Engineering Lab.
Kuers
Prop. & Vehicle Engr. Lab.
Lucas
Qual. & Relia. Assurance Lab.
Grau
Space Sciences Lab.
Stuhlinger
Test Lab.
Heimburg
Industrial Operations
O'Conner
Contracts
Hirsch
Project Logistics
Goodrum
Planning & Resources
Andressen
Saturn IB Program
Teir
Saturn V Program
James
Engine Program
Brown
Sat./Apollo Applications Program
Belew
Michoud Assembly Facility
Constan
Mission Operations
Speer
Mississippi Test Facility
Balch
Research & Development Opns. 3972
Industrial Opns. 1083
Total on Board 6363
MSFC 9/28/68

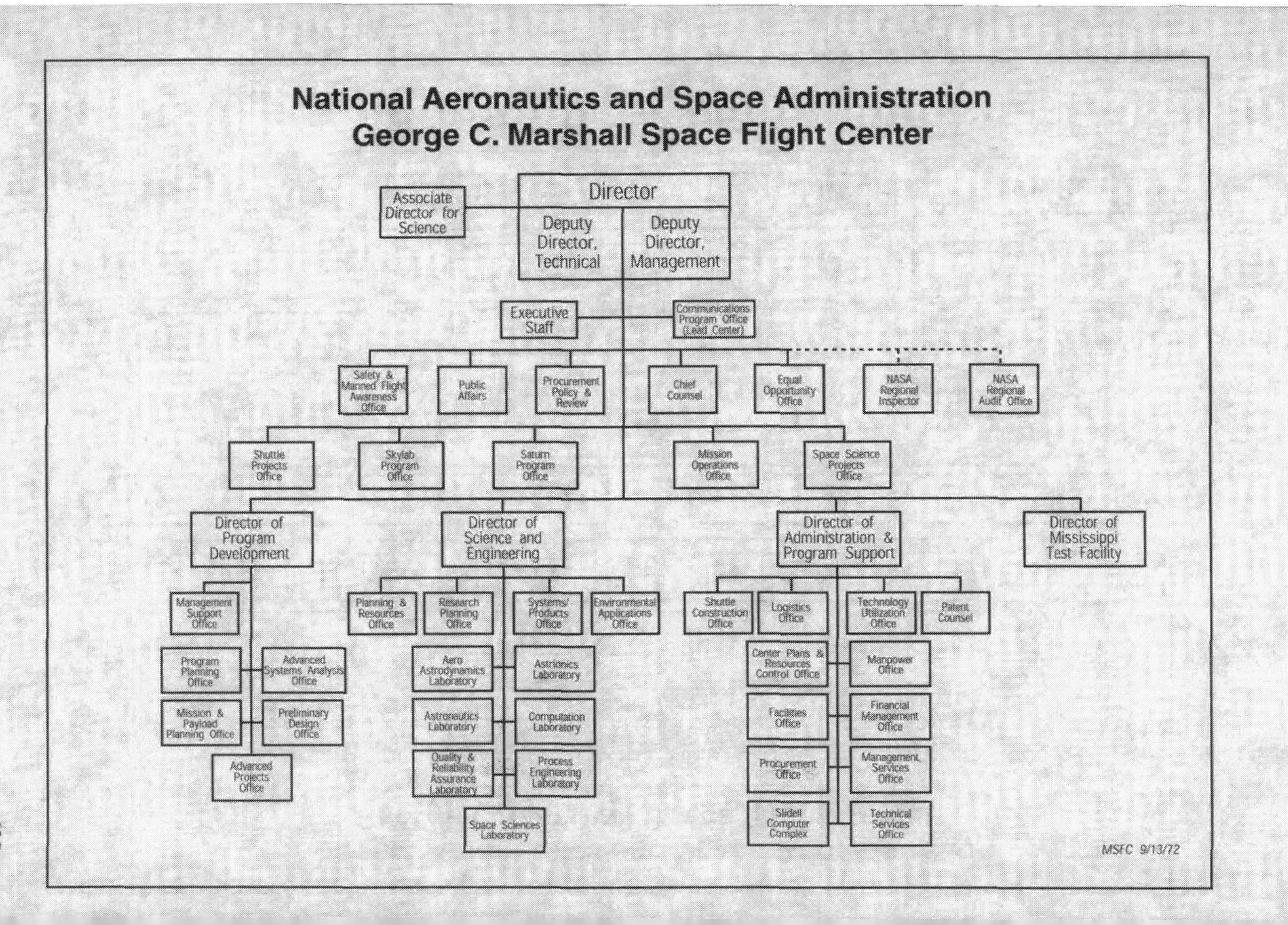
National Aeronautics and Space Administration
George C. Marshall Space Flight Center
Director
Deputy Director, Technical
Deputy Director, Management
Associate Director for Science
Executive Staff
Communications Program Office (Lead Center)
Safety & Manned Flight Awareness Office
Public Affairs
Procurement Policy & Review
Chief Counsel
Equal Opportunity Office
NASA Regional Inspector
NASA Regional Audit Office
Shuttle Projects Office
Skylab Program Office
Saturn Program Office
Mission Operations Office
Space Science Projects Office
Director of Program Development
Director of Science and Engineering
Director of Administration & Program Support
Director of Mississippi Test Facility
Management Support Office
Program Planning Office
Advanced Systems Analysis Office
Mission & Payload Planning Office
Preliminary Design Office
Advanced Projects Office
Planning & Resources Office
Research Planning Office
Systems/ Products Office
Environmental Applications Office
Aero Astrodynamics Laboratory
Astrionics Laboratory
Astronautics Laboratory
Computation Laboratory
Quality & Reliability Assurance Laboratory
Process Engineering Laboratory
Space Sciences Laboratory
Shuttle Construction Office
Logistics Office
Technology Utilization Office
Patent Counsel
Center Plans & Resources Control Office
Manpower Office
Facilities Office
Financial Management Office
Procurement Office
Management Services Office
Slidell Computer Complex
Technical Services Office
MSFC 9/13/72

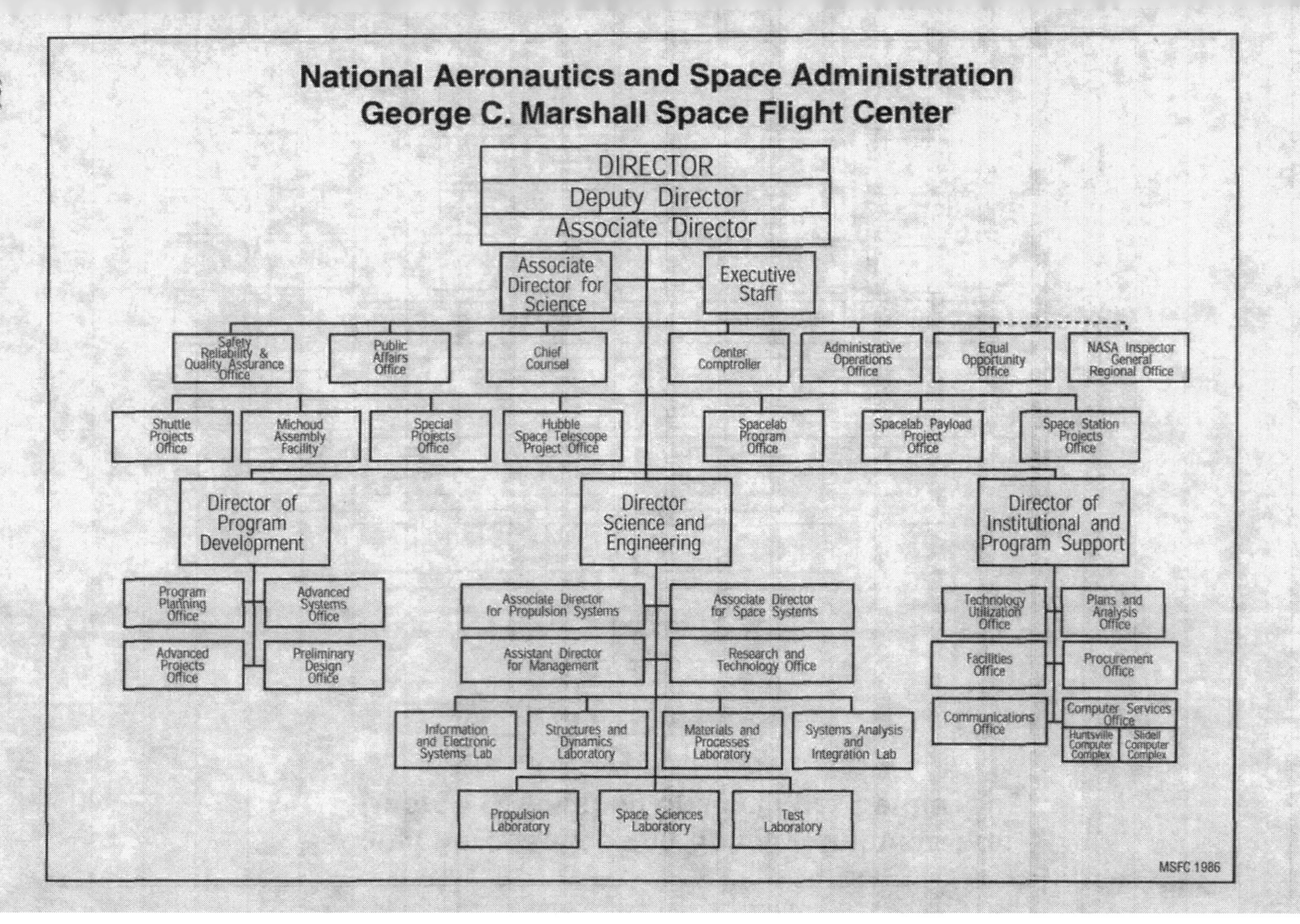
National Aeronautics and Space Administration
George C. Marshall Space Flight Center
DIRECTOR
Deputy Director
Associate Director
Associate Director for Science
Executive Staff
Safety Reliability & Quality Assurance Office
Public Affairs Office
Chief Counsel
Center Comptroller
Administrative Operations Office
Equal Opportunity Office
NASA Inspector General Regional Office
Shuttle Projects Office
Michoud Assembly Facility
Special Projects Office
Hubble Space Telescope Project Office
Spacelab Program Office
Spacelab Payload Project Office
Space Station Projects Office
Director of Program Development
Program Planning Office
Advanced Systems Office
Advanced Projects Office
Preliminary Design Office
Director Science and Engineering
Associate Director for Propulsion Systems
Associate Director for Space Systems
Assistant Director for Management
Research and Technology Office
Information and Electronic Systems Lab
Structures and Dynamics Laboratory
Materials and Processes Laboratory
Systems Analysis and Integration Lab
Propulsion Laboratory
Space Sciences Laboratory
Test Laboratory
Director of Institutional and Program Support
Technology Utilization Office
Plans and Analysis Office
Facilities Office
Procurement Office
Communications Office
Computer Services Office
Huntsville Computer Complex
Slidell Computer Complex
MSFC 1986

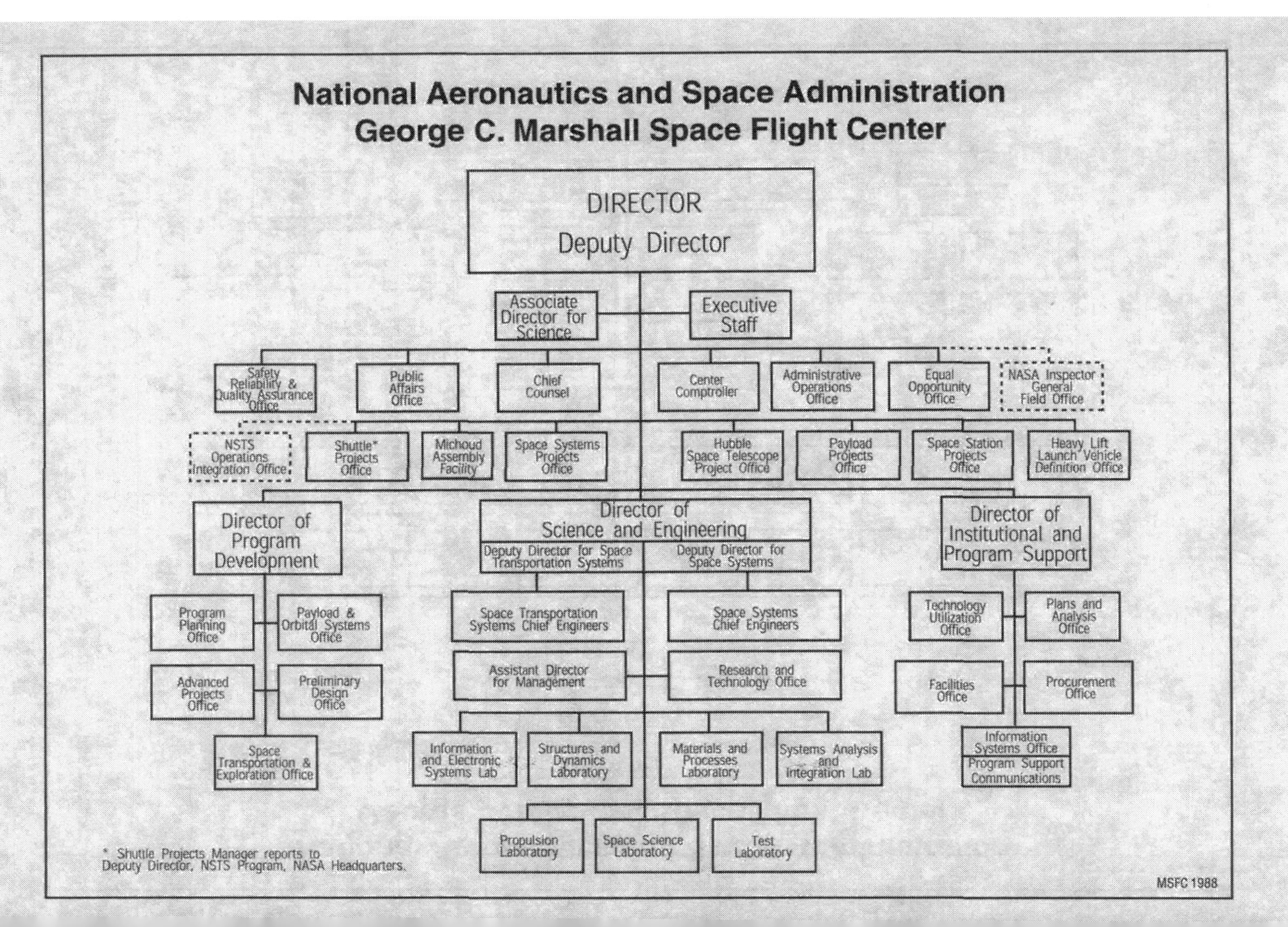
National Aeronautics and Space Administration
George C. Marshall Space Flight Center
DIRECTOR
Deputy Director
Associate Director for Science
Executive Staff
Safety Reliability & Quality Assurance Office
Public Affairs Office
Chief Counsel
Center Comptroller
Administrative Operations Office
Equal Opportunity Office
NASA Inspector General Field Office
NSTS Operations Integration Office
Shuttle* Projects Office
Michoud Assembly Facility
Space Systems Projects Office
Hubble Space Telescope Project Office
Payload Projects Office
Space Station Projects Office
Heavy Lift Launch Vehicle Definition Office
Director of Program Development
Director of Science and Engineering
Deputy Director for Space Transportation Systems
Deputy Director for Space Systems
Director of Institutional and Program Support
Program Planning Office
Payload & Orbital Systems Office
Advanced Projects Office
Preliminary Design Office
Space Transportation & Exploration Office
Space Transportation Systems Chief Engineers
Space Systems Chief Engineers
Assistant Director for Management
Research and Technology Office
Information and Electronic Systems Lab
Structures and Dynamics Laboratory
Materials and Processes Laboratory
Systems Analysis and Integration Lab
Propulsion Laboratory
Space Science Laboratory
Test Laboratory
Technology Utilization Office
Plans and Analysis Office
Facilities Office
Procurement Office
Information Systems Office
Program Support Communications
* Shuttle Projects Manager reports to Deputy Director, NSTS Program, NASA Headquarters.
MSFC 1988

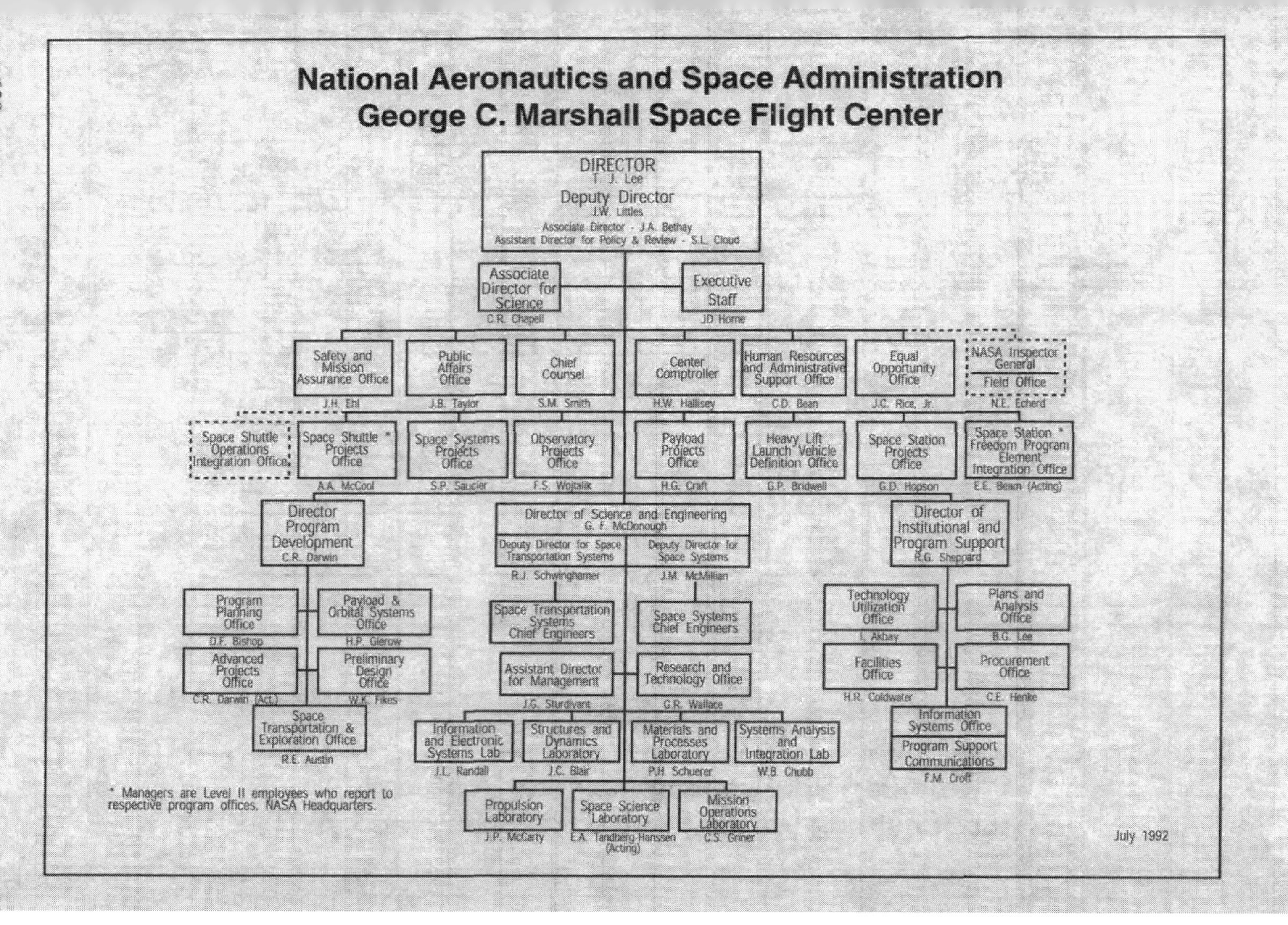
National Aeronautics and Space Administration
George C. Marshall Space Flight Center
DIRECTOR
T. J. Lee
Deputy Director
J.W. Littles
Associate Director - J.A. Bethay
Assistant Director for Policy & Review - S.L. Cloud
Associate Director for Science
C.R. Chapell
Executive Staff
JD Horne
Safety and Mission Assurance Office
J.H. Ehl
Public Affairs Office
J.B. Taylor
Chief Counsel
S.M. Smith
Center Comptroller
H.W. Hallisey
Human Resources and Administrative Support Office
C.D. Bean
Equal Opportunity Office
J.C. Rice, Jr.
NASA Inspector General
Field Office
N.E. Echerd
Space Shuttle Operations Integration Office
Space Shuttle * Projects Office
A.A. McCool
Space Systems Projects Office
S.P. Saucier
Observatory Projects Office
F.S. Wojtalik
Payload Projects Office
H.G. Craft
Heavy Lift Launch Vehicle Definition Office
G.P. Bridwell
Space Station Projects Office
G.D. Hopson
Space Station * Freedom Program Element Integration Office
E.E. Beam (Acting)
Director Program Development
C.R. Darwin
Director of Science and Engineering
G. F. McDonough
Deputy Director for Space Transportation Systems
R.J. Schwinghamer
Deputy Director for Space Systems
J.M. McMillian
Director of Institutional and Program Support
R.G. Sheppard
Program Planning Office
D.F. Bishop
Payload & Orbital Systems Office
H.P. Gierow
Advanced Projects Office
C.R. Darwin (Act.)
Preliminary Design Office
W.K. Fikes
Space Transportation & Exploration Office
R.E. Austin
Space Transportation Systems Chief Engineers
Space Systems Chief Engineers
Assistant Director for Management
J.G. Sturdivant
Research and Technology Office
G.R. Wallace
Information and Electronic Systems Lab
J.L. Randall
Structures and Dynamics Laboratory
J.C. Blair
Materials and Processes Laboratory
P.H. Schuerer
Systems Analysis and Integration Lab
W.B. Chubb
Propulsion Laboratory
J.P. McCarty
Space Science Laboratory
E.A. Tandberg-Hanssen (Acting)
Mission Operations Laboratory
C.S. Griner
Technology Utilization Office
I. Akbay
Plans and Analysis Office
B.G. Lee
Facilities Office
H.R. Coldwater
Procurement Office
C.E. Henke
Information Systems Office
Program Support Communications
F.M. Croft
* Managers are Level II employees who report to respective program offices, NASA Headquarters.
July 1992

Appendix E:

Budgets and Expenses

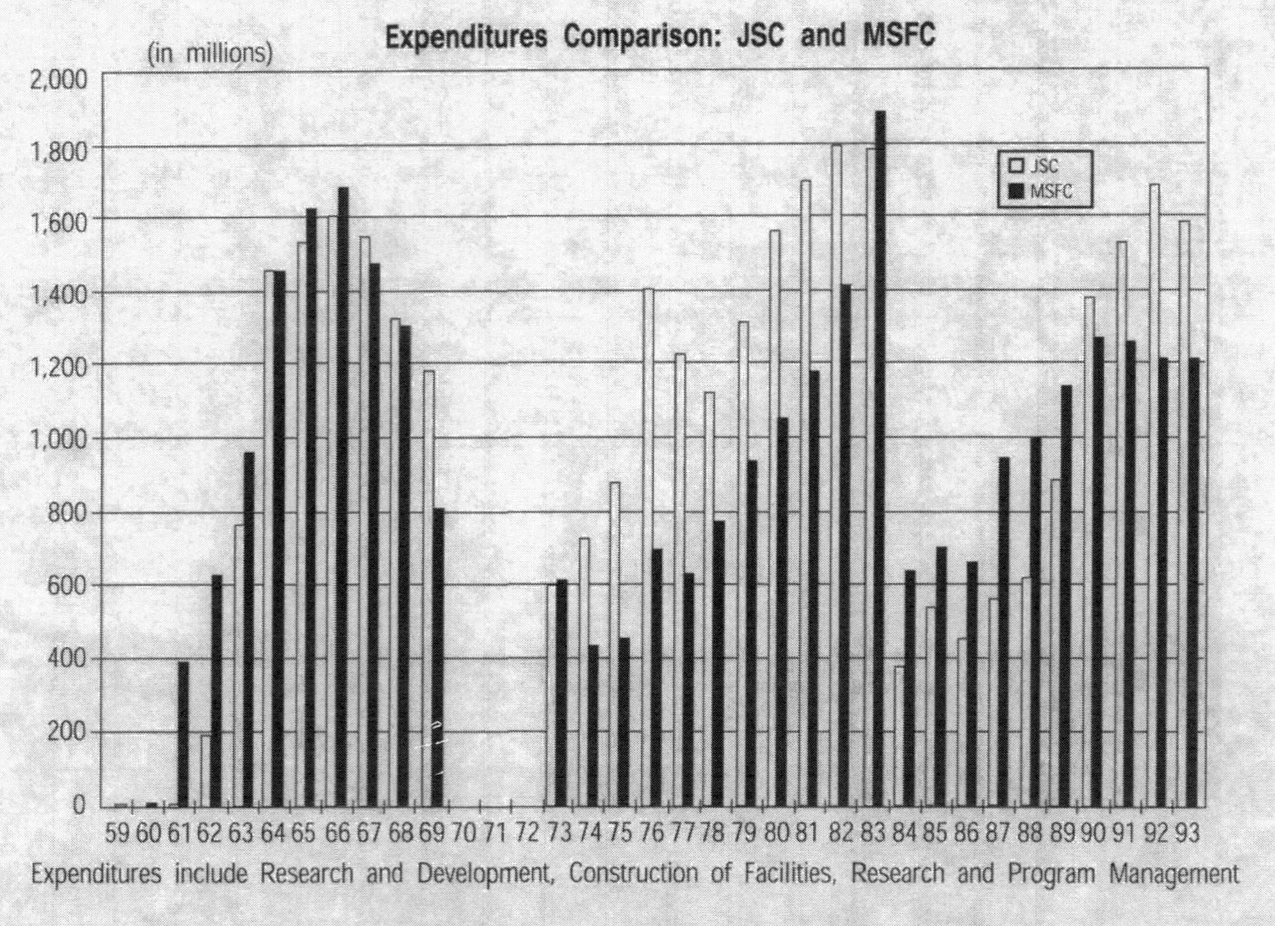
Expenditures Comparison: JSC and MSFC
(in millions)
2,000
1,800
1,600
1,400
1,200
1,000
800
600
400
200
0
JSC
MSFC
59 60 61 62 63 64 65 66 67 68 69 70 71 72 73 74 75 76 77 78 79 80 81 82 83 84 85 86 87 88 89 90 91 92 93
Expenditures include Research and Development, Construction of Facilities, Research and Program Management
Source: Pocket Statistics, 1979–1994 (Data unavailable for 1970–72)

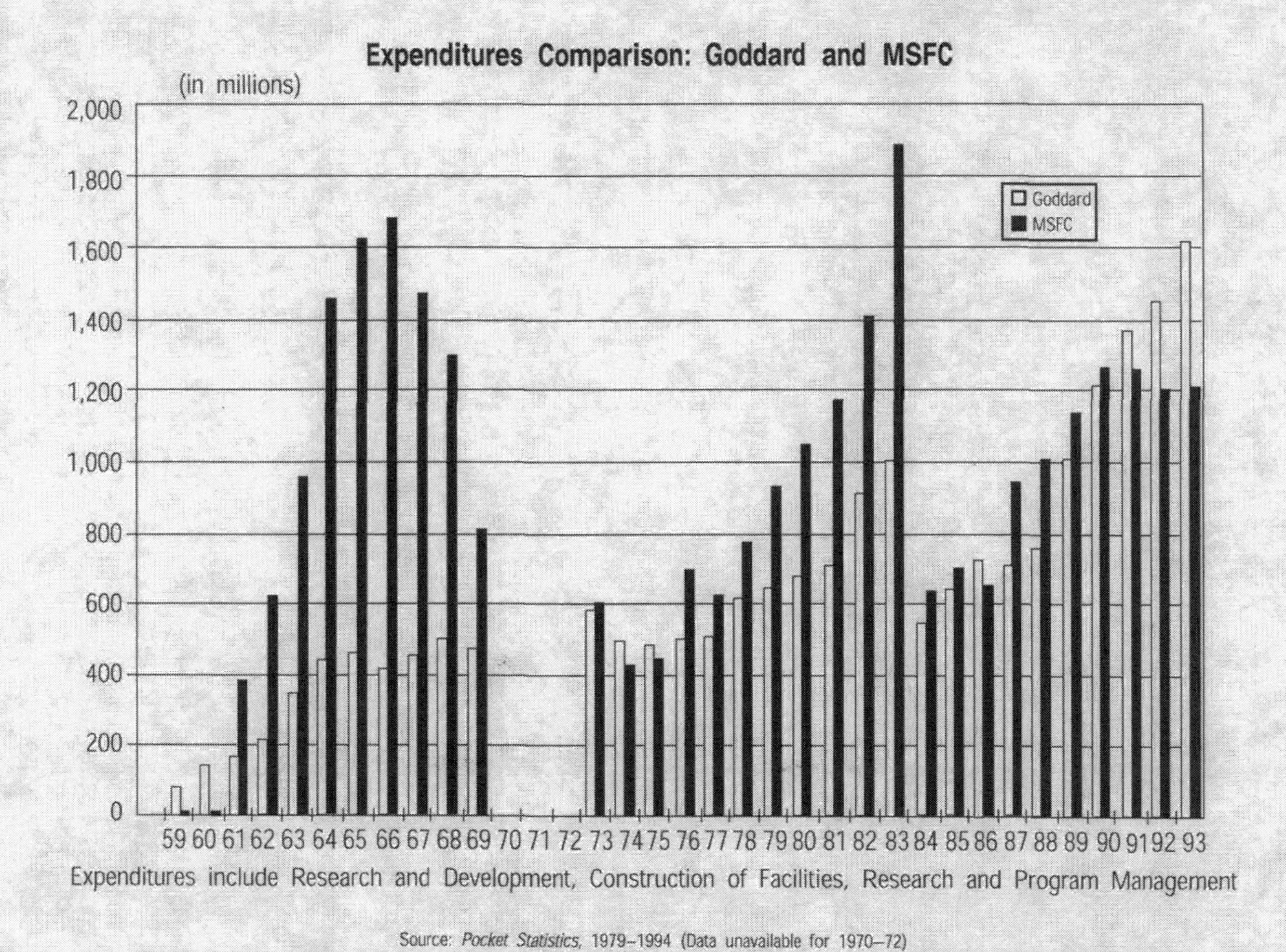

Source: *Pocket Statistics*, 1979–1994 (Data unavailable for 1970–72)

Appendix F:

Brief Chronology of Facilities Buildup Relating to History of Marshall Space Flight Center (Early 1950s through 1990)

1951: The Static Test Tower (Facility Number 4572) was constructed. It was initially used to conduct 487 tests involving the Army's Jupiter missile. It contained two test positions, and because of its appearance was sometimes called the "T-Tower." It was designed to test rocket systems with a maximum thrust of 500,000 lb. In 1961, the test stand was modified to permit static firing of the Saturn I and Saturn IB stages, which produced a total thrust of 1.6 million pounds. The name of the stand was then changed to the S–IB Static Test Stand and it has also been referred to as the Propulsion and Structural Test Facility. The west side of the stand was used to test the S–I stage. The east side was used to the test the S–IB stage. A total of 24 tests were performed on 10 S–I stages while 32 tests were performed on 12 S–IB stages. The west side was also used to test the F–1 engine; 75 F–1 engine tests were performed through July 1968. In 1984, the west side of the test stand was again modified to permit structural tests on the Space Shuttle solid rocket booster. Since its original construction and activation in 1951, a total of 649 tests have been conducted at the facility. The 140-foot-high facility was selected as a National Historic Landmark because it was the first test stand to fire rocket engines in a cluster. The name of the facility was later changed to the Hazardous Structural Test Complex. (MSFC Pamphlet "Propulsion Laboratory, Marshall Space Flight Center," 1989; Memorandum from Grady S. Jobe to Michael D. Wright, January 23, 1997, "Appendix for MSFC History"; Memorandum from B.R. McCullar to Michael Wright, March 10, 1997, "MSFC History, 1960 to 1990.)"

1952: The Redstone Interim Test Stand (Facility Number 4665), originally called the Ignition Test Stand, was constructed. Now a National Historic Landmark, this is the site where testing was conducted on the modified Redstone missile that launched America's first astronaut, Alan Shepard, into space. This dual position test structure was utilized as a Redstone vehicle center section cold flow facility on one side, and a vehicle hot firing position on the other. A total of 364 static firings were performed, including acceptance testing of Explorer I, Juno I, and Mercury-Redstone launch vehicle stage assemblies. The stand has its own control and instrumentation center which is housed in an earth-covered tank and trailer. The facility is noted for its simplicity when compared to test stands used to fire later generation of rocket engines. The stand is no longer active and was declared a National Historic Landmark in 1977. The steel support tower has a reinforced concrete base. The stand had a thrust capacity of 78,000 lb. (MSFC Pamphlet "Propulsion Laboratory, Marshall Space Flight Center," 1989; Memorandum from Grady S. Jobe to Michael D. Wright,

January 23, 1997, "Appendix for MSFC History"; Memorandum from B.R. McCullar to Michael Wright, March 10, 1997, "MSFC History, 1960 to 1990.")

1956: The Combustion Test Cells Facility (Facility Number 4583) was constructed to test liquid rocket engine components. Model rocket engines were fired in all the cells to develop design data for static and launch deflectors. Subscale 1–20 models used in testing included an RL–10 engine, an H–1 engine, an F–1 engine, and a J–2 engine. Tests were also conducted using a 1:56 scale F–1. Full-scale model tests were also conducted for the H–1 and S–3D engine. Modifications followed in 1983 and 1989. In 1987, Cell 103 was modified to support solid rocket ballistic testing. (MSFC Pamphlet "Propulsion Laboratory, Marshall Space Flight Center," 1989)

1956: The Cold Calibration Test Stand (Facility Number 4588) was constructed to cold flow test the Redstone engine and associated engine hardware. In 1957, a second test position was added to test the S–3D engine under cold flow conditions. In 1959, the stand was modified to add larger tanks to permit testing of the H–1 engine. At the same time, the north side of the stand was modified to conduct cold flow tests involving the Saturn I. (MSFC Pamphlet "Propulsion Laboratory, Marshall Space Flight Center," 1989)

1959: Facilities were completed that would house Marshall's **Structural Test Facilities** (Facility Number 4619). Large high- and low-bay facilities were constructed for structural static and dynamic tests of large and small vehicle components. A load test annex was constructed and later extended. The west end of the building was designed to include a Teleoperator and Robotic Evaluation Facility. Portions of the building have also housed a high-fidelity *Skylab* mock-up, an automated beam building machine, and a large vacuum chamber. The name of the Structural Test Facilities was changed to the Structural and Dynamics Research & Development Test Complex. (Memorandum from Grady S. Jobe to Michael D. Wright, January 23, 1997; MSFC 1996 Facilities Data Book, pages 50–52)

1963 and 1964: In 1963 the Marshall Center began construction on the first of a series of buildings in its **Headquarters Complex**. (Facility Numbers, 4200, 4201, 4202). In the early 1990s, construction began on a fourth portion of the complex. This facility would be designated as Building 4203. (1994 Facilities Data Book, p. 26)

1964: The Saturn V Dynamic Test Stand (Facility Number 4550) was constructed for low-frequency dynamic testing of the complete Saturn V launch vehicle to evaluate structural frequencies and assure decoupling from the vehicle control system. Various flight configurations were evaluated, including, the complete vehicle, the vehicle less the S–IC stage, S–II stage, etc. In the years that followed the tower was utilized to structurally qualify the *Skylab* orbital workshop and the meteoroid shield deployment for *Skylab*. The facility was modified in 1977 to perform low-frequency vibration tests on the mated Space Shuttle using the orbiter *Enterprise*. The facility was later modified to contain a drop tower and drop tube to provide a low-gravity environment for approximately three seconds. The overall height of the tower was 475 feet. The steel structure was 98 feet wide by 122 feet long by 360 feet high. The stiff leg overhead derrick was 115 feet high with a 200-ton capacity main hook and a 40-ton capacity auxiliary hook. The facility has been referred to as the Saturn V Dynamic Test Stand (Vacuum Drop Tube Facility/Low Gravity Materials Science Facility). (MSFC Pamphlet "Propulsion Laboratory, Marshall Space Flight Center," 1989; Memorandum from Grady S. Jobe to Michael D. Wright, January 23, 1997, "Appendix for MSFC History"; Memorandum from B.R. McCullar to Michael Wright, March 10, 1997, "MSFC History, 1960 to 1990")

1964: Test Stand 300 (Facility Number 4530) was constructed at the Marshall Center as a gas generator and heat exchanger test facility to support the Saturn/Apollo program. Deep space simulation was provided by a 1969 modification that added a thermal vacuum chamber and a 1981 modification that added a 12-foot vacuum chamber. The facility was again modified in 1989 when 3-foot- and 15-foot-diameter chambers were added to support *Space Station Freedom* and technology programs. The multiposition test stand was used to test a wide range of rocket engine components, systems, and subsystems. It was designed with the capability to simulate launch thermal and pressure profiles. The Marshall Center has used the stand in connection with solid rocket booster/external tank thermal protection system evaluations, solid rocket motor O-ring tests, Space Shuttle main engine injector evaluation tests, Space Station water electrolysis testing, and other programs and projects. (MSFC Pamphlet "Propulsion Laboratory, Marshall Space Flight Center," 1989)

1964: Test Stand 116 (Facility Number 4540) was constructed and activated as a an acoustical research technology model test facility. It was first used to support the Saturn/Apollo program, and then the Space Shuttle program. The stand was later modified to serve as a multiposition component test stand for programs requiring high-pressure (up to 15,000 psi) ambient and cryogenic propellants. (MSFC Pamphlet "Propulsion Laboratory, Marshall Space Flight Center," 1989)

1964: The Marshall Center constructed the **S–IC Stage Static Test Stand** (Facility Number 4670) to develop and test the first stage of the Saturn V launch vehicle which used five F–1 engines. Each F–1 engine developed 1.5 million pounds of thrust for a total lift-off thrust of 7.5 million pounds. The stand contains 12 million pounds of concrete in its base legs and could accept an engine configuration generating thrusts to that level. Eighteen tests were completed on the S–IC–T stage between April 1965 and August 1966. During 1966, testing was completed on the first three S–IC flight stages.

Modifications to the stand were initiated in 1974 to add a liquid hydrogen capability for testing liquid hydrogen tankage on the Space Shuttle external tank. These tests were completed in 1980. The facility was again modified in 1986 and its name was changed to the Advanced Engine Test Facility. These modifications were made to accommodate the technology test-bed engine that was intended to be a derivative of the Space Shuttle main engine. (MSFC Pamphlet "Propulsion Laboratory, Marshall Space Flight Center," 1989; Memorandum from Grady S. Jobe to Michael D. Wright, January 23, 1997, "Appendix for MSFC History"; Memorandum from B.R. McCullar to Michael Wright, March 10, 1997, "MSFC History, 1960 to 1990")

1964: Marshall completed and activated the **F–1 Turbopump Test Stand** (Facility Number 4696). The facility was used to perform checkout, calibration, qualification, and research and development tests on the F–1 engine turbopumps for the first stage of the Saturn V. A gas generator-driven F–1 turbopump was attached to an F–1 "bobtail" engine to constitute the test-bed for the reference testing. Testing continued on a regular basis through 1968 at which time the facility was placed on stand-by status. (MSFC Pamphlet "Propulsion Laboratory, Marshall Space Flight Center," 1989)

1965: The Marshall Center activated the **S–IVB Test Stand** (Facility Number 4520) in its East Test Area. The S–IVB served as the second stage of the Saturn IB and the third stage of the Saturn V. The stage utilized the J–2 engine which burned liquid hydrogen as fuel and liquid oxygen as the oxidizer. The S–IVB Test Stand was a liquid oxygen/liquid hydrogen facility designed to static fire the S–IVB stage in a vertical mode. The stand was used in 117 static firings. It was designed as an open steel structure capable of accepting stages 22 feet in diameter and 60 feet long, developing thrusts up to 300,000 lb. It was last used for the static firing in 1971 but later utilized for Space Shuttle external tank tests using a 10-foot-diameter tank for thermal protection system development. It was also used for inflatable nozzle technology tests. (MSFC Pamphlet "Propulsion Laboratory, Marshall Space Flight Center," 1989)

1966: Test Stand 500 was constructed to test liquid hydrogen/liquid oxygen turbopumps and combustion devices for the J–2 engine. The facility was modified in 1980 to support Space Shuttle main engine bearing testing. (MSFC Pamphlet "Propulsion Laboratory, Marshall Space Flight Center," 1989)

1968: The Center completed the **Neutral Buoyancy Simulator**. The facility was designed to provide a simulated zero-gravity environment in which engineers, designers, and astronauts could perform, for extended periods of time, the various phases of space development to gain a first-hand knowledge of design problems and operational characteristics. The tank is 75 feet in diameter and 40 feet deep and designed to hold 1.5 million gallons of water. There are four observation levels for underwater audio and video communications. The southwest corner of Building 4705 that houses the facility has a completely equipped test control center for directing, controlling, and monitoring the simulation activities. The simulator was used extensively to prepare astronauts for the *Skylab* missions. It was a vital element in defining rescue procedures for the crippled *Skylab* orbital workshop.

1968: In 1964, the Marshall Center instituted a Launch Information Exchange Facility (LIEF) linking Marshall and Kennedy Space Center. LIEF began operating in December 1964 to provide instantaneous launch data concerning the Saturn vehicle. By 1968, the Marshall Center had established the **Huntsville Operations Support Center** located in a portion of its Computation Laboratory (Facility Number 4663). During the Saturn/Apollo missions Marshall Center engineers were stationed at the facility to monitor data received from KSC. The data was evaluated and advice and guidance given through a series of engineering consoles. Engineers monitored the flights in order to deal with any malfunctions or failures in propulsion, navigation, or electrical control. The same facility was a critical element of mission support during the *Skylab* missions. Years later, extensive modifications were made to the facility in Building 4663 in order to support Space Shuttle missions. ("Marshall Historical Monograph Number 9, History of the George C. Marshall Space Flight Center, January 1–June 30, 1964"; *Marshall Star*, January 10, 1968, "Operations Support Center Helps in Saturn Launches"; Marshall Space Flight Center 25th Anniversary Report, p. 27)

1968: The Marshall Center built the **High Reynolds Number Wind Tunnel Facility** (Facility Number 4732) later known as the Wind Tunnel Complex. This facility was designed to simulate winds up to Mach 3.5. (1994 MSFC Facilities Data Book, 64; 1996 MSFC Facilities and Equipment Catalog)

1974: The Marshall Center was nearing completion of **the Space Shuttle Main Engine Hardware Simulation Laboratory** in Building 4436. The facility was designed to test and verify the SSME avionics and software, control system, and mathematical models. It would serve for years as an invaluable tool in the design and development of the SSME. (*Marshall Star,* October 9, 1974, "SSME Simulation Facility Being Prepared at MSFC"; MSFC Open House Brochure, May 3, 1997)

1975: The Hot Gas Facility (Facility Number 4554) was originally built for solid rocket booster and external tank thermal protection system material evaluations. The facility was designed to simulate flight vehicle environments of heating rates, pressures, shear, and other factors. The facility was modified in 1985 to extend the maximum run time from 60 to 180 seconds. Approximately 2,000 tests were performed in the qualification of external tank thermal protection system materials. These tests included testing MSFC sprayable ablative materials used on the solid rocket motors for thermal protection. (MSFC Pamphlet "Propulsion Laboratory, Marshall Space Flight Center," 1989)

April 1976: Marshall's original X-ray Test Facility was completed. The facility, the only one of its size and type at the time was used for x-ray verification testing and calibration of x-ray mirrors, telescope systems, and instruments. It was initially used to test instruments for Marshall's High Energy Astronomy Observatory (HEAO) program. The facility was designed with a 1,000-foot-long stainless steel x-ray path guide tube, almost 3 feet in diameter. The tube was connected to a chamber 20 feet in diameter to house the telescopes or other instruments to be tested. In the early 1990s, construction was completed on an improved **X-Ray Calibration Facility** for evaluating the mirrors for the Advanced X-Ray Astrophysics Facility. The facility is designated as Building

4718. (*Marshall Star*, March 31, 1976, "Huge X-Ray Test Facility to Be Completed Tomorrow"; Memorandum from Grady S. Jobe to Michael D. Wright, January 23, 1987, "Appendix for MSFC History"; Marshall Space Flight Center Master Plan 1992: MSFC Fact Sheet, August 1991, "Marshall X-Ray Calibration Facility")

1979: The Marshall Center began training science crews for missions involving Spacelab. The **Payload Crew Training Complex** in Building 4612 was designed to provide a simulated environment for training Spacelab payload crews in hands-on interaction with Spacelab systems and experiments including Spacelab command and data management systems operations and procedures and timeline verification. (*Marshall Star,* March 26, 1997, "Marshall's Payload Training Team…"; MSFC Fact Sheet, April 1990, "Payload Crew Training Complex Used for Spacelab Experiment Training"; "Marshall Space Flight Center Open House brochure, May 3, 1997)

1987: The Transient Pressure Test Article (Facility Number 4564) test stand was built to provide data to verify the sealing capacity of the redesigned solid rocket motor (SRM) field and nozzle joints. The facility was designed to apply pressure, temperature, and external loads to a short stack of solid rocket motor hardware. The simulated solid rocket motor ignition pressure and temperature transients were by firing approximately 500 pounds of specially configured solid propellant. Approximately 1 million pounds of dead weight on top of the test article simulated the weight of the other Shuttle elements. The steel structure was designed to be 14 feet wide, 26 feet long, and 33 feet high. (MSFC Pamphlet "Propulsion Laboratory, Marshall Space Flight Center", 1989; Memorandum from Grady S. Jobe to Michael D. Wright, January 23, 1997, "Appendix for MSFC History"; Memorandum from B.R. McCullar to Michael Wright, March 10, 1997, "MSFC History, 1960 to 1990")

1990: Teams of controllers and researchers began controlling all NASA Spacelab missions from Marshall's new **Spacelab Mission Operations Control Facility**. The new facility was located on two floors of Building 4663 at Marshall and replaced the payload operations control center formerly situated at the Johnson Space Center in Houston from which previous Spacelab missions were operated. (Marshall Space Flight Center Fact Sheet, "Spacelab Mission Operations Control Facility," May 1990)

Appendix G:

Major MSFC Patents

Source: Significant NASA Inventions. G.P.O., 1986.

Date Filed: 26 June 1970
Date Issued: 25 July 1972
Inventor: Bernard Rubin, et al.
Description: A process for the preparation of calcium phosphate salts for deposit from a gel medium onto the surface of a tooth. The gel diffusion process on the enamel aids repair of damaged tooth.

Date Filed: 16 January 1970
Date Issued: 4 January 1972
Inventor: John R. Rasquin, et al.
Description: A device to fabricate industrial-grade diamonds from common graphite by concentrated shock wave energy.

Date Filed: 24 March 1971
Date Issued: 15 July 1975
Inventor: Felix P. LaIacoma
Description: A process for the fabrication of a graphite-reinforced aluminum composite utilizing diffusion-bonding and nickel coating to produce a high-strength, low-density material.

Date Filed: 20 March 1972
Date Issued: 1 January 1974
Inventor: James M. Hoop
Description: A method of testing devices exposed to high voltage discharges utilizing ultrasonic energy. A high-frequency arc discharge through a coupling medium detects flaws.

Date Filed: 13 March 1975
Date Issued: 5 March 1974
Inventor: William Jabez Robinson, Jr.
Description: A microwave, remote, power transmission system automatically adjusted to increase or decrease the power output to a remote receiving station.

Date Filed: 11 March 1973
Date Issued: 24 June 1975
Inventor: Byron Hamilton Auker, et al.
Description: A boroaluminum silicate composite thermal coating for surfaces exposed to solar radiation, reentry heating, dust, and salt spray.

Date Filed: 11 July 1974
Date Issued: 9 December 1975
Inventor: James Albert Webster
Description: A fuel tank sealant composed of a polyimide that is strong and highly resistant to temperature extremes and is resistant to fuel corrosion.

Date Filed: 16 July 1974
Date Issued: 12 August 1975
Inventor: James Albert Webster
Description: A fuel tank sealant composed of tetracarboxylic acid and dianhydride.

Date Filed: 5 April 1975
Date Issued: 18 November 1975
Inventor: James Russell Lowery
Description: A panel for selectively absorbing solar energy for subsequent use in heating or cooling operations in a metal body.

Date Filed: 29 January 1976
Date Issued: 15 March 1977
Inventor: Lott W. Brantley, et al.
Description: A collector dish mount utilizing a rigid, angulated, axle that tracks the Sun both diurnally and seasonally.

Date Filed: 19 July 1976
Date Issued: 4 October 1977
Inventor: Frank J. Nola
Description: A power factor control system for alternate current induction motors that tests line voltage and regulates power to the motor.

Date Filed: 23 June 1976
Date Issued: 9 May 1978
Inventor: William Reynolds Feltner
Description: A method of making a field effect transistor from a semi-conductor through ion bombardment.

Date Filed: 8 June 1976
Date Issued: 18 July 1978
Inventor: Barbara Scott Askins
Description: An auto-radiography process for treating photographic film.

Date Filed: 24 February 1978
Date Issued: 4 March 1980
Inventor: John Kaufman
Description: A wind wheel electric power generator.

Date Filed: 12 March 1980
Date Issued: 5 January 1982
Inventor: William N. Myers, et al.
Description: A wind turbine utilizing two chambers rotating independently.

Date Filed: 23 October 1980
Date Issued: 21 February 1984
Inventor: Frank J. Nola
Description: A three-phase power factor controller for a three-phase induction motor.

Date Filed: 13 October 1981
Date Issued: 13 September 1983
Inventor: Frank J. Nola
Description: A reduced voltage starter utilizing a power factor controller.

Date Filed: 30 November 1981
Date Issued: 17 January 1984
Inventor: Frank J. Nola
Description: A trigger control circuit producing firing impulses through a power factor controller for preventing lags in current cycles of alternating current induction motors.

Date Filed: 23 April 1982
Date Issued: 11 October 1983
Inventor: John B. Tenney, Jr.
Description: A prosthetic device for use with tubular internal human organs.

Date Filed: 8 September 1983
Date Issued: 8 October 1985
Inventor: Glenn D. Craig
Description: A wide-range video camera.

Date Filed: 16 December 1982
Date Issued: 4 September 1984
Inventor: Frank J. Nola
Description: A three-phase power factor controller that contains an EMF sensing device for an alternating current induction motor.

Date Filed: 4 December 1982
Date Issued: 10 July 1984
Inventor: Frank J. Nola
Description: A phase detector for a three-phase power factor controller.

Date Filed: 23 July 1984
Date Issued: 1 November 1988
Inventor: Vernon W. Keller
Description: A warm fog dissipation device for airports by spraying large volumes of water.

Date Filed: 20 August 1987
Date Issued: 23 May 1989
Inventor: Daniel C. Carter
Description: A human serum albumin crystal for the production of new drugs.

Appendix H:

Huntsville Area Social and Economic Change

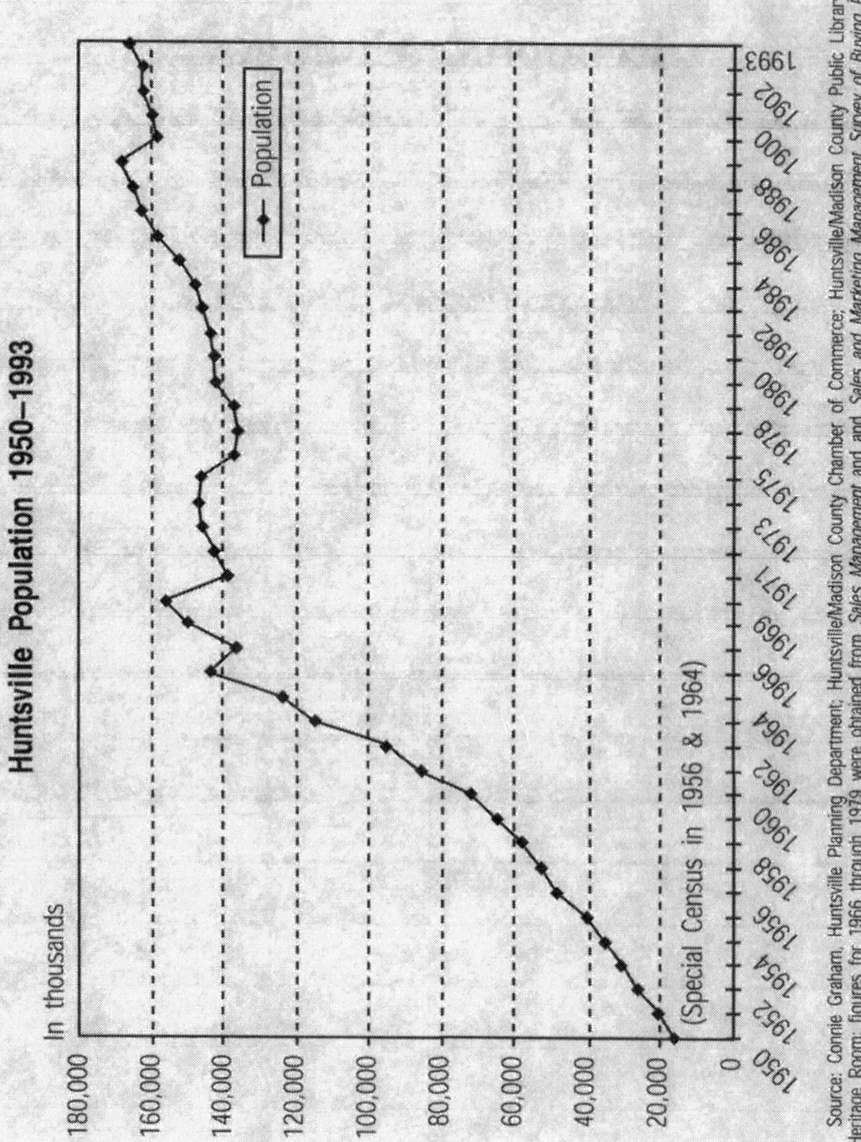
Huntsville Population 1950–1993
In thousands
180,000
160,000
140,000
120,000
100,000
80,000
60,000
40,000
20,000
0
Population
(Special Census in 1956 & 1964)
1950 1952 1954 1956 1958 1960 1962 1964 1966 1969 1971 1973 1975 1978 1980 1982 1984 1986 1988 1900 1902 1993
Source: Connie Graham, Huntsville Planning Department; Huntsville/Madison County Chamber of Commerce; Huntsville/Madison County Public Library, Heritage Room; figures for 1966 through 1979 were obtained from Sales Management and and Sales and Marketing Management Survey of Buying Power 1990 from Alabama Industrial Relations

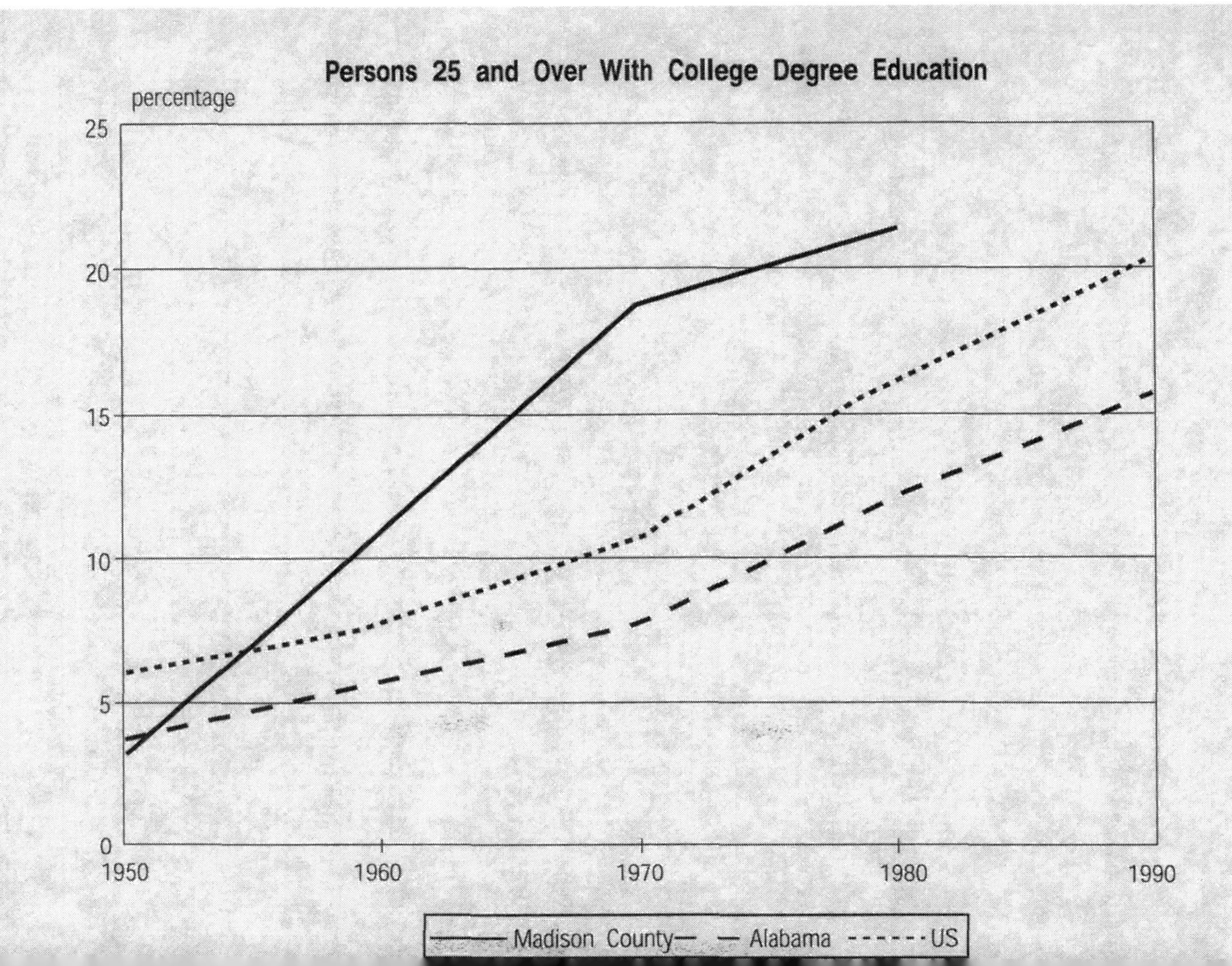
Persons 25 and Over With College Degree Education
percentage
25
20
15
10
5
0
1950
1960
1970
1980
1990
Madison County
Alabama
US

Sources and Research Materials

Marshall Space Flight Center's documentary collections relating its history are uneven, primarily because the Center had no history office from 1975 until 1986. When the office closed, Marshall sent many of these documents to the National Archives annex in Atlanta, and retrieval is complicated because shelf lists are incomplete and some of the documents have apparently been lost.

Today, Marshall's historical documents are in several collections. Most important are the History Archives, housed in Building 4203. This collection is built around a collection formerly used as a resource by the Office of the Center Director. It contains correspondence, Weekly Notes, official documents, and other records relating to projects, management, institutional issues, and other Centers, Headquarters, and other issues of interest to top management. The History Office, located in Building 4200, has a wide range of documents and other resources (including videotapes) collected since the office reopened in 1986, but covering all periods of Marshall's history. Many of the key documents are available on fiche. The histories of Marshall's involvement in Shuttle and Space Station have been documented by a contractor under the supervision of the history office. The Shuttle and Station materials include documents, annotated chronologies, and interviews.

Other collections on the Center and at the adjacent Redstone Arsenal have information on Marshall's history. The Marshall Document Repository houses technical documents on the Center's projects. The Redstone Scientific Information center is an Army regional library with a rich collection of documents, publications, and on-line retrieval systems. The Arsenal also has its own history office, which has information on pre-Marshall ABMA missile development in Huntsville.

Two other sites in Huntsville contain information on Marshall. The Space and Rocket Center holds documents of some retirees from Marshall, including many of the original German team. The Special Collections Department of the library at the University of Alabama in Huntsville houses a collection of materials assembled for Roger Bilstein's book on Saturn. The Saturn collection has interviews, technical and managerial documents, and brochures from prime contractors.

Because the two Centers have worked together on most of NASA's major human spaceflight projects, the history office at the Johnson Space Center in Houston has many resources relating to Marshall's history, including chronological document collections relating to Mercury, Apollo, Skylab (housed at the Fondren Library at Rice University), Shuttle, and Space Station. Several Houston projects over the years have conducted interviews, and many of these discuss Marshall. Many of the collections of the NASA Headquarters history office in Washington have information on Marshall. A vertical file containing an extensive clipping file and numerous documents includes biographical files on key NASA personnel, program files, and files on each of the NASA Centers. The office also has the papers of several NASA administrators and deputy administrators; the Fletcher and Myers papers in particular have material relating to Marshall. Management and administrative collections also bear on Marshall's history.

Other sites in Washington also have useful materials. The Space Division of the National Air and Space Museum has interviews Robert Smith conducted for his book on the Hubble Space Telescope. In addition, the National Archives houses the records of the presidential commission that investigated the *Challenger* accident. The commission records contain over one hundred interviews undertaken by the investigation staff and thousands of documents on over seventy reels of microfilm.

Most prominent among the publications dealing with the origins of Marshall Space Flight Center are those dealing with the Germans who came to Huntsville as a result of Operation Paperclip. The authors of most of these works were people who knew and worked with Wernher von Braun; these works comprise what historian Rip Bulkeley called the "Huntsville school" of aerospace history. The best of these works are Ernst Stuhlinger and Frederick I. Ordway III, *Wernher von Braun, Crusader for Space: A Biographical Memoir* (Malabar, Florida: Krieger Publishing Company, 1994) and Ordway and Mitchell R. Sharpe, *The Rocket Team* (New York: Thomas Y. Crowell, 1979). Michael Neufeld's *The*

Rocket and the Reich (New York: Basic Books, 1994), published as this book entered its final review process, is the most scholarly study of the German World War II missile program at Peenemünde. Neufeld shared many of his insights and allowed us to see chapters of his work in progress.

Roger Bilstein's *Orders of Magnitude: A History of the NACA and NASA, 1915–1990* (NASA SP–4406, 1989) is a useful overview. A more interpretive study of NASA's evolving culture is Howard E. McCurdy's *Inside NASA: High Technology and Organizational Change in the U. S. Space Program* (Baltimore: Johns Hopkins University Press, 1993). Other histories of NASA Centers in the NASA History Series that have information about Marshall include Charles D. Benson and William B. Faherty, *Moonport: A History of the Apollo Launch Facilities and Operations* NASA SP–4204 (Washington, 1978); Virginia P. Dawson, *Engines and Innovation: Lewis Laboratory and American Propulsion Technology* (NASA SP–4306, 1991); Henry C. Dethloff, *Suddenly, Tomorrow Came...: A History of the Johnson Space Center* NASA SP–4307 (Washington, DC, 1993); and James R. Hansen, *Engineer in Charge: A History of the Langley Aeronautical Laboratory, 1917–1958* (NASA SP–4305, 1987).

The political history of the origins of NASA, and of the absorption of ABMA by the Agency, are treated in Walter McDougall's *The Heavens and the Earth* (New York: Basic Books, Inc., 1985) and Robert A. Divine's *The Sputnik Challenge* (New York: Oxford University Press, 1993). Books by participants in these events with significant treatment of ABMA include J.D. Hunley (editor), *The Birth of NASA: The Diary of T. Keith Glennan* (Washington, DC: NASA History Series SP–4105) and Major General John B. Medaris with Arthur Gordon, *Countdown for Decision* (New York: G. P. Putnam's Sons, 1960). The best internal Marshall treatment of the Center's origins is David S. Akens, "Historical Origins of the George C. Marshall Space Flight Center," MSFC Historical Monograph No. 1 (Huntsville: MSFC, 1960). Most of the books that deal with the early space program concentrate on astronauts rather than engineers. An exception is Sylvia Doughty Fries, *NASA Engineers and the Age of Apollo* (NASA SP–4104, 1992).

Although many books discuss the Apollo Program, few cover MSFC in any detail. By far the most detailed history of the Center in the 1960s is Roger Bilstein's *Stages to Saturn: A Technological History of Apollo/Saturn* NASA SP–4206 (Washington, DC, 1980). The book is especially valuable because many of the documents that Bilstein used have since been lost. Other noteworthy works are: John A. Logsdon, *The Decision to Go to the Moon: Project Apollo and the National Interest* (Cambridge, Mass.: MIT Press, 1970); Charles Murray and Catherine Bly Cox, *Apollo: The Race to the Moon* (New York: Simon and Schuster, 1989); Dale Carter, *The Final Frontier: The Rise and Fall of the American Rocket State* (London: Verso, 1988); Courtney G. Brooks, James M. Grimwood, Loyd S. Swenson, *Chariots for Apollo: A History of Manned Lunar Spacecraft* NASA SP–4205 (Washington, DC, 1979); William D. Compton, *Where No Man Has Gone Before: A History of the Lunar Exploration Missions* NASA SP–4214 (Washington, DC, 1989); Norman Mailer offers colorful accounts of von Braun and the Apollo 11 launch in *Of a Fire on the Moon* (Boston, Little, Brown, 1969).

Several NASA publications provide statistics regarding institutional development of the Agency, with detailed information on individual Centers. The three volumes of the *NASA Historical Data Book* (Volume I, NASA Resources, 1958–1968 edited by Jane Van Nimmen, Leonard C. Bruno, and Robert L. Rosholt, SP–4011, 1976; Volume II, Programs and Projects, 1958–1968, and Volume III, Programs and Projects, 1969–1978, edited by Linda Neuman Ezell, SP–4012, 1988) are invaluable resources. Arnold S. Levine's *Managing NASA in the Apollo Era* (NASA SP–4102, 1982) analyzes NASA administration of budgets, planning, personnel, and interagency relations.

Books that shed light on Marshall's diversification include: W. David Compton and Charles D. Benson, *Living and Working in Space: A History of Skylab* NASA SP–4208 (Washington, DC, 1983); C. A. Lundquist, *Skylab's Astronomy and Space Sciences* NASA SP–404 (Washington, DC, 1979); J. A. Eddy, *A New Sun: The Solar Results from Skylab* NASA SP–402 (Washington, DC, 1979); Wallace H. Tucker, *The Star Splitters: The High Energy Astronomy Observatories* NASA SP–466 (Washington, DC, 1984); R. J. Naumann and H. W. Herring, *Materials Processing in Space: Early Experiments*, NASA SP–443 (Washington, DC, 1980); Douglas R. Lord, *Spacelab: An International Success Story* NASA SP–487 (Washington, DC, 1987). Robert W. Smith's *The Space Telescope: A Study of NASA, Science, Technology, and Politics* (Cambridge: Cambridge University Press, 1993) is a magisterial account of the Hubble Space Telescope.

Although it concentrates more on documents generated at Houston and Headquarters, the voluminous six-volume "Shuttle Chronology, 1964–1973" edited by John F. Guilmartin, Jr., and John Walker Mauer, JSC Management Analysis Office, 1988, has many documents that relate to Marshall's role in early Shuttle development.

The key source for the *Challenger* accident are the five volumes of the Presidential Commission (Washington, DC: US GPO, 6 June 1986). These volumes contain the Rogers Commission report, transcripts of hearings, reports of the NASA task groups, and many key documents from the Shuttle program. Neither the hearings or the published documents give a full record of pre-accident events. The secondary literature on the *Challenger* accident is extensive, but mainly follows the interpretation of the commission report. The main works are: Malcolm McConnell, *Challenger: A Major Malfunction, A True Story of Politics, Greed, and the Wrong Stuff* (Garden City, New York: Doubleday, 1987); Joseph Trento, *Prescription for Disaster* (New York: Crown, 1987); Diane Vaughn, "Autonomy, Interdependence, and Social Control: NASA and the Space Shuttle Challenger," *Administrative Science Quarterly* 35 (June, 1990); Frederick F. Lighthall, "Launching the Space Shuttle *Challenger*: Disciplinary Deficiencies in the Analysis of Engineering Data," *IEEE Transactions on Engineering Management* (February, 1991); Gregory Moorhead, Richard J. Ference, and Chris P. Neck, "Groupthink Decision Fiascoes Continue: Space Shuttle Challenger and a Revised Groupthink Framework," *Human Relations* 44 (June 1991); Phillip K. Tompkins, *Organizational Communication Imperatives: Lessons of the Space Program* (Los Angeles: Roxbury Publishing, 1993); Thomas F. Gieryn and Anne E. Figert, "Ingredients for a Theory of Science in Society: O-Rings, Ice Water, C-Clamp, Richard Feynman, and the Press" in Susan E. Cozzens and Thomas F. Gieryn, eds., *Theories of Science in Society* (Bloomington: University of Indiana Press, 1990). Commission member Richard P. Feynman describes his work on the investigation in *"What Do You Care What Other People Think?" Further Adventures of a Curious Character* (New York: Norton, 1988).

Howard E. McCurdy's *The Space Station Decision: Incremental Politics and Technological Choice* (Baltimore: The Johns Hopkins University Press, 1990) examines the political struggle to win approval for NASA's "next logical step." Adam L. Gruen's *The Port Unknown: A History of the Space Station Freedom Program* (NASA SP–4217, 1995) examines from a Washington perspective the politics, budgets, and configuration changes that characterized *Space Station Freedom's* developmental rollercoaster.

Index

A

E

G

The NASA History Series

Reference Works, NASA SP–4000:

Grimwood, James M. *Project Mercury: A Chronology.* (NASA SP–4001, 1963).

Grimwood, James M., and Hacker, Barton C., with Vorzimmer, Peter J. *Project Gemini Technology and Operations: A Chronology*. (NASA SP–4002, 1969).

Link, Mae Mills. *Space Medicine in Project Mercury*. (NASA SP–4003, 1965).

Astronautics and Aeronautics, 1963: Chronology of Science, Technology, and Policy. (NASA SP–4004, 1964).

Astronautics and Aeronautics, 1964: Chronology of Science, Technology, and Policy. (NASA SP–4005, 1965).

Astronautics and Aeronautics, 1965: Chronology of Science, Technology, and Policy. (NASA SP–4006, 1966).

Astronautics and Aeronautics, 1966: Chronology of Science, Technology, and Policy. (NASA SP–4007, 1967).

Astronautics and Aeronautics, 1967: Chronology of Science, Technology, and Policy. (NASA SP–4008, 1968).

Ertel, Ivan D., and Morse, Mary Louise. *The Apollo Spacecraft: A Chronology, Volume I, Through November 7, 1962*. (NASA SP–4009, 1969).

Morse, Mary Louise, and Bays, Jean Kernahan. *The Apollo Spacecraft: A Chronology, Volume II, November 8, 1962–September 30, 1964*. (NASA SP–4009, 1973).

Brooks, Courtney G., and Ertel, Ivan D. *The Apollo Spacecraft: A Chronology, Volume III, October 1, 1964–January 20, 1966*. (NASA SP–4009, 1973).

Ertel, Ivan D., and Newkirk, Roland W., with Brooks, Courtney G. *The Apollo Spacecraft: A Chronology, Volume IV, January 21, 1966–July 13, 1974*. (NASA SP–4009, 1978).

Astronautics and Aeronautics, 1968: Chronology of Science, Technology, and Policy. (NASA SP–4010, 1969).

Newkirk, Roland W., and Ertel, Ivan D., with Brooks, Courtney G. *Skylab: A Chronology*. (NASA SP–4011, 1977).

Van Nimmen, Jane, and Bruno, Leonard C., with Rosholt, Robert L. *NASA Historical Data Book, Volume I: NASA Resources, 1958–1968*. (NASA SP–4012, 1976, rep. ed. 1988).

Ezell, Linda Neuman. *NASA Historical Data Book, Volume II: Programs and Projects, 1958–1968*. (NASA SP–4012, 1988).

Ezell, Linda Neuman. *NASA Historical Data Book, Volume III: Programs and Projects, 1969–1978*. (NASA SP–4012, 1988).

Gawdiak, Ihor Y., with Fedor, Helen. Compilers. *NASA Historical Data Book, Volume IV: NASA Resources, 1969-1978*. (NASA SP–4012, 1994).

Astronautics and Aeronautics, 1969: Chronology of Science, Technology, and Policy. (NASA SP–4014, 1970).

Astronautics and Aeronautics, 1970: Chronology of Science, Technology, and Policy. (NASA SP–4015, 1972).

Astronautics and Aeronautics, 1971: Chronology of Science, Technology, and Policy. (NASA SP–4016, 1972).

Astronautics and Aeronautics, 1972: Chronology of Science, Technology, and Policy. (NASA SP–4017, 1974).

Astronautics and Aeronautics, 1973: Chronology of Science, Technology, and Policy. (NASA SP–4018, 1975).

Astronautics and Aeronautics, 1974: Chronology of Science, Technology, and Policy. (NASA SP–4019, 1977).

Astronautics and Aeronautics, 1975: Chronology of Science, Technology, and Policy. (NASA SP–4020, 1979).

Astronautics and Aeronautics, 1976: Chronology of Science, Technology, and Policy. (NASA SP–4021, 1984).

Astronautics and Aeronautics, 1977: Chronology of Science, Technology, and Policy. (NASA SP–4022, 1986).

Astronautics and Aeronautics, 1978: Chronology of Science, Technology, and Policy. (NASA SP–4023, 1986).

Astronautics and Aeronautics, 1979-1984: Chronology of Science, Technology, and Policy. (NASA SP–4024, 1988).

Astronautics and Aeronautics, 1985: Chronology of Science, Technology, and Policy. (NASA SP–4025, 1990).

Noordung, Hermann. *The Problem of Space Travel: The Rocket Motor*. Stuhlinger, Ernst, and Hunley, J.D., with Garland, Jennifer. Editor. (NASA SP–4026, 1995).

Astronautics and Aeronautics, 1986–1990: A Chronology. (NASA SP–4027, 1997).

Management Histories, NASA SP–4100:

Rosholt, Robert L. *An Administrative History of NASA, 1958–1963*. (NASA SP–4101, 1966).

Levine, Arnold S. *Managing NASA in the Apollo Era*. (NASA SP–4102, 1982).

Roland, Alex. *Model Research: The National Advisory Committee for Aeronautics, 1915–1958*. (NASA SP–4103, 1985).

Fries, Sylvia D. *NASA Engineers and the Age of Apollo*. (NASA SP–4104, 1992).

Glennan, T. Keith. *The Birth of NASA: The Diary of T. Keith Glennan*. Hunley, J.D. Editor. (NASA SP–4105, 1993).

Seamans, Robert C., Jr. *Aiming at Targets: The Autobiography of Robert C. Seamans, Jr.* (NASA SP–4106, 1996)

Project Histories, NASA SP–4200:

Swenson, Loyd S., Jr., Grimwood, James M., and Alexander, Charles C. *This New Ocean: A History of Project Mercury*. (NASA SP–4201, 1966).

Green, Constance McL., and Lomask, Milton. *Vanguard: A History*. (NASA SP–4202, 1970; rep. ed. Smithsonian Institution Press, 1971).

Hacker, Barton C., and Grimwood, James M. *On Shoulders of Titans: A History of Project Gemini*. (NASA SP–4203, 1977).

Benson, Charles D. and Faherty, William Barnaby. *Moonport: A History of Apollo Launch Facilities and Operations*. (NASA SP–4204, 1978).

Brooks, Courtney G., Grimwood, James M., and Swenson, Loyd S., Jr. *Chariots for Apollo: A History of Manned Lunar Spacecraft*. (NASA SP–4205, 1979).

Bilstein, Roger E. *Stages to Saturn: A Technological History of the Apollo/Saturn Launch Vehicles*. (NASA SP–4206, 1980).

SP–4207 not published.

Compton, W. David, and Benson, Charles D. *Living and Working in Space: A History of Skylab*. (NASA SP–4208, 1983).

Ezell, Edward Clinton, and Ezell, Linda Neuman. *The Partnership: A History of the Apollo-Soyuz Test Project*. (NASA SP–4209, 1978).

Hall, R. Cargill. *Lunar Impact: A History of Project Ranger.* (NASA SP–4210, 1977).

Newell, Homer E. *Beyond the Atmosphere: Early Years of Space Science*. (NASA SP–4211, 1980).

Ezell, Edward Clinton, and Ezell, Linda Neuman. *On Mars: Exploration of the Red Planet, 1958–1978*. (NASA SP–4212, 1984).

Pitts, John A. *The Human Factor: Biomedicine in the Manned Space Program to 1980*. (NASA SP–4213, 1985).

Compton, W. David. *Where No Man Has Gone Before: A History of Apollo Lunar Exploration Missions*. (NASA SP–4214, 1989).

Naugle, John E. *First Among Equals: The Selection of NASA Space Science Experiments*. (NASA SP–4215, 1991).

Wallace, Lane E. *Airborne Trailblazer: Two Decades with NASA Langley's Boeing 737 Flying Laboratory*. (NASA SP–4216, 1994).

Butrica, Andrew J. Editor. *Beyond the Ionosphere: Fifty Years of Satellite Communication* (NASA SP–4217, 1997).

Butrica, Andrews J. *To See the Unseen: A History of Planetary Radar Astronomy.* (NASA SP–4218, 1996).

Mack, Pamela E. Editor. *From Engineering Science to Big Science: The NACA and NASA Collier Trophy Research Project Winners*. (NASA SP–4219, 1998).

Reed, R. Dale. With Lister, Darlene. *Wingless Flight: The Lifting Body Story.* (NASA SP–4220, 1997).

Heppenheimer, T.A. *The Space Shuttle Decision: NASA's Quest for a Reusable Space Vehicle* (NASA SP–4221, 1999).

Center Histories, NASA SP–4300:

Rosenthal, Alfred. *Venture into Space: Early Years of Goddard Space Flight Center*. (NASA SP–4301, 1985).

Hartman, Edwin, P. *Adventures in Research: A History of Ames Research Center, 1940–1965*. (NASA SP–4302, 1970).

Hallion, Richard P. *On the Frontier: Flight Research at Dryden, 1946–1981*. (NASA SP–4303, 1984).

Muenger, Elizabeth A. *Searching the Horizon: A History of Ames Research Center, 1940–1976*. (NASA SP–4304, 1985).

Hansen, James R. *Engineer in Charge: A History of the Langley Aeronautical Laboratory, 1917–1958*. (NASA SP–4305, 1987).

Dawson, Virginia P. *Engines and Innovation: Lewis Laboratory and American Propulsion Technology*. (NASA SP–4306, 1991).

Dethloff, Henry C. *"Suddenly Tomorrow Came...": A History of the Johnson Space Center*. (NASA SP–4307, 1993).

Hansen, James R. *Spaceflight Revolution: NASA Langley Research Center from Sputnik to Apollo*. (NASA SP–4308, 1995).

Wallace, Lane E. *Flights of Discovery: 50 Years at the NASA Dryden Flight Research Center*. (NASA SP–4309, 1996).

Herring, Mack R. *Way Station to Space: A History of the John C. Stennis Space Center*. (NASA SP–4310, 1997).

Wallace, Harold D., Jr. *Wallops Station and the Creation of the American Space Program*. (NASA SP–4311, 1997).

Wallace, Lane E. *Dreams, Hopes, Realities: NASA's Goddard Space Flight Center, The First Forty Years* (NASA SP–4312, 1999).

General Histories, NASA SP–4400:

Corliss, William R. *NASA Sounding Rockets, 1958-1968: A Historical Summary*. (NASA SP–4401, 1971).

Wells, Helen T., Whiteley, Susan H., and Karegeannes, Carrie. *Origins of NASA Names*. (NASA SP–4402, 1976).

Anderson, Frank W., Jr. *Orders of Magnitude: A History of NACA and NASA, 1915–1980*. (NASA SP–4403, 1981).

Sloop, John L. *Liquid Hydrogen as a Propulsion Fuel, 1945–1959*. (NASA SP–4404, 1978).

Roland, Alex. *A Spacefaring People: Perspectives on Early Spaceflight*. (NASA SP–4405, 1985).

Bilstein, Roger E. *Orders of Magnitude: A History of the NACA and NASA, 1915–1990*. (NASA SP–4406, 1989).

Logsdon, John M. Editor. With Lear, Linda J., Warren-Findley, Jannelle, Williamson, Ray A., and Day, Dwayne A. *Exploring the Unknown: Selected Documents in the History of the U.S. Civil Space Program, Volume I, Organizing for Exploration*. (NASA SP–4407, 1995).

Logsdon, John M. Editor. With Day, Dwayne A., and Launius, Roger D. *Exploring the Unknown: Selected Documents in the History of the U.S. Civil Space Program, Volume II, Relations with Other Organizations*. (NASA SP–4407, 1996).

Logsdon, John M. Editor. With Launius, Roger D., Onkst, David H., and Garber, Stephen E. *Exploring the Unknown: Selected Documents in the History of the U.S. Civil Space Program, Volume III, Using Space*. (NASA SP–4407, 1998).